80116570
Feb '83

ARTHUR C. GUYTON, M.D.

Department of Physiology and Biophysics
University of Mississippi
School of Medicine
Jackson, Mississippi

Physiology of the Human Body

Fifth Edition

SAUNDERS COLLEGE PUBLISHING

Philadelphia

Saunders College Publishing
West Washington Square
Philadelphia, PA 19105

Library of Congress Cataloging in Publication Data

Guyton, Arthur C

Physiology of the human body.

First-4th ed. published under title: Function of the human body.

Includes bibliographies and index.

1. Human physiology.   I. Title.

QP34.5.G89 1979   612   77-25557

ISBN 0-7216-4378-7

Listed here is the latest translated edition of this book together
with the language of the translation and the publisher.

French *(4th Edition)* – Les Editions HRW Ltee., Montreal, Canada

Spanish *(4th Edition)* – NEISA, Mexico 4 D.F., Mexico

Portuguese *(4th Edition)* – Editora Interamericana LTD, Rio de Janeiro, Brazil

Physiology of the Human Body   ISBN   0-7216-4378-7

2   147   98765

PREFACE

This textbook, entitled *Physiology of the Human Body,* has been published in four previous editions under the title *Function of the Human Body.* This change in title was made simply to indicate that the text is used widely, as was intended, in physiology courses in a variety of schools throughout the world. However, along with the change in title there have been other changes as well, primarily complete revision of the figures and thorough and extensive updating of the text material itself. The figures are now presented in two colors, which has been very helpful in separating functional elements of the figures from background information. Revision of the text material has been made in literally every part of the book. It has been occasioned mainly by the fact that physiology continues in a dynamic stage of discovery, with new understandings of basic physiological concepts generated each day. For instance, it is only now that we are beginning to understand the intricate relationships between basic cellular function and function of the overall organ complex of the body. Therefore, a major share of the revision is concerned with explaining these interrelationships.

My original purpose in writing this text was to present the basic philosophy of human bodily function, with the hope that I could pass on to others my own love for the intrinsic functional beauty that underlies life itself. I have hoped to present the human being as a thinking, sensing, active creature capable of living almost automatically and yet capable of the immense diversity that characterizes higher forms of life. There is no machine yet devised or ever to be devised that can have more excitement or more majesty than the human body. Therefore, I hope that it will be with pleasure as well as with enthusiasm that the reader learns how his body functions.

Because the field of human physiology is extremely extensive, the character of a textbook in this field is necessarily determined by the choice of material that is presented. In this text a special attempt has been made to choose those aspects of human physiology that will lead the reader to understand basic principles and concepts. Yet, because some aspects of human physiology are still only partly understood, a strong effort has been made to distinguish fact from theory, but not to burden the reader with intricate and insignificant details that more properly belong in a reference textbook.

A text such as this requires work by many different people, not the least of whom are the teachers who send suggestions to the author. Feedback of this type has helped immensely in making the earlier editions progressively better and I hope also in making this Fifth Edition still much better as well. I wish also to express my appreciation to Mrs. Billie Howard for her superb secretarial help in preparing this edition, to Mrs. Carolyn Hull and Ms. Tomika Mita for drawing most of the figures, and to the staff of the W. B. Saunders Company for their continued excellence in production of this book.

ARTHUR C. GUYTON

CONTENTS

INTRODUCTION: CELL PHYSIOLOGY

I

INTRODUCTION TO HUMAN PHYSIOLOGY

1

What Is Physiology? We could spend the remainder of our lives attempting to define the word "physiology," for physiology is the study of life itself. It is the study of function of all parts of living organisms, as well as of the whole organism. It attempts to discover answers to such questions as: How and why do plants grow? What makes bacteria divide? How do fish obtain oxygen from the sea and in what way do they utilize it once they have obtained it? How is food digested? And what is the nature of the thinking process in the brain?

Even small viruses weighing one-millionth of a single bacterium have the characteristics of life, for they feed on their surroundings, they grow and reproduce, and they excrete by-products. These very minute living structures are the subject of the simplest type of physiology, *viral physiology*. Physiology becomes progressively more complicated and vast as it extends through the study of higher and higher forms of life such as cells, plants, lower animals, and, finally, human beings. It is obvious, then, that the subject of this book, "human physiology," is but a small part of the vast discipline of physiology.

As small children we begin to wonder what enables people to move, how it is possible for them to talk, how they can see the expanse of the world and feel the objects about them, what happens to the food they eat, how they derive from food the energy needed for exercise and other types of bodily activity, by what process they reproduce other beings like themselves so that life goes on, generation after generation. All these and other human activities make up *life*. Physiology attempts to explain them and hence to explain life itself.

ROLE OF THE CELL IN THE HUMAN BODY

The basic functional unit of the body is the cell, and 75 trillion cells make up the human body. Each of these is a living organism in itself, capable of existing, performing chemical reactions, and contributing its part to the overall function of the body — also capable in most instances of reproducing itself to replenish the cells that die.

The cells are the building blocks of the organs, and each of the organs performs its own specialized function. One will appreciate the importance of the cell when he realizes that many more millions of years went into the evolutionary development of the cell than into evolution from the cell to the human being. Therefore, before one can understand how any one of the organs functions or how the organs function together to maintain life, it is necessary that he understand the inner workings of the cell itself. The next few chapters will be devoted entirely to discussion of basic cellular function, and throughout the remainder of this book we will refer many times again to cellular function as the basis of organ and system operation.

THE INTERNAL ENVIRONMENT AND HOMEOSTASIS

All cells of the body live in a bath of fluid, fluid that weaves its way through the minute spaces between the cells, that moves in and out of the blood vessels, and that is carried in the blood from one part of the body to another. This mass of fluid that constantly bathes the outsides of the cells is called the *extracellular fluid*.

3

For the cells of the body to continue living, there is one major requirement: The composition of the extracellular fluid must be controlled very exactly from moment to moment and day to day, with no single important constituent ever varying more than a few per cent. Indeed, cells can live even after removal from the body if they are placed in a fluid bath that contains the same constituents and has the same physical conditions as those of the extracellular fluid. Claude Bernard, the great nineteenth century physiologist who originated much of our modern physiologic thought, called the extracellular fluids that surround the cells the *milieu intérieur,* the "internal environment," and Walter Cannon, another great physiologist of the first half of this century, referred to the maintenance of constant conditions in these fluids as *homeostasis.*

Thus, at the very outset of our discussion of physiology of the human body, we are beset with a major problem. How does the body maintain the required constancy of the internal environment, that is, the constancy of the extracellular fluid? The answer to this is that almost every organ plays some role in the control of one or more of the fluid constituents. For instance, the *circulatory system,* composed of the *heart* and *blood vessels,* transports blood throughout the body; and water and dissolved substances diffuse back and forth between the blood and the fluids that surround the cells. Thus, the circulatory system keeps the extracellular fluid in all parts of the body constantly mixed. This function of the circulatory system is so effective that hardly any portion of fluid in any part of the body remains unmixed with the other fluid more than a few minutes at a time.

The *respiratory system* transfers oxygen from the air to the blood, and the blood in turn transports the oxygen to all the tissue fluids surrounding the cells, thus maintaining the level of oxygen that is required for life by all the cells. The carbon dioxide excreted by the cells enters the tissue fluids, then becomes mixed with the blood and is finally removed through the lungs.

The *digestive system* performs a similar function for other nutrients besides oxygen; it processes nutrients which are then absorbed into the blood and are rapidly transported throughout the body fluids where they can be used by the cells. The *liver,* the *endocrine glands,* and some of the other organs participate in what is collectively known as *intermediary metabolism,* which converts many of the nutrients absorbed from the gastrointestinal tract into substances that can be used directly by the cells. The *kidneys* remove the remains of the nutrients after their energy has been extracted by the cells, and other organs provide for *hearing, feeling, tasting, smelling,* and *seeing,* all of which aid the animal in his search for and selection of food and also help him to protect himself from dangers so that he can perpetuate the almost Utopian internal environment in which his cells continue their life processes.

Thus, we can emphasize once again that organ functions of the body depend on individual functions of cells, and sustained life of the cells depends on maintenance of an appropriate environment in the extracellular fluids. In turn, the organs and the cells, in their own ways, play individual roles in maintaining constancy of this internal fluid environment, the process that we call homeostasis.

ORGANS AND SYSTEMS OF THE HUMAN BODY

For those students who have not yet learned the basic structure of the human body, we need now to retreat for a few moments and review its major functional components.

THE SKELETON AND ITS ATTACHED MUSCLES

Figure 1–1 illustrates the skeleton with some of its attached muscles. Each joint of the skeleton is enveloped by a loose *capsule,* and the space within the capsule and between the two respective bones is the *joint cavity.* In the joint cavities is a thick, slippery fluid containing a mucus-like substance, hyaluronic acid, which lubricates the joints, promoting ease of movement. On the sides of each capsule are strong fibrous *ligaments* that keep the joints from pulling apart. Often the ligaments are only on two sides of the joint, which allows the joint to move freely in one direction but not so freely in another direction. Other joints, particularly those of the spine, hips, and shoulders, not having very restrictive ligaments, can move in almost any direction; that is, they can bend forward, backward, and to either side, or they can even be rotated. In these instances, loose ligaments merely limit the degree of motion to prevent excessive movement in any one direction.

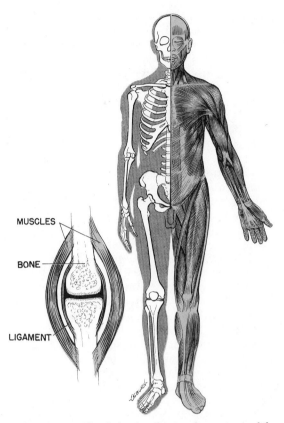

MUSCLES

BONE

LIGAMENT

Figure 1-1. The skeletal and muscular systems of the human body.

Muscles move the limbs and other parts of the body in the directions allowed by the ligaments. In the case of movement at the knee joint, for instance, one major muscle is on the front and several muscles are on the back of the joint. There is a similar arrangement of muscles anteriorly and posteriorly about the ankle, except that the ligaments of the ankle allow the ankle joint to move also from side to side, and additional muscles are available to provide the sidewise movements. The muscles of the spine are especially interesting because, contrary to what might be expected, the back muscles are not just a few very large muscles, but are composed of about one hundred different individual muscles each one of which performs a specific function: One rotates an adjacent vertebra while a second one flexes the vertebra forward, a third backward, and so on. This is analogous to the arrangement of the centipede, for each segment can bend independently of all the others. The joint connecting the head to the spinal column is supplied with many additional muscles arranged on all sides so that the head can be rotated from side to side or bent in any direction.

In summary, then, the skeleton is literally a bag of bones which can be contorted into many different configurations. Each bone has its own function, and the limitations of angulation of each joint are decreed by the ligaments. The knee joint bends in only one direction, the ankle joint in two, the hip joint in two directions plus an additional rotary motion; and, in general, at least two opposing muscles are available for each motion that the ligaments of a joint allow.

The muscles themselves are composed of long *muscle fibers*. Usually many thousands of these fibers are oriented side by side like the threads in a skein of wool. At each end of the muscle, the muscle fibers fuse with strong *tendon fibers* that form a bundle called the *muscle tendon*. The muscle tendons in turn penetrate and fuse with the bones on the two sides of the respective joints so that any pull exerted will effect appropriate movement.

All muscles are not exactly alike in size and appearance; for instance, the smallest skeletal muscle of the body, the stapedius, is a minute muscle in the middle ear only a few millimeters long, while the longest muscle, the sartorius, extends almost 2 feet down the entire length of the thigh, connecting the pelvic bone with the lower leg. Some muscles, such as those of the abdominal wall, are arranged in thin sheets, while others are round, cigar-shaped structures, for example, the biceps which lifts the lower arm and the gastrocnemius which pulls the foot down when one wishes to stand on tiptoes.

The precise method by which muscle fibers contract is still not completely clear, but we do know that signals arriving in the muscles through nerves cause each fiber to shorten for a brief instant, allowing the entire muscle belly to contract and thereby to perform its function. This will be discussed in detail in Chapter 10.

THE NERVOUS SYSTEM

The nervous system, illustrated in Figure 1-2, is composed of the brain, the spinal cord, and the peripheral nerves that extend throughout the body. A major function of the nervous system is to control many of the bodily activities, especially those of the muscles, but to exert this control intelligently the brain must be apprised continually of the body's surroundings. Therefore, to perform

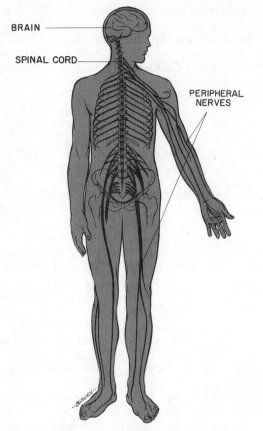

BRAIN

SPINAL CORD

PERIPHERAL NERVES

Figure 1-2. The nervous system.

these varied activities, the nervous system is composed of two separate portions, the *sensory portion* which reports and analyzes the nature of conditions around and inside the body, and the *motor portion* which controls the muscles and glandular secretion.

The sensory portion operates through the senses of sight, hearing, smell, taste, and feel. The sense of feel is actually many different senses, for one can feel light touch, pin pricks, pressure, pain, vibration, position of the joints, tightness of the muscles, and tension on the tendons.

Once information has been relayed to the brain from all the senses, the brain then determines what movement, if any, is most suitable, and the muscles are called into action to implement the decision.

One of the most important functions of the nervous system is to control walking. In walking, the body must be supported against gravity, the legs must move rhythmically in a walking motion, equilibrium must be maintained, and the direction of movement of the limbs must be guided. Therefore, the initia-

tion and control of locomotion are very complex functions of the nervous system and require the services of major portions of the brain.

The Autonomic Nervous System. The autonomic nervous system, which is really part of the motor portion of the nervous system, controls many of the internal functions of the body. It operates principally by causing contraction or relaxation of a type of muscle called *smooth muscle,* which constitutes major portions of many of the internal organs. Smooth muscle fibers are much smaller than skeletal muscle fibers, and they usually are arranged in large muscular sheets. For instance, the gastrointestinal tract, the urinary bladder, the uterus, the biliary ducts, and the blood vessels are all composed mainly of smooth muscle sheets rolled into tubular or spheroid structures. Some of the autonomic nerves cause the muscles of these organs to contract while others cause relaxation.

The autonomic nerves also control secretion by many of the glands in the gastrointestinal tract and elsewhere in the body, and at times their nerve endings even secrete hormones that can increase or decrease the rates of chemical reactions in the body's tissues.

Finally, the autonomic nervous system helps to control the heart, which is composed of *cardiac muscle,* a type of muscle intermediate between smooth muscle and skeletal muscle. Stimulation of the so-called *sympathetic fibers* causes the rate and force of contraction of the heart to increase, while stimulation of the *parasympathetic* fibers causes the opposite effects.

In summary, the autonomic nervous system helps to control most of the body's internal functions.

THE CIRCULATORY SYSTEM

The circulatory system, illustrated in Figure 1-3, is composed mainly of the heart and blood vessels. The heart consists of two separate pumps arranged side by side. The first pump pumps blood into the lungs. From here, the blood returns to the second pump to be pumped then into the *systemic arteries* which transport it through the body. From the arteries it flows into the *capillaries,* then into the *veins* and finally back to the heart, thus making a complete circuit. Circulating around and around through the body, the blood acts as a transportation system for conducting various substances from one place to another. It is the circulatory system that carries nu-

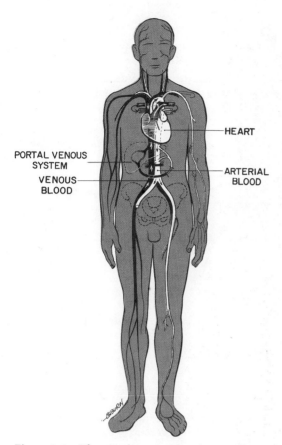

PORTAL VENOUS SYSTEM

VENOUS BLOOD

HEART

ARTERIAL BLOOD

Figure 1–3. The circulatory system: heart and major vessels.

trients to the tissues and then carries excretory products away from the tissues.

The capillaries are porous, allowing fluid and nutrients to diffuse into the tissues and excreta from the cells to reenter the blood.

The Lymphatic System. Large particles that appear for any reason in the tissue spaces, such as old debris of dead tissues, protein molecules, and dead bacteria, cannot pass from the tissues through the small pores of the capillaries. A special accessory circulatory system known as the *lymphatic system* takes care of these materials. Lymph vessels originate in small *lymph capillaries* which lie beside the blood capillaries. And *lymph,* which is fluid derived from the spaces between the cells, flows along the lymph vessels up to the neck where these vessels empty into the neck veins. The lymph capillaries are extremely porous so that large particles can enter the lymph vessels and be transported by the lymph. At several points along the course of the lymph vessels, the lymph passes through *lymph nodes* where most large parti-

cles are filtered out and where bacteria are engulfed and digested by special cells called *reticuloendothelial* cells.

THE RESPIRATORY SYSTEM

Figure 1–4 illustrates the respiratory system, showing the two fundamental portions of this system: (1) the air passages and (2) the blood vessels of the lungs. Air is moved in and out of the lungs by contraction and relaxation of the respiratory muscles, and blood flows continually through the vessels. Only a very thin membrane separates the air from the blood, and since this membrane is porous to gases it allows free passage of oxygen into the blood and of carbon dioxide from the blood into the air.

Oxygen is one of the nutrients needed by the body's tissues. It is carried by the blood and tissue fluids to the cells where it combines chemically with foods to release energy. This energy, in turn, is used to promote muscle contraction, secretion of digestive juices, conduction of signals along nerve fibers, and synthesis of many substances needed for growth and function of the cells.

When oxygen combines with foods to liberate energy, carbon dioxide is formed. This diffuses through the tissue fluids into the blood and is then carried by the blood to the lungs. Then, the carbon dioxide diffuses from the blood into the lung air to be breathed out into the atmosphere.

THE GASTROINTESTINAL SYSTEM

The gastrointestinal system is illustrated in Figure 1–5. Food, after being swallowed, enters the stomach, then the duodenum, the jejunum, the ileum, and the large intestine, finally to be defecated through the anus. However, during this passage through the gastrointestinal tract, the food is *digested* and those portions of the food valuable to the body are *absorbed* into the blood. Along the entire extent of the gastrointestinal tract, special substances are secreted into the gut, especially when food is present. These secretions contain *digestive enzymes* which cause the foods to split into chemicals small enough to pass through the pores of the intestinal membrane into underlying blood and lymphatic capillaries. Thence the digestive products enter the circulating blood to be transported and

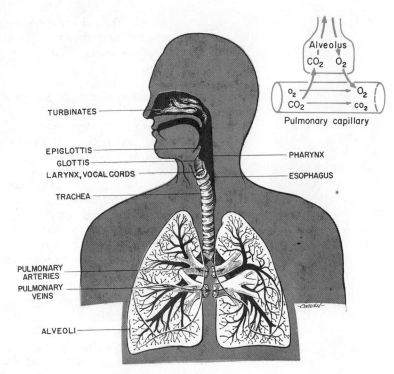

Figure 1–4.   The respiratory system.

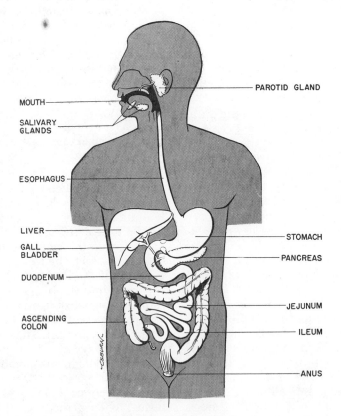

Figure 1–5.   The gastrointestinal system.

used wherever in the body they may be needed.

METABOLIC SYSTEMS

Metabolism and Growth. The term *metabolism* means simply the chemical reactions that occur in the animal organism. These reactions occur inside the individual cells which make up the tissues, and their functions are to provide energy to perform the bodily activities and to build new structures. It is because of the metabolic processes that the cells grow larger and more numerous. The metabolism of special cells allows them to form structures such as bones and fibrous tissue, enlarging the entire animal. Thus, metabolism is the basis not only for the energy needed by the body but also for growth itself.

INTERMEDIARY METABOLISM. Many of the foods entering the blood from the digestive tract can be used by the tissue cells without alteration, but some tissues require special chemicals which are not normally found in the food. To supply these, much of the absorbed food passes to special organs where it is changed into new substances needed by the cells. This process is called *intermediary metabolism.*

The Liver. The liver is one of the internal organs especially adapted for intermediary metabolism and storage. It can split fats and proteins into smaller substances so that the cells of other tissues can then use them for energy or for synthesizing specially needed cellular chemicals. The liver also forms products needed for blood coagulation, for transport of fat, for immunity to infection, and for many other purposes. And the liver is capable of storing large quantities of fats, carbohydrates, and even proteins and then later releasing these foods into the blood when the tissues need them. An animal can live for only a few hours without a liver.

Control of Metabolism by the Hormones. Metabolism is an inherent function of every cell of the body. However, the rate of metabolism in each respective cell is very often increased or decreased by the controlling action of *hormones* secreted by *endocrine glands* in different parts of the body. The *thyroid gland,* located in the neck, secretes *thyroxine* which acts on all cells of the body to increase the rates of most metabolic reactions. *Epinephrine* and *norepinephrine,* two hormones secreted by the *adrenal medullas*

located at the upper poles of the two kidneys, also increase the rate of metabolism in all cells. The *ovaries* secrete *estrogens* and *progesterone,* and the *testes* secrete *testosterone,* which help to control metabolism in the sex organs of the female and male, respectively. *Insulin,* secreted by the *pancreas,* a gland located behind and beneath the stomach, increases the utilization of carbohydrates and decreases the utilization of fats in all the tissues. *Adrenocortical hormones,* secreted by the two *adrenal cortices* located at the upper poles of the kidneys, help to convert proteins to carbohydrates, and they control the passage of proteins, salts, and perhaps other substances through cell walls. *Parathyroid hormone,* secreted by four minute *parathyroid glands* located behind the thyroid gland in the neck, help to control the concentration of calcium in the blood and extracellular fluid by removing calcium from the bones when it is needed by the fluids. Finally, at least eight different hormones secreted by the *pituitary gland* located at the base of the brain control a great host of bodily functions such as growth, rates of secretion of many of the other hormones, sexual functions, and excretion of water and electrolytes by the kidneys.

THE EXCRETORY SYSTEM

The *kidneys,* illustrated in Figure 1–6, constitute an excretory system for ridding the blood of unwanted substances. Most of the substances are the end-products of metabolic reactions, including mainly urea, uric acid, creatinine, phenols, sulfates, and phosphates. If they were allowed to collect in the blood in large quantities, these "ashes" of the cellular fires would soon "smother the flames" so that no further metabolic reactions could take place. For this reason it is important that the kidneys remove these unwanted substances.

The kidneys have another very valuable function besides that of excreting the waste products: They regulate the concentrations of most of the ions in the body fluids. A very large proportion of these ions is sodium chloride or common table salt. The kidneys continually adjust the concentrations of both sodium and chloride in the blood and tissue fluids; and they also regulate very precisely the concentrations of potassium, magnesium, phosphates, and many other substances. The kidneys perform this function by allowing the unwanted substances such as urea to pass easily into the urine while retaining the

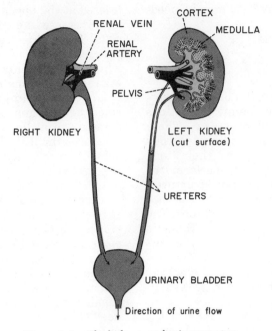

Figure 1-6.   The kidneys and urinary system.

wanted substances such as glucose. Likewise, if sodium is already present in the blood in too large a concentration, it becomes an unwanted substance, and much of it is excreted by the kidneys. However, if the concentration of sodium is too low, it then becomes a wanted substance instead, and the kidney's special properties, which will be described in a future chapter, then prevent sodium loss from the blood.

THE REPRODUCTIVE SYSTEMS

All the functions and systems of the body that maintain life would be useless were it not for those that provide for life's reproduction. The reproductive systems of the female and male are shown in Figure 1–7. The female (Fig. 1–7A) provides the *ovum* (egg) from which a new human being is to develop, but this ovum cannot begin developing until it is fertilized by a *sperm* from the male (Fig. 1–7B). The fertilized ovum derives half of its developmental characteristics from the mother and half from the fertilizing sperm of the father, so that the offspring owes its characteristics equally to each of the parents.

After the ovum has been fertilized it is at first still a single cell, but soon it divides into two cells, then four cells, and finally into many cells, thus becoming an *embryo* and then a *fetus*. Gradually, the newly developing cells *differentiate* into the organs of the body.

The mother provides nutrition for the growing fetus by means of the *placenta,* a structure that attaches to the inner wall of her *uterus* and through which *fetal blood* flows from the fetus. Nutrients diffuse from the mother's blood into the baby's blood through the *placental membrane,* which is very much like the membrane of the respiratory system. In turn, excretory products from the baby pass into the mother's blood. Thus, the fetus is nurtured through a period of nine months in the mother's body until it becomes capable of sustaining life on its own in the outer world. At that time the mother's uterus expels the baby.

COMMUNAL ORGANIZATION OF THE BODY

If this chapter has succeeded in presenting a résumé of the functioning systems of the human body, it should by now be obvious that no single part of the body can live by itself, but that each of the body's functions is neces-

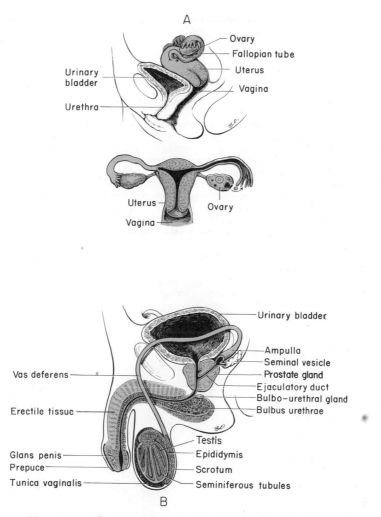

Figure 1–7. The reproductive systems: (A) female, (B) male.

sary for continuous operation of the others. The human animal is a sensing, thinking, and motile organism which, by virtue of nervous and hormonal systems of control, can adapt itself to most surroundings offered by Earth. Its activities are initiated and controlled in part involuntarily, in part by intuition, and in part by reasoning. In the framework of the organs and other tissues are about 75 trillion individual *cells,* each one of which is a living structure. It is the magic of these cells that makes the human body possible. The next few chapters will describe function of the cell itself.

GENERAL REFERENCES IN THE FIELD OF PHYSIOLOGY

American Journal of Physiology. This journal, published monthly, contains articles by authors throughout the world on current research projects.

Annual Review of Physiology. Palo Alto, Calif., Annual Reviews, Inc. (One book each year.) Each book contains review articles that cover the literature of the preceding year in almost the entire field of animal physiology.

Brobeck, J. R.: *Best and Taylor's Physiological Basis of Medical Practice.* 9th ed. Baltimore, The Williams & Wilkins Company, 1973. This text is written principally for the medical student and the postgraduate student.

Guyton, A. C.: *Textbook of Medical Physiology.* 5th ed. Philadelphia, W. B. Saunders Company, 1976. This text presents physiology at the level of the medical student. In general it covers the same material as that in the present text, but in much greater detail and with more emphasis on the medical aspects of human physiology.

International Review of Physiology. (Published every two years in eight volumes.) This review covers the entire field of physiology. In general, the articles are designed for persons with only moderate backgrounds in physiology.

Mountcastle, V. B.: *Medical Physiology.* 13th ed. St. Louis, The C. V. Mosby Company, 1973. This text is most useful for postgraduate students.

Physiological Reviews. This journal, published four times a year, contains reviews of most subjects in the field of physiology every few years.

Ruch, T. C., and Patton, H. D.: *Medical Physiology and*

Biophysics. 20th ed. Philadelphia, W. B. Saunders Company, 1973. This text is written at the level of the postgraduate student.

QUESTIONS

1. What is the internal environment, and what is the relationship of the extracellular fluid to this?
2. What is homeostasis, and what is its importance to functions of the body?
3. What are the interrelationships between the skeleton, the ligaments, and the muscles? Why are two muscles required for each range of motion at a joint?
4. What are the roles of the sensory and motor portions of the nervous system?
5. What is the function of the autonomic nervous system, and why is the smooth muscle of the body so closely associated with this system?
6. Describe the function of the circulatory system in mixing the body fluids.
7. Describe the transport of oxygen from the air to the peripheral cells and carbon dioxide from these cells back to the air.
8. What type of substances are secreted in the gastrointestinal tract, and how are these related to digestion and absorption?
9. What roles do the liver and hormones play in the overall function of the body?
10. Besides the excretion of waste products through the kidneys, what other important function do the kidneys subserve?

THE CELL AND ITS COMPOSITION

2

The human body contains about 75 trillion cells, each of which is a living structure. Several hundred basic types of cells exist in the body, and each type plays a special role in bodily function. Yet, despite the differences between cells, they all have some functions in common, for instance, their abilities to live, grow, and, in most instances, reproduce. The purpose of this chapter is mainly to emphasize these similarities of cells and their functions. Yet, first, let us describe some of the different cell types and their related structures.

SOME REPRESENTATIVE TYPES OF CELLS

Figure 2–1 illustrates five types of tissue that subserve different functions in the body. Example A depicts *connective tissue* which holds the different structures of the body together. This tissue contains cells called *fibroblasts* that are enmeshed in *collagenous* and *elastic* fibers. The fibroblasts secrete chemical substances that later polymerize to form the fibers, and the fibers in turn provide tensile strength to the tissues, thereby holding them together.

Example B illustrates several *red* and *white blood cells.* The red cells carry oxygen in the circulating blood from the lungs to the tissues and carbon dioxide from the tissues back to the lungs, while the white blood cells cleanse the blood and tissues of unwanted materials

such as bacteria, debris from degenerating tissues, and so forth.

Example C shows a *nerve cell* in the brain. The long projection of the nerve cell is its *axon* that occasionally extends as long as one meter. Electrochemical impulses travel over the surface of the nerve cell and along the membrane of the axon to transmit information from one part of the body to another.

Example D illustrates *muscle cells* which can also transmit electrochemical impulses over their membranes but which are different from nerve cells in that they contain long myofibrils that extend the entire length of the muscle and contract when an electrochemical impulse travels over the surface of the muscle cell.

Example E illustrates several different types of cells and structures in the *kidney.* Several *kidney tubules* are shown, lined by *epithelial cells;* these structures help to form the urine, as is explained in a later chapter. Also, two small *blood vessels* are illustrated in cross-section in this figure; these vessels are filled with red blood cells, and connective tissue is present throughout the kidney to hold the different structures together.

The representative tissues and cells shown in Figure 2–1 are but a few of the many types found in the body, but they show the wide variability between different cells. These dissimilarities allow cells to perform different functions. The remainder of this chapter, on the other hand, presents the *similarities* between cells rather than their dissimilarities.

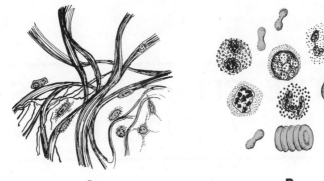
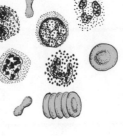

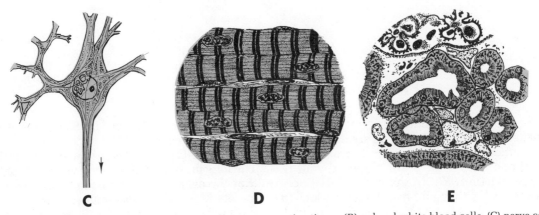

Figure 2–1. Examples of different types of cells: (A) connective tissue, (B) red and white blood cells, (C) nerve cell, (D) muscle cells, (E) kidney tissue.

In future chapters, many different types of specialized cells are described in detail, and their functions are presented.

ORGANIZATION AND CHEMICAL COMPOSITION OF THE CELL

A typical cell, as seen by the light microscope, is illustrated in Figure 2–2. The cell is filled with a viscid, colloid type of material called *protoplasm,* and this is divided between two separate compartments, the nuclear compartment that contains *nucleoplasm* and a second compartment outside the nucleus containing *cytoplasm.* The *nucleoplasm* is separated from the cytoplasm by the *nuclear membrane,* and the cytoplasm is separated from the surrounding fluids by the *cell membrane.*

Protoplasm is composed mainly of five basic substances: *water, ions* (also called electrolytes), *proteins, lipids,* and *carbohydrates.*

Water.   Water is present in protoplasm in a concentration between 70 and 85 per cent. Many cellular chemicals are dissolved in the water, while others are suspended in small particulate form. Chemical reactions take place between the dissolved chemicals or at the surfaces of the suspended particles in the water. The fluid nature of water allows both the dissolved and suspended substances to diffuse or flow to different parts of the cell, thereby providing transport of the substances from one part of the cell to another.

Ions.   The most important ions in the cell are *potassium, magnesium, phosphate, sulfate, bicarbonate,* and small quantities of *sodium* and *chloride.* These will be discussed in much greater detail in Chapters 8 and 16, which will consider the interrelationships between the different fluids of the body.

The ions are dissolved in the water of protoplasm, and they provide inorganic chemicals for cellular reactions. Also, they are necessary for operation of some of the cellular control mechanisms. For instance, ions acting at the cell membrane allow transmission of electro-

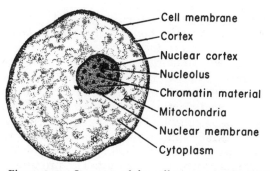

Cell membrane
Cortex
Nuclear cortex
Nucleolus
Chromatin material
Mitochondria
Nuclear membrane
Cytoplasm

Figure 2–2. Structure of the cell as seen by the light microscope.

chemical impulses in nerve and muscle fibers, and the intracellular ions determine the activity of different enzymatically catalyzed reactions that are necessary for cellular metabolism.

Proteins. Next to water, the most abundant substance in most cells is proteins, which normally constitute 10 to 20 per cent of the cell mass. These can be divided into two different types, *structural proteins* and *enzymes.*

To get an idea of what is meant by *structural proteins,* one needs only to note that leather is composed principally of structural proteins, and that hair is almost entirely a structural protein. Proteins of this type are present in the cell membrane, in the nuclear membrane, and in membranes of special intracellular structures, such as of the endoplasmic reticulum and the mitochondria, which will be discussed later in this chapter. Thus structural proteins hold the structures of the cell together. Most structural proteins are *fibrillar;* that is, the individual protein molecules are polymerized into long fibrous threads. The threads in turn provide tensile strength for the cellular structures.

Enzymes, on the other hand, are an entirely different type of protein, composed usually of individual protein molecules or at most of aggregates of a few molecules in a *globular* form rather than a fibrillar form. These proteins, in contrast to the fibrillar proteins, are often dissolved in the fluid of the cell or are adsorbed to the surfaces of membranous structures inside the cell. The enzymes come into direct contact with other substances inside the cell and catalyze chemical reactions. For instance, the chemical reactions that split glucose into its component parts and then combine these with oxygen to form carbon dioxide and water are catalyzed by a series of protein enzymes.

Thus, enzyme proteins control the metabolic functions of the cell.

Special types of proteins are present in different parts of the cell. Of particular importance are the *nucleoproteins,* present both in the nucleus and in the cytoplasm. The nucleoproteins of the nucleus contain deoxyribonucleic acid (DNA), which constitutes the *genes,* and these control the overall function of the cell as well as the transmission of hereditary characteristics from cell to cell. These substances are so important that they will be considered in detail in Chapter 4. In addition, the chemical nature of proteins will be considered in Chapter 31, and the different structural and enzymatic functions of proteins will be subjects of discussion at numerous points throughout this text.

Lipids. Lipids are several different types of substances that are grouped together because of their common property of being soluble in fat solvents. The most abundant lipid of animal tissues is *triglyceride,* also called *neutral fat.* However, in addition to neutral fat two other types of lipids, *phospholipids* and *cholesterol,* are very common throughout cells.

The usual cell contains 2 to 3 per cent lipids which are dispersed throughout the cell but are present in especially high concentrations in the cell membrane, the nuclear membrane, and the membranes lining intracytoplasmic organelles, such as the endoplasmic reticulum and the mitochondria.

The special importance of lipids in the cell is that they are either insoluble or only partially soluble in water. They combine with structural proteins to form the different membranes that separate the different water compartments of the cell from each other. The lipids of each membrane form a boundary between the solutions on the two sides of the membrane, making it impervious to many dissolved substances.

The chemical natures of the different types of lipids and their functions in the body will be discussed in Chapter 31.

Carbohydrates. In general, carbohydrates have very little structural function in the cell, but they play a major role in nutrition of the cell. Most human cells do not maintain large stores of carbohydrates, usually averaging about 1 per cent of their total mass. However, carbohydrate, in the form of glucose, is always present in the surrounding extracellular fluid so that it is readily available to the cell. The small amount of carbohydrates stored in the cells is almost entirely in the form of *glycogen,* which is an insoluble polymer of glucose.

PHYSICAL STRUCTURE OF THE CELL

The cell is not merely a bag of fluid, enzymes, and chemicals; it also contains highly organized physical structures called *organelles,* which are as important to the function of the cell as the cell's chemical constituents. For instance, without one of the organelles, the *mitochondria,* more than 95 per cent of the energy supply of the cell would cease immediately. Some principal organelles of the cell are the *cell membrane, nuclear membrane, endoplasmic reticulum, mitochondria,* and *lysosomes.* These are illustrated in Figure 2–3. Others not shown in the figure are the *Golgi complex, centrioles, cilia,* and *microtubules.*

MEMBRANES OF THE CELL

Essentially all physical structures of the cell are lined by a membrane composed primarily of lipids and proteins. All of the different membranes are similar in structure, and this common type of structure is called the "unit membrane," which will be described below in relation to the cell membrane. The different membranes include the *cell membrane,* the *nuclear membrane,* the *membrane of the endoplasmic reticulum,* and the *membranes of the mitochondria, lysosomes, Golgi complexes,* and so forth.

The Cell Membrane. The cell membrane is thin—approximately 75 to 100 Ångstroms* (7.5 to 10 nanometers), or 1/100,000 mm. — and elastic. It is composed almost entirely of proteins and lipids, with a percentage composition approximately as follows: proteins, 62 per cent; lipids, 35 per cent; polysaccharides, perhaps 3 per cent. The proteins in the cell membrane are mainly proteins called *tektins,* insoluble structural proteins having elastic properties. The lipids are approximately 60 per cent phospholipids, 25 per cent cholesterol, and 15 per cent other lipids.

The precise molecular organization of the cell membrane is unknown, but the classical description of its structure is that illustrated in Figure 2–4A. This shows a central layer of lipids covered by protein layers and a thin mucopolysaccharide layer on the outside surface. The presence of protein and mucopolysaccharide on the surfaces makes the membrane *hydrophilic,* which means that water adheres easily to the membrane. The lipid center of the membrane makes the membrane mainly impervious to lipid-insoluble substances. The small knobbed structures lying at the bases of the protein molecules in Figure 2–4A are phospholipid molecules; the fat portion of the phospholipid molecule is attracted to the central lipid phase of the cell membrane, and the polar (ionized) portion of the molecule protrudes toward either surface, where it is physically adherent to the inner and outer layers of protein. The thin layer of mucopolysaccharide on the outside of the cell membrane helps to make the outside different from the inside, thus polarizing the membrane so that the chemical reactivities of the cell's inner surface are different from those of the outer surface.

Another interpretation of the cell membrane, and one that much recent evidence supports, is that the basic structure of the membrane is almost entirely lipid, with very little protein on the two surfaces. Instead, the protein seems to be dispersed throughout the lipid where it performs many important functions, including (1) contributing to the structural strength of the membrane; (2) acting as enzymes to promote chemical reactions; (3) acting as carrier proteins for transport of substances through the membrane; and

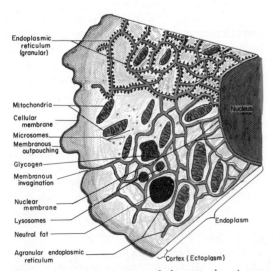

Endoplasmic
reticulum
(granular)

Mitochondria

Cellular
membrane

Microsomes

Membranous
outpouching

Glycogen

Membranous
invagination

Nuclear
membrane

Lysosomes

Neutral fat

Agranular endoplasmic
reticulum

Nucleus

Endoplasm

Cortex (Ectoplasm)

Figure 2–3. Organization of the cytoplasmic compartment of the cell.

*1 Ångstrom equals 10^{-7} mm. and 0.1 nanometer.

CELL MEMBRANE

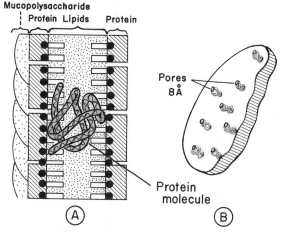

Figure 2–4. (A) Postulated molecular organization of the cell membrane. (B) Pores in the cell membrane.

(4) providing molecular breaks in the lipid substances and, therefore, providing pores through the membrane.

ENZYMES IN THE MEMBRANE. A number of protein enzymes either are dissolved in the cell membrane or are adherent to it. Many of these are specifically present in the inner protein layer of the membrane and, therefore, function at the boundary between the inner surface of the membrane and the cytoplasm to catalyze many chemical reactions. For instance, it will be pointed out in Chapter 8 that enzyme-catalyzed reactions of this type are responsible for transporting many important substances through the cell membrane. These enzymatic characteristics of the membrane also help to polarize the membrane, making the chemical reactivities of the inner surface different from those of the outer surface.

THE "UNIT" MEMBRANE. The triple layer membrane described above, with lipid in the center and two layers of protein on the two sides of the lipid, is called the *unit membrane.* This same basic structure, though in slightly different forms, is found in all the different membranes of the cell, such as the nuclear membrane and the membranes of the mitochondria, endoplasmic reticulum, lysosomes, and so forth; these will be discussed in subsequent paragraphs.

PORES IN THE MEMBRANE. The membrane is believed to have many minute pores that pass from one side to the other as shown in Figure 2–4B. These pores have never been demonstrated even with an electron micro-

scope, but it is likely that the pores in the cell membrane are caused by the presence of large protein molecules that interrupt the lipid membrane structure and extend all the way through the membrane from one side to the other, thus providing direct watery passage through the interstices of the protein molecules. One such protein molecule is illustrated in Figure 2–4A. Furthermore, when we discuss the "pores" in nerve membranes in Chapter 9, we shall see that there are probably at least two varieties of pores — and perhaps still more — each allowing free movement of different types of substances.

Functional experiments to study the movement of molecules of different sizes between the extra- and intracellular fluids have demonstrated free diffusion of molecules up to a size of approximately 8 Ångstroms (0.8 nanometer). It is through these pores that lipid-insoluble substances of very small sizes, such as water and urea molecules, are believed to pass with relative ease between the interior and exterior of the cell.

The Nuclear Membrane. The nuclear membrane is actually two "unit" membranes having a wide space in between. Each unit membrane is almost identical to the cell membrane, having lipids in its center and protein on its two surfaces, but having no mucopolysaccharide layer. Very large holes, or "pores," of several hundred Ångstroms diameter are present in the nuclear membrane as illustrated in Figure 2–10 so that almost all dissolved substances can move with ease between the fluids of the nucleus and those of the cytoplasm.

The Endoplasmic Reticulum. Figure 2–3 illustrates in the cytoplasm a network of tubular and vesicular structures, constructed of a system of unit membranes, called the *endoplasmic reticulum.* The detailed structure of this organelle is illustrated in Figure 2–5. The space inside the tubules and vesicles is filled with *endoplasmic matrix,* a fluid medium that is different from the fluid outside the endoplasmic reticulum.

Electron micrographs show that the space inside the endoplasmic reticulum is connected with the space between the two membranes of the double nuclear membrane. Also, this space is continuous with the inner chambers of the Golgi complex, to be described below. In some instances the endoplasmic reticulum connects directly through small openings with the exterior of the cell. Substances formed in different parts of the cell enter the

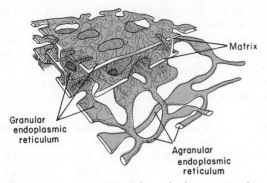

Figure 2–5. Structure of the endoplasmic reticulum. (Modified from De Robertis, Saez, and De Robertis: Cell Biology, 6th ed. W. B. Saunders Co., 1975.)

spaces of this vesicular system and are then conducted to other parts of the cell.

RIBOSOMES AND THE GRANULAR ENDOPLASMIC RETICULUM.   Attached to the outer surfaces of many parts of the endoplasmic reticulum are large numbers of small granular particles called *ribosomes*. Where these are present, the reticulum is frequently called the *granular endoplasmic reticulum*. The ribosomes are composed mainly of ribonucleic acid, which functions in the synthesis of protein in the cells, as discussed in Chapter 4.

THE AGRANULAR ENDOPLASMIC RETICULUM.   Part of the endoplasmic reticulum has no attached ribosomes. This part is called the *agranular*, or *smooth, endoplasmic reticulum.*

The agranular reticulum helps in the synthesis of lipid substances and probably also plays an important role in glycogen resorption.

Golgi Complex.   The Golgi complex, illustrated in Figure 2–6, is a specialized derivative of the endoplasmic reticulum. It has membranes similar to those of the agranular endoplasmic reticulum and is usually composed of four or more layers of thin vesicles. Electron micrographs show direct connections between the endoplasmic reticulum and parts of the Golgi complex.

The Golgi complex is very prominent in secretory cells; in these, it is located on the side of the cell from which substances will be secreted. Its function is believed to be temporary storage and condensation of secretory substances and preparation of these substances for final secretion. In addition, the Golgi complex also synthesizes carbohydrates and then combines these with protein in the Golgi cisternae to form glycoproteins, which are important secretory substances of many cells; especially important are the mucopolysaccharides that are major constituents of (1) mucus, (2) the ground substance of the interstitial spaces, and (3) the ground substance of both cartilage and bone.

Finally, the Golgi complex is involved in the formation of the lysosomes, cytoplasmic organelles that are important for digesting intracellular substances, as will be explained in a subsequent section of this chapter.

THE FLUID OF THE CYTOPLASM, AND THE DISPERSED ORGANELLES

The cytoplasm is filled with both minute and large dispersed particles and organelles ranging in size from a few Ångstroms to 3 microns (micrometers) in diameter. The clear fluid portion of the cytoplasm in which the particles are dispersed is called *hyaloplasm;* this contains mainly dissolved proteins, electrolytes, glucose, and small quantities of phospholipids, cholesterol, and esterified fatty acids.

The portion of the cytoplasm immediately beneath the cell membrane is frequently gelled into a semisolid called the *cortex,* or *ectoplasm.* The cytoplasm between the cortex and the nuclear membrane is liquefied and is called the *endoplasm.*

Among the large dispersed particles in the cytoplasm are neutral fat globules, glycogen

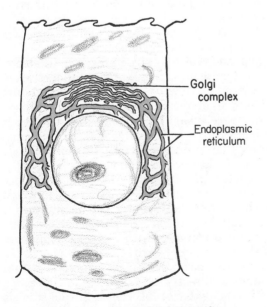

Figure 2–6. A typical Golgi complex.

granules, ribosomes, secretory granules, and two important organelles — the *mitochondria* and *lysosomes* — which are discussed below.

Colloidal Nature of the Cytoplasm. All the particles dispersed in the cytoplasm, whether the large lysosomes and mitochondria or the small granules, are hydrophilic — that is, attracted to water — because of electrical charges on their surfaces. The lining membranes of the mitochondria and lysosomes have charges on their surfaces because of the protein molecules on the membrane surfaces. The particles remain dispersed in the cytoplasm mainly because of mutual repulsion of the charges on the different particles. Thus, the cytoplasm is actually a colloidal solution.

Mitochondria. Mitochondria extract energy from the nutrients and oxygen and in turn provide most of the energy needed elsewhere in the cell for performing the cellular functions. These organelles are present in the cytoplasm of all cells, as illustrated in Figure 2–3, but the number per cell varies from a few hundred to many thousand, depending on the amount of energy required by each cell. Mitochondria are also very variable in size and shape; some are only a few hundred millimicrons (nanometers) in diameter and globular in shape while others are as large as 1 micron (micrometer) in diameter, as long as 7 microns, and filamentous in shape.

The basic structure of the mitochondrion is illustrated in Figure 2–7, which shows it to be composed mainly of two layers of unit membrane: an *outer membrane* and an *inner membrane*. Many infoldings of the inner membrane form *shelves* onto which the oxidative enzymes of the cell are attached. In addition,

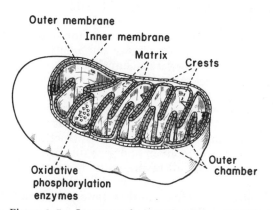

Figure 2–7. Structure of a mitochondrion. (Modified from De Robertis, Saez, and De Robertis: Cell Biology, 6th ed. W. B. Saunders Co., 1975.)

the inner cavity of the mitochondrion is filled with a gel *matrix* containing large quantities of dissolved enzymes that are necessary for extracting energy from nutrients. These enzymes operate in association with the oxidative enzymes on the shelves to cause oxidation of the nutrients, thereby forming carbon dioxide and water. The liberated energy is used to synthesize a high energy substance called *adenosine triphosphate (ATP)*. ATP is then transported out of the mitochondrion, and it diffuses throughout the cell to release its energy wherever it is needed for performing cellular functions. The function of ATP is so important to the cell that it is discussed in detail in the following chapter.

Mitochondria are probably self-replicative, which means that one mitochondrion can probably form a second one, a third one, and so on whenever there is need in the cell for increased amounts of ATP. Indeed, the mitochondria contain a special type of deoxyribonucleic acid similar to, but different from, that found in the nucleus. In Chapter 4 we shall see that deoxyribonucleic acid is the basic substance that controls replication of the entire cell; it is reasonable, therefore, to suspect that this substance might play a similar role in the mitochondrion.

Lysosomes. The lysosomes provide an intracellular digestive system that allows the cell to digest and thereby remove unwanted substances and structures, especially damaged or foreign structures, such as bacteria. The lysosome, illustrated in Figure 2–3, is a spherical organelle 250 to 750 millimicrons (nanometers) in diameter, and is surrounded by a unit membrane. It is filled with large numbers of small granules 55 to 80 Ångstroms (5.5 to 8 nanometers) in diameter which are protein aggregates of hydrolytic (digestive) enzymes. A hydrolytic enzyme is capable of splitting an organic compound into two or more parts by combining hydrogen from a water molecule with part of the compound and by combining the hydroxyl portion of the water molecule with the other part of the compound. For instance, protein is hydrolyzed to form amino acids, and glycogen is hydrolyzed to form glucose. More than a dozen different *acid hydrolases* have been found in lysosomes, and the principal substances that they digest are proteins, nucleic acids, mucopolysaccharides, and glycogen.

Ordinarily, the membrane surrounding the lysosome prevents the enclosed hydrolytic enzymes from coming in contact with other substances in the cell. However, many different conditions of the cell will break the membranes of some of the lysosomes, allowing release of the enzymes. These enzymes then split the organic substances with which they come in contact into small, highly diffusible substances, such as amino acids and glucose. Some of the more specific functions of lysosomes are discussed in the following chapter.

Microtubules. Located in many cells are fine tubular structures that are approximately 250 Å (25 nanometers) in diameter and that have lengths of 1 to many microns (micrometers). These structures, called *microtubules,* are very thin in relation to their length, but they are usually arranged in bundles, which gives them, en masse, considerable structural strength. Furthermore, microtubules are usually stiff structures that break if bent too severely. Figure 2–8 illustrates some typical microtubules that have been teased from the flagellum of a sperm. Another example of microtubules is the tubular filaments that give cilia their structural strength, radiating upward from within the cell cyto-

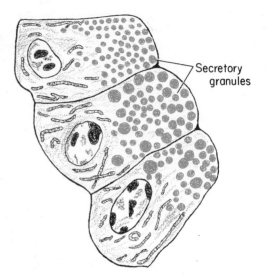

Figure 2–9. Secretory granules in acinar cells of the pancreas.

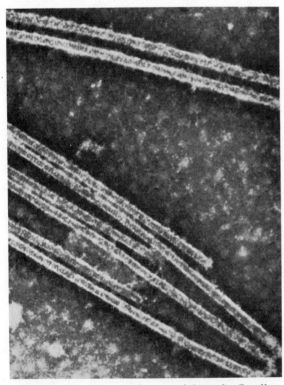

Figure 2–8. Microtubules teased from the flagellum of a sperm. (From Porter, K. R.: Ciba Foundation Symposium: Principles of Biomolecular Organization. Little, Brown and Co., 1966.)

plasm to the tip of the cilium; these will also be discussed in the following chapter.

The primary function of microtubules appears to be to act as a *cytoskeleton,* providing rigid physical structures for certain parts of cells such as the cilia just mentioned. However, the tubular nature of their structure also suggests that substances might be transported through the tubules. Indeed, cytoplasmic streaming has been observed in the vicinity of microtubules, indicating that these tubular structures might play a role in causing movement of cytoplasm.

Secretory Granules. One of the most important functions of many cells is secretion of special substances. The secretory substances are usually formed inside the cell and are held there until an appropriate time for release to the exterior. The storage depots within the cells are called secretory granules; these are illustrated by the colored spots in Figure 2–9, which shows pancreatic acinar cells that have formed and stored enzymes that will be secreted later into the intestinal tract.

Many secretory granules lie inside the tubules and vesicles of the endoplasmic reticulum and Golgi complex, while others are free in the cytoplasm and probably are formed in most if not all instances by the Golgi complex.

Other Structures and Organelles of the Cytoplasm. The cytoplasm of each cell contains two pairs of *centrioles,* which are small cylindrical structures that play a major role in cell division, as will be discussed in Chapter 4. Also, most cells contain small *lipid droplets* and *glycogen granules* that play important roles in energy metabolism of the cell. And certain cells contain highly specialized structures such as the *cilia* of ciliated cells which are actually outgrowths from the cytoplasm, and the *myofibrils* of muscle cells which are the contractile component of these cells. All of these are discussed in detail at different points in this text.

THE NUCLEUS

The nucleus is the control center of the cell. It controls both the chemical reactions that occur in the cell and reproduction of the cell. Briefly, the nucleus contains large quantities of *deoxyribonucleic acid,* which we have called *genes* for years. The genes control the formation of the protein enzymes of the cytoplasm, and in this way control cytoplasmic activities. On the other hand, to control reproduction, the genes first reproduce themselves, and after this is accomplished the cell splits by a special process called *mitosis* to form two daughter cells, each of which receives one of the two new sets of genes. These activities of the nucleus are considered in detail in Chapter 4.

The appearance of the nucleus under the microscope does not give much of a clue to the mechanisms by which it performs its control activities. Figure 2–10 illustrates the light microscopic appearance of the interphase nucleus (period between mitoses), showing darkly staining *chromatin material* throughout the *nuclear sap*. During mitosis, the chromatin material becomes readily identifiable as part of the highly structured *chromosomes,* which can then be seen easily with the light microscope. Even during the interphase of cellular activity, the chromatin material is still organized into fibrillar chromosomal structures, but this is impossible to see except in a few types of cells.

Nucleoli. The nuclei of many cells contain one or more lightly staining structures called nucleoli. The nucleolus, unlike most of the organelles that we have discussed, does not have a limiting membrane. Instead, it is simply a protein structure that contains a large amount of *ribonucleic acid* of the type found in ribosomes. The nucleolus becomes considerably enlarged when a cell is actively synthesizing proteins. The reason for this is that the genes of one of the chromosomes synthesize great quantities of ribonucleic acid that is immediately stored in the nucleolus. This is at first a loose fibrillar RNA that later condenses to form the granular ribosomes. These in turn migrate into the cytoplasm where most of them become attached to the endoplasmic reticulum.

COMPARISON OF THE ANIMAL CELL WITH PRECELLULAR FORMS OF LIFE

Many of us think of the cell as the lowest level of animal life. However, the cell is a very complicated organism, which probably required several billion years to develop after the earliest form of life, some organism similar to the present-day *virus,* first appeared on Earth. Figure 2–11 illustrates the relative sizes of the smallest known virus, a large

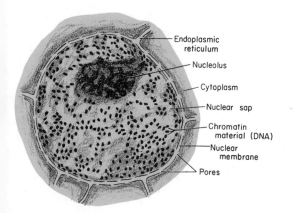

Figure 2–10. Structure of the nucleus.

Endoplasmic reticulum
Nucleolus
Cytoplasm
Nuclear sap
Chromatin material (DNA)
Nuclear membrane
Pores

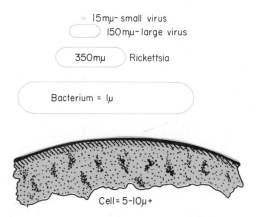

15mμ- small virus
150mμ- large virus
350mμ   Rickettsia
Bacterium = 1μ
Cell = 5-10μ+

Figure 2–11. Comparison of sizes of subcellular organisms with that of the average cell in the human body.

virus, a *rickettsia,* a *bacterium,* and a cell, showing that the cell has a diameter about 1000 times that of the smallest virus, and, therefore, a volume about 1 billion times that of the smallest virus. Correspondingly, the functions and anatomic organization of the cell are also far more complex than those of the virus.

The principal constituent of the very small virus is a *nucleic acid* embedded in a coat of protein. This nucleic acid is similar to that of the cell, and it is capable of reproducing itself if appropriate nutrients are available. Thus, the virus is capable of propagating its lineage from generation to generation, and, therefore, is a living structure in the same way that the cell and the human being are living structures.

As life evolved, other chemicals besides nucleic acid and simple proteins became integral parts of the organism, and specialized functions began to develop in different parts of the virus. A membrane formed around the virus, and inside the membrane a fluid matrix appeared. Specialized chemicals developed inside the matrix to perform special functions; many protein enzymes appeared which were capable of catalyzing chemical reactions, and, therefore, of controlling the organism's activities.

In still later stages, particularly in the rickettsial and bacterial stages, *organelles* developed inside the organism, these representing aggregates of chemical compounds that perform functions in a more efficient manner than can be achieved by dispersed chemicals throughout the fluid matrix. And, finally, in the cell, still more complex organelles developed, the most important of which is the *nucleus*. The nucleus distinguishes the cell from all lower forms of life; this structure provides a control center for all cellular activities, and it also provides for very exact reproduction of new cells generation after generation, each new cell having essentially the same structure as its progenitor.

REFERENCES

See references at end of Chapter 3.

QUESTIONS

1. Describe the physical parts of the typical cell.
2. List the chemical components of protoplasm.
3. What are some of the important functions of proteins in cells?
4. What are some of the important functions of lipids in cells?
5. Describe the cell membrane.
6. What is the function of enzymes in the cell membrane?
7. What are the functions of the endoplasmic reticulum?
8. What is the structure of the mitochondrion, and what does it contain?
9. Describe the relationship of the Golgi apparatus to the endoplasmic reticulum.
10. What types of enzymes are found in lysosomes, and what are their functions?
11. What is the most important functional substance found in the nucleus?

FUNCTIONAL SYSTEMS OF THE CELL

3

Each organelle of the cell performs its own specialized functions. For instance, the cell membrane has special abilities to transport selected substances into the cell. The lysosomes digest foreign particles or dead portions of the cell, and the digested end-products are then excreted from the cell through the cell membrane or are reused by the cell as nutrients. The mitochondria are called the "powerhouses" of the cell because they burn oxygen with the nutrients and produce the energy required for promoting most chemical reactions in the cell. The endoplasmic reticulum performs such varied functions as synthesis of substances required in the cell and transport of substances from one part of the cell to another. The Golgi complex aids the endoplasmic reticulum in some of these operations. Finally, the nucleus is the control center of the cell, controlling the types and rates of chemical reactions within the cell, the formation of new substances in the cell, and even reproduction of the cell.

In this chapter we will discuss most of the functional systems of the cell, but two of these systems are so important that they deserve special chapters of their own. First, function of the nucleus and its genes in controlling protein synthesis, intracellular chemical reactions, and reproduction will be presented in the following chapter. Second, movement of substances through the cell membrane will be discussed in Chapter 8.

INGESTION BY THE CELL

PINOCYTOSIS

If a cell is to live and grow, it must obtain nutrients and other substances from the surrounding fluids. Substances can pass through a cell membrane in three separate ways: (1) by *diffusion* through the pores in the membrane or through the membrane matrix itself; (2) by *active transport* through the membrane, a mechanism in which enzyme systems and special carrier substances "carry" the substances through the membrane; and (3) by *pinocytosis,* a mechanism by which the membrane actually engulfs some of the extracellular fluid and its contents. Transport of substances through the membrane is such an important subject that it will be considered in detail in Chapter 8, but one of these mechanisms of transport, pinocytosis, is a specialized cellular function that deserves mention here as well.

Figure 3–1 illustrates the mechanism of pinocytosis. Figure 3–1A shows three molecules of protein in the extracellular fluid approaching the surface of the cell. In Figure 3–1B these molecules have become attached to the surface, presumably by the simple process of adsorption. The presence of these proteins causes the surface tension properites of the cell surface to change in such a way that the membrane invaginates and

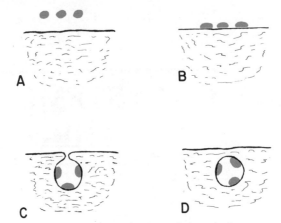

Figure 3–1. Mechanism of pinocytosis.

closes over the proteins as shown in Figure 3–1C. Immediately thereafter, the invaginated portion of the membrane breaks away from the surface of the cell, forming a *pinocytic vesicle* (Fig. 3–1D) which then penetrates deep into the cytoplasm away from the cell membrane.

Pinocytosis occurs only in response to certain types of substances that contact the cell membrane, the two most important of which are *proteins* and *strong solutions of ions*. It is especially significant that proteins cause pinocytosis, because pinocytosis is the only means by which proteins can pass through the cell membrane.

PHAGOCYTOSIS

Phagocytosis means the ingestion of large particulate matter by a cell, such as the ingestion of (a) a bacterium, (b) some other cell, or (c) particles of degenerating tissue. Though pinocytosis was discovered only after the advent of the electron microscope because pinocytic vesicles are smaller than the resolution of the light microscope, phagocytosis has been known to occur from the earliest studies using the light microscope.

The mechanism of phagocytosis is almost identical with that of pinocytosis. The particle to be phagocytized must have a surface that can become adsorbed to the cell membrane. Many bacteria have membranes that, on contact with the cell membrane, actually become miscible with the cell membrane so that the cell membrane simply spreads around the bacterium and invaginates to form a *phagocytic vesicle* containing the bac-

terium, a vesicle that is essentially the same as the pinocytic vesicle shown in Figure 3–1D but much larger. Phagocytosis will be discussed at further length in Chapter 5 in relation to function of the white blood cells.

THE DIGESTIVE ORGAN OF THE CELL — THE LYSOSOMES

Almost immediately after a pinocytic or phagocytic vesicle appears inside a cell, one or more lysosomes become attached to the vesicle and empty their digestive enzymes, the hydrolases, into the vesicle, as illustrated in Figure 3–2. Thus, a *digestive vesicle* is formed in which the hydrolases begin digesting the proteins, glycogen, nucleic acids,

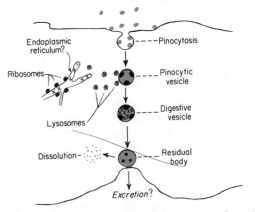

Figure 3–2. Digestion of substances in pinocytic vesicles by enzymes derived from lysosomes. (Modified from C. De Duve: Lysosomes, ed. by Reuck and Cameron. Little, Brown and Co., 1963.)

mucopolysaccharides, and other substances in the vesicle. The products of digestion are small molecules of amino acids, glucose, phosphates, and so forth that can then diffuse through the membrane of the vesicle into the cytoplasm. What is left of the digestive vesicle, called the *residual body,* is then either excreted or undergoes dissolution inside the cytoplasm. Thus, the lysosomes may be called the *digestive organs* of the cells.

Regression of Tissues and Autolysis of Cells. Often, tissues of the body regress to a much smaller size than previously. For instance, this occurs in the uterus following pregnancy, in muscles during long periods of inactivity, and in mammary glands at the end of a period of lactation. Lysosomes are probably responsible for much if not most of this regression, for one can show that the lysosomes become very active at this time. However, the mechanism by which the lack of activity in a tissue causes the lysosomes to increase their activity is completely unknown.

Another very special role of the lysosomes is the removal of damaged cells or damaged portions of cells from tissues — cells damaged by heat, cold, trauma, chemicals, disease, or any other factor. Damage to the cell causes lysosomes to rupture, and the released hydrolases begin immediately to digest the surrounding organic substances. If the damage is slight, only a portion of the cell will be removed, followed by repair of the cell. However, if the damage is severe the entire cell will be digested, a process called *autolysis.* In this way, the cell is completely removed and a new cell of the same type ordinarily is formed by mitotic reproduction of an adjacent cell to take the place of the old one.

EXTRACTION OF ENERGY FROM NUTRIENTS — FUNCTION OF THE MITOCHONDRIA

The principal nutrients from which cells extract energy are oxygen and the carbohydrates, fats, and proteins. In the human body essentially all carbohydrates are converted into glucose before they reach the cell, the proteins are converted into amino acids, and the fats are converted into fatty acids. Figure 3–3 shows oxygen and the foodstuffs — glucose, fatty acids, and amino

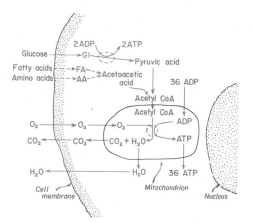

Figure 3–3. Formation of adenosine triphosphate in the cell, showing that most of the ATP is formed in the mitochondria.

acids — all entering the cell. Inside the cell, mainly inside the mitochondria, the foodstuffs react chemically with the oxygen under the influence of various enzymes that control their rates of reactions and channel the energy that is released in the proper direction.

ADENOSINE TRIPHOSPHATE (ATP)

The energy released from the nutrients is used to form adenosine triphosphate, generally called ATP, the formula for which is illustrated in Figure 3–4. Note that ATP is a nucleotide composed of the nitrogenous base *adenine,* the pentose sugar *ribose,* and three *phosphate radicals.* The last two phosphate radicals are connected with the remainder of the molecule by so-called *high energy phosphate bonds,* which are represented by the symbol $\sim$. Each of these bonds contains about 8000 calories of energy per mole of ATP under the physical conditions of the body (7000 calories under standard conditions), which is much greater than the energy stored in the average chemical bond of other organic compounds, thus giving rise to the term "high energy" bond. Furthermore, the high energy phosphate bond can be formed by the metabolic systems of the cell. The cells utilize these bonds as a means for temporary storage of the energy derived from the nutrients.

Another characteristic of ATP is that it can be split very rapidly by the functional systems of the cell, thereby deriving energy

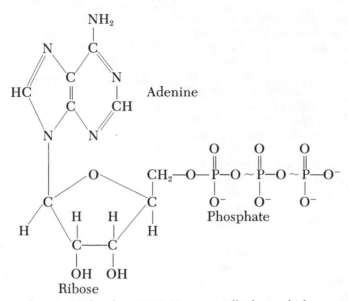

Figure 3–4. Adenosine triphosphate (ATP). Note especially the two high energy bonds (~).

as it is required to promote other cellular reactions.

ATP has a special importance in the cell because it can be used to supply energy to the various cellular functional systems, and then it can be reformed again. The ATP releases its energy by splitting one or both the high-energy phosphate bonds, as illustrated to the far right in the molecule in Figure 3–4. When the first of these bonds is split, the phosphoric acid radical is released into the surrounding fluid, and the remaining molecule becomes *adenosine diphosphate* (ADP). When the second bond is split, another phosphoric acid radical is released, and the remaining molecule then becomes *adenosine monophosphate* (AMP). As each of the high-energy phosphate bonds is split, 8000 calories of energy are released, and it is this energy that is used to perform the other functions in the cell.

Once ADP and AMP have been formed in the cell, they are then rapidly reconverted into ATP by the metabolic systems of the cell, utilizing energy derived mainly from the combination of oxygen with the various foodstuffs. This process of reconverting AMP and ADP to ATP occurs almost entirely in the mitochondria, which explains the extreme importance of these organelles to the cell.

Note especially that ATP can be degraded, then reformed again, then degraded, this process cycling again and again. Each time it is degraded, it releases energy that can be used immediately by the functional components of the cell. For this reason, ATP is frequently called the "energy currency" of the cell. Therefore, one can readily see the overall importance of ATP to cellular function. Furthermore, a major share of the chemical processes of the cell is committed to the formation of new ATP every time the old is used. Some of the more important of these chemical processes for formation of ATP are the following:

Chemical Processes for the Formation of ATP. On entry into the cells, glucose, fatty acids, and amino acids are subjected to enzymes in the cytoplasm or the nucleoplasm that convert glucose into *pyruvic acid* (a process called *glycolysis*), and convert fatty acids and most amino acids into *acetic acid*. The chemical reactions of these conversions will be presented in Chapter 31. A small amount of ADP is changed into ATP by energy released during the conversion of glucose to pyruvic acid, but this amount plays only a minor role in the overall energy metabolism of the cell.

ROLE OF THE MITOCHONDRIA. By far the major portion of the ATP formed in the cell is formed in the mitochondria. The pyruvic and acetic acids are both converted into the compound *acetyl co-A* in the cytoplasm, and this is transported along with oxygen through the mitochondrial membrane into the matrix of the mitochondrion. Here this substance is acted upon by a series of enzymes and undergoes dissolution in a sequence of chemical reactions called the *tricarboxylic acid cycle,* or *Krebs cycle.*

In the tricarboxylic acid cycle, acetyl co-A is split into its component parts, hydrogen atoms and carbon dioxide. The carbon dioxide in turn diffuses out of the mitochondria and eventually out of the cell. The hydrogen atoms combine with carrier substances and are carried to the surfaces of the shelves that protrude into the mitochondria, as shown in Figure 2–7 of the previous chapter. Attached to these shelves are the so-called *oxidative enzymes.* These, by a series of sequential reactions, cause the hydrogen atoms to combine with oxygen. The enzymes are arranged on the surfaces of the shelves in such a way that the products of one chemical reaction are immediately relayed to the next enzyme, then to the next, and so on until the complete sequence of reactions has taken place. During the course of these reactions, the energy released from the combination of hydrogen with oxygen is used to manufacture tremendous quantities of ATP from ADP and AMP. The ATP is then transported out of the mitochondrion into all parts of the cytoplasm and nucleoplasm where its energy is used to energize the functions of the cell.

ANAEROBIC FORMATION OF ATP DURING GLYCOLYSIS. Note in Figure 3–3 that no oxygen is required for the splitting of glucose to form pyruvic acid, but, even so, a small amount of ATP is formed during this process. Therefore, this mechanism (glycolysis) for providing energy to the cell is called *anaerobic energy metabolism.*

OXIDATIVE ENERGY METABOLISM IN THE MITOCHONDRIA. Note, on the other hand, that tremendous quantities of ATP are formed by the chemical reactions in the mitochondria. Because these reactions require oxygen, this mechanism for providing energy is called *oxidative energy metabolism.* Approximately 90 per cent of all ATP formed in the cell is formed in the mitochondria, in comparison with only 10 per cent outside the mitochondria. Because of the tremendous quantity of ATP formed in the mitochondria and because of the need for ATP to supply energy for all cellular functions, the mitochondria are called the *powerhouses* of the cell.

USES OF ATP FOR CELLULAR FUNCTION

ATP is used to promote three major categories of cellular functions: (1) *membrane transport,* (2) *synthesis of chemical compounds* inside the cell, and (3) *mechanical work.* Examples of these three different uses of ATP are illustrated in Figure 3–5: (a) to supply energy for the transport of glucose through the membrane, (b) to promote protein synthesis by the ribosomes, and (c) to supply the energy needed during muscle contraction.

Membrane Transport. In addition to membrane transport of glucose, energy from ATP is required for transport of sodium and potassium ions and, in certain cells, calcium ions, phosphate ions, chloride ions, hydrogen ions, and still many other special substances. Membrane transport is so important to cellular function that some cells utilize as much as 50 per cent of the ATP formed in the cells for this purpose alone.

Synthesis. In addition to synthesizing proteins, cells also synthesize phospholipids, cholesterol, purines, pyrimidines, and a great host of other substances. Synthesis of almost any chemical compound requires energy. For instance, a single protein molecule might be composed of as many as several thousand amino acids attached to each other by peptide linkages; the formation of each of these linkages requires the breakdown of three ATP molecules; thus many thousand

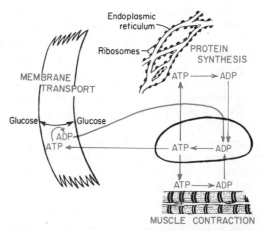

Figure 3–5. Use of adenosine triphosphate to provide energy for three of the major cellular functions: (1) membrane transport, (2) protein synthesis, and (3) muscle contraction.

ATP molecules must release their energy as each protein molecule is formed. Indeed, cells often utilize as much as 75 per cent of all the ATP formed in the cell simply to synthesize new chemical compounds; this is particularly true during the growth phase of cells.

Muscle Contraction. The final major use of ATP is to supply energy for special cells to perform mechanical work. We shall see in Chapter 10 that each contraction of a muscle fibril requires expenditure of tremendous quantities of ATP. This is true whether the fibril is in skeletal muscle, smooth muscle, or cardiac muscle. Other cells perform mechanical work in two additional ways, by *ciliary* or *ameboid motion,* both of which will be described later in this chapter. The source of energy for all these types of mechanical work is ATP.

In summary, therefore, ATP is always available to release its energy rapidly and almost explosively wherever in the cell it is needed. To replace the ATP used by the cell, other much slower chemical reactions break down carbohydrates, fats, and proteins and use the energy derived from these to form new ATP.

FUNCTIONS OF THE ENDOPLASMIC RETICULUM

The endoplasmic reticulum exists in many different forms in different cells, sometimes highly granular with a large number of ribosomes on its surface, sometimes agranular, sometimes tubular, sometimes vesicular with large shelflike surfaces, and so forth. Therefore, simply from anatomic considerations alone, it is certain that the endoplasmic reticulum performs a very large share of the cell's functions. Yet, understanding of its functions has been extremely slow to develop because of the difficulty of studying function with the electron microscope.

On the vast surfaces of the endoplasmic reticulum are adsorbed many of the protein enzymes of the cell, and perhaps still other enzymes are actually integral parts of the reticulum itself. Many of these enzymes synthesize substances in the cells, and others undoubtedly act to transport substances through the membrane of the endoplasmic reticulum from the hyaloplasm of the cytoplasm into the matrix of the reticulum or in the opposite direction. Some of the proved functions of the endoplasmic reticulum are the following:

SECRETION OF PROTEINS BY SECRETORY CELLS

Many cells, particularly cells in the various glands of the body, form special proteins that are secreted to the outside of the cells. The mechanism for this involves the endoplasmic reticulum and Golgi complex as illustrated in Figure 3–6. The ribosomes on the surface of the endoplasmic reticulum synthesize the protein that is to be secreted. This protein is either discharged directly into the tubules of the endoplasmic reticulum by the ribosomes or is immediately transported into the tubules to form small protein granules. These granules then move slowly through the tubules toward the Golgi complex, arriving there a few minutes to an hour or more later. In the Golgi complex the protein may be conjugated with carbohydrates to form glycoproteins which are a common secretory product. Also, the granules are condensed into *coalesced granules* that then are evaginated outward through the membrane of the Golgi complex into the cytoplasm of the cell to form *secretory granules*. Each of these granules carries with it part of the membrane of the Golgi apparatus which provides a membrane around the secretory granule and prevents it from dispersing in the cytoplasm. Gradually, the secretory granules move toward the surface of the cell where their membranes become miscible with the membrane of the cell itself,

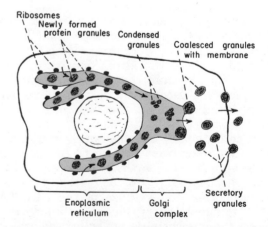

Figure 3–6. Function of the endoplasmic reticulum and Golgi complex in secreting proteins.

and in some way not completely understood they expel their substances to the exterior. It is in this way that protein digestive enzymes, for instance, are secreted by the glands of the gastrointestinal tract.

LIPID SECRETION

Almost exactly the same mechanism applies to lipid secretion. The one major difference is that lipids are synthesized by the agranular portion of the endoplasmic reticulum. The Golgi complex also provides much the same function in lipid secretion as in protein secretion, for here lipids are stored for long periods of time before finally being extruded into the cytoplasm as lipid droplets and thence to the exterior of the cells.

RELEASE OF GLUCOSE FROM GLYCOGEN STORES OF CELLS

Electron micrographs show that glycogen is stored in cells as minute granules lying in close apposition to agranular tubules of the endoplasmic reticulum. The actual chemical reactions that cause polymerization of glucose to form glycogen granules probably occur in the hyaloplasm and not in the wall of the endoplasmic reticulum, but the reticulum seems to play a role in the breakdown of glycogen (glycogenolysis) because some of the enzymes for this process are found in the endoplasmic reticulum.

OTHER POSSIBLE FUNCTIONS OF THE ENDOPLASMIC RETICULUM

The functions presented thus far for the endoplasmic reticulum have involved, first, synthesis of substances and, second, transport of these substances to the exterior of the cell. Many substances are also synthesized by the endoplasmic reticulum and are then secreted to the interior of the cell rather than to the exterior. The Golgi complex also plays a role in these internal secretory processes; indeed, large Golgi complexes are found in some cells, such as nerve cells, that do not perform any external secretory function.

Because of the multitude of different anatomic forms of the endoplasmic reticulum, it is certain that we will discover many more

functional roles that it plays besides those few that have been discussed here.

CELL MOTION

By far the most important type of cell motion that occurs in the body is that caused by the specialized muscle cells in skeletal, cardiac, and smooth muscle, which comprise almost 50 per cent of the entire body mass. The specialized functions of these cells will be discussed in Chapter 10. However, two other types of motion occur in other cells, *ameboid motion and ciliary motion.*

AMEBOID MOTION

Ameboid motion means movement of an entire cell in relation to its surroundings, such as the movement of white blood cells through tissues. Typically ameboid motion begins with protrusion of a *pseudopodium* from one end of the cell. The pseudopodium projects far out away from the cell body, and then the remainder of the cell moves toward the pseudopodium. Formerly, it was believed that the protruding pseudopodium attached itself far away from the cell and then pulled the remainder of the cell toward it. However, in recent years, new studies have changed this idea to a "streaming" concept, as follows: Figure 3–7 illustrates diagrammatically an elongated cell, the right-hand end of which is a protruding pseudopodium. The membrane of this end of the cell is continually moving forward, while the membrane at the left-hand end of the cell is continually following along as the cell moves.

It is postulated that ameboid movement is

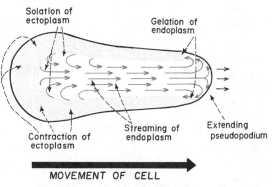

Figure 3–7.   Ameboid motion by a cell.

caused in the following way: The outer portion of the cytoplasm is in a *gel* state and is called the *ectoplasm,* whereas the central portion of the cytoplasm is in a *sol* state and is called *endoplasm.* In the gel is a contractile protein called *myxomyosin,* which contracts in the presence of ATP and calcium ions. Therefore, normally there is a continual tendency for the ectoplasm to contract. However, in response to a "chemotaxic" stimulus the ectoplasm at one end of the cell becomes thin, causing a pseudopodium to bulge outward in the direction of the chemotaxic source. Thus, the pseudopodium moves progressively forward. However, this also allows contraction and thickening of the ectoplasm at the opposite end of the cell. As some of this ectoplasm is compressed toward the center of the cell, the gel "solates." Then the dissolved ectoplasm "streams" forward through the endoplasm toward the pseudopodium. Thus, the combination of contraction of the ectoplasm, as well as solation of its inner layers, forces a stream of endoplasm forward toward the pseudopodium, pushing the membrane of the pseudopodium progressively forward. On reaching the pseudopodial end of the cell, the endoplasm turns toward the sides of the cell and then becomes gelated to form new ectoplasm. Therefore, at the tail end of the cell, ectoplasm is continually being solated while new ectoplasm is being formed at the sides of the pseudopodial end. The continuous repetition of this process makes the cell move in the direction in which the pseudopodium projects. One can readily see that this streaming movement inside the cell is analogous to the revolving movement of the track of a Caterpillar tractor.

Types of Cells That Exhibit Ameboid Motion. The most common cells to exhibit ameboid motion in the human body are the *white blood cells* moving out of the blood into the tissues in the form of *tissue macrophages* or *microphages.* However, many other types of cells can move by ameboid motion under certain circumstances. For instance, fibroblasts will move into any damaged area to help repair the damage, and even some of the germinal cells of the skin, though ordinarily completely sessile cells, will move by ameboid motion toward a cut area to repair the rent. Finally, ameboid motion is especially important in the development of the fetus, for embryonic cells often migrate long distances from the primordial sites of origin to new areas during the development of special structures.

Control of Ameboid Motion — "Chemotaxis." The most important factor that usually initiates ameboid motion is the appearance of certain chemical substances in the tissues. This phenomenon is called *chemotaxis,* and the chemical substance causing it to occur is called a *chemotaxic substance.* Most ameboid cells move toward the source of the chemotaxic substance — that is, from an area of lower concentration toward an area of higher concentration — which is called *positive chemotaxis.* However, some cells move away from the source, which is called *negative chemotaxis.*

Though the cause of chemotaxis is not known, it is believed that the chemotaxic substance alters either the elastic or the electrical properties of the cell membrane, and this in turn causes one side of the membrane and its sublying ectoplasm to thin while the other side contracts.

MOVEMENT OF CILIA

A second type of cellular motion, *ciliary movement,* is the bending of cilia along the surface of cells in the respiratory tract and in the fallopian tubes of the reproductive tract. As illustrated in Figure 3–8, a cilium looks like a minute, sharp-pointed hair that projects 3 to 4 microns (micrometers) from the surface of the cell. Many cilia can project from a single cell.

The cilium is covered by an outer layer of cell membrane, and it is supported by 11 microtubular filaments, 9 double tubular filaments located around the periphery of the cilium and 2 single tubular filaments down the center, as shown in the cross-section illustrated in Figure 3–8. Each cilium is an outgrowth of a structure that lies immediately beneath the cell membrane called the *basal body* of the cilium.

In the inset of Figure 3–8 movement of the cilium is illustrated. The cilium moves forward with a sudden rapid stroke 10 to 17 times per second, bending sharply where it projects from the surface of the cell. Then it moves backward slowly in a whiplike manner. The rapid forward movement pushes the fluid lying adjacent to the cell in the direction that the cilium moves, then the slow whiplike movement in the other direction has almost no effect on the fluid. As a re-

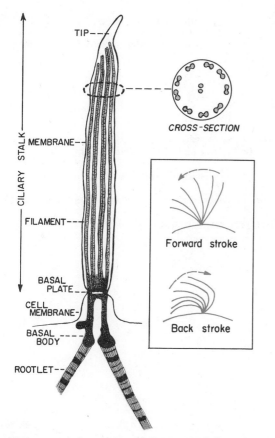

TIP

CROSS-SECTION

CILIARY STALK

MEMBRANE

FILAMENT

Forward stroke

BASAL PLATE

CELL MEMBRANE

BASAL BODY

Back stroke

ROOTLET

Figure 3–8. Structure and function of the cilium. (Modified from Satir: *Sci. Amer.*, 204 [2]:108, 1961.)

sult, fluid is continually propelled in the direction of the forward stroke. Since most ciliated cells have large numbers of cilia on their surfaces, and since ciliated cells on a surface are oriented in the same direction, this is a very satisfactory means for moving fluids from one part of the surface to another, for instance, for moving mucus out of the lungs or for moving the ovum along the fallopian tube.

Mechanism of Ciliary Movement. Though the precise way in which ciliary movement occurs is unknown, we do know the following: First, the nine double tubules and the two single tubules are all linked to each other by a complex system of protein cross-linkages; this total complex of tubules and cross-linkages is called the *axoneme.* Second, even after removal of the membrane and destruction of other elements of the cilium besides the *axoneme,* the cilium can still beat under appropriate conditions. Third, there are two necessary conditions for continued beating of the axoneme after removal of the other structures of the

cilium: (1) the presence of ATP, and (2) appropriate ionic conditions especially including appropriate concentrations of magnesium and calcium. Fourth, the cilium will continue to beat even after it has been removed from the cell body. Fifth, when the cilium bends forward, the tubules on the front edge of the bending cilium slide outward toward the tip of the cilium while the tubules on the back edge of the cilium remain in place. Sixth, three protein arms with ATPase activity project from each set of peripheral tubules toward the next set.

Given the above basic information, it has been postulated that the release of energy from ATP in contact with the ATPase arms causes the arms to "crawl" along the surfaces of the adjacent pair of tubules. If this occurs simultaneously on the two sides of the axoneme in a synchronized manner, the front tubules will crawl outward while the back tubules will remain stationary. Because of the elastic structure of the axoneme, this obviously will cause bending.

Since many cilia on a cell surface contract simultaneously in a wavelike manner, it is presumed that some synchronizing signal — perhaps an electrochemical signal over the cell surface — is transmitted from cilium to cilium. The ATP required for the ciliary movements is provided by mitochondria near the bases of the cilia from which the ATP diffuses into the cilia.

Reproduction of Cilia. Cilia have the peculiar ability to reproduce themselves. This is achieved by the *basal body,* which is almost identical to the centriole, an important structure in the reproduction of whole cells, as we shall see in the next chapter. The basal body, like the centriole, has the ability to reproduce itself by means not yet understood. After it does reproduce itself, each new basal body then grows an additional cilium from the surface of the cell.

REFERENCES

Capaldi, R. A.: A dynamic model of cell membranes. *Sci. Amer.,* 230(3):26, 1974.
De Robertis, E. D. P., Nowinski, W. W., and Saez, F. A.: Cell Biology. 4th ed. Philadelphia, W. B. Saunders Company, 1965.
Fawcett, D. W.: The Cell. Philadelphia, W. B. Saunders Company, 1966.
Giese, A. C.: Cell Physiology. 4th ed. Philadelphia, W. B. Saunders Company, 1973.
Hinkle, P. C., and McCarty, R. E.: How cells make ATP. *Sci. Amer.,* 238(3):104, 1978.

Rasmussen, H., and Goodman, D. B. P.: Relationships between calcium and cyclic nucleotides in cell activation. *Physiol. Rev., 57*:421, 1977.

Satir, B.: The final steps in secretion. *Sci. Amer., 233(4)*:28, 1975.

Satir, P.: How cilia move. *Sci. Amer., 231(4)*:44, 1974.

Stephens, R. E., and Edds, K. T.: Microtubules: structure, chemistry, and function. *Physiol. Rev., 56*:709, 1976.

QUESTIONS

1. Describe pinocytosis and phagocytosis.
2. How do lysosomes enter into the formation of the digestive vesicle?
3. Describe the function of adenosine triphosphate in the cell and also its relationship to the mitochondria.
4. What is the difference between anaerobic energy metabolism and oxidative energy metabolism?
5. What are the roles of the endoplasmic reticulum and Golgi complex in the secretion of proteins?
6. What functions does the endoplasmic reticulum perform in relation to lipid secretion and glucose release from glycogen?
7. Describe the mechanisms of ameboid and ciliary motion.

GENETIC CONTROL OF CELL FUNCTION — PROTEIN SYNTHESIS AND CELL REPRODUCTION

4

Almost everyone knows that the genes control heredity from parents to children, but most persons do not realize that the same genes control the reproduction of and the day-by-day function of all cells. The genes control function of the cell by determining what substances will be synthesized within the cell — what structures, what enzymes, what chemicals. Figure 4–1 illustrates the general schema by which the genes control cellular function. The gene, which is a nucleic acid called *deoxyribonucleic acid (DNA)*, automatically controls the formation of another nucleic acid, *ribonucleic acid (RNA)*, which then spreads throughout the cell and controls the formation of the different proteins. Some of these proteins are *structural proteins* which, in association with various lipids, form the structures of the various organelles that were discussed in the preceding chapter. But by far the majority of the proteins are *enzymes* that catalyze the different chemical reactions in the cells. For instance, enzymes promote all the oxidative reactions that supply energy to the cell, and they promote the synthesis of various chemicals such as lipids, glycogen, adenosine triphosphate, and others.

THE GENES

The genes are contained in long, double-stranded, helical molecules of *deoxyribonucleic acid (DNA)* having molecular weights usually measured in the millions. A very short segment of such a molecule is illustrat-

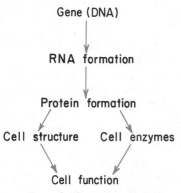

Gene (DNA)

RNA formation

Protein formation

Cell structure   Cell enzymes

Cell function

Figure 4–1. General schema by which the genes control cell function.

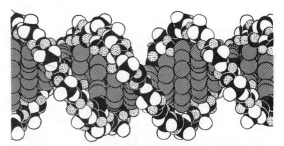

Figure 4–2. The helical, double-stranded structure of the gene. The outside strands are composed of phosphoric acid and the sugar deoxyribose. The internal molecules connecting the two strands of the helix are purine and pyrimidine bases; these determine the "code" of the gene.

33

PHOSPHORIC ACID:

DEOXYRIBOSE:

BASES:

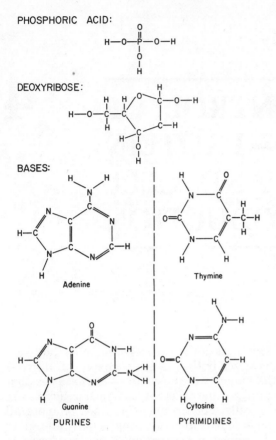

Adenine

Thymine

Guanine

Cytosine

PURINES

PYRIMIDINES

Figure 4–3. The basic building blocks of DNA.

PAIR #1

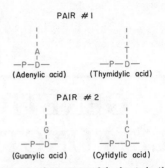

A   T

—P—D—   —P—D—

(Adenylic acid)   (Thymidylic acid)

PAIR #2

G   C

—P—D—   —P——D—

(Guanylic acid)   (Cytidylic acid)

Figure 4–5. Combinations of the basic building blocks of DNA to form nucleotides. (A, adenine; C, cytosine; D, deoxyribose; G, guanine; P, phosphoric acid; T, thymine.) Note that there are four basic nucleotides that make up DNA, and these always occur together in two pairs.

ed in Figure 4–2. This molecule is composed of several simple chemical compounds arranged in a regular pattern explained in the following few paragraphs.

The Basic Building Blocks of DNA. Figure 4–3 illustrates the basic chemical compounds involved in the formation of DNA. These include *phosphoric acid,* a sugar called *deoxyribose,* and four nitrogenous bases (two purines, *adenine* and *guanine,* and two pyrim-

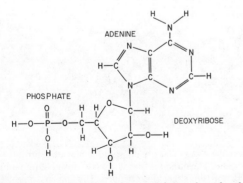

ADENINE

PHOSPHATE

DEOXYRIBOSE

Figure 4–4. Adenylic acid, one of the nucleotides that make up DNA.

idines, *thymine* and *cytosine*). The phosphoric acid and deoxyribose form the two helical strands of DNA, and the bases lie between the strands and connect them together, as illustrated by the colored circles making up the core of the DNA molecule in Figure 4–2.

The Nucleotides. The first stage in the formation of DNA is the combination of one molecule of phosphoric acid, one molecule of deoxyribose, and one of the four bases to form a nucleotide. One such nucleotide, adenylic acid, is illustrated in Figure 4–4. Four separate nucleotides are thus formed, one for each of the four bases: *adenylic, thymidylic, guanylic,* and *cytidylic acids.* Figure 4–5 illustrates simple symbols for all the four basic nucleotides that form DNA.

Note also in Figure 4–5 that the nucleotides are separated into two *complementary pairs.* Adenylic acid and thymidylic acid form one pair. Guanylic acid and cytidylic acid form the other pair. The bases of each pair can attach loosely (by hydrogen bonding) to each other, thus providing the means by which the two strands of the DNA helix are bound together: One nucleotide of a pair is on one strand of DNA, the other nucleotide of that pair is in a corresponding position on the other strand, and these are bound together by loose and reversible bonds between the nucleotide bases.

Organization of the Nucleotides to Form DNA. Figure 4–6 illustrates the manner in which multiple numbers of nucleotides are bound together to form DNA. Note that these are combined in such a way that phosphoric acid and deoxyribose alternate with each other in the two separate strands, and these

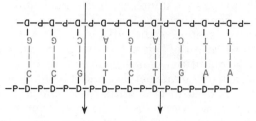

Figure 4–6. Combination of deoxyribose nucleotides to form DNA.

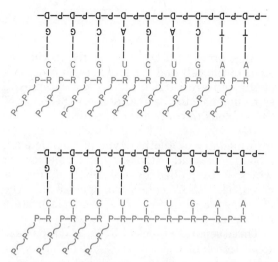

Figure 4–7. Combination of ribose nucleotides with a strand of DNA to form a molecule of ribonucleic acid (RNA) that carries the DNA code from the gene to the cytoplasm. The wavy lines represent high energy phosphate bonds that energize formation of the RNA strand.

strands are held together by the respective complementary pairs of bases. Thus, in Figure 4–6 the sequence of complementary pairs of bases is CG, CG, GC, TA, CG, TA, GC, AT, and AT. However, the bases are bound together by very loose hydrogen bonding, represented in the figure by dashed lines. Because of the looseness of these bonds, the two strands can pull apart with ease, and they do so many times during the course of their function in the cell.

Now, to put the DNA of Figure 4–6 into its proper physical perspective, one needs merely to pick up the two ends and twist them into a helix. Ten pairs of nucleotides are present in each full turn of the helix in the DNA molecule, as illustrated in Figure 4–2.

THE GENETIC CODE

The importance of DNA lies in its ability to control the formation of other substances in the cell. It does this by means of the so-called genetic code. When the two strands of a DNA molecule are split apart, this exposes a succession of purine and pyrimidine bases projecting to the side of each strand. It is these projecting bases that form the code.

Research studies in the past few years have demonstrated that so-called *code words* are present on the separated DNA strand. These consist of "triplets" of bases — that is, each three successive bases are a code word. The successive code words control the sequence of amino acids in a protein molecule during its synthesis in the cell. Note in Figure 4–6 that each of the two strands of the DNA molecule carries its own genetic code. For instance, the top strand, reading from left to right, has the genetic code GGC, AGA, CTT, the code words being separated from each other by the arrows. As we follow this genetic code through Figures 4–7 and 4–8, we shall see

that these three code words are responsible for placement of the three amino acids, *proline, serine,* and *glutamic acid,* in a molecule of protein. Furthermore, these three amino acids will be lined up in the protein molecule in exactly the same way that the genetic code is lined up in this strand of DNA.

It is also important that some code words do not cause amino acids to incorporate into proteins but, instead, perform other functions in the synthesis of protein molecules, such as initiating and stopping the formation of a protein molecule. For instance, some DNA molecules of viruses have molecular weights of 40 million or more and cause the formation of many more than a single protein molecule — 8, 10, 20, or more molecules. The code words that do not cause incorporation of amino acids into a protein molecule signal when to stop the formation of one protein and to start the formation of another.

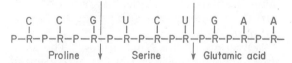

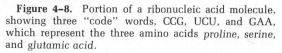

Figure 4–8. Portion of a ribonucleic acid molecule, showing three "code" words, CCG, UCU, and GAA, which represent the three amino acids *proline, serine,* and *glutamic acid.*

RIBONUCLEIC ACID (RNA)

Since almost all DNA is located in the nucleus of the cell and yet most of the functions of the cell are carried out in the cytoplasm, some means must be available for the genes of the nucleus to control the chemical reactions of the cytoplasm. This is done through the intermediary of another type of nucleic acid, ribonucleic acid (RNA), the formation of which is controlled by the DNA of the nucleus. The RNA is then transported into the cytoplasmic cavity where it controls protein synthesis.

Three separate types of RNA are important to protein synthesis: *messenger RNA, transfer RNA,* and *ribosomal RNA.* Before we describe the functions of these different RNA's in the synthesis of proteins, let us see how DNA controls the formation of RNA.

Synthesis of RNA.   One strand of the DNA molecule, which contains the genes, acts as a template for synthesis of RNA molecules. (The other strand of the DNA has no genetic function but does function for replication of the gene itself, which will be discussed later in the chapter.) The code words in DNA cause the formation of *complementary* code words (or *codons*) in RNA. The stages of RNA synthesis are as follows:

The Basic Building Blocks of RNA.   The basic building blocks of RNA are almost the same as those of DNA except for two differences. First, the sugar deoxyribose is not used in the formation of RNA. In its place is another sugar of very slightly different composition, *ribose.* The second difference in the basic building blocks is that thymine is replaced by another pyrimidine, *uracil.*

Formation of RNA Nucleotides.   The basic building blocks of RNA first form nucleotides exactly as described above for the synthesis of DNA. Here again, four separate nucleotides are used in the formation of RNA. These nucleotides contain the bases *adenine, guanine, cytosine,* and *uracil,* respectively, the uracil replacing the thymine found in the four nucleotides that make up DNA. Also, uracil takes the place of thymine in pairing with the purine adenine, as we shall see in the following paragraphs.

Activation of the Nucleotides.   The next step in the synthesis of RNA is activation of the nucleotides. This occurs by addition to each nucleotide of two phosphate radicals to form triphosphates. These last two phosphates are combined with the nucleotide by *high energy phosphate bonds* derived from the energy system of the cell.

The result of this activation process is that large quantities of energy are made available to each of the nucleotides, and it is this energy that is used in promoting the chemical reactions that eventuate in the formation of the RNA chain.

Combination of the Activated Nucleotides with the DNA Strand.   The next stage in the formation of RNA is the splitting apart of the two strands of the DNA molecule. Then, activated nucleotides become attached to the bases on the DNA strand that contains the genes, as illustrated by the top panel in Figure 4–7. Note that ribose nucleotide bases always combine with the deoxyribose bases in the following combinations:

DNA base	RNA base
guanine	cytosine
cytosine	guanine
adenine	uracil
thymine	adenine

Polymerization of the RNA Chain.   Once the ribose nucleotides have combined with the DNA strand as shown in the top panel of Figure 4–7, they are then lined up in proper sequence to form code words (codons) that are complementary to those in the DNA molecule. At this time an enzyme, *RNA polymerase,* causes the two extra phosphates on each nucleotide to split away and, at the same time, to liberate enough energy to cause bonds to form between the successive ribose and phosphoric acid radicals of the adjacent nucleotides. As this happens, the RNA strand automatically separates from the DNA strand and becomes a free molecule of RNA. This bonding of the ribose and phosphoric acid radicals and the simultaneous splitting of the RNA from the DNA is illustrated in the lower part of Figure 4–7.

Once the RNA molecules are formed, they diffuse into all parts of the cytoplasm where they perform further functions that eventually lead to protein formation.

MESSENGER RNA—THE PROCESS OF "TRANSCRIPTION"

The type of RNA that carries the genetic code to the cytoplasm for formation of pro-

Table 4–1. RNA Codons for the Different Amino Acids and for Start and Stop

Amino Acid			RNA Codons			
Alanine	GCU	GCC	GCA	GCG		
Arginine	CGU	CGC	CGA	CGG	AGA	AGG
Asparagine	AAU	AAC				
Aspartic acid	GAU	GAC				
Cysteine	UGU	UGC				
Glutamic acid	GAA	GAG				
Glutamine	CAA	CAG				
Glycine	GGU	GGC	GGA	GGG		
Histidine	CAU	CAC				
Isoleucine	AUU	AUC	AUA			
Leucine	CUU	CUC	CUA	CUG	UUA	UUG
Lysine	AAA	AAG				
Methionine	AUG					
Phenylalanine	UUU	UUC				
Proline	CCU	CCC	CCA	CCG		
Serine	UCC	UCC	UCA	UCG		
Threonine	ACU	ACC	ACA	ACG		
Tryptophan	UGG					
Tyrosine	UAU	UAC				
Valine	GUU	GUC	GUA	GUG		
Start (CI)	AUG					
Stop (CT)	UAA	UAG	UGA			

teins is called *messenger RNA*. Molecules of messenger RNA are usually composed of several hundred to several thousand nucleotides in a *single strand,* and this strand contains codons that are exactly complementary to the code words of the genes. Figure 4–8 illustrates a small segment of a molecule of messenger RNA. Its codons are CCG, UCU, and GAA. These are the codons for proline, serine, and glutamic acid. If we now refer back to Figure 4–7, we see that the events recorded in this previous figure represent transfer of this particular genetic code from the DNA strand to the RNA strand. This process of transferring the genetic code from DNA to messenger RNA is called *transcription.*

Messenger RNA molecules are long, straight strands that remain suspended in the cytoplasm. These molecules migrate to the ribosomes where protein molecules are manufactured, which is explained below.

RNA Codons. Table 4–1 gives the RNA codons for the 20 common amino acids found in protein molecules. Note that several of the amino acids are represented by more than one codon; and, as was pointed out before, some codons represent such signals as "start manufacturing a protein molecule" or "stop manufacturing a protein molecule." In Table 4–1, these two codons are designated CI for "chain-initiating" and CT for "chain-terminating."

TRANSFER RNA

Another type of RNA that plays a prominent role in protein synthesis is called *transfer RNA* because it transfers amino acid molecules to protein molecules as the protein is synthesized. There is a separate type of transfer RNA for each of the 20 amino acids that are to be incorporated into proteins. Furthermore, each type of transfer RNA combines specifically with only one type of amino acid — one and no more. Transfer RNA then acts as a *carrier* to transport its specific type of amino acid to the ribosomes where protein molecules are to be formed. In the ribosomes, each specific type of transfer RNA recognizes a particular code word on the messenger RNA, as is described below, and thereby delivers the appropriate amino acid to the appropriate place in the chain of the newly forming protein molecule.

Transfer RNA contains only about 80 nucleotides and therefore is a relatively small molecule in comparison with messenger RNA. It is a folded chain of nucleotides with a cloverleaf appearance similar to that illustrated in Figure 4–9. At one end of the molecule is always an adenylic acid nucleotide; it is to this that the transported amino acid attaches to a hydroxyl group of the ribose in the adenylic acid. A specific enzyme causes this attachment for each specific type of transfer RNA; this enzyme also determines the type of amino acid that will attach to the respective type of transfer RNA.

Since the function of transfer RNA is to cause attachment of a specific amino acid to a forming protein chain, it is essential that

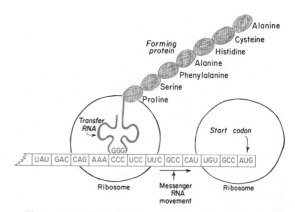

Figure 4–9. Postulated mechanism by which a protein molecule is formed in ribosomes in association with messenger RNA and transfer RNA.

each type of transfer RNA also have specificity for a particular codon in the messenger RNA. The specific prosthetic group in the transfer RNA that allows it to recognize a specific codon is called the *anticodon,* and this is located approximately in the middle of the transfer RNA molecule (at the bottom of the cloverleaf configuration illustrated in Figure 4–9). During formation of a protein molecule, the anticodon bases combine loosely by hydrogen bonding with the codon bases of the messenger RNA. In this way the respective amino acids are lined up one after another along the messenger RNA chain, thus establishing the appropriate sequence of amino acids in the protein molecule.

RIBOSOMAL RNA

The third type of RNA in the cell is that found in the ribosomes; it constitutes between 40 and 50 per cent of the ribosome. The remainder of the ribosome is protein.

The ribosomal RNA exists, along with the ribosomal protein, in particles of two different sizes. The transfer RNA, along with its attached amino acid, first binds with the smaller particle. Then as the messenger RNA passes through the ribosome, the amino acid is released to the forming protein while the transfer RNA is released back into the cytoplasm to combine again with another molecule of amino acid of the same type. The larger ribosomal particle provides the enzymes that promote peptide linkage between the successive amino acids, and various fractions of the ribosomal protein also enter into some of the intermediate chemical reactions that occur in the overall process. Thus the ribosome acts as a manufacturing plant in which the protein is formed.

Formation of Ribosomes in the Nucleolus. In the previous chapter it was pointed out that the *nucleolus* is a specialized structure of the nucleus that functions in association with one of the chromosomes. Also, the nucleolus contains a moderate concentration of RNA, which is formed by the process of transcription by the DNA of the associated chromosome. In the nucleolus the RNA is complexed with different types of proteins to form granular condensation products. These products are primordial forms of the *ribosomes.* The ribosomes are then released from the nucleolus and pass through the large

pores of the nuclear membrane to migrate to almost all parts of the cytoplasm. In the cytoplasm most of the ribosomes become attached to the outer surfaces of the endoplasmic reticulum, and it is here that they perform their function of manufacturing protein molecules. It is believed that many, if not most, of the protein molecules that are manufactured pass directly into the cisternae of the endoplasmic reticulum, then to be transported within the endoplasmic reticulum tubules to other parts of the cell, as was described in the previous chapter.

FORMATION OF PROTEINS IN THE RIBOSOMES – THE PROCESS OF "TRANSLATION"

When a molecule of messenger RNA comes in contact with a ribosome, it travels through the ribosome, and as it travels through, a protein molecule is formed — a process called *translation.* This is illustrated in Figure 4–9. Thus, the ribosome reads the code of the messenger RNA in much the same way that a tape is "read" as it passes through the playback head of a tape recorder. When a "stop" (or "chain-terminating") codon slips past the ribosome, the end of a protein molecule is signaled, and the entire molecule is freed into the cytoplasm.

It is especially important to note that a messenger RNA can cause the formation of a protein molecule in any ribosome, and that there is no specificity of ribosomes for given types of protein. The ribosome seems to be simply the physical structure in which or on which the chemical reactions take place.

Peptide Linkage. The successive amino acids in the newly forming protein chain combine with each other according to the following typical reaction:

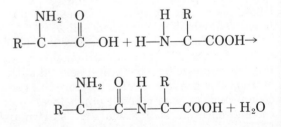

In this chemical reaction, a hydroxyl radical is removed from the COOH portion of one amino acid while a hydrogen of the NH₂ portion of the other amino acid is removed. These combine to form water, and the two reactive sites left on the two successive amino acids combine. This process is called *peptide linkage.*

SYNTHESIS OF OTHER SUBSTANCES IN THE CELL

Many hundreds or thousands of different types of protein enzymes formed in the manner just described control essentially all the other chemical reactions that take place in cells. These enzymes promote synthesis of lipids, glycogen, purines, pyrimidines, and hundreds of other substances. We will discuss many of these synthetic processes in relation to carbohydrate, lipid, and protein metabolism in Chapter 31. It is by means of all these different substances that the many functions of the cells are performed.

CONTROL OF GENETIC FUNCTION AND BIOCHEMICAL ACTIVITY IN CELLS

There are basically two different methods by which the biochemical activities in the cell are controlled. One of these can be called *genetic regulation,* in which the activities of the genes themselves are controlled, and the other can be called *enzyme regulation,* in which the rates of activities of the enzymes within the cell are controlled.

GENETIC REGULATION

Figure 4–10 illustrates the general plan by which function of the genes is regulated. At the top of the figure is a *regulatory gene* that has the ability to regulate the degree of activity of other genes. It does this by controlling the formation of a *repressor substance,* a compound of small molecular weight which in turn represses the activities of other genes. The repressor substance does not act directly on these other genes but instead on a small part of the DNA strand called an *operator,* which lies adjacent to a group of several related genes that are to be controlled. When not repressed the genetic operator excites the genes, but when the operator is repressed by the repressor substance, the genes become inactive.

The genes controlled by the operator are

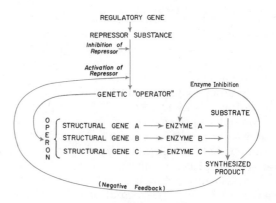

Figure 4–10. Control system that regulates function of genes. See text for detailed explanation.

called *structural genes;* it is they that cause the formation of RNA and therefore also cause the eventual formation of enzymes in the cell. The group of structural genes under the influence of each genetic operator is known as an *operon.* In Figure 4–10, the structural genes are shown to initiate the formation of enzymes A, B, and C, which in turn control biochemical reactions in the cell leading to the formation of some specific chemical product.

As noted in Figure 4–10, the repressor substance from the regulatory gene can be either *inhibited* or *activated.* If inhibited, the genetic operator is no longer repressed and therefore becomes very active. As a result, the biochemical synthesis proceeds unabated. On the other hand, if the repressor substance is activated, it immediately represses the genetic operator, and the events leading to the synthesized product cease within minutes or hours.

Control of Cellular Constituents by Feedback Repression of the Genes. The importance of the genetic regulatory system of Figure 4–10 is that it allows the concentrations of different substances in the cell to be controlled. The mechanism of this is the following: The synthesized product often can *activate* the repressor substance, and the repressor substance then *inactivates* the genetic operator, leading to decreased or ceased production of the enzymes required for forming the synthesized product. Thus, *negative feedback* occurs in such a way that the concentration of the synthesized product controls the product's rate of synthesis. When its concentration becomes too great, the rate of synthesis falls; when its concentration becomes too low, its rate of synthesis increases.

Genetic control systems of this type are

especially important in controlling the intracellular concentrations of amino acids, amino acid derivatives, and some of the intermediary substances of carbohydrate, lipid, and protein metabolism.

Induction of Genetic Activity and Enzyme Formation. Note in Figure 4–10 that the repressor substance can be inhibited as well as activated. When it is inhibited, the repressor no longer holds the operon in abeyance, and the structural genes operate to their full capacity in forming enzymes. Therefore, a simple way to "induce" the formation of a particular group of enzymes is to inhibit the repressor. This is one of the means by which hormones act to increase or decrease specific cellular activities. For instance, the steroid hormone aldosterone increases the quantities of enzymes required for transport of sodium through membranes, and the steroid sex hormones induce the formation of enzymes required for protein synthesis in the sex organs. These effects will be discussed in later chapters in relation to the functions of these hormones.

CONTROL OF ENZYME ACTIVITY

In the same way that inhibitors and activators can affect the genetic regulatory system, so also can the enzymes themselves be directly controlled by other inhibitors or activators. This, then, represents a second mechanism by which cellular biochemical functions can be controlled.

Enzyme Inhibition. A great many of the chemical substances formed in the cell have a direct feedback effect on the respective enzyme systems that synthesize them. Thus, in Figure 4–10 the synthesized product is shown to act directly back on enzyme A to inactivate it. If enzyme A is inactivated, then none of the substrate will begin to be converted into the synthesized product. Almost always in enzyme inhibition, the synthesized product acts on the first enzyme in a sequence, rather than on the subsequent enzymes. One can readily recognize the importance of inactivating this first enzyme: It prevents buildup of intermediary products that will not be utilized.

This process of enzyme inhibition is another example of negative feedback control; it is responsible for controlling the intracellular concentrations of some of the amino acids that are not controlled by the genetic mechanism as well as the concentrations of many of the purines, pyrimidines, vitamins, and other substances.

Enzyme Activation. Enzymes that are either normally inactive or that have been inactivated by some inhibitor substance can often be activated. An example of this is the action of cyclic adenosine monophosphate (AMP) in activating the enzymes that split glycogen so that the released glucose molecules can be used to form high energy ATP, as discussed in the previous chapter. When the adenosine triphosphate is used for energy in a cell, a considerable amount of cyclic AMP begins to be formed as a breakdown product of the ATP; the presence of this cyclic AMP in the cell indicates that the reserves of ATP have approached a low ebb. The cyclic AMP immediately activates the glycogen-splitting enzyme phosphorylase, liberating glucose molecules that are rapidly used for replenishment of the ATP stores. Thus, in this case the cyclic AMP acts as an enzyme activator and thereby helps to control intracellular ATP concentration.

Another interesting instance of both enzyme inhibition and enzyme activation occurs in the formation of the purines and the pyrimidines. These substances are needed by the cell in approximately equal quantities for formation of DNA and RNA. When purines are formed, they inhibit the enzymes that are required for formation of additional purines, but they activate the enzymes for formation of the pyrimidines. Conversely, the pyrimidines inhibit their own enzymes but activate the purine enzymes. In this way there is continual cross-feed between the synthesizing systems for these two substances, resulting in almost exactly equal amounts of the two substances in the cells at all times.

Summary. In summary, there are two different methods by which the cells control proper proportions and proper quantities of different cellular constituents: (1) the mechanism of genetic regulation and (2) the mechanism of enzyme regulation. The genes can be either activated or inhibited, and, likewise, the enzymes can be either activated or inhibited. Furthermore, it is principally the substances synthesized by the enzymes that cause activation or inhibition; but on occasion, substances from without the cell (especially some of the hormones which will be discussed later in this text) also control the intracellular biochemical reactions.

CELL REPRODUCTION

Most cells are continually growing and reproducing. The new cells take the place of the old ones that die, thus maintaining a complete complement of cells in the body.

Also, one can remove most types of cells from the human body and grow them in tissue culture where they will continue to grow and reproduce so long as appropriate nutrients are supplied and so long as the end-products of the cells' metabolism are not allowed to accumulate in the nutrient medium. Thus, the life lineage of most cells is indefinite.

As is true of almost all other events in the cell, reproduction also begins in the nucleus itself. The first step is *replication (duplication) of all DNA in the chromosomes.* The next step is division of the two sets of DNA between two separate nuclei. And the final step is splitting of the cell itself to form two new daughter cells, a process called *mitosis.*

The complete life cycle of a cell that is not inhibited in some way is about 10 to 30 hours from reproduction to reproduction, and the period of mitosis lasts for approximately one-half hour. The period between mitoses is called *interphase.* However, in the body there are almost always inhibitory controls that slow or stop the uninhibited life cycle of the cell and give cells life cycle periods that vary from as little as 10 hours for stimulated bone marrow cells to an entire lifetime of the human body for nerve cells.

REPLICATION OF THE DNA

The DNA is reproduced several hours before mitosis takes place, and the duration of DNA replication is only about one hour. The DNA is duplicated only once. The net result is two exact duplicates of all DNA, which respectively become the DNA in the two new daughter cells that will be formed at mitosis. Following replication of the DNA, the nucleus continues to function normally for several hours before mitosis begins abruptly.

Chemical and Physical Events. The DNA is duplicated in almost exactly the same way that RNA is formed from DNA. First, the two strands of the DNA helix of the gene pull apart. Second, each of these strands combines with deoxyribose nucleotides of the four types described early in the chapter, and complementary DNA strands are formed. The only difference between this formation of the new strands of DNA and the formation of an RNA strand is that the new strands of DNA remain attached to the old strands that have formed them, thus forming two new double-stranded DNA helixes.

THE CHROMOSOMES AND THEIR REPLICATION

The chromosomes consist of two major parts: DNA and protein. The protein consists of many small molecules of *histones.* The DNA is bound loosely with the protein and sometimes during the life cycle of the cell seems to become separated from the protein. The combination of the two is known as *nucleoprotein.*

Recent experiments indicate that all the DNA of a particular chromosome is arranged in one long double helix and that the genes are attached end-on-end to each other. Such a molecule in the human being, if spread out linearly, would be approximately 7.5 cm. long or several thousand times as long as the diameter of the nucleus itself; but the experiments also indicate that this long double helix is folded or coiled like a spring and is held in this position by its linkages to protein molecules. The protein has nothing to do with the genetic potency of the chromosome, and the protein molecules can be replaced by new protein molecules without any alteration in the functions of the genes.

Replication of the chromosomes follows as a natural result of replication of the DNA strand. When the new double helix separates from the original double helix, it presumably carries some of the old protein with it or combines with new protein, the DNA acting as the backbone of the newly replicated chromosome and the protein acting only as an accessory to the chromosomal structure.

Number of Chromosomes in the Human Cell. Each human cell contains 46 chromosomes arranged in 23 pairs. In general, the genes in the two chromosomes of each pair are almost identical with each other, so that it is usually stated that the different genes exist in pairs, though this is not always the case, especially for the sex chromosomes, as explained in Chapter 37.

MITOSIS

The actual process by which the cell splits into two new cells is called mitosis. Once the genes have been duplicated and each chromosome has split to form two new chromosomes, mitosis follows automatically, almost without fail, within a few hours.

The Mitotic Apparatus. The first event of mitosis takes place in the cytoplasm, occurring during the latter part of interphase in the small structures called *centrioles*. As illustrated in Figure 4–11, two pairs of centrioles lie close to each other near one pole of the nucleus. Each centriole is a small cylindrical body about 0.4 micron (micrometer) long and about 0.15 micron in diameter, consisting mainly of nine parallel, tubular structures arranged around the inner wall of the cylinder. The two centrioles of each pair lie at right angles to each other.

Insofar as is known, the two pairs of centrioles remain dormant during interphase until shortly before mitosis is to take place. At that time, the two pairs begin to move apart from each other. This is caused by protein microtubules growing between the respective pairs and actually pushing them apart. At the same time, microtubules grow radially away from each of the pairs. Some of these penetrate the nucleus. The set of microtubules connecting the two centriole pairs is called the *spindle,* and the entire set of microtubules plus the two pairs of centrioles is called the *mitotic apparatus*.

Prophase. The first stage of mitosis, called *prophase,* is shown in Figure 4–11 *A, B,* and *C*. While the spindle is forming, the *chromatin material* of the nucleus (the DNA), which in interphase consists of long loosely coiled strands, becomes shortened into well-defined chromosomes.

Prometaphase. During this stage (Fig. 4–11*D*) the nuclear envelope dissolutes, and microtubules from the forming mitotic apparatus become attached to the chromosomes. This attachment always occurs at the same point on each chromosome, at a small condensed portion called the *centromere*.

Metaphase. During metaphase (Fig. 4–11*E*) the centriole pairs are pushed far apart by the growing spindle, and the chromosomes are thereby pulled tightly by the attached microtubules to the very center of the cell, lining up in the equatorial plane of the mitotic spindle.

Anaphase. With still further growth of the spindle, each pair of chromosomes is now broken apart, a stage of mitosis called anaphase (Fig. 4–11*F*). A microtubule connect-

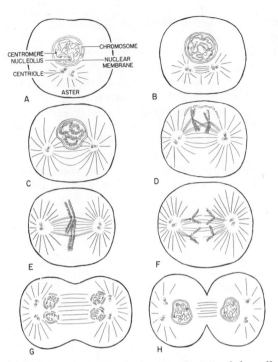

Figure 4–11. Stages in the reproduction of the cell. A and B, late interphase; C and D, prophase; E, metaphase; F, anaphase; G and H, telophase. (Redrawn from Mazia: *Sci. Amer.* 205[3]:102, 1961.)

ing with one pair of centrioles pulls one chromosome, and a microtubule connecting with the other centriole pair pulls the opposite chromosome. Thus, all 46 pairs of chromosomes are separated, forming 46 daughter chromosomes that are pulled toward one mitotic spindle and another 46 duplicate chromosomes that are pulled toward the other mitotic spindle.

Telophase. In telophase (Figs 4–11*G* and *H*) the mitotic spindle grows still longer, pulling the two sets of daughter chromosomes completely apart. Then the mitotic apparatus dissolutes and a new nuclear membrane develops around each set of chromosomes, this membrane perhaps being formed from portions of the endoplasmic reticulum that are already present in the cytoplasm. Shortly thereafter, the cell pinches in two midway between the two nuclei, for reasons totally unexplained at present.

Note also, that each of the two pairs of centrioles is replicated during telophase, the mechanism of which is not understood. These new pairs of centrioles remain dormant through the next interphase until a mitotic apparatus is required for the next cell division.

CONTROL OF CELL GROWTH AND REPRODUCTION

Cell growth and reproduction usually go together; growth normally leads to replication of the DNA of the nucleus, followed a few hours later by mitosis.

In the normal human body, regulation of cell growth and reproduction is mainly a mystery. We know that certain cells grow and reproduce all the time, such as the blood-forming cells of the bone marrow, the germinal layers of the skin, and the epithelium of the gut. However, many other cells, such as muscle cells, do not reproduce for many years. And a few cells, such as the neurons, do not reproduce after birth during the entire life of the person.

If there is an insufficiency of certain types of cells in the body, these will often grow and reproduce very rapidly until appropriate numbers of them are again available. For instance, seven-eighths of the liver can be removed surgically, and the cells of the remaining one-eighth will grow and divide until the liver mass returns almost to normal. The same process occurs for almost all glandular cells, for cells of the bone marrow, the subcutaneous tissue, the intestinal epithelium, and for cells of many other tissues except highly differentiated cells such as nerve and muscle cells.

We know very little about the mechanisms that maintain proper numbers of the different types of cells in the body. It is assumed that control substances are secreted by the different cells that cause feedback effects to stop or slow their growth when too many of them have been formed, though only a few such substances have been found. We know that cells of any type removed from the body and grown in tissue culture can grow and reproduce rapidly and indefinitely if the medium in which they grow is continually replenished. Yet they will stop growing when even small amounts of their own secretions are allowed to collect in the medium, which supports the idea that control substances limit cellular growth.

CANCER

Cancer can occur in any tissue of the body. It results from a change in certain cells that allows them to disrespect normal growth limits, no longer obeying the feedback controls that normally stop cellular growth and reproduction after a given number of such cells have developed. As pointed out above, even normal cells when removed from the body and grown in tissue culture can grow and proliferate indefinitely if the growth medium is continually changed. Therefore, in tissue culture, normal tissue cells behave exactly as cancer cells; but in the body, normal tissue cells behave differently, for they are subject to limits, whereas cancer cells are not.

What is the difference between the cancer cell and the normal tissue cell that allows the cancer cell to grow and reproduce unabated? The answer to this question is not known, but researchers have found that the genetic make-up of most, if not all, cancer cells is different from that of normal cells — that is, they have abnormal chromosomes or sometimes an excess number of chromosomes. This has led to the idea that cancer almost invariably results from mutation of part of the genetic system in the nucleus (sometimes spontaneous but other times caused by irritants, irradiation, or the presence of a virus in the cell). The mutated "genome" eliminates the feedback mechanisms that normally limit growth and reproduction of the cell. Once even a single such cell is formed, it obviously might grow and proliferate indefinitely, its number increasing exponentially.

Cancerous tissue competes with normal tissues for nutrients, and because cancer cells continue to proliferate indefinitely, their number multiplying day-by-day, one can readily understand that the cancer cells will soon demand essentially all the nutrition available to the body. As a result, the normal tissues gradually suffer nutritive death.

REFERENCES

Calvin, M.: Chemical evolution. Am. Sci., *63*:169, 1975.

Cohen, S. N.: The manipulation of genes. *Sci. Amer., 233(1)*:24, 1975.

Croce, C. M., and Koprowski, H.: The genetics of human cancer. *Sci. Amer., 238(2)*:117, 1978.

Fraser, G. R., and Mayo, O.: Textbook of Human Genetics. Philadelphia, J. B. Lippincott Company, 1975.

Friedman, D. L.: Role of cyclic nucleotides in cell growth and differentiation. *Physiol. Rev., 56*:652, 1976.

Grobstein, C.: The recombinant-DNA debate. *Sci. Amer., 237(1)*:22, 1977.

Maniatis, T., and Ptashne, M.: A DNA operator-repressor system. *Sci. Amer., 234(1)*:64, 1976.

Meerson, F. Z.: Role of synthesis of nucleic acids and protein in adaptation to the external environment. *Physiol. Rev., 55*:79, 1975.

Rich, A., and Kim, S. H.: The three-dimensional structure of transfer RNA. *Sci. Amer., 238(1)*:52, 1978.

Stein, G. S., Stein, J. S., and Kleinsmith. Chromosomal proteins and gene regulation. *Sci. Amer., 232(2)*:46, 1975.

Wolpert, L., and Lewis, J. H.: Towards a theory of development. *Fed. Proc., 34*:14, 1975.

QUESTIONS

1. What are the basic building blocks of DNA?
2. How are nucleotides combined to form DNA?
3. How does DNA control the formation of RNA?
4. Describe the role of messenger RNA in the formation of proteins.
5. Describe the function of transfer RNA and the ribosomes in the formation of protein.
6. How are the concentrations of cellular constituents controlled by feedback repression of the genes?
7. Describe the control of enzyme activity in the cell.
8. In cell reproduction what role do replication of DNA and replication of chromosomes play?
9. Describe the stages of mitosis and the events that take place during each of these stages.
10. What is the cause of cancer?

BLOOD CELLS, IMMUNITY, AND BLOOD CLOTTING

II

THE BLOOD CELLS, HEMOGLOBIN, AND RESISTANCE TO INFECTION

5

Almost all the cells in the blood are *red blood cells*. However, approximately 1 out of every 500 cells is a *white blood cell,* or *leukocyte,* which is called *"white"* because it is not colored by hemoglobin. White blood cells have several different functions, but the most important of these is to protect the body against invasion by disease organisms.

The blood also contains large numbers of *platelets,* which are often classified as white blood cells. However, the platelets are not really cells but instead very small fragments of a special type of bone marrow cell called the *megakaryocyte.* The platelets are essential for the clotting of blood, which will be explained in Chapter 7.

THE RED BLOOD CELLS

The major function of red blood cells is to transport hemoglobin, which in turn carries oxygen from the lungs to the tissues

Normal red blood cells are biconcave disks, as illustrated in Figure 5–1, having a mean diameter of approximately 8 microns (micrometers) and a thickness at the thickest point of 2 microns and in the center of 1 micron or less. The shapes of red blood cells can change remarkably as the cells pass through capillaries. Actually, the red blood cell is a "bag" that can be deformed into almost any shape. Furthermore, because the normal cell has a great excess of cell membrane for the quantity of material inside, deformation does not stretch the membrane, and consequently does not rupture the cell as would be the case with many other cells.

The average number of red blood cells per cubic millimeter of blood is about 5 million; women have a few per cent below this value, and men a few per cent above it. The altitude at which the person lives and the degree of exercise affect the number of red blood cells; these are discussed later.

Quantity of Hemoglobin in the Cells and Transport of Oxygen. Normal red blood cells contain approximately 34 grams of hemoglobin per 100 ml. of cells. However, when hemoglobin formation is deficient in the bone marrow, the percentage of hemoglobin in the cells may fall to as low as 15 grams per 100 ml. or less.

When the hematocrit (defined as the percentage of the blood that is cells — normally about 40 per cent) and the quantity of hemoglobin in each respective cell are normal, the whole blood contains an average of 15 grams of hemoglobin per 100 ml. As will be discussed in more detail in connection with respiration and the transport of oxygen in Chapter 19, each gram of hemoglobin is capable of combining with approximately 1.33 ml. of oxygen. Therefore, 20 ml. of oxygen can normally be carried in combination with hemoglobin in each 100 ml. of blood.

GENESIS OF THE RED BLOOD CELL

The red blood cells are produced in the bone marrow and are derived from a cell known as

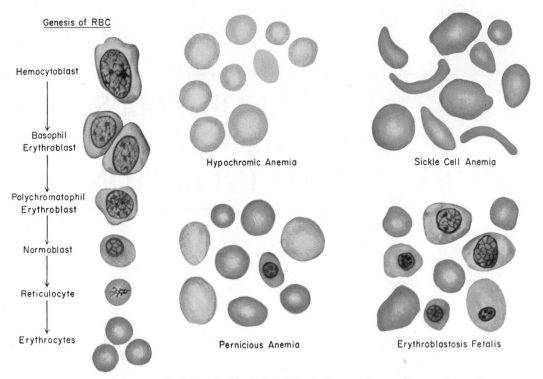

Figure 5-1.   Genesis of red blood cells, and the blood pictures of several types of anemia.

the *hemocytoblast,* illustrated in Figure 5–1.

The hemocytoblast passes through several stages of development, becoming first a *basophil erythroblast,* then a *polychromatophil erythroblast,* a *normoblast,* a *reticulocyte,* and finally an *erythrocyte.* During the earlier stages the cells divide many times, and they change color, as illustrated in Figure 5–1, owing to progressive formation of more and more hemoglobin. In the normoblastic stage, the nucleus degenerates, becomes extremely small, and is extruded, thus forming the reticulocyte. Then the cell usually leaves the bone marrow in the reticulocyte stage. The reticulocyte still contains small strands of basophilic reticulum mixed in with the hemoglobin in the cytoplasm. This reticulum is chiefly the remains of the endoplasmic reticulum, and it continues to produce small amounts of hemoglobin in the early cell even after it begins to circulate in the blood. However, this reticulum is usually dissolved within 2 days after release of the reticulocyte from the bone marrow, and the cell becomes the mature red blood cell, the erythrocyte. The erythrocyte then circulates in the blood for a period of approximately 120 days before it is destroyed. The proportion of the early red

blood cells, the reticulocytes, in the circulating blood is usually slightly less than 1 per cent.

REGULATION OF RED BLOOD CELL PRODUCTION

The total number of red blood cells in the circulatory system is regulated within very narrow limits, so that an adequate number of cells is always available to provide sufficient tissue oxygenation and, yet, so that the cells are not overly concentrated to the extent that they impede blood flow. Thus, when a person becomes extremely *anemic* as a result of hemorrhage or any other condition, the bone marrow immediately begins to produce large quantities of red blood cells. Also, at very *high altitudes,* where the quantity of oxygen in the air is greatly decreased, insufficient oxygen is transported to the tissues, and red cells are then also produced so rapidly that their number in the blood increases considerably. Finally, the degree of physical activity of a person determines to a slight extent the rate at which red blood cells will be produced. The athlete will often have a red blood cell count as high as 5.5 million per cubic millimeter, whereas the asthenic person will have a count as low as 4.5 million per cubic millimeter.

Erythropoietin, Its Response to Hypoxia, and Its Function in Regulating Red Blood Cell Production. Despite the very marked effect of hypoxia on red blood cell production, hypoxia does not have a direct effect on the bone marrow. Instead, the hypoxia first causes formation of a hormone called *erythropoietin*. This hormone in turn stimulates the bone marrow to produce greatly increased numbers of red blood cells.

Erythropoietin is a glycoprotein having a molecular weight of about 39,000. Its mechanism of formation is the following: Hypoxia has an effect in the kidneys to cause release of an enzyme called *renal erythropoietic factor*. This is secreted into the blood where it acts within a few minutes on one of the plasma proteins, a globulin, to split away the glycoprotein erythropoietin molecule. Erythropoietin, in turn, circulates in the blood, lasting for about one day, and during this time it acts on the bone marrow to cause red cell formation. Only minute quantities of erythropoietin can be formed in the whole body in the complete absence of the kidneys. Therefore, persons whose kidneys have been destroyed and who are being kept alive by use of the artificial kidney usually have very severe anemia, usually with red blood cell counts of less than one-half normal.

Though erythropoietin begins to be formed almost immediately upon placing an animal or person in an atmosphere of low oxygen, very few new red blood cells appear in the circulation for the first 2 days; and it is only after 5 or more days that the maximum rate of new red cell production is reached. Thereafter, cells continue to be produced as long as the person remains in the low oxygen state or until he has produced enough red blood cells to carry adequate amounts of oxygen to his tissues despite the low oxygen.

The action of the erythropoietin is mainly to cause the formation of large numbers of hemocytoblasts in the bone marrow and also to cause proliferation of the hemocytoblasts themselves. The other stages of erythrogenesis are also accelerated, but this probably results mainly from the initial stimulatory effect of the erythropoietin at the early levels. In the complete absence of erythropoietin, very few red blood cells are formed by the bone marrow. At the other extreme, when extreme quantities of erythropoietin are formed, the rate of red blood cell production can rise to as high as 8 to 10 times normal.

VITAMINS NEEDED FOR FORMATION OF RED BLOOD CELLS

The Maturation Factor — Vitamin B_{12} (Cyanocobalamin). Vitamin B_{12} is an essential nutrient for all cells of the body, and growth of tissues in general is greatly depressed when this vitamin is lacking. This results from the fact that vitamin B_{12} is required for conversion of ribose nucleotides into deoxyribose nucleotides, one of the essential steps in DNA formation. Therefore, lack of this vitamin causes failure of nuclear maturation and division, and greatly inhibits the rate of red blood cell production.

MATURATION FAILURE CAUSED BY POOR ABSORPTION OF VITAMIN B_{12} — PERNICIOUS ANEMIA. The most common cause of maturation failure is not a lack of vitamin B_{12} in the diet but instead failure to absorb vitamin B_{12} from the gastrointestinal tract. This often occurs in the disease called *pernicious anemia,* in which the basic abnormality is an *atrophic gastric mucosa* that fails to secrete normal gastric secretions. These secretions contain a substance called *intrinsic factor,* which combines with vitamin B_{12} of the food and makes the B_{12} available for absorption by the gut. It does this by protecting the B_{12} from digestion by the gastrointestinal enzymes until it can be absorbed.

Once vitamin B_{12} has been absorbed from the gastrointestinal tract, it is stored in large quantities in the liver and then released slowly as needed to the bone marrow and other tissues of the body. The total amount of vitamin B_{12} required each day to maintain normal red cell maturation is less than 1 microgram, and the normal store in the liver is about 1000 times this amount.

Relationship of Folic Acid (Pteroylglutamic Acid) to Red Cell Formation. Occasionally maturation failure anemia results from folic acid deficiency instead of vitamin B_{12} deficiency. Folic acid, like B_{12}, is required for formation of DNA but in a different way. It promotes the methylation of deoxyuridylate to form deoxythymidylate, one of the nucleotides required for DNA synthesis.

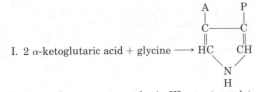

I. 2 α-ketoglutaric acid + glycine ⟶ (pyrrole structure)

II. 4 pyrrole ⟶ protoporphyrin III   (pyrrole)

III. protoporphyrin III + Fe ⟶ heme

IV. 4 heme + globin ⟶ hemoglobin

Figure 5–2. Formation of hemoglobin.

FORMATION OF HEMOGLOBIN

Synthesis of hemoglobin begins in the erythroblasts and continues through the reticulocyte stage. Even when young red blood cells, the reticulocytes, leave the bone marrow and pass into the blood stream, they continue to form hemoglobin for another day or so.

Figure 5–2 gives the basic chemical steps in the formation of hemoglobin. From tracer studies with isotopes it is known that hemoglobin is synthesized mainly from *acetic acid* and *glycine,* which together form a *pyrrole* compound. In turn, four pyrrole compounds combine to form a protoporphyrin compound. One of the protoporphyrin compounds, known as *protoporphyrin III,* then combines with iron to form the *heme* molecule. Finally, four heme molecules combine with one molecule of

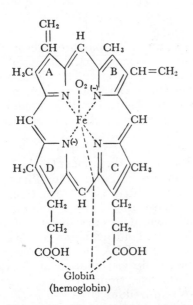

Figure 5–3. The hemoglobin molecule.

globin, a globulin, to form hemoglobin, the formula for which is shown in Figure 5–3. Hemoglobin has a molecular weight of 68,000.

Combination of Hemoglobin with Oxygen. The most important characteristic of the hemoglobin molecule is its ability to combine loosely and reversibly with oxygen. This ability will be discussed in detail in Chapter 19 in relation to respiration, for the primary function of hemoglobin in the body depends upon its ability to combine with oxygen in the lungs and then to release this oxygen readily in the tissue capillaries where the gaseous tension of oxygen is much lower than in the lungs.

Oxygen *does not* combine at the two positive valences of the ferrous iron in the hemoglobin molecule. Instead, it binds loosely at two of the six "coordination" valences of the iron atom. This is an extremely loose bond, so that the combination is easily reversible.

IRON METABOLISM

Because iron is important for formation of hemoglobin, myoglobin, and other substances such as the cytochromes, cytochrome oxidase, peroxidase, and catalase, it is essential to understand the means by which iron is utilized in the body.

The total quantity of iron in the body averages about 4 grams, approximately 65 per cent of which is present in the form of hemoglobin. About 4 per cent is present in the form of myoglobin, 1 per cent in the form of the various heme compounds that control intracellular oxidation, 0.1 per cent in the form of transferrin in the blood plasma, and 15 to 30 per cent stored mainly in the form of ferritin in the liver.

Transport and Storage of Iron. A schema for transport, storage, and metabolism of iron in the body is illustrated in Figure 5–4, which may be explained as follows: When iron is absorbed from the small intestine, it immediately combines in the blood plasma with a beta globulin, *transferrin.* The iron is very loosely combined with this compound, and consequently can be released to any of the tissue cells at any point in the body. Excess iron in the blood is deposited in all cells of the body *but especially in the liver cells,* where about 60 per cent of the excess is stored. There it combines with a protein of very large molecular weight, *apoferritin,* to form *ferritin.*

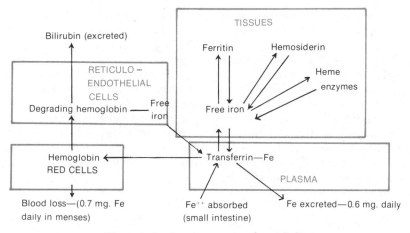

Figure 5–4.   Iron transport and metabolism.

This iron stored in the form of ferritin is called *storage iron.*

When the quantity of iron in the plasma falls very low, iron is removed from ferritin quite easily. The iron is then transported to the portions of the body where it is needed.

When red blood cells have lived their life span and are destroyed, the hemoglobin released from the cells is ingested by the reticuloendothelial cells (the white blood cells and certain tissue cells — these will be discussed later in this chapter). There free iron is liberated, and it can then either be restored in the ferritin pool or be reused for formation of hemoglobin.

Daily Loss of Iron.   About 0.6 mg. of iron is excreted each day by the male, mainly into the feces. Additional quantities of iron are lost whenever bleeding occurs. Thus, in the female, the menstrual loss of blood brings the average iron loss to a value of approximately 1.3 mg. per day.

Obviously, the average quantity of iron derived from the diet each day must at least equal that lost from the body.

Absorption of Iron from the Gastrointestinal Tract.   Iron is absorbed almost entirely in the upper part of the small intestine, primarily in the duodenum. It is absorbed by an active process, though the precise mechanism of this active absorption is unknown. Also, only the *ferrous* form of iron is absorbed in significant amounts. Since a large share of the iron in foodstuffs is in the *ferric* form rather than the ferrous form, this becomes an important consideration in the selection of foods and drugs for the treatment of iron deficiency anemias.

Regulation of the Total Body Iron — Role of Intestinal Absorption.   As is true of almost all essential substances in the body, it is exceedingly important to maintain an appropriate amount — not too little, not too much — of iron in the body. This is achieved in the following way. When the apoferritin in the liver and other organs where iron is stored becomes saturated with iron, transferrin then cannot release iron from the plasma into the storage depots. Consequently, the transferrin also becomes saturated with iron, and the active transport processes of the intestinal mucosal cells then cannot transport iron into the blood. Therefore, the iron handling system simply turns off the intestinal absorption of iron so that most of the iron in the food fails to be absorbed and instead is excreted in the feces.

On the other hand, whenever the iron stores become depleted, this mechanism reverses itself completely, and the intestinal absorptive process transports iron actively from the intestinal lumen into the blood. Thus, the system is always geared to keep the iron storage depots full of iron but not overly filled, except when the person develops a nutritional iron deficiency.

DESTRUCTION OF RED BLOOD CELLS

When red blood cells are delivered from the bone marrow into the circulatory system they normally circulate an average of 120 days before being destroyed. Even though red cells do not have a nucleus, they do still have cytoplasmic enzymes for metabolizing glucose and other substances and for the utilization of oxygen, but many of these metabolic systems become progressively less active with time. As the cells become older they also become progressively more fragile, presumably because their life processes simply wear out.

Once the red cell membrane becomes very fragile, it may rupture during passage through some tight spot of the circulation. Many of the red cells fragment in the spleen where the cells squeeze through the red pulp of the spleen. When the spleen is removed, the number of abnormal cells and old cells circulating in the blood increases considerably.

Destruction of Hemoglobin. The hemoglobin released from the cells when they burst is phagocytized and digested almost immediately by reticuloendothelial cells, releasing iron back into the blood to be carried by transferrin either to the bone marrow for production of new red blood cells or to the liver and other tissues for storage in the form of ferritin. The heme portion of the hemoglobin molecule is converted by the reticuloendothelial cell, through a series of stages, into the bile pigment *bilirubin,* which is released into the blood and later secreted by the liver into the bile; this will be discussed in relation to liver function in Chapter 30.

ANEMIA

Anemia means a deficiency of red blood cells, which can be caused either by too rapid loss or by too slow production of red blood cells. Some of the common types of anemia were illustrated in Figure 5–1. These are often caused by:

1. *Blood loss.*
2. *Bone marrow destruction.* Common causes of this are drug poisoning or gamma ray irradiation — for instance, exposure to radiation from a nuclear bomb blast.
3. *Failure of red blood cells to mature* because of lack of vitamin B_{12} or folic acid, as was previously explained and as occurs in pernicious anemia.
4. *Hemolysis of red cells* — that is, rupture of the cells — resulting from many possible causes, such as (a) drug poisoning, (b) hereditary diseases that make the red cell membranes friable (for example, sickle cell anemia), and (c) erythroblastosis fetalis, a disease of the newborn in which antibodies from the mother destroy red cells in the baby.

As long as an anemic person's rate of activity is low, he often can live without fatal hypoxia of the tissues, even though his concentration of red blood cells may be reduced to one-fourth normal. However, when he begins to exercise, his heart will not be capable of pumping enough blood to the tissues to supply the needed oxygen. Consequently, during exercise, which greatly increases the demands for oxygen, extreme tissue hypoxia results, and acute heart failure also ensues.

WHITE BLOOD CELLS (LEUKOCYTES), AND RESISTANCE OF THE BODY TO INFECTION

The body is constantly exposed to bacteria, these occurring especially in the mouth, the respiratory passageways, the colon, the mucous membranes of the eyes, and even the urinary tract. Many of these bacteria are capable of causing disease if they invade the deeper tissues. In addition, a person is intermittently exposed to highly virulent bacteria and viruses from outside the body which can cause specific diseases such as pneumonia, streptococcal infections, and typhoid fever.

On the other hand, a group of tissues including the *white blood cells* and the *reticuloendothelial system* constantly combat any infectious agent that tries to invade the body. These tissues function in two different ways to prevent disease: (1) by actually destroying invading agents by the process of phagocytosis and (2) by forming *antibodies* against the invading agent, the antibodies in turn destroying the invader; this process is called *immunity*. The present discussion is concerned with phagocytic destruction of the invading agents, while the following chapter is concerned with immunity.

The white blood cells, also called *leukocytes,* are the *mobile units* of the body's protective system. They are formed partially in the bone marrow and partially in the lymph nodes, but after formation they are transported in the blood to the different parts of the body where they are to be used.

The Types of White Blood Cells. Five different types of white blood cells are normally found in the blood. These are *polymorphonuclear neutrophils, polymorphonuclear eosinophils, polymorphonuclear basophils, monocytes,* and *lymphocytes.* In addition, there are large numbers of *platelets,* which are fragments of a sixth type of white blood cell, the

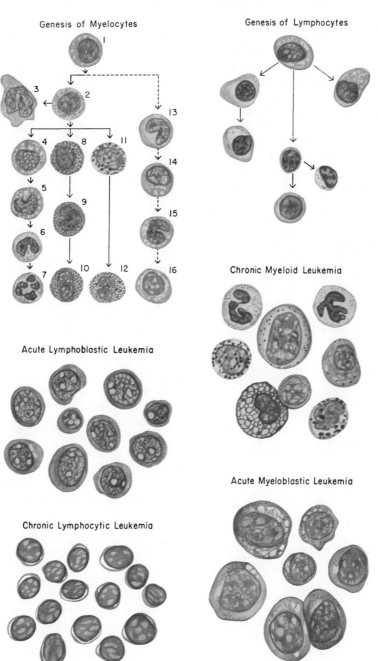

Figure 5–5. Genesis of the white blood cells, and the blood picture in several types of leukemia. Cell number 1 is a *myeloblast*, and number 3 is a *megakaryocyte*. Cells 4, 5, 6, and 7 illustrate the formation of *neutrophils*; 8, 9, and 10, the formation of *eosinophils*; 11 and 12, the formation of *basophils*; and 13, 14, 15, and 16, the formation of *monocytes*. (Redrawn in part from Piney: *A Clinical Atlas of Blood Diseases.* The Blakiston Co.)

megakaryocyte. The three types of polymorphonuclear cells have a granular appearance, as illustrated in Figure 5–5, for which reason they are called *granulocytes,* or in clinical terminology they are often called simply "polys."

The granulocytes and the monocytes protect the body against invading organisms by ingesting them — that is, by the process of *phagocytosis.* Also, a function of another type of white cell, the lymphocyte, is to attach to specific invading organisms and to destroy them; this is part of the immunity system and will be discussed in the following chapter. Finally, the function of platelets is to activate the blood clotting mechanism. All these functions are protective mechanisms of one type or another.

Concentrations of the Different White Blood Cells in the Blood. The adult human being has approximately 7000 white blood cells per cubic millimeter of blood. The normal percentages of the different types of white blood cells are approximately the following:

Polymorphonuclear neutrophils	62.0%
Polymorphonuclear eosinophils	2.3%
Polymorphonuclear basophils	0.4%
Monocytes	5.3%
Lymphocytes	30.0%

The number of platelets in each cubic millimeter of blood is normally about 300,000.

GENESIS OF THE WHITE BLOOD CELLS

The upper left portion of Figure 5–5 illustrates the stages in the development of the white blood cells. The polymorphonuclear cells and monocytes are normally formed only in the bone marrow. On the other hand, lymphocytes are produced in the various lymphogenous organs, including the lymph glands, the spleen, the thymus, the tonsils, and various lymphoid rests in the gut and elsewhere.

Some of the white blood cells formed in the bone marrow, especially the granulocytes, remain stored within the marrow until they are needed in the circulatory system. Then when the need arises, various factors that are discussed later cause them to be released.

As also illustrated in Figure 5–5, megakaryocytes are also formed in the bone marrow and are part of the myelogenous group of bone marrow cells. These megakaryocytes break up into very small fragments in the bone marrow, the small fragments known as *platelets* passing then into the blood.

PROPERTIES OF WHITE BLOOD CELLS

Phagocytosis. The most important function of the *neutrophils* and *monocytes* and, to a lesser extent, of some of the other white blood cells is phagocytosis.

Obviously, the phagocytes must be selective in the material that is phagocytized, or otherwise some of the structures of the body itself would be ingested. Whether or not phagocytosis will occur depends on three selective procedures. First, if the surface of a particle is rough, the likelihood of phagocytosis is increased. Second, most natural substances of the body have electronegative surface charges that repel the phagocytes, which also carry electronegative surface charges. On the other hand, dead tissues and foreign particles are frequently electropositive and are therefore subject to phagocytosis. Third, the body has a means for promoting phagocytosis of specific foreign materials by first combining them with protein "antibodies" called *opsonins*. After the opsonin has combined with the particle, it allows the phagocyte to adhere to the surface of the particle, and this promotes phagocytosis. The special features of opsonization and its relationship to immunity are discussed in the following chapter.

Enzymatic Digestion of the Phagocytized Particles. Once a foreign particle has been phagocytized, lysosomes immediately come in contact with the phagocytic vesicle, and their membranes fuse with those of the vesicle, thereby dumping the digestive enzymes of the lysosomes into the vesicle. Thus, the phagocytic vesicle now becomes a *digestive vesicle*, and digestion of the phagocytized particle begins immediately. This was discussed in more detail in Chapter 3.

Diapedesis. White blood cells can move out of the blood into the tissue spaces. They do this by squeezing through the capillary pores and even through holes in some endothelial cells of the blood vessels by the process of diapedesis. That is, even though a pore is much smaller than the size of the cell, a small portion of the cell slides through the pore at a time, the portion sliding through being momentarily constricted to the size of the pore, as illustrated in Figure 5–6.

Ameboid Motion. Once the cells have entered the tissue spaces, the polymorphonuclear leukocytes especially, and the large lymphocytes and monocytes to a lesser degree, move through the tissues by ameboid motion, which was described in Chapter 3. Some of the cells can move through the tissues at rates as great as 40 microns (micrometers) per minute — that is, they can move at least three times their own length each minute.

Chemotaxis. Different chemical substances in the tissues cause the leukocytes to

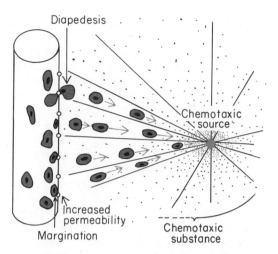

Figure 5–6. Movement of neutrophils by the process of *chemotaxis* toward an area of tissue damage.

move either toward or away from the source of the chemical. This phenomenon, known as chemotaxis, is illustrated in Figure 5–6. Degenerative products of inflamed tissues, especially tissue polysaccharides and also one of the reaction products of a complex of substances called "complement" (discussed in the following chapter), can *cause neutrophils and monocytes to move toward the area of inflammation*. In addition, a number of bacterial toxins can cause chemotaxis of leukocytes.

THE RETICULOENDOTHELIAL SYSTEM

In addition to the white blood cells, another group of cells distributed widely throughout the tissues and lining some of the blood and lymph channels also helps to protect the body against foreign invaders. This group of cells is mainly nonmotile and is collectively called the *reticuloendothelial system*. However, this term is used differently by different persons. Most frequently, the term includes two types of cells: (1) cells derived mainly from monocytes that have enlarged to become *tissue macrophages* — these are present in the various tissues and are also adherent to the walls of blood and lymph channels; and (2) *lymphocytic cells* which either are wandering through the tissues or are entrapped in special lymphoid tissue, such as the lymph nodes.

Most of the functions of the lymphocytes, especially those functions related to immunity, will be discussed in the following chapter.

THE RETICULOENDOTHELIAL CELLS DERIVED FROM MONOCYTES

Many monocytes, on entering the tissues, enlarge greatly, become fixed to the tissues, and perform phagocytic activity. Some of these types of cells are the following:

Tissue Macrophages. Monocytes that wander into the tissues often become fixed in the tissues and then swell to become fixed *tissue macrophages*. During the course of inflammation, these can divide in situ and form more macrophages. Frequently, they proliferate and form giant cell capsules around foreign particles that cannot be digested, such as particles of silica dust, carbon, and so forth, thus effectively isolating these particles from the remaining tissue. This "walling off" process also frequently occurs in response to certain chronic infections — tuberculosis, for instance — and therefore is an important mechanism for preventing spread of disease.

The Macrophages of the Lymph Nodes. Almost no particulate matter that enters the tissues can be absorbed directly through the capillary membranes into the blood. Instead, if the particles are not destroyed locally in the tissues, they enter the lymph and flow through the lymphatic vessels to the lymph nodes located intermittently along the course of the lymphatics. The foreign particles are trapped there in a meshwork of sinuses that are lined by tissue macrophages.

Figure 5–7 illustrates the general organization of the lymph node, showing lymph entering by way of the *afferent lymphatics,* flowing through the *medullary sinuses,* and finally passing out of the *hilus* into the *efferent lym-*

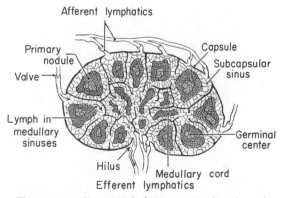

Figure 5–7. Functional diagram of a lymph node. (Redrawn from Ham: Histology. J. B. Lippincott Co., 1971.)

phatics. Large numbers of tissue macrophages line the sinuses, and if any particles enter the sinuses, these cells phagocytize them and prevent their general dissemination throughout the body.

The Alveolar Macrophages. Another route by which invading organisms frequently enter the body is through the respiratory system. Fortunately, large numbers of tissue macrophages are present as integral components of the respiratory surfaces. These can phagocytize particles that become entrapped in the alveoli.

The Tissue Macrophages (Kupffer Cells) in the Liver Sinuses. Still another favorite route by which bacteria invade the body is through the gastrointestinal tract. Large numbers of bacteria constantly pass through the gastrointestinal mucosa into the portal blood. However, before this blood enters the general circulation, it passes through the sinuses of the liver; these sinuses are lined with tissue macrophages called *Kupffer cells,* illustrated in Figure 5–8. These cells form such an effective particulate filtration system that almost none of the bacteria from the gastrointestinal tract succeeds in passing from the portal blood into the general systemic circulation. Indeed, motion pictures of phagocytosis by Kupffer cells have demonstrated phagocytosis of single bacteria in less than 1/100 second.

The Macrophages of the Spleen and Bone Marrow. If any invading organism does succeed in entering the general circulation, there still remain other lines of defense by the reticuloendothelial system, especially by reticuloendothelial cells of the spleen and bone marrow. In both of these tissues, macrophages have become entrapped by the reticular meshworks of the two organs, and when foreign particles come in contact with them the particles are phagocytized.

The spleen is similar to the lymph nodes, except that blood, instead of lymph, flows through the substance of the spleen. Figure 5–9 illustrates the general structure of the spleen, showing a small peripheral segment of the spleen. Note that a small artery penetrates from the splenic capsule into the *splenic pulp,* and terminates in small capillaries. The capillaries are highly porous, allowing large numbers of whole blood cells to pass out of the capillaries into the *cords of the red pulp.* These cells then gradually *squeeze* through the tissue substance of the cords and eventually return to the circulation through the endothelial walls of the *venous sinuses.* The cords of the red pulp are loaded with macrophages, and in addition the venous sinuses are also lined with macrophages. This peculiar passage of blood through the cords of the red pulp provides an exceptional means for phagocytosis of unwanted debris in the blood, especially old and abnormal red blood cells. The spleen is also an important organ for phagocytic removal of abnormal platelets, blood parasites, and any bacteria that might succeed in entering the general circulating blood.

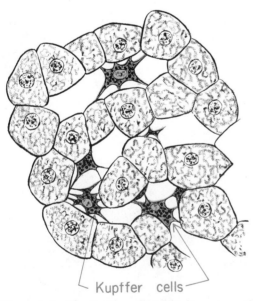

Figure 5–8. The Kupffer cells of the liver sinusoids. These are typical reticuloendothelial cells. (Redrawn from Smith and Copenhaver: *Bailey's Textbook of Histology.* The Williams & Wilkins Co.)

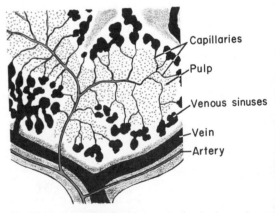

Figure 5–9. Functional structures of the spleen. (Modified from Bloom and Fawcett: Textbook of Histology. W. B. Saunders Co., 1975.)

In a similar way, macrophages of the bone marrow also help to remove unwanted debris and pathologic agents from the blood.

THE FUNCTION OF WHITE BLOOD CELLS IN INFLAMED TISSUES

THE PROCESS OF "INFLAMMATION"

Inflammation is a complex of sequential changes in the tissues in response to injury. When tissue injury occurs, whether it be caused by bacteria, trauma, chemicals, heat, or any other phenomenon, the substance *histamine,* along with other humoral substances, is liberated by the damaged tissue into the surrounding fluids. This increases the local blood flow and also increases the permeability of the capillaries, allowing large quantities of fluid and protein to leak into the tissues. Local extracellular edema results, and the extracellular fluid and lymphatic fluid both clot because of the coagulating effect of tissue exudates on one of the leaked proteins, fibrinogen. Thus, *brawny edema* develops in the spaces surrounding the injured cells.

The "Walling Off" Effect of Inflammation. It is clear that one of the first results in inflammation is to "wall off" the area of injury from the remaining tissues. The tissue spaces and the lymphatics in the inflamed area are blocked by fibrinogen clots so that fluid barely flows through the spaces. Therefore, walling off the area of injury delays the spread of bacteria or toxic products.

Attraction of Neutrophils to the Area of Inflammation. When tissues are damaged, several effects occur to cause movement of neutrophils into the damaged area. First, the neutrophils stick to the walls of the damaged capillary, causing the process known as margination, which was shown in Figure 5–6. Gradually, the cells pass by diapedesis into the tissue spaces.

The second effect is chemotaxis of the neutrophils toward the damaged area; this is caused by bacterial or cellular products that attract the neutrophils. Thus, within a few hours after damage begins, the area becomes well supplied with neutrophils.

Neutrophilia During Inflammation — Leukocytosis-Promoting Factor. The term *neutrophilia* means an increase above normal in the number of neutrophils in the blood, and the term *leukocytosis* means an excess total number of white blood cells.

A substance known as *leukocytosis-promoting factor* is liberated by inflamed tissues. This factor diffuses into the blood and finally to the bone marrow where it has two actions: First, it causes large numbers of granulocytes, especially neutrophils, to be released within a few minutes to a few hours into the blood from the storage areas of the bone marrow, thus increasing the total number of neutrophils per cubic millimeter of blood sometimes to as high as 20,000 to 30,000, four to six times the normal number. Second, the rate of granulocyte production by the bone marrow increases either as a direct result of the factor or as an indirect result of the bone marrow release of the granulocytes. Within a day or two after onset of the inflammation, the bone marrow becomes hyperplastic and then continues to produce large numbers of granulocytes as long as leukocyte-promoting factor continues to be formed in the inflamed tissues.

The Macrophage Response in Chronic Inflammation. The monocytic cells — including the tissue macrophages and the blood monocytes — also play a major role in protecting the body against infection. First, the tissue macrophages develop ameboid motion and migrate chemotaxically toward the area of inflammation. These cells provide the first line of defense against infection within the first hour or so, but their numbers are not very great. Within the next few hours, the neutrophils become the primary defense, reaching their maximum effectiveness in about 6 to 12 hours. By that time large numbers of monocytes have begun to enter the tissues from the blood. They change their characteristics drastically during the first few hours. They start to swell, to form greatly increased quantities of cytoplasmic lysosomes, to exhibit increased ameboid motion, and to move chemotaxically toward the damaged tissues.

The macrophages are several times as phagocytic as the neutrophils. Furthermore, they are large enough that they can engulf large quantities of necrotic tissue, including dead neutrophils themselves.

Formation of Pus. When the neutrophils and macrophages engulf large amounts of bacteria and necrotic tissue, they themselves eventually die. After several days, a cavity is often excavated in the inflamed tissues containing varying portions of necrotic tissue, dead neutrophils, and dead macrophages. Such a mixture is commonly known as *pus.*

Ordinarily, pus formation continues until all infection is suppressed. Sometimes the pus cavity eats its way to the surface of the body or into an internal cavity and in this way empties itself. At other times the pus cavity remains enclosed even after tissue destruction has ceased. When this happens the dead cells and necrotic tissue in the pus gradually autolyze over a period of days, and the end-products of autolysis are usually absorbed into the surrounding tissues until most of the evidence of tissue damage is gone.

THE EOSINOPHILS

The function of the eosinophils is almost totally unknown even though they normally comprise 1 to 3 per cent of all the leukocytes. Eosinophils are weak phagocytes, and they exhibit chemotaxis, but in comparison with the neutrophils, it is doubtful that the eosinophils are of significant importance in protection against usual types of infection.

Eosinophils enter the blood in large numbers following foreign protein injection, during allergic reactions, and during infection with parasites. The function of eosinophils in these conditions is unknown, though it is supposed that they might remove toxic substances from the tissues.

THE BASOPHILS

The basophils comprise only 0.4% of the circulating white blood cells, but they are very similar to the large *mast* cells located immediately outside many of the capillaries in the body. The mast cells liberate *heparin* into the blood, a substance that can prevent blood coagulation and that can also speed the removal of fat particles from the blood after a fatty meal. Therefore, it is probable that the basophils in the circulating blood also secrete heparin within the blood stream, or it is possible that the blood transports basophils to tissues where they then become mast cells and perform the function of heparin liberation.

THE LYMPHOCYTES

Until recently it was believed that the lymphocytes represent a homogeneous group of cells. However, it is now known that the "lymphocyte" comprises a number of different types of cells, all of which have essentially the same staining characteristics. A few of these cells, probably produced primarily in the bone marrow, seem to be multipotential cells that are similar, if not identical, to the primordial *stem cell* from which almost any other type of cell can be formed. These multipotential cells can be changed into erythroblasts, myeloblasts, monocytes, plasma cells, other types of small lymphocytes, fibroblasts, and so forth.

Many of the other circulatory "lymphocytes" play special roles in the process of immunity. These will be discussed in detail in the following chapter.

AGRANULOCYTOSIS

A clinical condition known as agranulocytosis occasionally occurs, in which the bone marrow stops producing white blood cells, leaving the body unprotected against bacteria and other agents that might invade the tissues. The cause is usually drug poisoning or irradiation following a nuclear bomb blast. Within two days after the bone marrow stops producing white blood cells, ulcers appear in the mouth and colon, or the person develops some form of severe respiratory infection. Bacteria from the ulcers then rapidly invade the surrounding tissues and the blood. Without treatment, death usually ensues 3 to 6 days after agranulocytosis begins.

THE LEUKEMIAS

Uncontrolled production of white blood cells can be caused by cancerous mutation of a myelogenous or a lymphogenous cell, causing leukemia, which means greatly increased numbers of white blood cells in the circulating blood. Ordinarily, leukemias are divided into two general types: the *lymphogenous leukemias* and the *myelogeous leukemias*.

The first effect of leukemia is metastatic growth of leukemic cells in abnormal areas of the body. The leukemic cells of the bone marrow may reproduce so greatly that they invade the surrounding bone, causing a tendency to easy fracture. Almost all leukemias spread to the spleen, the lymph nodes, the liver, and other especially vascular regions, regardless of whether the origin of the leukemia is in the bone marrow or in the lymph nodes. In each of these areas the rapidly growing cells invade the surrounding tissues, consequently causing tissue destruction.

Perhaps the most important effect of leukemia on the body is the excessive use of metabolic substrates by the growing cancerous cells. The leukemic tissues reproduce new cells so rapidly that tremendous demands are made for foodstuffs, especially the amino acids and vitamins. Consequently, the energy of the person is greatly depleted, and the excessive utilization of amino acids causes rapid deterioration of the normal protein tissues of the body. Thus, while the leukemic tissues grow, the other tissues are decimated. Obviously, after metabolic starvation has continued long enough, this alone is sufficient to cause death.

REFERENCES

Adamson, J. W., and Finch, C. A.: Hemoglobin function, oxygen affinity, and erythropoietin. *Ann. Rev. Physiol., 37*:351, 1975.

Bellanti, J. A., and Dayton, D. H. (eds.): The Phagocytic Cell in Host Resistance. New York, Raven Press, 1975.

Custer, R. P.: An Atlas of the Blood and Bone Marrow. 2nd ed. Philadelphia, W. B. Saunders Company, 1975.

Ersley, A. J.: Renal biogenesis of erythropoietin. *Am. J. Med., 58*:25, 1975.

Ersley, A. J., and Gabuzda, T. G.: Pathophysiology of Blood. Philadelphia, W. B. Saunders Company, 1975.

Forth, W., and Rummel. W.: Iron absorption. *Physiol. Rev., 53*:724, 1973.

Golde, D. W., and Cline, M. J.: Regulation of granulopoiesis. *N. Engl. J. Med., 291*:1388, 1974.

Munro, H. N., and Linder, M. C.: Ferritin: structure, biosynthesis, and role in iron metabolism. *Physiol. Rev., 58*:317, 1978.

Nathan, D. G., and Oski, F. A.: Hematology of Infancy and Childhood. Philadelphia, W. B. Saunders Company, 1975.

Surgenor, D. M.: The Red Blood Cell. 2nd ed. Vol. 21. New York, Academic Press, Inc., 1975.

Van Arman, C. G. (ed.): White Cells in Inflammation. Springfield, Ill., Charles C Thomas, Publisher, 1974.

QUESTIONS

1. What is the significance and function of the "baglike" structure of the red blood cell?
2. Discuss the genesis of red blood cells.
3. Give the mechanism for regulating the total number of red blood cells in the circulatory system.
4. What are the roles of vitamin B_{12} and folic acid in the formation of red blood cells?
5. Discuss iron metabolism and the role of iron in the formation of hemoglobin.
6. What causes the ultimate destruction of red blood cells?
7. List four different causes of anemia.
8. Contrast the genesis of white blood cells with that of red blood cells.
9. Which white blood cells have the greatest capability for phagocytizing bacteria?
10. Discuss the role of white blood cells in the inflammatory process.
11. Describe the function of macrophages in the different tissues of the body.
12. What is the cause of the leukemias?

6 IMMUNITY AND ALLERGY

IMMUNITY

In the previous chapter we discussed innate roles of white blood cells and of the reticuloendothelial system to protect the body against organisms that tend to damage tissues and organs. However, the body has another entirely different system for protection, a system that operates not only against organisms but against many toxins as well. This is called the system of *acquired immunity*. It results from the formation of *antibodies* and *sensitized lymphocytes* that attack and destroy the invading organisms or toxins.

The immune system does not resist the initial invasion by organisms. Nor does it resist the effects of toxins upon first exposure. However, within a few days to a few weeks after initial exposure, the immune system develops an extremely powerful resistance to the invader that is generally specific for the particular invader and for no other one. For instance, protection against the paralytic toxin of botulinum bacteria or the tetanizing toxin of tetanus bacteria can be produced, once immunity has developed, at doses as high as 100,000 times those that would be lethal without immunity. This is the reason that the process known as *vaccination* is so extremely important in protecting human beings against disease and toxins, as will be explained in the course of this chapter.

Unfortunately, the immune process does not always work exactly the way that it should. Sometimes elements of the immune system attack the person's own tissues rather than a specific invader. Under these conditions, serious damaging effects can occur as a result of *autoimmunity* or *allergy*, both of which also will be explained later in the chapter.

TWO BASIC TYPES OF IMMUNITY

Two basic, but closely allied, types of immunity occur in the body. In one of these the body develops circulating *antibodies*, which are globulin molecules that are capable of attacking the invading agent. This type of immunity is called *humoral immunity*. The second type of immunity is achieved through the formation of large numbers of highly specialized lymphocytes that are specifically sensitized against the foreign agent. These *sensitized lymphocytes* have the special capability to attach to the foreign agent and to destroy it. This type of immunity is called *cellular immunity* or, sometimes, *lymphocytic immunity*.

We shall see shortly that both the antibodies and the sensitized lymphocytes are formed in the lymphoid tissue of the body. First, let us discuss the initiation of the immune process by *antigens*.

ANTIGENS

Each toxin or each type of organism contains one or more specific chemical compounds in its make-up that are different from all other compounds. In general, these are proteins, large polysaccharides, or large lipoprotein complexes, and it is one or more of these compounds that cause the immunity. These substances are called *antigens*.

Essentially all toxins secreted by bacteria are also proteins, large polysaccharides, or mucopolysaccharides, and they are highly antigenic. Also, the bodies of bacteria or viruses usually contain several antigenic chemical compounds. Likewise, animal tissues such as a transplanted heart from other human

60

beings also contain numerous antigens that can elicit the immune process and cause subsequent destruction.

For a substance to be antigenic it usually must have a high molecular weight, 8000 or greater. Furthermore, the process of antigenicity probably depends upon regularly recurring prosthetic radicals on the surface of the large molecule, which perhaps explains why proteins and many polysaccharides are antigenic, for they both have this characteristic.

ROLE OF LYMPHOID TISSUE IN IMMUNITY

Immunity is absolutely dependent on function of the body's lymphoid tissue. In persons who have a genetic lack of lymphoid tissue, no acquired immunity whatsoever can develop. And almost immediately after birth such a person dies of fulminating infection unless treated by heroic measures.

The lymphoid tissue is located most extensively in the lymph nodes, but it is also found in special lymphoid tissue such as in the spleen, in submucosal areas of the gastrointestinal tract, and, to a slight extent, in the bone marrow. The lymphoid tissue is distributed very advantageously in the body to intercept the invading organisms or toxins before they can spread too widely.

TWO TYPES OF LYMPHOCYTES THAT PROMOTE, RESPECTIVELY, CELLULAR IMMUNITY AND HUMORAL IMMUNITY

Though most of the lymphocytes in normal lymphoid tissue look alike when studied under the microscope, these cells are distinctly divided into two separate populations. One of the populations is responsible for forming the sensitized lymphocytes that provide cellular immunity and the other for forming the antibodies that provide humoral immunity.

Both of these types of lymphocytes are derived originally in the embryo from *lymphocytic stem cells in the bone marrow*. The descendants of the stem cells eventually migrate to the lymphoid tissue. Before doing so, however, those lymphocytes that are eventually destined to form sensitized lymphocytes first migrate to and are preprocessed in the *thymus* gland, for which reason they are called "T" lymphocytes. These are responsible for cellular immunity.

The other population of lymphocytes—those that are destined to form antibodies—is processed in some unknown area of the body, possibly the liver and spleen. However, this population of cells was first discovered in birds in which the preprocessing occurs in the *bursa of Fabricius,* a structure not found in mammals. For this reason this population of lymphocytes is called the "B" lymphocytes, and they are responsible for humoral immunity.

Figure 6–1 illustrates the two separate lymphocyte systems for the formation, respectively, of the sensitized lymphocytes and the antibodies.

Role of the Thymus Gland for Preprocessing the "T" Lymphocytes. Most of the preprocessing of the "T" lymphocytes of the thymus gland occurs shortly before birth of the baby and for a few months after birth. Therefore, beyond this period of time, removal of the thymus gland usually will not seriously impair the "T" lymphocytic immunity system, the system necessary for cellular immunity. However, removal of the thymus several months before birth can completely prevent the development of all cellular immunity. Since it is the cellular type of immunity that is mainly responsible for rejection of transplanted organs such as transplanted hearts and kidneys, one can transplant organs with little likelihood of rejection if the thymus is removed from an animal a reasonable period of time before birth.

Role of the Bursa of Fabricius for Preprocessing "B" Lymphocytes in Birds. It is during the latter part of fetal life that the bursa of Fabricius preprocesses the "B" lymphocytes and prepares them to manufacture antibodies. Here again, this process continues for a while after birth. In mammals, recent experiments indicate that it is lymphoid tissue in the fetal liver, and perhaps to a slight extent lymphoid tissue in the spleen, that performs this same function.

Spread of Processed Lymphocytes to the Lymphoid Tissue. After formation of processed lymphocytes in both the thymus and the bursa, these first circulate freely in the blood and gradually filter into the tissues. Then they enter the lymph and are carried to the lymphoid tissue. The lymphoid tissue contains reticulum cells that form a fine reticulum meshwork. This filters the lymphocytes from the lymph, thereby entrapping them in the lymphoid tissue. Thus, the lymphocytes do not originate primordially in the

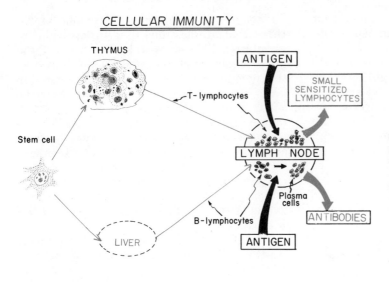

Figure 6-1. Formation of antibodies and sensitized lymphocytes by a lymph node in response to antigens. This figure also shows derivation of "T" and "B" lymphocytes responsible for the cellular and humoral immune processes of the lymph nodes.

lymphoid tissue, but instead are transported to this tissue by way of the preprocessing areas of the thymus and probably the fetal liver.

MECHANISMS FOR DETERMINING SPECIFICITY OF SENSITIZED LYMPHOCYTES AND ANTIBODIES—LYMPHOCYTE CLONES

Earlier in the chapter it was pointed out that the lymphocytes of the lymphoid tissue can form sensitized lymphocytes and antibodies that are highly specific against particular types of invading agents. This effect is believed to occur in the following way:

It is thought that literally hundreds or thousands of different types of precursor lymphocytes preexist in the lymph nodes. Each one of these is theoretically capable of forming a specific type of sensitized lymphocyte or a specific antibody. However, this sensitized lymphocyte or antibody will not be formed in significant quantity until the lymph node is exposed to the appropriate antigen. But when it is exposed, the corresponding precursor lymphocyte for that particular antigen begins to proliferate madly, forming tremendous numbers of progeny; and these in turn lead to the formation of large quantities of antibodies if the precursor lymphocyte is a "B" lymphocyte, or to the formation of numerous sensi-

tized lymphocytes if the precursor lymphocyte is a "T" lymphocyte.

The large mass of new lymphocytes, all of the same type, that are formed in response to a single specific antigen is called a *clone* of lymphocytes. Thus, the lymphoid tissue is capable of forming literally hundreds or thousands of different types of *clones,* each of which is capable of forming a specific antibody or a specific type of sensitized lymphocyte.

SPECIFIC ATTRIBUTES OF HUMORAL IMMUNITY—THE ANTIBODIES

Formation of Antibodies by Plasma Cells. Prior to exposure to a specific antigen, the "B" precursor lymphocytes remain dormant in the lymphoid tissue. However, upon entry of a foreign antigen, the lymphocytes specific for that antigen immediately enlarge and take on the appearance of a *plasmablast.* Some of these then further differentiate to form in four days a total population of about 500 cells for each original plasmablast. The mature plasma cell then produces gamma globulin antibodies at an extremely rapid rate — about 2000 molecules per second for each cell. The antibodies are secreted into the lymph and are carried to the circulating blood. This process continues for several days until death of the plasma cells.

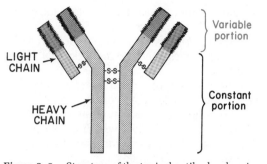

Figure 6–2. Structure of the typical antibody, showing it to be composed of two heavy polypeptide chains and two light polypeptide chains. The antigen binds at two different sites on the variable portions of the chains.

The Nature of the Antibodies

The antibodies are gamma globulins called *immunoglobulins,* and they have molecular weights between approximately 150,000 and 900,000.

All of the immunoglobulins are composed of combinations of *light* and *heavy polypeptide chains,* and most are a combination of two light and two heavy chains, as illustrated in Figure 6–2.

This figure also shows a designated end of each of the light and each of the heavy chains called the "variable portion." The remainder of each chain is called the "constant portion."

Specificity of Antibodies. Each antibody that is specific for a particular antigen has a different organization of amino acid residues in the variable portions of both the light and heavy chains. These have a specific steric shape for each antigen specificity so that when an antigen comes in contact with it, the prosthetic radicals of the antigen fit as a mirror image with those of the antibody, thus allowing a rapid and tight chemical bond between the antibody and the antigen.

Note, especially, in Figure 6–2 that there are two variable sites on the antibody, which allows the antibody to attach to two separate antigen molecules. Thus, most antibodies are *bivalent.* However, a small proportion of the antibodies, which have high molecular weight combinations of light and heavy chains, have more than two reactive sites.

Mechanisms of Action of Antibodies

Antibodies can act in three different ways to protect the body against invading agents:

(1) by direct attack on the invader, (2) by activation of the complement system that then destroys the invader, or (3) by activation of the anaphylactic system that changes the local environment around the invading antigen and in this way reduces its virulence.

Direct Action of Antibodies on Invading Agents. Because of the bivalent nature of the antibodies and the multiple antigen sites on most invading agents, the antibodies can inactivate the invading agent in one or more of several ways as follows:

1. *Agglutination,* in which multiple antigenic agents are bound together into a clump by the antibodies.

2. *Precipitation,* in which the complex of antigen and antibody becomes insoluble and precipitates.

3. *Neutralization,* in which the antibodies cover the toxic sites of the antigenic agent.

4. *Lysis,* in which some very potent antibodies are capable of directly attacking membranes of cellular agents and thereby causing rupture of the cells.

However, the direct actions of antibodies attacking the antigenic invaders probably, under normal conditions, are not strong enough to play a major role in protecting the body against the invader. Most of the protection probably comes through the *amplifying* effects of the complement and anaphylactic effector systems described below.

The Complement System for Antibody Action. Complement is a system of nine different enzyme precursors (designated C-1 through C-9) which are found normally in the plasma and other body fluids, but the enzymes are normally inactive. However, when an antibody combines with an antigen, the antigen-antibody complex then becomes an activator of the complement system. Figure

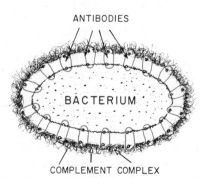

Figure 6–3. Attachment of antibodies to a bacterium and envelopment of the bacterium by complement complex.

6–3, for instance, illustrates a bacterium bound to a few antibodies and these in turn activating the complement complex. Only a few antigen-antibody combinations are required to activate large numbers of enzyme precursor molecules in the first stage of the complement system, and the enzymes thus formed then activate still far more of the enzymes in the later stages of the system. The activated enzymes then attack the invading agent in several different ways, and they also initiate local tissue reactions that provide protection against damage by the invader. Among the more important effects that occur are the following:

1. *Lysis.* The proteolytic enzymes of the complement system digest portions of the cell membrane, thus causing rupture of cellular agents such as bacteria or other types of invading cells.

2. *Opsonization and phagocytosis.* The complement enzymes attack the surfaces of bacteria and other antigens, making these highly susceptible to phagocytosis by neutrophils and tissue macrophages. This process is called *opsonization.* It often enhances the number of bacteria that can be destroyed many hundredfold.

3. *Chemotaxis.* One or more of the complement products cause chemotaxis of neutrophils and macrophages, thus greatly enhancing the number of these phagocytes in the local region of the antigenic agent.

4. *Agglutination.* The complement enzymes also change the surfaces of some of the antigenic agents so that they adhere to each other, thus causing agglutination.

5. *Neutralization of viruses.* The complement enzymes frequently attack the molecular structures of viruses and thereby render them nonvirulent.

6. *Inflammatory effects.* The complement products elicit a local inflammatory reaction, leading to hyperemia, coagulation of proteins in the tissues, and other aspects of the inflammation process, thus preventing movement of the invading agent through the tissues.

Activation of the Anaphylactic System by Antibodies. Some of the antibodies, particularly a special type called IgE antibodies, attach to the membranes of cells in the tissues and blood. Among the important such cells are the *mast cells* in tissues surrounding the blood vessels and the *basophils* circulating in the blood. When an antigen reacts with one of the antibody molecules attached to the cell, there is an immediate swelling and then rupture of the cell, with the release of a large number of factors that affect the local environment. These factors include:

1. *Histamine,* which causes local vasodilatation and increased permeability of the capillaries.

2. *Slow-reacting substance of anaphylaxis,* which causes prolonged contraction of certain types of smooth muscle such as the bronchi.

3. *Chemotaxic factor,* which causes chemotaxis of neutrophils and macrophages into the area of antigen-antibody reaction.

4. *Lysosomal enzymes,* which elicit a local inflammatory reaction.

These anaphylactic reactions can frequently be very harmful to the body, often causing the harmful reactions of allergy, as will be discussed subsequently. However, it is also known that in persons who are genetically unable to respond with the anaphlactic reaction, many types of infection spread much more rapidly through the body than when the reaction can take place. Therefore, this reaction presumably helps to immobilize the antigenic invader.

SPECIAL ATTRIBUTES OF CELLULAR IMMUNITY

Release of Sensitized Lymphocytes from Lymphoid Tissue. On exposure to proper antigens, sensitized lymphocytes are released from lymphoid tissue in ways that parallel antibody release. The only real differences are that the "T" lymphocytes rather than the "B" lymphocytes are stimulated to multiply, and that instead of releasing antibodies, whole sensitized lymphocytes are formed and released into the lymph. These then pass into the circulation where they remain at most a few minutes to a few hours, after which they filter out of the circulation into all the tissues of the body.

Persistence of Cellular Immunity. An important difference between cellular immunity and humoral immunity is its persistence. Humoral antobidies rarely persist more than a few months, or at most a few years. On the other hand, sensitized lymphocytes probably have an indefinite life span and seem to persist until they eventually come in contact with their specific antigen. There is reason to believe that such sensitized lymphocytes might persist as long as 10 years in some instances.

Types of Organisms Resisted by Sensitized Lymphocytes. Although the humoral antibody mechanism for immunity is especially efficacious against more acute bacterial diseases, the cellular immunity system is activated much more potently by the more slowly developing becterial diseases such as tuberculosis, brucellosis, and so forth. Also, this system is active against cancer cells, cells of transplanted organs, and fungus organisms, all of which are far larger than bacteria. And, finally, the system is very active against some viruses.

Therefore, cellular immunity is especially important in protecting the body against some virus diseases, in destroying many early cancerous cells before they can cause cancer, and unfortunately in causing rejection of tissues transplanted from one person to another.

Mechanism of Action of Sensitized Lymphocytes

The sensitized lymphocyte, on coming in contact with its specific antigen, combines with the antigen. This combination in turn leads to a sequence of reactions whereby the sensitized lymphocytes destroy the invader. As is also true of the humoral immunity system, the sensitized lymphocyte destroys the invader either directly or indirectly.

Direct Destruction of the Invader. Figure 6–4 illustrates sensitized lymphocytes that have become bound with antigens in the membrane of an invading cell such as a cancer cell, a heart transplant cell, or a parasitic cell of another type. The immediate effect of this attachment is swelling of the sensitized lymphocyte and release of cytotoxic substances from the lymphocyte to attack the invading cell. The cytotoxic substances are probably lysosomal enzymes manufactured in the lymphocytes. However, these direct effects of the sensitized lymphocyte in destroying the invading cell are probably relatively weak in comparison with two indirect effects, as follows.

Release of Transfer Factor. The sensitized lymphocytes release a polypeptide substance with a molecular weight of less than 10,000, called *transfer factor*. This then reacts with other small lymphocytes in the tissues that are of the nonsensitized variety. On entering these lymphocytes, transfer factor causes them to take on the same characteristics as the sensitized lymphocytes. Thus, transfer factor recruits additional lymphocytes having the same capability for causing the same cellular immunity reaction as the original sensitized lymphocytes. Furthermore, the newly sensitized lymphocytes are as specific for the original antigen as were the original sensitized lymphocytes. Thus, this mechanism multiplies the effect of the sensitized lymphocytes.

Attraction and Activation of Macrophages. A second product of the activated sensitized lymphocyte is a *macrophage chemotaxic factor* that causes as many as 1000 macrophages to enter the vicinity of each activated sensitized lymphocyte. Then, another substance increases the phagocytic activity of the macrophages. Therefore, the macrophages play a major role in removing the foreign antigenic invader.

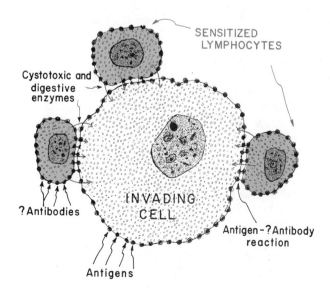

Figure 6–4. Destruction of an invading cell by sensitized lymphocytes.

Thus, it is by a combination of a weak direct effect of the sensitized lymphocytes on the antigen invader and much more powerful indirect reactions that the cellular immunity system destroys the invader.

TOLERANCE OF THE ACQUIRED IMMUNITY SYSTEM TO ONE'S OWN TISSUES – ROLE OF THE THYMUS AND THE BURSA

Obviously, if a person should become immune to his own tissues, the process of acquired immunity would destroy his own body. Fortunately, the immune mechanism normally "recognizes" a person's own tissues as being completely distinctive from those of invaders, and his immunity system forms neither antibodies nor sensitized lymphocytes against his own antigens. This phenomenon is known as *tolerance* to the body's own tissues.

Mechanism of Tolerance. It is possible that tolerance to one's own tissues is determined genetically — that is, via absence of genes to form sensitized lymphocytes and antibodies against the person's own tissues. However, there is much reason to believe that tolerance develops during the processing of the lymphocytes in the thymus and in the "B" lymphocyte processing area. The reason for this belief is that injecting a strong antigen into a fetus at the time that the lymphocytes are being processed in these two areas will prevent the development of precursor lymphocytes in the lymphoid tissue that are specific for the injected antigen.

Therefore, it is believed that during the processing of lymphocytes in the thymus and in the "B" lymphocyte processing area, all those precursor lymphocytes that are specific for the body's own tissues are self-destroyed because of their exposure to the body's antigens.

Failure of the Tolerance Mechanism — Autoimmune Diseases. Unfortunately, persons frequently lose some of their immune tolerance to their own tissues. This usually results from destruction of some of the body's tissues, which releases antigens that then for the first time circulate in the body fluids in significant quantity. For instance, the proteins of the cornea do not seem to circulate in the fluids of the fetus; this is also true of the thyroglobulin molecule of the thyroid; therefore, tolerance to these never develops. But when damage occurs to either of these two tissues in later life, these protein molecules can then elicit immunity, and the immunity in turn can attack the cornea or the thyroid gland to cause corneal opacity or destructive thyroiditis.

Other diseases that result from autoimmunity include: *rheumatic fever,* in which the body becomes immunized against tissues in the heart and joints following exposure to a specific type of streptococcal toxin; *acute glomerulonephritis,* in which the person becomes immunized against the glomeruli, resulting from exposure to another specific type of streptococcal toxin; *myasthenia gravis,* in which immunity develops against muscles and thereby causes paralysis; and *lupus erythematosus,* in which the person becomes immunized against many different body tissues at the same time, a disease that occasionally causes extensive damage and rapid death.

VACCINATION

The process of vaccination has been used for many years to cause acquired immunity against specific diseases. A person can be vaccinated by injecting dead organisms that are no longer capable of causing disease though still have their chemical antigens. This type of vaccination is used to protect against typhoid fever, whooping cough, diphtheria, and many other types of bacterial diseases. Also, immunity can be achieved against toxins that have been treated with chemicals so that their toxic nature has been destroyed even though their antigens for causing immunity are still intact. This procedure is used in vaccinating against tetanus, botulism, and other similar toxic diseases. And, finally, a person can be vaccinated by infecting him with live organisms that have been "attenuated." That is, these organisms either have been grown in special culture mediums or have been passed through a series of animals until they have mutated enough that they will not cause disease but will still carry the specific antigens. This procedure is used to protect against poliomyelitis, yellow fever, measles, smallpox, and many other viral diseases.

PASSIVE IMMUNITY

Thus far, all the acquired immunity that we have discussed has been *active immunity.* That is, the person's body develops either antibodies or sensitized lymphocytes in response to invasion of the body by a foreign antigen. However, it is possible also to achieve temporary immunity in a person without injecting any antigen whatsoever. This is done by infusing antibodies, sensitized lymphocytes, or both from someone else or from some other animal that has been actively immunized against the antigen. The immune protection that is thus achieved is called *passive immunity*. The antibodies will last for 2 to 3 weeks, and during that time the person is protected against the invading disease. Sensitized lymphocytes will last for a few weeks if transfused from another person, and for a few hours to a few days if transfused from an animal.

ALLERGY

One of the important side effects of immunity is the development, under some conditions, of allergy. There are several different types of allergy, some of which can occur in any person, and others that occur only in persons who have a specific allergic tendency.

ALLERGIES THAT OCCUR IN NORMAL PEOPLE

A delayed-reaction type of allergy frequently causes skin eruptions even in normal so-called nonallergic persons. These often occur in response to certain drugs or chemicals, particularly some cosmetics and household chemicals, to which one's skin is often exposed. Another example of such an allergy is the skin eruption caused by exposure to poison ivy.

Delayed-reaction allergy is caused by sensitized lymphocytes and not by antibodies. In the case of poison ivy, the toxin of poison ivy in itself does not cause much harm to the tissues. However, upon repeated exposure it does cause the formation of sensitized lymphocytes. Then, following subsequent exposure to the poison ivy toxin, within a day or so the sensitized lymphocytes diffuse in sufficient numbers into the skin to combine with the poison ivy toxin and elicit a cellular immunity type of reaction. Remembering that cellular immunity can cause release of many toxic substances from the sensitized lymphocytes, as well as extensive invasion of the tissues by macrophages that cause extensive

subsequent effects, one can well understand that the eventual result of some delayed-reaction allergies can be serious tissue damage.

ALLERGIES IN THE "ALLERGIC" PERSON

Some persons have an "allergic" tendency. This tendency is genetically passed from parent to child, and it is caused by the inherited characteristic of these persons to form large quantities of an abnormal type of antibody called a *reagin*. The specific antigens that react with the reagins are called *allergens* because every time one of these allergens comes in contact with the reagin, an allergic reaction occurs.

Reagins have a special propensity to attach to cells throughout the body. Therefore, the reaction of an allergen with these antibodies takes place in direct contact with the cells and damages them. Many allergic reactions result from the fact that large numbers of reagins are often attached to eosinophils and basophils. The reagin-allergen reaction causes rupture of these cells, allowing the release of large numbers of very toxic substances into the tissues, including especially *histamine* and large numbers of *lysosomal digestive enzymes*.

Some of the serious allergic reactions that can result from such reactions are:

Anaphylaxis. When a specific allergen is injected directly into the circulation it can react in widespread areas of the body with the basophils and eosinophils of the blood and the mast cells located immediately outside the small blood vessels. Therefore an anaphylactic type of reaction occurs everywhere. The histamine released into the circulation causes widespread peripheral vasodilatation as well as increased permeability of the capillaries and marked loss of plasma from the circulation. Often, persons experiencing this reaction die of circulatory shock within a few minutes. But also released from the cells are other substances that sometimes cause spasm of the smooth muscle of the bronchioles, eliciting an asthmalike attack.

Hay Fever. In hay fever, the allergen-reagin reaction occurs in the nose. *Histamine* released in response to this causes local vascular dilatation with resultant increased capillary pressure, and it also causes increased capillary permeability. Both of these effects cause rapid fluid leakage into the tissues of the nose, and the nasal linings become swollen and secretory. Use of antihistaminic drugs can prevent this swelling reaction. However, other products of the allergen-reagin reaction still cause irritation of the nose, eliciting the typical sneezing syndrome despite drug therapy.

Asthma. In asthma, the allergen-reagin reaction occurs in the bronchioles of the lungs. Here, the most important product released from the mast cells seems to be a substance called *slow-reacting substance of anaphylaxis,* which causes spasm of the bronchiolar smooth muscle. Consequently, the person has difficulty breathing until the reactive products of the allergic reaction have been removed. Unfortunately, administration of antihistaminics has little effect on the course of asthma, because histamine does not appear to be the major factor eliciting the asthmatic reaction.

REFERENCES

Burke, D. C.: The status of interferon. *Sci. Amer., 236(4)*:42, 1977.

Capra, J. D., and Edmundson, A. B.: The antibody combining site. *Sci. Amer., 236(1)*:50. 1977.

Carpenter, P. L.: Immunology and Serology. 3rd ed. Philadelphia. W. B. Saunders Company, 1975.

Edelman, G. M. (ed.): Cellular Selection and Regulation in the Immune Response. New York. Raven Press, 1974.

Klinman, N. R., and Press. J. L.: Expression of specific clones during B cell development. *Fed. Proc., 34*:47, 1975.

Nisonoff, A., Hopper, J. E., and Spring, S. B.: The Antibody Molecule. New York, Academic Press, Inc., 1975.

Old, L. J.: Cancer immunology. *Sci. Amer., 236(5)*:62, 1977.

Raff, M. C.: Cell-surface immunology. *Sci. Amer., 234(5)*: 30, 1976.

Taylor, G.: Immunology in Medical Practice. Philadelphia, W. B. Saunders Company, 1975.

QUESTIONS

1. What is meant by immunity?
2. How do antigens activate the immunity mechanism?
3. What are the two different types of lymphocytes that play important roles in the immune process?
4. What is the function of lymphocyte clones in determining the specificity of sensitized lymphocytes and antibodies?
5. What are antibodies, and how are they formed?
6. How do antibodies react with antigens?
7. What are some of the functions of the complement system?
8. What are some of the functions of the anaphylactic system?
9. Describe the mechanisms by which sensitized lymphocytes destroy invading organisms.
10. What is meant by immune tolerance?
11. How is vaccination used to protect a person against disease?
12. Distinguish between the types of allergies that can occur in normal people and those that occur in the "allergic" person.

BLOOD COAGULATION, TRANSFUSION, AND TRANSPLANTATION OF ORGANS

7

HEMOSTASIS

The term hemostasis means prevention of blood loss. Whenever a vessel is severed or ruptured, hemostasis is achieved by a succession of different mechanisms including (1) vascular spasm, (2) formation of a platelet plug, (3) blood coagulation, and (4) growth of fibrous tissues into the blood clot to close the hole in the vessel permanently.

VASCULAR SPASM

Immediately after a blood vessel is cut or ruptured, the trauma to the vessel wall itself causes the vessel to contract; this instantaneously reduces the flow of blood from the vessel rupture. The more the vessel is traumatized, the greater is the degree of spasm; this means that the sharply cut blood vessel usually bleeds much more than does the vessel ruptured by crushing. This local vascular spasm lasts for as long as 20 minutes to half an hour, during which time ensuing processes of platelet plugging and blood coagulation can take place.

FORMATION OF A PLATELET PLUG

The second event in hemostasis is an attempt by the platelets to plug the rent in the vessel. To understand this it is important that we first understand the nature of platelets themselves.

Platelets are minute round or oval discs about 2 microns (micrometers) in diameter. They are fragments of *megakaryocytes,* which are extremely large white blood cells formed in the bone marrow. The megakaryocytes disintegrate into platelets while they are still in the bone marrow and release the platelets into the blood. The normal concentration of platelets in the blood is between 200,000 and 400,000 per cubic millimeter.

Mechanism of the Platelet Plug. Platelet repair of vascular openings is based on several important functions of the platelet itself: When platelets come in contact with a *wettable* surface, such as the collagen fibers in the vascular wall, they immediately swell and become sticky so that they stick to the collagen fibers. They secrete large quantities of ADP, and the ADP, in turn, acts on nearby platelets to activate them as well, and the stickiness of these additional platelets causes them also to adhere to the originally activated platelets. Thus, very large numbers of platelets accumulate to form a *platelet plug.* If the rent in a vessel is small, the platelet plug by itself can stop blood loss completely, but if there is a large hole, a blood clot in addition to the platelet plug is required to stop the bleeding.

The platelet plugging mechanism is extremely important to close the minute ruptures in very small blood vessels that occur

69

hundreds of times daily. A person who has very few platelets develops literally hundreds of small hemorrhagic areas under his skin and throughout his internal tissues, but this does not occur in the normal person.

CLOTTING IN THE RUPTURED VESSEL

The third mechanism for hemostasis is formation of the blood clot. The clot begins to develop in 15 to 20 seconds if the trauma of the vascular wall has been severe, and in 1 or 2 minutes if the trauma has been minor. Activator substances both from the traumatized vascular wall and from platelets and blood proteins adhering to the collagen of the traumatized vascular wall initiate the clotting process. The physical events of this process are illustrated in Figure 7–1 and the chemical events will be discussed in detail below.

Within 3 to 6 minutes after rupture of a vessel, the entire cut or broken end of the vessel is filled with clot. After 20 minutes to an hour, the clot retracts; this closes the vessel still further.

FIBROUS ORGANIZATION OF THE BLOOD CLOT

Once a blood clot has formed, it usually becomes invaded by fibroblasts, which sub-

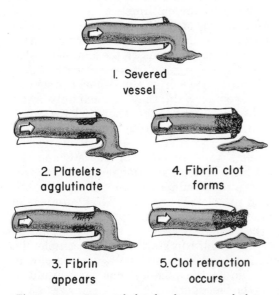

I. Severed vessel

2. Platelets agglutinate

3. Fibrin appears

4. Fibrin clot forms

5. Clot retraction occurs

Figure 7–1. Stages of clot development and clot retraction following injury to a blood vessel. (Redrawn from Seegers and Sharp: Hemostatic Agents. Charles C Thomas.)

sequently form fibrous tissue all through the clot. This begins within a few hours after the clot is formed, and fibrous closure of the hole in the vessel is complete within approximately 7 to 10 days.

MECHANISM OF BLOOD COAGULATION

Almost all research workers in the field of blood coagulation agree that clotting takes place in three essential steps:

First, a substance called *prothrombin activator* **is formed in response to rupture of the vessel or damage to the blood itself.**
Second, the prothrombin activator catalyzes the conversion of prothrombin into *thrombin.*
Third, the thrombin acts as an enzyme to convert fibrinogen into *fibrin threads* **that enmesh red blood cells and plasma to form the clot itself.**

Let us first discuss the latter two of these steps — that is, those involving the formation of the clot itself, beginning with the conversion of prothrombin to thrombin; then we will come back to the initiating stages in the clotting process.

CONVERSION OF PROTHROMBIN TO THROMBIN

After prothrombin activator has been formed as a result of rupture of the blood vessel or as a result of damage to substances in the blood, this activator will cause conversion of prothrombin to thrombin, which in turn causes polymerization of fibrinogen molecules into fibrin threads within the next 10 to 15 seconds.

Prothrombin is a plasma protein having a molecular weight of 68,700. It is an unstable protein that can split easily into smaller compounds, one of which is *thrombin,* having a molecular weight of 33,700, almost exactly half that of prothrombin.

Prothrombin is formed continually by the liver, and if the liver fails to produce prothrombin, its concentration in the plasma falls too low within 24 hours to provide normal blood coagulation. Vitamin K is required by the liver for this formation of prothrombin; therefore, either lack of vitamin K or the presence of liver disease that prevents normal

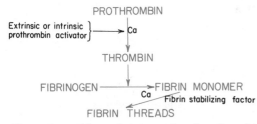

Figure 7–2. Schema for conversion of prothrombin to thrombin, and polymerization of fibrinogen to form fibrin threads.

prothrombin formation can often decrease the prothrombin level so low that a bleeding tendency results.

Effect of Prothrombin Activator to Form Thrombin from Prothrombin. Figure 7–2 illustrates the conversion of prothrombin to thrombin under the influence of prothrombin activator and calcium ions. The rate of formation of thrombin from prothrombin is almost directly proportional to the quantity of prothrombin activator available, which in turn is approximately proportional to the degree of trauma to the vessel wall or to the blood. And the rapidity of the clotting process is proportional to the quantity of thrombin formed.

CONVERSION OF FIBRINOGEN TO FIBRIN—FORMATION OF THE CLOT

Fibrinogen. Fibrinogen is a high molecular weight protein (340,000), most if not all of which is formed in the liver. Liver disease occasionally decreases the concentration of circulating fibrinogen, as it does the concentration of prothrombin which was pointed out previously.

Action of Thrombin on Fibrinogen to Form Fibrin. Thrombin is a protein *enzyme* with proteolytic capabilities. It acts on fibrinogen to activate each molecule, thus forming a molecule of *fibrin monomer* which has the automatic capability of polymerizing with other fibrin monomer molecules. Therefore, many fibrin monomer molecules polymerize within seconds into *long fibrin threads* that form the *reticulum* of the clot. Then another plasma globulin factor, *fibrin stabilizing factor,* acts as an enzyme to cause stronger bonding between the fibrin monomer molecules and also bonding between adjacent polymer chains. This adds tremendously to the strength of the threads.

The Blood Clot. The clot is composed of the meshwork of fibrin threads running in all directions with entrapped blood cells, platelets, and plasma. The fibrin threads adhere to damaged surfaces of blood vessels; therefore, the blood clot prevents blood loss.

INITIATION OF COAGULATION: FORMATION OF PROTHROMBIN ACTIVATOR

Now that we have discussed the clotting process itself, we can return to the more complex mechanisms that initiate the activation of the prothrombin. Clotting can be initiated by (1) trauma to the tissues, (2) trauma to the blood, or (3) contact of the blood with special substances such as collagen outside the blood vessel endothelium. In each instance, they lead to the formation of *prothrombin activator,* which then causes the actual clotting.

There are two basic ways in which prothrombin activator is formed: (1) by the *extrinsic pathway* that begins with trauma to the tissues outside the blood vessels, or (2) by the *intrinsic pathway* that begins in the blood itself.

The Extrinsic Mechanism for Initiating Clotting

The extrinsic mechanism for initiating the formation of prothrombin activator begins with blood coming in contact with traumatized tissues and occurs according to the basic steps illustrated in Figure 7–3.

1. *Release of Tissue Factor and Tissue Phospholipids.* The traumatized tissue releases two factors that set the clotting process into motion. These are (a) *tissue factor,* which is a proteolytic enzyme, and (b) *tissue phospholipids,* which are mainly phospholipids of the tissue cell membranes.

2. *Conversion of Protein Clotting Factors in Plasma to Form Prothrombin Activator.* The tissue factor and tissue phospholipids released in the first step now react with several protein *clotting factors* that are present in the plasma, *factor V, factor VII,* and *factor X.* The product of this reaction is *prothrombin activator.*

As has already been discussed, the prothrombin activator can then cause conversion of prothrombin to thrombin, a complete blood clot developing within the next 10 to 15

EXTRINSIC PATHWAY

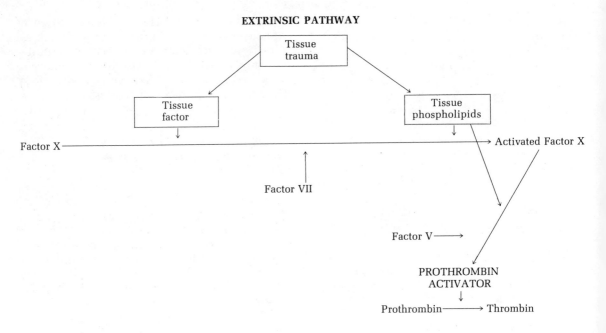

Figure 7–3.   The extrinsic pathway for initiating blood clotting.

INTRINSIC PATHWAY

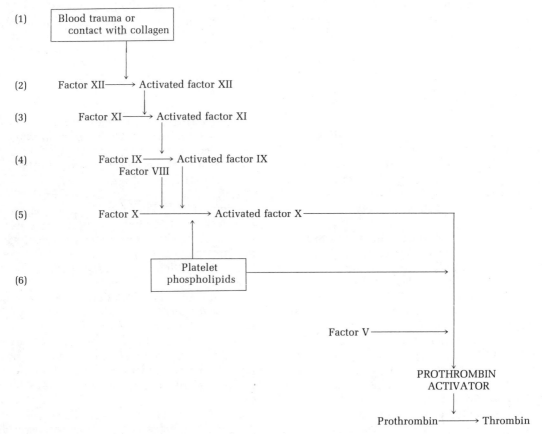

Figure 7–4.   The intrinsic pathway for initiating blood clotting.

seconds. Thus, any time that a blood vessel is ruptured, the trauma to the wall of the vessel itself can initiate the clotting process and thereby stop the bleeding.

The Intrinsic Mechanism for Initiating Clotting

The second mechanism for initiating the formation of prothrombin activator, and therefore for initiating clotting, begins with trauma to the blood itself and continues through a series of cascading reactions, as illustrated in Figure 7–4.

Here again, a number of protein clotting factors in the plasma are involved: *factor V, factor VIII, factor IX, factor XI,* and *factor XII.* Damage of almost any type to the blood or contact of the blood with collagen activates one or more of these factors, and this in turn activates the next, then the next, until eventually the final product is again prothrombin activator. And, as is true for the extrinsic pathway of blood clotting, the prothrombin activator then causes the actual clotting process.

Role of Calcium Ions in the Intrinsic and Extrinsic Pathways

Except for the first two steps in the intrinsic pathway, calcium ions are required for promoting all of the reactions in both clotting pathways. Therefore, in the absence of calcium ions, blood clotting will not occur. Fortunately, in the living body the calcium ion concentration never falls low enough to affect significantly the kinetics of blood clotting.

Summary of Blood Clotting Initiation

It is clear from the above schemas of the intrinsic and extrinsic systems for initiating blood clotting that clotting is initiated after rupture of blood vessels by both of the pathways. The tissue factor and tissue phospholipids initiate the extrinsic pathway, while contact of factor XII and the platelets with collagen in the vascular wall initiates the intrinsic pathway. In contrast, when blood is removed from the body and maintained in a test tube, it is the intrinsic pathway alone that must elicit the clotting.

An especially important difference between the extrinsic and intrinsic pathways is that the extrinsic pathway is explosive in nature; once initiated, its speed of occurrence is limited only by the amount of tissue factor and tissue phospholipids released from the traumatized tissues, and by the quantities of factors X, VII, and V in the blood. With severe tissue trauma, clotting can occur in as little as 15 seconds. On the other hand, the intrinsic pathway is much slower to proceed, usually requiring 1 to 3 minutes to cause clotting.

PREVENTION OF BLOOD CLOTTING IN THE NORMAL VASCULAR SYSTEM

Obviously, it is important that blood not clot in the normal circulation. Fortunately, several important anticlotting mechanisms prevent this from occurring. The three most important are the following:

Endothelial Surface Factors. Probably the most important factor for preventing clotting in the normal vascular system is the nature of the endothelial lining of the blood vessels themselves. The endothelial cells are extremely smooth, which prevents contact activation of the clotting system. But perhaps even more important, a thin molecular layer of negatively charged protein is adsorbed to the inner surface of the endothelium, and this repels the negatively charged protein clotting factors as well as the platelets. This, too, prevents the onset of clotting. However, when the endothelial wall becomes damaged, both of these effects are lost, and clotting usually begins immediately.

Antithrombin. A small amount of protein called *antithrombin* is present in the normal circulating plasma. And when tiny amounts of prothrombin are converted to thrombin, the resulting small quantities of thrombin are destroyed by the antithrombin before clotting can occur. Therefore, clotting will not occur unless the rate of conversion of prothrombin to thrombin rises above a certain critical level.

Heparin. An anticlotting factor called *heparin* is secreted by the *mast cells,* located in the pericapillary connective tissue, especially in the liver and lungs; also, heparin is secreted by circulating basophil cells which are identical with the mast cells. Heparin blocks several of the different steps in the clotting process, especially the action of thrombin to cause conversion of fibrinogen to fibrin threads. Therefore, this substance also serves to prevent clotting under normal conditions.

CONDITIONS THAT CAUSE EXCESSIVE BLEEDING IN HUMAN BEINGS

Excessive bleeding can result from deficiency of any one of the many different clotting factors. However, deficiency of a few of these is especially likely to cause clinically serious bleeding tendencies as follows:

Hemophilia. Hemophilia results from genetic lack of *factor VIII, factor IX,* or *factor XI* in the blood, usually factor VIII. Many persons with hemophilia die in early life because of severe bleeding. However, if a person with hemophilia is given a transfusion of normal plasma, this will provide the missing clotting factor, and clotting will occur normally for several days thereafter. This is the method for treating the serious episodes of bleeding in these patients.

Thrombocytopenia. Thrombocytopenia means lack of adequate numbers of platelets in the circulating blood. This, too, is sometimes a genetically determined disease, but it often also results from poisoning by toxins or drugs. The patient with thrombocytopenia usually has large numbers of *minute hemorrhages* both in the skin and in the deep tissues because the platelet plugging method for stopping small bleeding points in the vasculature becomes deficient. The hemorrhages in the skin cause *purplish blotches* all over the surface of the body, usually about 0.5 cm. in diameter. The bleeding tendency can be stopped for a period of a few days by transfusing fresh whole blood into the patient or by transfusing separated platelets, though both of these are very difficult procedures.

Bleeding in Persons with Liver Disease. In liver disease, several of the clotting factors fail to be produced, especially *prothrombin* and *factors VII, IX,* and *X.* Therefore, any serious liver disease is likely to lead also to serious bleeding tendency.

PREVENTION OF CLOTTING OF BLOOD REMOVED FROM THE BODY

When blood is removed for transfusion, it is necessary to prevent its clotting. This can be done for a few hours by adding a large amount of *heparin.* However, for more prolonged periods of time, the usual procedure is to remove the calcium ions from the blood. As discussed earlier, calcium ions are necessary to promote most of the steps in the clotting process. The usual method for removing calcium ions is to add either *citrate ions* or *oxalate ions.* The citrate ions combine with the calcium ions to form calcium citrate, a soluble but un-ionized substance, and this obviously prevents clotting thereafter. The oxalate combines with the calcium to form a precipitated calcium oxalate, and this too prevents clotting.

When blood is to be used for transfusion, citrate ion instead of oxalate ion is used because the recipient's body can metabolize citrate and remove it from the transfused blood, whereas oxylate is very toxic. After the citrate has been metabolized, the blood will again clot normally.

BLOOD COAGULATION TESTS

Bleeding Time. When a sharp knife is used to pierce the tip of the finger or lobe of the ear, bleeding ordinarily lasts three to six minutes.

Clotting Time. A method widely used for determining clotting time is to collect blood in a chemically clean glass test tube and then to tip the tube back and forth approximately every 30 seconds until the blood has clotted. By this method, the normal clotting time ranges between 5 and 8 minutes.

Prothrombin Time. A test to determine the quantity of prothrombin in the blood is performed by first rendering the blood incoagulable with oxalate, which precipitates the ionic calcium. Then large amounts of calcium, thromboplastin, accelerator globulin, and proconvertin are all mixed suddenly with the sample of blood. The only additional factor needed for the formation of thrombin that is not added to the blood is prothrombin. Therefore, the length of time required for the blood to clot depends on the amount of prothrombin already in the blood. The longer the time required, the less the amount of prothrombin available. The normal prothrombin time is approximately 12 seconds — that is, this is the time required for the clot to appear after mixing has occurred. If the prothrombin time is as long as 20 to 30 seconds, the person can be expected to be a bleeder.

The quantities of proconvertin, accelerator

globulin, and several of the other factors that enter into blood coagulation can be estimated by similar procedures.

TRANSFUSION

Often a person loses so much blood that to save his life he must be given a transfusion of new blood immediately. At other times transfusions are given to treat anemia or some other blood deficiency such as hemophilia or thrombocytopenia. Unfortunately, the bloods of different people are not all exactly alike, and failure to transfuse the appropriate type of blood is likely to cause death of the recipient. For this reason it is very important to understand the physiology of blood grouping and blood matching prior to giving transfusions.

The principal reason one person's blood may not be suitable for another person is that the recipient may be immune to some of the proteins in the blood cells of the donor. Antibodies in the recipient's plasma can cause *agglutination* (clumping) and *hemolysis* (rupture) of the injected cells, thus plugging some of the vessels and releasing large quantities of hemoglobin out of the cells into the circulation.

O-A-B BLOOD GROUPS

The A and B Antigens — the Agglutinogens. Two different but related antigens — type A and type B — occur in the cells of different persons. Because of the way these antigens are inherited, a person may have neither of them in the cells, or may have one or both simultaneously.

As will be discussed below, some bloods also contain strong antibodies called agglutinins that react specifically with either the type A or type B antigens in the cells, causing agglutination and hemolysis. Because the type A and type B antigens in the cells make the cells susceptible to agglutination, these antigens are called *agglutinogens*.

THE FOUR MAJOR O-A-B BLOOD GROUPS. In transfusing blood from one person to another, the bloods of donors and recipients are normally classified into four major O-A-B groups, as illustrated in Table 7–1, depending on the presence or absence of the two agglutinogens. When neither A nor B agglutinogen is present, the blood group is *group O*.

Table 7–1. The Blood Groups and Their Constituent Agglutinogens and Agglutinins

BLOOD GROUPS	AGGLUTINOGENS	AGGLUTININS
O	–	Anti-A and Anti-B
A	A	Anti-B
B	B	Anti-A
AB	A and B	–

When only type A agglutinogen is present, the blood is *group A*. When only type B agglutinogen is present, the blood is *group B*. And when both A and B agglutinogens are present, the blood is *group AB*.

The Agglutinins. When type A agglutinogen *is not present* in a person's red blood cells, antibodies known as "anti-A" agglutinins develop in the plasma. Also, when type B agglutinogen *is not present* in the red blood cells, antibodies known as "anti-B" agglutinins develop in the plasma.

Thus referring once again to Table 7–1, it will be observed that group O blood, though containing no agglutinogens, does contain both *anti-A* and *anti-B agglutinins*, whereas group A blood contains type A agglutinogens and *anti-B agglutinins*, and group B blood contains type B agglutinogens and *anti-A agglutinins*. Finally, group AB blood contains both A and B agglutinogens but no agglutinins at all.

Blood Typing. Prior to giving a transfusion, it is necessary to determine the blood group of the recipient and the group of the donor blood so that the bloods will be appropriately matched.

The usual method of blood typing is the slide technique illustrated in Figure 7–5. In using this technique a drop or more of blood is removed from the person to be typed. This is then diluted approximately 50 times with saline to provide a suspension of red blood cells. Two separate drops of this suspension are placed on a microscope slide, and a drop of anti-A agglutinin serum is mixed with one of the drops of cell suspension while a drop of anti-B agglutinin serum is mixed with the second drop. After allowing several minutes for the agglutination process to take place, the slide is observed under microscope to determine whether or not the cells have clumped. If they have clumped, one knows

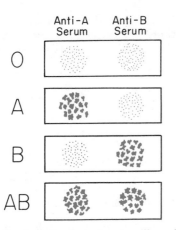

Figure 7–5. Agglutination of some cells and lack of agglutination of other cells in the process of blood typing.

that an immune reaction has resulted between the serum and the cells.

Table 7–2 illustrates the reactions that occur with each of the four different types of blood. Group O red blood cells have no agglutinogens and, therefore, do not clump with either the anti-A or the anti-B serum. Group A blood has A agglutinogens and therefore does clump with anti-A agglutinins. Group B blood has B agglutinogens and clumps with the anti-B serum. Group AB blood has both A and B agglutinogens and clumps with both types of serum.

THE Rh BLOOD TYPES

In addition to the O-A-B blood group system, several other systems are sometimes important in the transfusion of blood, the most important of which is the Rh system. The one major difference between the O-A-B system and the Rh system is the following: In the O-A-B system, the agglutinins responsible for causing transfusion reactions develop spontaneously, whereas in the Rh system anti-Rh antibodies almost never occur spontaneously. Instead, persons must first be massively exposed to the Rh factor, usually by transfusion of blood into them, before they will develop enough agglutinins to cause significant transfusion reaction.

The Rh Agglutinogens. There are at least eight different types of Rh agglutinogens, each of which is called an *Rh factor*. Certain ones of these are very likely to cause transfusion reactions. If one of them is present in the blood cells, the person is said to be Rh positive.

When the red blood cells do not contain one of the potent Rh agglutinogens, the person is said to be Rh negative.

The Rh Immune Response. FORMATION OF ANTI-RH AGGLUTININS. When red blood cells containing Rh factor are injected into an Rh negative person, anti-Rh agglutinins develop very slowly, the maximum concentration of agglutinins occurring approximately 2 to 4 months later. This immune response occurs to a much greater extent in some people than in others. On multiple exposure to the Rh factor, the Rh negative person eventually becomes strongly "sensitized" to the Rh factor, and thereafter he will have a serious transfusion reaction if transfused with additional Rh positive blood.

ERYTHROBLASTOSIS FETALIS. Erythroblastosis fetalis is a disease of the fetus and newborn infant characterized by progressive destruction of the baby's blood.

This disease occurs in many babies when the mother is Rh negative and the father Rh positive. If the baby inherits the Rh positive characteristic from the father, the mother is likely to become immunized against the baby's Rh positive blood cells. This is especially true when the mother has had several Rh positive babies one after another. Once the mother becomes immunized against the Rh factor, *anti-Rh agglutinins* formed by the mother's immune system then diffuse through the placenta from the mother into the baby and begin to destroy the baby's cells. In many instances this kills the baby prior to birth, and the baby is stillborn. In other instances the baby is born extremely anemic because of destruction of the red blood cells. Also, the baby is severely *jaundiced,* which means that its skin has a very yellow color

Table 7–2.   **Blood Typing – Showing Agglutination of Cells of the Different Blood Groups with Anti-A and Anti-B Agglutinins**

RED BLOOD CELLS	SERA	
	Anti-A	*Anti-B*
O	–	–
A	+	–
B	–	+
AB	+	+

because the reticuloendothelial cells convert hemoglobin released from the destroyed red blood cells into *bilirubin,* which is a yellow pigment that colors the skin.

The treatment for the newborn jaundiced and anemic child is to replace its blood with Rh negative blood that also does not have anti-Rh agglutinins. Otherwise, the baby is still very likely to die.

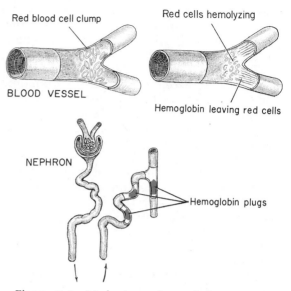

Figure 7-6. Mechanism of transfusion reactions caused by mismatched blood.

TECHNIQUE FOR TRANSFUSION

The usual technique for transfusion is to remove blood from a donor into a sterile bottle containing a citrate solution. The citrate deionizes the calcium so that the blood remains unclotted. This blood can then be kept for several weeks at 4° Celsius until it is needed. Also, fresh blood can be transfused by using heparin as an anticoagulant.

TRANSFUSION REACTIONS RESULTING FROM MISMATCHED BLOOD GROUPS

Hemolysis of Red Cells Following Transfusion Reactions. When there is an antibody-antigen reaction involving one of the blood groups, the usual effect is for the red blood cells to *agglutinate.* Clumps of agglutinated cells then become lodged in the small capillaries throughout the body as shown in Figure 7-6. Over a period of hours these cells gradually break up and release their cellular contents, mainly hemoglobin, into the circulating blood. Therefore, the net result of the transfusion reaction is almost always to release large amounts of hemoglobin into the circulating blood. Sometimes the antigen-antibody reaction is so strong that it actually causes direct hemolysis of the red cells without first causing agglutination; this direct hemolysis occurs within minutes rather than over a period of hours.

Acute Kidney Shutdown Following Transfusion Reactions. The large amount of circulating hemoglobin following a transfusion reaction can often cause acute kidney shutdown, and this in turn sometimes leads to death within a week to 12 days as a result of *uremia.*

The kidney shutdown is caused mainly by blockage of the kidney tubules with hemoglobin, as shown in the lower part of Figure 7-6.

When the hemoglobin filters through the glomerulus into the renal tubules, it is in the dissolved form. However, large quantities of fluid are then reabsorbed back into the blood in the tubules, and this concentrates the hemoglobin so greatly that it precipitates. Consequently, the tubules become totally plugged with hemoglobin, causing acute renal failure.

TRANSPLANTATION OF TISSUES AND ORGANS

In this modern age of surgery, many attempts are being made to transplant tissues and organs from one person to another, or, occasionally, from lower animals to the human being. Many of the different antigenic proteins of red blood cells that cause transfusion reactions plus still many more are present in the other cells of the body. Consequently, any foreign cells transplanted into a recipient can cause immune responses and immune reactions. In other words, most recipients are just as able to resist invasion by cells from another person as to resist invasion by bacteria.

In the case of transplants from one person to another, except when special drug therapy is employed, immune reactions almost always occur, causing death of all the cells in the transplant within 3 to 10 weeks after transplantation. This process is called "rejection" of the transplant.

Transplants from one identical twin to another are an exception in which transplants are usually successful without drug therapy. The reason for this is that the antigenic proteins of both twins are determined by identical genes derived originally from the single fertilized ovum.

Some of the different tissues and organs that have been transplanted either experimentally or for temporary benefit from one person to another are skin, kidney, heart, liver, glandular tissue, bone marrow, and lung. When special procedures have been employed to prevent the antibody reactions that cause rejection, kidney transplants have been successful for as long as ten years, a rare liver and heart transplant for 1 to 8 years, and lung transplants for 1 month.

ATTEMPTS TO OVERCOME REJECTION OF TRANSPLANTED TISSUE

Tissue Typing.   In the same way that red blood cells can be typed to prevent reactions between recipient and donor, so also is it possible to "type" tissues to help prevent graft rejection, though thus far this procedure has met with far less success than has been achieved in red blood cell typing. There are 25 to 30 antigens that are especially important in transplant rejection and for which the tissues must be typed with appropriate antisera.

Use of Anti-Lymphocyte Serum.   It was pointed out in Chapter 6 that transplanted tissues are usually destroyed by lymphocytes that have become sensitized against the transplant. These cells invade the transplant and gradually destroy the cells. The cells of the transplant begin to swell, their membranes become very permeable, and finally they rupture. Simultaneously, macrophages move in to clean up the debris. Within a few days to a few weeks after this process begins, the tissue is completely destroyed.

Therefore, one of the most effective procedures to prevent rejection of transplanted tissues has been to inoculate the recipient with *anti-lymphocyte serum*. This serum is made in horses by injecting human lymphocytes into them; the antibodies that develop in these animals will then attack and destroy human lymphocytes. When serum from one of these animals is injected into the transplanted recipient, the number of circulating small lymphocytes can be decreased to as little as 5 to 10 per cent of the normal number, and there is corresponding decrease in the likelihood of the rejection reaction.

Suppression of Antibody Formation.   Occasionally, a human being has naturally suppressed antibody formation resulting from a hereditary depression of his immune system. Transplants of tissues into such individuals are occasionally successful, or at least their destruction is delayed. Also, irradiative destruction of most of the lymphoid tissue by either x-rays or gamma rays renders a person much more susceptible than usual to a transplant. And treatment with certain drugs, such as azathioprine (Imuran) which suppresses antibody formation, also increases the likelihood of success with transplants. Unfortunately, all of these procedures also leave the person unprotected by the immune system against disease.

To summarize, transplantation of living tissues in human beings up to the present has been mainly an experiment, though some degree of success is now being recorded, especially for kidney transplants. But when someone succeeds in blocking the rejection response of the recipient to a donor organ without at the same time destroying the recipient's specific immunity for disease, this story will change overnight.

REFERENCES

Cerottini, J. C., and Brunner, K. T.: Cell-mediated cytotoxicity, allograft rejection, and tumor immunity. *Adv. Immunol., 18*:67, 1974.

Cooper, H. A., Mason, R. G., and Brinkhous, K. M.: The platelet: membrane and surface reactions. *Ann. Rev. Physiol., 38*:501, 1976.

Cunningham, B. A.: The structure and function of histocompatibility antigens. *Sci. Amer., 234(4)*:96, 1977.

Davidson, J. F., Samama, M. M., and Desnoyers, P. C. (eds.): Progress in Chemical Fibrinolysis and Thrombolysis. New York. Raven Press, 1975.

Lyle, L. R., and Parker, C. W.: New approaches to immunosuppression. *Fed. Proc., 33*:1889, 1974.

Minna, J. D., Robboy, S. J., and Colman, R. W.: Disseminated Intravascular Coagulation in Man. Springfield, Ill., Charles C Thomas, Publisher, 1974.

Quick, A. J.: The Hemorrhagic Diseases and the Pathology of Hemostasis. Springfield, Ill., Charles C Thomas, Publisher, 1974.

Suttie, J. W., and Jackson, C. M.: Prothrombin structure, activation, and biosynthesis. *Physiol. Rev., 57*:1, 1977.

Terasaki, P. I., Opelz, G., and Mickey, M. R.: Histocompatibility and clinical kidney transplants. *Transplant. Proc., 6(Suppl. I)*:33, 1974.

QUESTIONS

1. What role does vascular spasm play in hemostasis?
2. What role do platelet plugs play in hemostasis?
3. Give the functions of thrombin and fibrinogen in the clotting process.
4. Describe the extrinsic mechanism for initiating clotting; what is the role of tissue damage in this mechanism?
5. What is the difference between the intrinsic mechanism and the extrinsic mechanism for initiating clotting?
6. How is clotting prevented in the normal vascular system?
7. How do the following conditions cause bleeding tendencies: hemophilia, thrombocytopenia, and liver disease?
8. What are the agglutinogens?
9. What are the agglutinins?
10. How does the Rh factor sometimes cause transfusion reactions and at other times cause erythroblastosis fetalis?
11. Discuss the problems in the transplantation of organs and tissues.

8 FLUID ENVIRONMENT OF THE CELL AND TRANSPORT THROUGH THE CELL MEMBRANE

The fluid inside the cells of the body, called *intracellular fluid,* is very different from that outside the cells, called *extracellular fluid.* The extracellular fluid circulates in the spaces between the cells and also mixes freely with the fluid of the blood through the capillary walls. Thus, it is the extracellular fluid that supplies the cells with nutrients and other substances needed for cellular function. But before the cell can utilize these substances, they must be transported through the cell membrane.

DIFFERENCES BETWEEN EXTRACELLULAR FLUID AND INTRACELLULAR FLUID

Figure 8–1 illustrates the compositions of extracellular fluid and intracellular fluid. Both of these fluids contain reasonable amounts of the usual nutrients required by the cells for metabolism, including such substances as glucose, amino acids, and the lipids cholesterol, phospholipid, and neutral fat. Also, both fluids contain oxygen and carbon dioxide, and the hydrogen ion concentration (which indicates the degree of acidity of the fluids) is only slightly different between the two fluids. On the other hand, some of the ions are distributed quite differently between the extracellular fluid and the intracellular fluid. Note especially in Figure 8–1 that the sodium ion concentration is very high in the extra-

cellular fluid though very low intracellularly. In contrast, the potassium ion concentration is very high inside the cell but very low outside. Similarly, calcium is high outside and low inside; magnesium high inside, low outside; chloride high outside, low inside; bicarbonate high outside, low inside; phosphates high inside, low outside; and sulfates low in both fluids but higher inside the cell than in the extracellular fluid.

There are two major reasons for the dif-

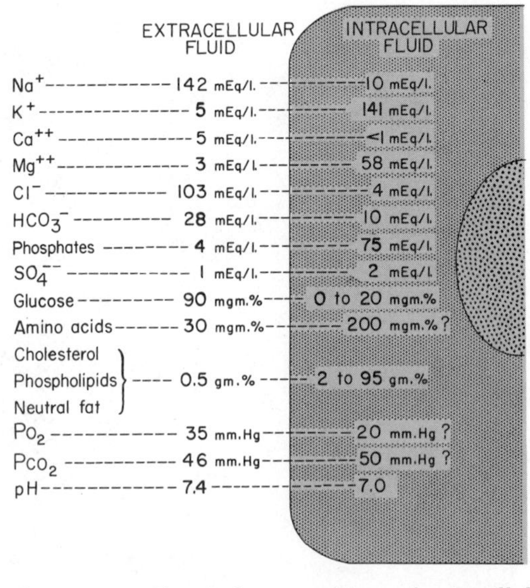

	EXTRACELLULAR FLUID	INTRACELLULAR FLUID
Na^+	142 mEq/l.	10 mEq/l.
K^+	5 mEq/l.	141 mEq/l.
Ca^{++}	5 mEq/l.	<1 mEq/l.
Mg^{++}	3 mEq/l	58 mEq/l.
Cl^-	103 mEq/l.	4 mEq/l.
HCO_3^-	28 mEq/l.	10 mEq/l.
Phosphates	4 mEq/l.	75 mEq/l.
SO_4^{--}	1 mEq/l.	2 mEq/l.
Glucose	90 mgm.%	0 to 20 mgm.%
Amino acids	30 mgm.%	200 mgm.%?
Cholesterol Phospholipids Neutral fat	0.5 gm.%	2 to 95 gm.%
Po_2	35 mm.Hg	20 mm.Hg ?
Pco_2	46 mm.Hg	50 mm.Hg ?
pH	7.4	7.0

Figure 8–1. Chemical compositions of extracellular and intracellular fluids.

80

ferent concentrations between extracellular fluids and intracellular fluids: First, some substances that enter the cells are utilized so rapidly that their concentrations become reduced inside the cell in comparison with the outside. For instance, the concentrations of both glucose and oxygen are lower inside than outside because both are continually being used in the metabolic reactions of the cell. These metabolic reactions inside the cell also create new substances, thereby causing the concentrations of these to be greater inside than outside. For instance, large quantities of carbon dioxide are formed in the cells, which means that the carbon dioxide concentration is normally somewhat higher inside than outside. Likewise, other end-products of cellular metabolism, such as urea, creatinine, sulfates, and so forth, are all present in the intracellular fluid in considerably higher concentration than in extracellular fluid.

The second reason for major differences between the concentrations of extracellular and intracellular fluid is selectivity of transport of substances through the cell membrane. The cell membrane is highly permeable to some substances and very poorly permeable to others, which obviously allows some substances to enter or to leave the cells more easily than others. But, in addition to differences in permeability, the cell membrane has the capability of *actively* transporting many substances through the membrane. That is, they can be carried through the membrane by a chemical mechanism and released on the opposite side of the membrane. Furthermore, active transport can occur even when the concentration of the substance is less on the first side of the membrane than on the second. This property of the cell membrane is responsible for many of the body's most important functions, as we shall discuss throughout this text. However, to give a simple example, it is active transport of sodium and potassium ions through the membrane that causes the concentration differences of these two most important ions on the inside versus the outside of the membrane, as illustrated in Figure 8–1. In turn, these concentration differences are the cause of the electrical potentials that occur in nerve and muscle fibers. And, finally, the electrical potentials are responsible for transmission of nerve impulses and for control of muscle contraction. Therefore, the simple process of selective transport of substances through the cell membrane can lead, first, to differences in the compositions of the

intracellular and extracellular fluids and then in turn to many other important functions of the human body.

MOVEMENT AND MIXING OF THE EXTRACELLULAR FLUIDS THROUGHOUT THE BODY

Fortunately, the extracellular fluids are continually mixed. Were it not for this, the cells would remove all the nutrients from the fluids in their immediate vicinity until none would be left, and cellular excreta would accumulate locally until they would kill the cells. Yet, two different mechanisms provide for continual mixing and movement of the extracellular fluid throughout the body; these are (1) circulation of the blood and (2) the phenomenon of diffusion.

Figure 8–2 illustrates the general plan of

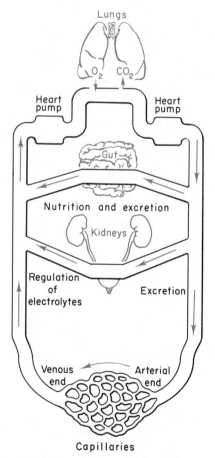

Figure 8–2. General schema of fluid flow in the circulation.

circulation, showing continuous flow of blood around the circulatory system, through the lungs and through the different tissues. The blood picks up oxygen in the lungs and picks up various nutrients in the gut and then carries these to all other areas of the body. On passing through the tissues the blood picks up carbon dioxide, urea, and other excreta from the cells and transports these to the lungs and kidneys to be removed from the body. Thus, the circulatory system provides long-distance transport of the extracellular fluids, in this way keeping the fluids in the different parts of the body mixed with each other.

The phenomenon of diffusion will be described in more detail later in the chapter in relation to transport of substances through the cell membrane. However, the basic principle of diffusion is simply that all molecules in the fluid, including both the molecules of water and those of the dissolved substances, are continually moving among each other. This motion allows continual mixing of all substances within the blood and also in the fluids of the tissue spaces, the so-called *interstitial fluid*. Furthermore, since there are large numbers of small openings in the capillaries, the *capillary pores*, the molecules are continually moving both out of the capillaries into the interstitial fluid and then back again through the pores into the blood. And because the blood is continually moving through all parts of the body, one can readily see that all the extracellular fluids of the body become mixed with each other. Indeed, this process is so effective that almost every portion of the extracellular fluid, even that in the minutest tissue space, becomes mixed with all the other extracellular fluid of the body at least once every 10 to 30 minutes. Because of this rapidity of mixing, the concentrations of substances in the extracellular fluid of one portion of the body are rarely more than a few per cent different from those anywhere else. It is in this way that nutrients from the gut, hormones from the endocrine glands, and oxygen from the lungs are transported to the cells, and it is also in this way that the excreta from the cells are carried either to the lungs or kidneys to be removed from the body.

TRANSPORT OF SUBSTANCES THROUGH THE CELL MEMBRANE

Substances are transported through the cell membrane by two major processes, *diffusion* and *active transport*. Though there are many different variations of these two basic mechanisms, as we shall see later in this chapter, basically, diffusion means movement of substances in a random fashion caused by the normal kinetic motion of matter, whereas active transport means movement of substances as a result of chemical procsses that impart energy to cause the movement.

DIFFUSION

All molecules and ions in the body fluids, including both water molecules and dissolved substances, are in constant motion, each particle moving its own separate way. Motion of these particles is what physicists call heat — the greater the motion, the higher is the temperature — and the motion never ceases. When a moving particle, A, bounces against a stationary particle, B, its energy of impact repels particle B, momentarily adding some of this energy to particle B. Consequently, particle B gains kinetic energy of motion while particle A slows down, losing some of it kinetic engery. Thus, Figure 8–3 shows the pathway of a single molecule in solution bouncing among the other molecules first in one direction, then another, then another, and so forth, bouncing hundreds to millions of times each second. At times it travels a far distance before striking the next molecule, but at other times only a short distance.

This continual movement of molecules among each other in liquids, or in gases, is called *diffusion*. Ions diffuse in exactly the same manner as whole molecules, and even suspended colloid particles diffuse in a similar manner, except that because of their very

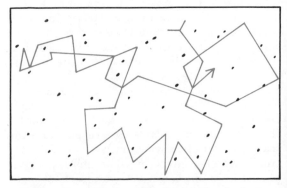

Figure 8–3. Diffusion of a fluid molecule during a fraction of a second.

large sizes they diffuse far less rapidly than molecular substances.

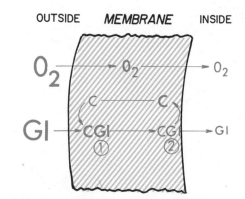

Figure 8-4. Diffusion of substances through the lipid matrix of the membrane. The upper part of the figure shows *free diffusion* of oxygen through the membrane, and the lower part shows *facilitated diffusion* of glucose.

KINETICS OF DIFFUSION — EFFECT OF CONCENTRATION DIFFERENCE

If we consider all the different factors that affect the rate of diffusion of a substance from one area to another, they are the following: (1) The greater the concentration differences between the areas, the greater the rate of diffusion. (2) The less the molecular weight, the greater the rate of diffusion. (3) The shorter the distance, the greater the rate. (4) The greater the cross-section of the diffusion pathway, the greater is the rate of diffusion. (5) The greater the temperature, the greater is the molecular motion and also the greater is the rate of diffusion. All these can be placed in the approximate formula for diffusion in solutions shown at the bottom of this page.

DIFFUSION THROUGH THE CELL MEMBRANE

The cell membrane is essentially a sheet of lipid material, called the *lipid matrix,* probably covered on each surface by a layer of protein, the detailed structure of which was discussed in Chapter 2 and shown diagrammatically in Figure 2-4. The fluids on each side of the membrane are believed to penetrate the protein portion of the membrane with ease so that any dissolved substance can diffuse into this portion of the cell membrane without any impediment. However, the lipid portion of the membrane is an entirely different type of fluid medium, acting as a limiting membrane between the extracellular and intracellular fluids.

However, two different methods by which the substances can diffuse through the membrane are: (a) by becoming dissolved in the lipid and then diffusing through it in the same way that diffusion occurs in water, and (b) by diffusing through minute pores that pass directly through the membrane.

Effect of Lipid Solubility on Diffusion Through the Lipid Matrix. The primary factor that determines how rapidly a substance can diffuse through the lipid matrix of a cell membrane is its solubility in lipids. If it is very lipid-soluble, it becomes dissolved in the membrane very easily and therefore also passes through easily. On the other hand, a substance that dissolves very poorly in lipids will be greatly retarded. For instance, oxygen, carbon dioxide, alcohol, and fatty acids are all very soluble, and they pass through the lipid matrix with ease, as shown for oxygen in Figure 8-4; whereas water, which is almost completely insoluble in lipids, passes through the lipid matrix almost not at all.

CARRIER-MEDIATED FACILITATED DIFFUSION. Some substances are very insoluble in lipids and yet can still pass through the lipid matrix by a process called *facilitated diffusion.* This is the means by which different sugars in particular cross the membrane. The most important of these sugars is glucose, the membrane transport of which is illustrated in the lower part of Figure 8-4. This shows that glucose Gl combines with a *carrier* substance C at point 1 to form the compound CGl. This combination is soluble in the lipid so that it can diffuse to the other side of the membrane, where the glucose then breaks away from the carrier (point 2) and passes to the inside of the cell while the carrier moves back to the outside surface of the membrane to pick up still more glucose and transport it also to the inside. Thus, the effect of the carrier is to make the glucose soluble in the membrane; without it, glucose cannot pass through the mem-

$$\text{Diffusion rate} \propto \frac{\text{Concentration difference} \times \text{Cross-sectional area} \times \text{Temperature}}{\sqrt{\text{Molecular weight}} \times \text{Distance}}$$

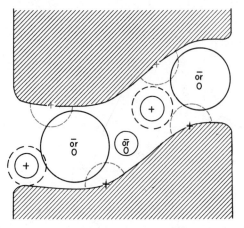

Figure 8–5. Postulated structure of the pore in the mammalian red cell membrane, showing the sphere of influence exerted by charges along the surface of the pore. (Modified from Solomon: *Sci. Amer.,* 203 [6]:146, 1960.)

brane. The carrier itself is probably a small protein or lipoprotien.

Diffusion Through the Membrane Pores. Some substances, such as water and many of the dissolved ions, go through holes in the cell membrane called simply the cell *pores*. The nature of these pores is unknown. They might be round holes, small slits in the cell membrane, or most likely some large hydrophilic protein molecule penetrating all the

way through the cell membrane and providing a water-filled pathway for movement of water-soluble substances along the axis of the molecule. At any rate, these so-called "pores" in the cell membrane behave as if they were minute round holes appproximately 8 Å (0.8 nanometer) in diameter and as if the total area of the pores should equal approximately 1/1600 of the total surface area of the cell. Despite this very minute total area of the pores, both molecules and ions diffuse so rapidly that the entire volume of fluid in a cell can easily pass through the pores within a few hundredths of a second.

Figure 8–5 illustrates a postulated structure of a pore, indicating that its surface is probably lined with positively charged prosthetic groups. This figure shows several small particles passing through the pore and also shows that the maximum size of the particle that can pass through is approximately equal to the size of the pore itself.

EFFECT OF PORE SIZE ON DIFFUSION THROUGH THE PORE. Table 8–1 gives the effective diameters of various substances in comparison with the diameter of the pore. Note that some substances, such as the water molecule, urea molecule, and chloride ion, are considerably smaller in size than the pore. All these pass through the pore with great ease. For instance, the rate per second of diffusion of water in each direction through the pores of

Table 8–1. **Relationship of Effective Diameters of Different Substances to Pore Diameter and Relative Permeabilities***

SUBSTANCE	DIAMETER Å	RATIO TO PORE DIAMETER	APPROXIMATE RELATIVE PERMEABILITY
Water molecule	3	0.38	50,000,000
Urea molecule	3.6	0.45	1,500,000
Hydrated chloride ion			
(red cell)	3.86	0.48	500,000
(nerve membrane)	–	–	0.06
Hydrated potassium ion			
(red cell)	3.96	0.49	1.1
(nerve membrane)	–	–	0.03
Hydrated sodium ion			
(red cell)	5.12	0.64	1
(nerve membrane)	–	–	0.0003
Lactate ion	5.2	0.65	?
Glycerol molecule	6.2	0.77	?
Ribose molecule	7.4	0.93	?
Pore size	8 (Ave.)	1.00	–
Galactose	8.4	1.03	?
Glucose	8.6	1.04	0.4
Mannitol	8.6	1.04	?
Sucrose	10.4	1.30	?
Lactose	10.8	1.35	?

*These data have been gathered from different sources but relate primarily to the red cell membrane. Other cell membranes have different characteristics.

a cell is about 100 times as great as the volume of the cell itself. It is fortunate that an identical amount of water normally diffuses in each direction, which keeps the cell from either swelling or shrinking despite the rapid rate of diffusion. The rates of diffusion of urea and chloride ions through the membrane are somewhat less than that of water, which is in keeping with the fact that their effective diameters are greater than that of water.

Table 8-1 also shows that most of the surgars, including glucose, have effective diameters that are slightly greater than that of the pores. Obviously, not even a single molecule as large as these could go through a pore that is smaller than its size. For this reason essentially none of the sugars can pass through the pores; instead most of these must pass through the lipid matrix by the process of facilitated diffusion.

EFFECTS OF ELECTRICAL CHARGE ON TRANSPORT OF IONS THROUGH THE MEMBRANE. Positively charged ions, such as sodium ions, pass through the cell membrane with extreme difficulty, as shown in Table 8-1. The reason for this is believed to be the presence of positive charges of proteins or adsorbed positive ions, such as calcium ions, lining the pores. Figure 8-5 illustrates that each positive charge causes a sphere of electrostatic space charge to protrude into the lumen of the pore. A positive ion attempting to pass through a pore also exerts a sphere of positive electrostatic charge so that the two positive charges repel each other. This repulsion, therefore, blocks or greatly impedes movement of the positive ion through the pore.

In contrast to the positive ions, negatively charged ions ordinarily pass through mammalian cell membrane pores much more easily. Thus, for the nerve membrane, chloride ions permeate the membrane about 2 times as easily as potassium ions and 100 to 200 times as easily as sodium ions. Moreover, for the red cell membrane, chloride ions permeate the membrane still another half million times as easily. It is believed that these differences are caused by lack of, or few numbers of, negative charges lining the pores in contrast to the large numbers of positive charges.

EFFECT OF CONCENTRATION DIFFERENCE ON NET DIFFUSION RATE

Figure 8-6 illustrates a cell membrane with a substance in high concentration on the outside and low concentration on the

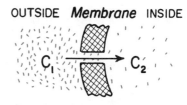

OUTSIDE *Membrane* INSIDE

Figure 8-6. Effect of concentration gradient on diffusion of molecules and ions through a cell membrane pore.

inside. The rate at which the substance diffuses *inward* is proportional to the concentration of molecules on the outside, for this concentration determines how many of the molecules strike the pore each second. On the other hand, the rate at which the molecules diffuse *outward* is proportional to their concentration inside the membrane. Obviously, therefore, the rate of net diffusion is proportional to the concentration on the outside *minus* the concentration on the inside, or

$$\text{Net diffusion} \propto P(C_1 - C_2)$$

in which C_1 is the concentration on the outside, C_2 is the concentration on the inside, and P is the permeability of the membrane for the substance.

NET DIFFUSION OF WATER ACROSS CELL MEMBRANES—OSMOSIS

By far the most abundant substance to diffuse through the cell membrane is water. Enough water ordinarily diffuses in each direction through the red cell membrane per second to equal about *100 times the volume of the cell itself.* Yet, *normally,* the amount that diffuses in the two directions is so precisely balanced that not even the slightest *net* diffusion of water occurs. Therefore, the volume of the cell normally remains constant. However, a *concentration difference for water* can develop across a membrane, just as concentration differences for other substances can also occur. When this happens, net diffusion of water—that is, more diffusion in one direction than in the other direction—does occur across the cell membrane, causing the cell either to swell or to shrink, depending on the direction of the net diffusion. This process of net diffusion of water caused by a water concentration difference is called *osmosis.*

To give an example of osmosis, let us assume that we have the conditions shown in Figure 8-7, with pure water on one side of the cell membrane and a solution of sodium chlo-

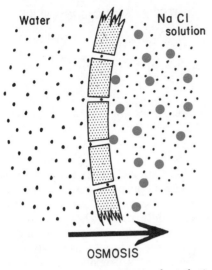

Figure 8-7. Osmosis occurring at a hypothetical cellular membrane when a sodium chloride solution is placed on one side of the membrane and water on the other side.

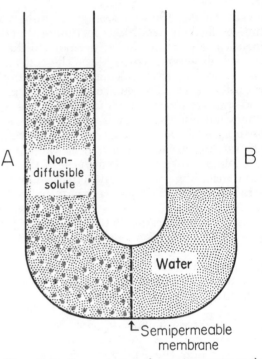

Figure 8-8. Demonstration of osmotic pressure on the two sides of the semipermeable membrane.

ride on the other side. Water molecules pass through the cell membrane with extreme ease while sodium ions pass through only with difficulty. And the chloride ions cannot pass through the membrane because the positive charge of the sodium ions holds the negatively charged chloride ions back. Therefore, sodium chloride solution is actually a mixture of diffusible water molecules and nondiffusible sodium ions and chloride ions. Yet the presence of the sodium and chloride has reduced the concentration of water molecules in a given volume of solution below that of pure water. That is, it is said that the *chemical potential* of the water molecules has been reduced. As a result, in the example of Figure 8-7, more water molecules strike the pores on the left side where there is pure water than on the right side where the water concentration has been reduced. Thus, osmosis occurs from left to right.

Osmotic Pressure. If in Figure 8-7 physical pressure were applied to the sodium chloride solution, osmosis of water into this solution could be slowed or even stopped. The amount of pressure required to stop osmosis completely is called the *osmotic pressure* of the sodium chloride solution.

The principle of a pressure difference opposing osmosis is illustrated in Figure 8-8, which shows a semipermeable membrane separating two separate columns of fluid, one containing water and the other containing a solu-

tion of water and some solute that will not penetrate the membrane. Osmosis of water from chamber B into chamber A causes the levels of the fluid columns to become farther and farther apart, until eventually a pressure difference is developed that is great enough to oppose the osmotic effect. The pressure difference across the membrane at this time is said to be the osmotic pressure of the solution containing the nondiffusible solute.

OSMOTIC PRESSURE OF THE EXTRACELLULAR AND INTRACELLULAR FLUIDS. The ions dissolved in the extracellular and intracellular fluids, as well as other substances such as glucose, amino acids, free fatty acids, and so forth, can all cause osmotic pressure at the cell membrane. The overall concentration of substances in the extracellular and intracellular fluids is enough to create a total osmotic pressure of approximately 5400 mm. Hg; that is, if clear water should be placed on one side of the cell membrane while either extracellular or intracellular fluid should be placed on the other side, it would require a pressure of 5400 mm. Hg to stop osmosis of water into the fluid. This amount of pressure is equal to about 7 atmospheres of pressure, or approximately equal to the pressure exerted by a column of water 230 feet high. Therefore, one can understand the extreme force that sometimes develops for movement of water through the pores of cell membranes by the phenomenon of osmosis.

LACK OF EFFECT OF MOLECULAR AND IONIC MASS ON OSMOTIC PRESSURE. The osmotic pressure exerted by nondiffusible particles in a solution, whether they be molecules or ions, is determined by the *numbers* of dissolved particles per unit volume of fluid and not the mass of the particles. This holds true whether the particles be ions, small urea molecules, or large protein molecules. The reason for this is that each particle in a solution, regardless of its mass, has almost exactly the same amount of kinetic energy of motion and therefore exerts, on the average, the same amount of pressure against the membrane.

Isotonicity, Hypertonicity, and Hypotonicity of Solutions. A solution that, when placed on the outside of cells, has exactly the same osmotic pressure as the intracellular fluid and therefore causes no osmosis through the cell membrane in either direction is said to be *isotonic* with the body fluids. For instance, a 0.9 per cent solution of sodium chloride is isotonic. On the other hand, a solution that causes osmosis of fluid out of the cell into the solution is said to be *hypertonic*. Thus, a sodium chloride solution having a concentration greater than 0.9 per cent is hypertonic. Finally, a solution that causes osmosis into cells is *hypotonic,* for instance, a sodium chloride solution of less than 0.9 per cent concentration. In other words, cells placed in an isotonic solution will maintain their volumes at a constant level, cells placed in a hypertonic solution will shrink, and cells placed in a hypotonic solution will swell.

OSMOTIC EQUILIBRIUM OF EXTRACELLULAR AND INTRACELLULAR FLUIDS

Except for very short periods of time, usually measured in seconds, the intracellular and extracellular fluids remain in constant osmotic equilibrium; that is, the osmotic concentrations of nondiffusible substances on the two sides of the membranes remain almost exactly equal. The reason for this is that, if ever the concentrations become unequal, osmosis of water through the cell membranes occurs so rapidly that equilibrium will be reestablished within a few seconds. This effect is illustrated in Figure 8–9. Figure 8–9A shows a cell sud-

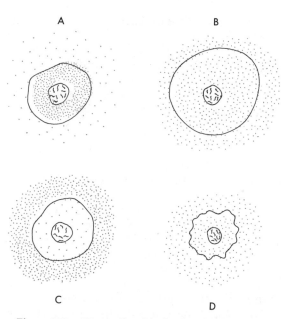

Figure 8–9.   (A) A cell suddenly placed in a hypotonic solution. (B) The same cell after osmotic equilibrium has been established, showing swelling of the cell and equilibration of the concentration of extra- and intracellular fluids. (C) A cell placed in a hypertonic solution. (D) The same cell after osmotic equilibrium has been established, showing shrinkage of the cell and equilibration of fluid concentrations.

denly placed in a very dilute solution *(hypotonic solution)*. Within a few seconds, water passes by osmosis through the cell membrane to the inside of the cell, causing: (1) increase in the intracellular fluid volume, and thus swelling of the cell, (2) decrease in the extracellular fluid volume, (3) dilution of the dissolved substances in the intracellular fluid, and (4) concentration of the dissolved substances in the extracellular fluid. Once these two fluids have reached the same concentrations of osmotically active substances, osmosis through the membrane ceases. Thus, within a few seconds, a new state of osmotic equilibrium has been reestablished, as shown in Figure 8–9B.

Figure 8–9C shows exactly the opposite situation, in which a cell having a dilute intracellular fluid is placed in a concentrated extracellular fluid *(hypertonic solution)*. Water passes by osmosis out of the cell, decreasing the intracellular fluid volume and increasing the extracellular fluid volume. This also concentrates the intracellular fluids while diluting the extracellular fluids. Therefore, within a few seconds, the concentrations of the two fluids reach equilibrium, but in the meantime the cell has decreased in size as shown in Figure 8–9D. To illustrate how rapidly osmosis can take place at the cell membrane, a red blood cell placed in pure water gains an amount of water equal approximately to its own volume in 2 to 3 seconds. Therefore, one would expect almost complete osmotic equilibrium to take place within 15 to 20 seconds.

ACTIVE TRANSPORT

Often only a minute concentration of a substance is present in the extracellular fluid, and yet a large concentration of the substance is required in the intracellular fluid for function of the cell. For instance, this is true of potassium ions. Conversely, other substances frequently enter cells and must be removed even though the concentration inside is far less than that outside. This is true of sodium ions.

From the discussion thus far it is evident that *no substances can diffuse against a concentration difference*, or, as is often said, "uphill." To cause movement of substances uphill, energy must be imparted to the substance. This is analogous to the compression of air by a pump. After compression, the concentration of the air molecules is far greater than before compression, but to create this greater concentration, energy had to be expended by the piston of the pump as it compressed the air molecules. Likewise, as molecules are transported through a cell membrane from a dilute solution to a concentrated solution, energy must be expended. The process of moving molecules "uphill" against a concentration difference is called *active transport*. Among the different substances that are actively transported through the cell membrane in at least some parts of the body are sodium ions, potassium ions, calcium ions, iron ions, hydrogen ions, chloride ions, iodide ions, ureate ions, several different sugars, and most of the amino acids.

BASIC MECHANISM OF ACTIVE TRANSPORT

The mechanism of active transport is believed to be similar for all substances and to be dependent on transport by *carriers*. Figure 8–10 illustrates the basic mechanism, showing a substance S entering the outside surface

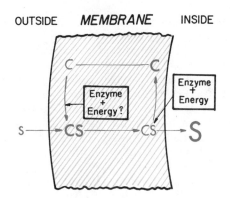

Figure 8–10.   Basic mechanism of active transport.

of the membrane where it combines with carrier C. At the inside surface of the membrane, S separates from the carrier and is released to the inside of the cell. C then moves back to the outside to pick up more S.

One will immediately recognize the similarity between this mechanism of active transport and that of facilitated diffusion discussed earlier in the chapter. The difference, however, is that energy is imparted to the system in the course of active transport, so that transport can occur *against a concentration difference*.

Unfortunately, little is known about the mechanisms by which energy is utilized in the transport mechanism. The energy could cause combination of the carrier with the substance, or it could cause splitting of the substance away from the carrier. Regardless of which of these is true, the energy is probably supplied to the transport system at the inside surface of the cell membrane, because it is inside the cell that large quantities of the energy-giving substance ATP are available for promoting chemical reactions.

It is also evident from Figure 8–10 that the chemical reactions are believed to be promoted by specific enzymes which catalyze the chemical reactions. Indeed, the carrier might itself be an enzyme as well as a carrier.

Chemical Nature of the Carrier. Carrier substances are believed to be either proteins or lipoproteins, the protein moiety providing a specific site for attachment of the substance to be transported and the lipid moiety providing solubility in the lipid phase of the cell membrane.

Several different carrier systems exist in cell membranes, each of which transports only certain specific substances. One carrier system, for instance, transports sodium to the outside of the membrane and transports potassium to the inside at the same time. Another system actively transports sugars through the membranes of certain cells, while still other specific carrier systems transport different ones of the amino acids.

The specificity of active transport systems for substances is determined either by the chemical nature of the carrier, which allows it to combine only with certain substances, or by the nature of the enzymes that catalyze the specific chemical reactions. Unfortunately, very little is known about the precise function of these two components of the carrier systems.

Energetics of Active Transport. In terms of calories, the amount of energy required to concentrate 1 mol of substance 10-fold is about 1400 calories. Thus, one can see that the energy expenditure for concentrating substances in cells or for removing substances from cells against a concentration gradient can be tremendous. Some cells, such as those lining the renal tubules, expend as much as 50 per cent of their energy for this purpose alone.

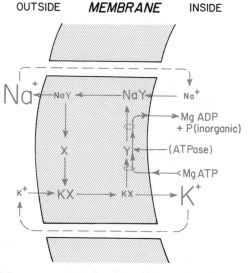

Figure 8–11. Postulated mechanism for active transport of sodium and potassium through the cell membrane, showing coupling of the two transport mechanisms and delivery of energy to the system at the inner surface of the membrane.

ACTIVE TRANSPORT OF SODIUM, POTASSIUM, AND OTHER ELECTROLYTES

Because minute quantities of sodium and potassium can leak through the pores of the cell, the concentrations of the two ions would eventually become equal inside and outside the cell unless there were some means to remove the sodium from the inside and to transport potassium back in.

Fortunately, a system for active transport of sodium and potassium is present in all cells of the body. The mechansim has been postulated to be that illustrated in Figure 8–11, which shows sodium (Na) inside the cell combining with carrier Y at the membrane surface to form large quantities of the combination NaY. This then moves to the outer surface where sodium is released and the carrier Y changes its chemical composition slightly to become carrier X. This carrier then combines with potassium (K) to form KX, which moves to the inner surface of the membrane where energy is provided to split K from X under the influence of the enzyme ATPase, the energy being derived from MgATP.

The transport mechanism in Figure 8–11 is

believed to be more effective in transporting sodium than in transporting potassium, usually transporting about three sodium ions for every two potassium ions. The carrier is probably a lipoprotein, and it is likely that this same lipoprotein molecule acts as the enzyme ATPase to release the energy required for transport.

The sodium transport mechanism is so important to many different functioning systems of the body — such as to nerve and muscle fibers for transmission of impulses, various glands for the secretion of different substances, and all cells of the body to prevent cellular swelling — that it is frequently called the *sodium pump*. We will discuss the sodium pump at many places in this text.

Other Electrolytes. Calcium and magnesium are probably transported by all cell membranes in much the same manner that sodium and potassium are transported, and certain cells of the body have the ability to transport still other ions. For instance, the glandular cell membranes of the thyroid gland can transport large quantities of iodide ion; the epithelial cells of the intestine can transport sodium, chloride, calcium, iron, bicarbonate, and probably many other ions; and the epithelial cells of the renal tubules can transport hydrogen, calcium, sodium, potassium, and a number of other ions.

ACTIVE TRANSPORT OF SUGARS

Facilitated diffusion of glucose and certain other sugars occurs in essentially all cells of the body, but active transport of sugars occurs in only a few places in the body. For instance, in the intestine and renal tubules, glucose and several other monosaccharides are continually transported into the blood even though their concentrations be minute in the lumens. Therefore, essentially none of these sugars is lost either in the intestinal excreta or the urine.

Though not all sugars are actively transported, almost all monosaccharides that are important to the body *are* actively transported, including *glucose, galactose, fructose, mannose, xylose, arabinose,* and *sorbose.* On the other hand, the disaccharides such as sucrose, lactose, and maltose are not actively transported at all.

As is true of essentially all other active transport mechanisms, the precise carrier system and chemical reactions responsible for transport of the monosaccharides are yet unknown.

Transport of the monosaccharides will not occur when active transport of sodium is blocked. Because of this, it is believed that active transport of sodium ions, by means of coupled reactions with the monosaccharide transport mechanisms, provides the energy required to move the monosaccharides through the membrane. Therefore, it is stated that monasaccharide transport is *secondary active transport,* in contrast to *primary active transport* represented by the transport of sodium through the membrane.

ACTIVE TRANSPORT OF AMINO ACIDS

Amino acids and the closely related substances, the amines, are actively transported through the membranes of all cells that have been studied. Active transport through the epithelium of the intestine, the renal tubules, and many glands is especially important. Probably at least four different carrier systems exist for transporting different groups of amino acids. It is especially interesting that amino acid transport in the intestine, like sugar transport, is also dependent on energy from the sodium transport mechanism. One of the few known features of the amino acid carrier systems is that transport of at least

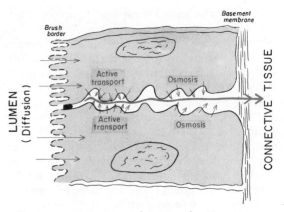

Figure 8–12. Basic mechanism of active transport through a layer of cells.

some amino acids depends on pyridoxine (vitamin B_6). Therefore, deficiency of this vitamin causes protein deficiency.

ACTIVE TRANSPORT THROUGH CELLULAR SHEETS

In many places in the body, substances must be transported through an entire *cellular layer* instead of simply through the cell membrane itself. Transport of this type occurs through the intestinal epithelium, the epithelium of the renal tubules, the epithelium of all exocrine glands, the membrane of the choroid plexus of the brain, and many other membranes.

The general mechanism of active transport through a sheet of cells is illustrated in Figure 8–12. This figure shows two adjacent cells in a typical epithelial membrane as found in the intestine, in the gallbladder, and in certain other areas of the body. On the luminal surface of the cells is a brush border that is highly permeable to both water and solutes, allowing both of these to diffuse readily from the lumen of the intestine to the interior of the cell. Once inside the cell, some of the solutes are actively transported into the space between the cells. This space is closed at the brush border of the epithelium but is wide open at the base of the cells where they rest on the basement membrane. Furthermore, the basement membrane is extremely permeable. Therefore, substances transported into this channel between the cells flow toward the connective tissue.

One of the most important substances actively transported in this manner is sodium;

and when sodium ions are transported into the space between the cells, their positive electrostatic charges pull negatively charged chloride ions through the membrane as well. Then, when the concentration of sodium and chloride ions increases in this intercellular space, this in turn causes osmosis of water out of the cell and into the space. In consequence, the water, along with the sodium and chloride, flows into the connective tissue behind the basement membrane where the water and ions diffuse into the blood capillaries. These same principles of transport apply generally wherever such transport occurs through cellular sheets, whether in the intestine, gallbladder, kidneys, or elsewhere.

REFERENCES

Andreoli, T. E., and Schaefer, J. A.: Mass transport across cell membranes: the effects of antidiuretic hormone on water and solute flows in epithelia. *Ann. Rev. Physiol., 38*:451, 1976.

Finn, A. L.: Changing concepts of transepithelial sodium transport. *Physiol. Rev., 56*:453, 1976.

Glynn. I. M., and Karlish, S. J. D.: The sodium pump. *Ann Rev. Physiol., 37*:43, 1975.

Korenbrot, J. I.: Ion transport in membranes: incorporation of biological ion-translocating proteins in model membrane systems. *Ann. Rev. Physiol., 39*:19, 1977.

Korn, E. D.: Transport. New York, Plenum Publishing Corp., 1975.

Macknight, A. D. C., and Leaf, A.: Regulation of cellular volume. *Physiol. Rev., 57*:510, 1977.

Montal, M.: Experimental membranes and mechanisms of bioenergy transductions. *Ann. Rev. Biophys. Bioeng., 5*:119, 1976.

Packer, I.: Biomembranes: Architecture, Biogenesis, Bioenergetics, and Differentiation. New York, Academic Press, Inc., 1975.

Van Winkle, W. B., and Schwartz, A.: Ions and inotropy. *Ann. Rev. Physiol., 38*:247, 1976.

Wright, E. M., and Diamond, J. M.: Anion selectivity in biological systems. *Physiol. Rev., 57*:109, 1977.

QUESTIONS

1. What are the ionic differences between the extracellular and intracellular fluids?
2. Give the theory of the diffusion process.
3. What is the difference between facilitated diffusion and simple diffusion?
4. What substances are transported by diffusion through membrane pores, and what by diffusion through the cell membrane matrix?
5. How do electric charges on ions and concentration differences affect the net diffusion rate?
6. Explain the principles of osmosis across cell membranes.
7. Why do extracellular and intracellular fluids remain continually in osmotic equilibrium?
8. What is the difference between active transport and facilitated diffusion?
9. Discuss the active transport of sodium and potassium across the cell membrane.
10. What is meant by secondary active transport of monosaccharides?

NERVE AND MUSCLE III

MEMBRANE POTENTIALS AND NERVE TRANSMISSION

9

All cells of the human body have an electrical potential across the cell membrane. Under resting conditions, this potential is more negative inside the membrane than outside. The potential is caused by differences between the ion compositions of intracellular and extracellular fluids. Especially important is the fact that the intracellular fluid contains a very high concentration of potassium while the extracellular fluid contains a very low concentration, and the extracellular fluid contains a very high concentration of sodium while the intracellular fluid contains a very low concentration. One of the purposes of the present chapter will be to explain why these ionic concentration differences cause the cell membrane potentials.

Membrane potentials also play an extremely important role in the transmission of signals by nerves, as well as in the control of muscle contraction, glandular secretion, and probably still many other cell functions. Therefore, in addition to discussing the basic cause of membrane potentials themselves, we will discuss also the development of membrane potentials in nerve fibers and their employment for signal transmission. With this in mind, let us first discuss the physiologic anatomy of the nerve fiber itself.

PHYSIOLOGIC ANATOMY OF THE NERVE FIBER

Figure 9–1A illustrates a neuron with its long fiber extension called an *axon,* and Figure 9–1B shows a cross section of the axon. It is the axon that is commonly called the "nerve fiber." One sees that the axon contains in its center a gelled substance called *axoplasm,* and this is surrounded by a *membrane* that separates the axoplasm from the interstitial fluid. This membrane has the same functions as any other cellular membrane, except that it is specifically adapted for transmission of electrochemical impulses.

Figure 9–1A also shows that some nerve fibers are surrounded on their outsides by an additional layer, the *myelin sheath,* which is discontinuous at the *nodes of Ranvier.* The myelin sheath is an electrical insulator, the function of which is explained later in the chapter.

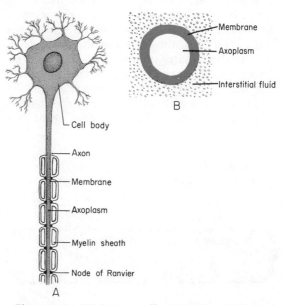

Figure 9–1. (A) A nerve cell with its threadlike axon, showing the nerve membrane, the axoplasm, and the myelin sheath. (B) Cross-section of an axon.

MEMBRANE POTENTIALS

The first essential for conduction of a nerve impulse is the establishment of a basic electric potential across the axon membrane. This potential is called a *membrane potential*. It results from concentration differences of certain ions between the two sides of the membrane.

Ionic Concentration Differences Across the Nerve Membrane. The membrane of the axon, like other cellular membranes, actively transports some ions from the extracellular fluid to the inside of the fiber, and others from inside back into the extracellular fluid. For nerve conduction, the most important of these transfer processes is the active transport of sodium out of the fiber into the extracellular fluid, which is often referred to as the *sodium pump*. This pump, by making the quantity of sodium inside the fiber very little, causes a membrane potential to develop as follows:

In Figure 9–2, the concentration of sodium inside the nerve fiber is shown to be approximately 10 milliequivalents, in contrast with its concentration of 142 milliequivalents in the extracellular fluid. On the other hand, the potassium concentration inside the nerve fiber is appproximately 28 times its concentration in the extracellular fluid. The high concentration of potassium on the inside of the fiber is caused partially by a potassium pump that pumps potassium to the inside of the fiber, but this is not a very important effect because the membrane is so permeable to potassium that this potassium leaks back out of the fiber almost immediately. Instead, by far the greater majority of the high potassium concentration inside the fiber is a secondary effect of the sodium pump as follows: When sodium ions, which are positive ions, are pumped to the outside a great void

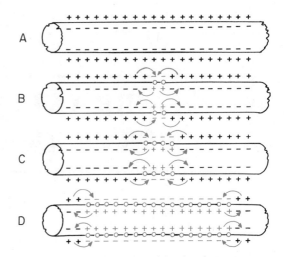

Figure 9–3. Transmission of the depolarization wave, the initial event in the nerve impulse.

of positive ions is left inside the membrane. As a result, the inside becomes strongly electronegative, approximately −85 millivolts more negative than the outside of the fiber, and since potassium ions, which are also positive ions, are diffusible through the membrane, the electronegativity pulls a large quantity of these ions to the inside, thus mainly filling the deficit left by the expulsed sodium ions.

The electronegative potential created inside the fiber as a result of the sodium pump is the *resting membrane potential*. In Figure 9–3A the resting membrane potential is illustrated by a series of positive and negative charges, positivity outside the membrane and negativity inside. This is the normal resting state of the nerve fiber, and the actual voltage of the resting membrane potential *inside the fiber* normally is about −0.085 volt or −85 millivolts.

THE ACTION POTENTIAL AND THE NERVE IMPULSE

When a signal is transmitted over a nerve fiber, the membrane potential goes through a series of changes called the action potential. Before the action potential begins, the resting membrane potential is very negative, but at the outset of the action potential the membrane potential suddenly becomes positive, followed a few ten-thousandths of a second later by return to the very negative resting level. This sudden rise of the membrane potential to positivity and then its return back to its normal negative state *is* the action po-

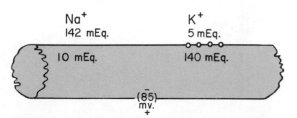

Figure 9–2. Concentration gradients of sodium and potassium at the axon membrane, showing that the membrane in the resting state is permeable only to potassium.

tential; it is also called the *nerve impulse*. The impulse (or action potential) spreads along the nerve fiber, and by means of such impulses the nerve fiber transmits information from one part of the body to another.

Action potentials can be elicited in nerve fibers by any factor that suddenly increases the permeability of the membrane to sodium ions. This membrane, in the normal resting state, is relatively impermeable to sodium ions though quite permeable to potassium ions. When the fiber is suddenly made permeable to sodium ions, the positively charged sodium ions leak to the inside of the fiber and make this positive, thus initiating the action potential. This first stage of the action potential, that is, the initial positive change in the membrane potential, is called *depolarization*. The subsequent return of the potential to its resting level is called *repolarization*.

DEPOLARIZATION OF THE MEMBRANE, REVERSAL POTENTIAL, AND TRANSMISSION OF THE NERVE IMPULSE

Figure 9–3B illustrates at its central-most point an area that has suddenly become so permeable that even sodium ions can now diffuse through the membrane with ease. Because the concentration of sodium ions outside the fiber is 14 times as great as inside, the positively charged sodium ions rush to the inside, causing the membrane to become suddenly positive inside and negative outside. This is opposite to the usual resting state of the membrane, and it is called the *reversal potential*. This sudden reversal is also called *depolarization* because the normal polarized state, with positivity on the outside and negativity on the inside, no longer exists.

The Depolarization Wave or Nerve Impulse. In figures 9–3C and D, the area of depolarization in the middle of the fiber has extended in both directions, and the area of increased permeability has also extended. The cause of this extension is the flow of electrical current from the original depolarized area to adjacent areas. This passage of electrical current, for reasons not yet completely understood, makes the adjacent areas also permeable to sodium. Therefore, sodium ions now flow through the membrane at these new areas, depolarizing these areas as well and causing still more electrical current to spread along the fiber. Thus, the area of increased

permeability to sodium ions spreads still farther along the membrane in both directions away from the area of original depolarization. And this process repeats itself over and over again. This spread of increased sodium permeability and electrical current along the membrane is called a *depolarization wave* or a *nerve impulse*.

It is obvious from the preceding discussion that once a single point anywhere on the nerve becomes depolarized a nerve impulse travels away from that point in both directions, and each impulse keeps on traveling until it comes to both ends of the fiber. In other words, a nerve fiber can conduct an impulse either toward the nerve cell or away from it.

Repolarization of the Nerve Fiber. Immediately after a depolarization wave has traveled along a nerve fiber, the inside of the fiber has become positively charged as illustrated in Figure 9–3D because of the large number of sodium ions that have diffused to the inside. This positivity stops further flow of sodium ions to the inside of the fiber, and this cessation of flow causes the membrane to become impermeable to sodium ions again, though the exact mechanism of this also is not understood. Yet, potassium can still diffuse through the pores, and indeed for a short period of time the pores become more permeable than ever to potassium. And because of the high concentration of potassium on the inside many potassium ions now diffuse outward carrying positive charges with them. This once again creates electronegativity inside the membrane and positivity outside, a process called *repolarization* because it reestablishes the normal polarity of the membrane. Repolarization usually begins at the same point in the fiber where depolarization had originally begun, and it spreads along the fiber in the manner illustrated in Figure 9–4. Repolarization occurs a few ten-thousandths of a second after depolarization. That is, the whole cycle of depolarization and repolarization takes place in large nerve fibers in less than one-thousandth of a second, and the fiber is then ready to transmit a new impulse.

REFRACTORY PERIOD. When an impulse is traveling a nerve fiber, the nerve fiber cannot transmit a second impulse until the fiber membrane has been repolarized. For this reason the fiber is said to be in a *refractory* state, and the time that the fiber remains refractory is called the *refractory period*. This varies

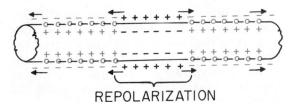

REPOLARIZATION

Figure 9–4. The repolarization process, the concluding event in the nerve impulse.

from about $1/2500$ second for large nerve fibers up to $1/500$ second for small fibers.

REESTABLISHMENT OF IONIC CONCENTRATION DIFFERENCES AFTER NERVE IMPULSES ARE CONDUCTED

After the nerve fiber has become repolarized, the sodium ions that have leaked to the inside of the membrane and the potassium ions that have leaked to the outside must be returned to their original sides of the membrane. This is accomplished by the sodium pump which was discussed previously. That is, this pump pumps the extra sodium ions inside the fiber back to the outside. This creates an extra deficit of positive charges inside the membrane so that the membrane interior becomes temporarily even more negative than usual. The excessive electronegativity inside the membrane now pulls potassium ions back to the inside (and a few potassium ions are also pumped to the inside by the potassium pump). Thus, this process restores the ionic differences to their original levels.

However, it must be emphasized that even when the sodium pump suddenly fails to act, a hundred thousand or more impulses can still be transmitted over the nerve fiber before transmission will cease. The reason for this is that only a few trillionths of a mole of sodium enter the fiber and approximately the same amount of potassium leaves each time a single impulse is transmitted, so that it takes a hundred thousand or more impulses for the concentration differences across the membrane to be dissipated and therefore to cause cessation of impulse transmission. It is evident, then, that the sodium pump is not necessary for the initial repolarization of the membrane after each nerve impulse; this is accomplished by the diffusion of potassium outward through the membrane pores. Instead, the sodium pump simply plods along

slowly, reestablishing ionic concentration differences across the membrane whenever a large number of impulses tend to alter these.

SUMMARY OF THE STEPS IN THE ACTION POTENTIAL

Let us now summarize the essential steps in membrane potential generation and in action potential transmission:

1. The sodium pump causes a high concentration of sodium on the outside of the membrane and a low concentration on the inside. This creates a negative potential inside the fiber.

2. The negative potential inside the fiber pulls positively charged potassium ions to the inside, and the potassium pump pumps a small amount of additional potassium ions to the inside. These two effects cause a high concentration of potassium inside the fiber and a low concentration outside.

3. A sudden increase in permeability of the membrane to sodium ions initiates the action potential. Sodium ions rush to the inside of the fiber, carrying positive charges, and this creates positivity inside the membrane at the local point where the membrane has become highly permeable. This is called the *depolarization* process.

4. The positive electricity entering the nerve fiber moves along the inside of the fiber. This has an effect on the adjacent membrane to cause it also to become highly permeable to sodium. Therefore, sodium leaks in here also, and the process is repeated again and again along the nerve fiber. Thus, the nerve impulse travels along the fiber.

5. After the fiber becomes fully depolarized, the membrane now suddenly becomes impermeable to sodium again and slightly excessively permeable to potassium. Because of the high concentration of potassium in the fiber, large quantities of positively charged potassium ions diffuse to the outside. Loss of these positive charges makes the inside become negative again. This is the *repolarization* process. The nerve fiber is now ready to transmit another impulse.

6. During the period when the fiber had been depolarized, a small number of sodium ions had moved to the inside of the fiber; and during the initial process of repolarization, a small number of potassium ions had moved to the outside. The sodium pump now begins to

work again, even between action potentials, and pumps the sodium ions outward; at the same time potassium ions diffuse inward, and a small quantity of potassium ions is pumped inward by the potassium pump. This reestablishes the appropriate concentration differences between the inside and the outside.

RECORDING ACTION POTENTIALS

A method for recording potentials between the inside and outside of the nerve fiber is illustrated in Figure 9–5. This figure shows a minute glass pipet having a tip less than 1 micron (micrometer) in diameter and filled with a strong solution of potassium chloride that conducts electricity from the tip. The tip pierces the nerve membrane to make contact with the fluid in the center of the fiber. On the outside of the fiber, located anywhere in contact with the extracellular fluid, is another electrode illustrated by the dark rectangle. These two electrodes are connected to a recording meter.

The changing potentials across the membrane are illustrated schematically in Figure 9–6. Section A of this figure shows that during the original resting state, a potential of −85 millivolts is recorded inside the fiber with respect to the outside. In Figure 9–6B a depolarization wave has traveled down the fiber until it is directly at the electrodes. At this instant, the rapid influx of positive sodium ions to the inside of the fiber reverses the potential, causing positivity inside the fiber and negativity outside. Then, soon after the impulse has passed the electrodes, repolarization occurs (Fig. 9–6C) and the potential returns again to approximately the original resting potential of −85 millivolts because of diffusion of positive potassium ions to the outside.

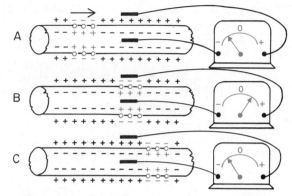

Figure 9–6. Principles of recording a monophasic action potential.

Figure 9–7 illustrates a continuous recording of the potential changes in a large nerve fiber as a nerve impulse passes the electrodes. This record is called the *monophasic action potential*. It begins with a normal resting membrane potential of −85 millivolts on the inside of the fiber and during the peak of the action potential reverses to become about +35 millivolts. Within approximately one-half millisecond, the normal membrane potential recovers, and the record returns almost to its original level. Actually, it "overshoots" a little, one or two millivolts, because the sodium pump immediately begins to pump positive sodium ions to the outside, which makes the potential even more negative than the normal value of −85 millivolts. This slight overshoot is called an *after potential*. It lasts for as long as 50 to 100 milliseconds; during this time,

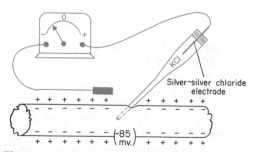

Figure 9–5. Measurement of membrane potentials and action potentials of a nerve fiber by means of a microelectrode.

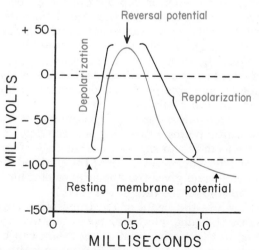
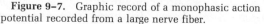

Figure 9–7. Graphic record of a monophasic action potential recorded from a large nerve fiber.

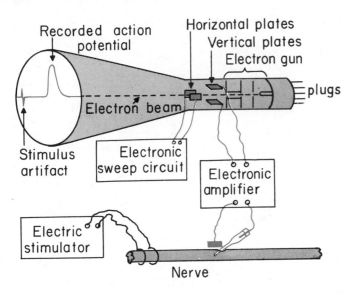

Figure 9–8. Diagram of an oscilloscope for recording action potentials from nerves.

the sodium that has entered the fiber during the action potential is removed from the fiber.

Use of the Oscilloscope to Record Action Potentials. Mechanical recording apparatus cannot function quickly enough to record the rapidly transient voltages of nerve action potentials. Therefore, a special instrument called the *oscilloscope,* which is similar to a television receiver, is normally used for this purpose. The principal components of the oscilloscope are shown in Figure 9–8. The basic part of this instrument is a cathode ray tube. An *electron gun* at the base of this tube shoots toward the face of the tube a fine beam of electrons having a diameter of about 1 mm. The beam passes between four metal plates, two of which are placed on the two sides of the beam and the other two of which are above and below the beam. On the face of the cathode ray tube is a fluorescent material that glows brightly when the electron beam strikes. The beam, in turn, can be moved back and forth or up and down across the face of the tube by applying electrical potentials to the horizontal and vertical plates.

An *electronic sweep circuit,* connected to the horizontal plates, causes the electron beam to move from left to right across the face of the tube. The fluorescence along the path of the moving beam gives the appearance of a horizontal line on the face of the tube.

If, while the beam of electrons is moving across the tube, electrical potentials are applied to the vertical plates, the beam can be made to move up or down. In Figure 9–8 the beam of electrons deviates from the line

slightly at the point called *stimulus artifact,* and then it deviates again at the point called *recorded action potential.* The electrical voltages across the nerve membrane are responsible for these deviations. When the nerve is stimulated by an electrical stimulator, some of the electrical current from the stimulator spreads through the fluids surrounding the nerve fiber to the pickup electrodes and causes the stimulus artifact. At the same time, an action potential begins traveling down the nerve fiber toward the two pickup electrodes. Then, when the action potential reaches the electrodes, the amplified signal applied to the vertical plates makes the electron beam move first upward and then back down again to record the potential.

Most action potentials of nerves have a total duration of not more than a few ten-thousandths to a few thousandths of a second. The cathode ray oscilloscope fortunately can record an electrical potential that lasts for this short period of time. Therefore, it is quite capable of giving a true record of nerve action potentials.

TYPES OF STIMULI THAT CAN EXCITE THE NERVE FIBER

In the living body, nerve fibers normally are stimulated by both physical and chemical means. For example, pressure on certain of the nerve endings in the skin mechanically stretches these endings, in this way opening the membrane pores to sodium ions and thereby setting off impulses. Heat and cold

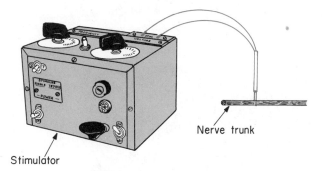

Figure 9-9. A laboratory stimulator.

Nerve trunk

Stimulator

affect other nerve endings to elicit impulses, and damage to the tissues, such as cutting the skin or stretching the tissues too much, can generate pain impulses.

In the central nervous system, impulses are transmitted from one neuron to another mainly by chemical means. The nerve ending of the first neuron secretes a chemical substance called a "transmitter" substance that in turn excites the second neuron. In this way, impulses are sometimes passed through many hundred neurons before stopping. This will be discussed in detail in Chapter 22.

In the laboratory, nerve fibers are usually stimulated electrically. A typical stimulator is illustrated in Figure 9-9. It emits electrical impulses of any desired voltage either singly or in rapid succession. Usually, a small probe having two wires at its end is placed on either side of a nerve trunk, and the electrical stimulus is applied so that electrical current flows through the fibers. As the current passes through the fiber membranes the permeability is altered, eliciting nerve impulses.

The All or None Law. From the preceding discussion of the nerve impulse it is evident that if a stimulus is strong enough to cause a nerve impulse at all, it causes the entire fiber to fire. In other words, a weak stimulus does not cause only part of the nerve to depolarize; the stimulus is strong enough to depolarize either the entire fiber or not any of it. This is called the *all or none law*.

TRANSMISSION OF SIGNALS OVER NERVE TRUNKS

TYPES OF FIBERS IN THE NERVE BUNDLE

Figure 9-10 shows a nerve bundle with fibers of different sizes and of two different types. The large black dots surrounded by

white areas are *myelinated fibers;* the white area is a tube of the insulator material *myelin.* In addition to the myelinated fibers there are about twice as many small fibers without myelin sheaths. These are called *unmyelinated fibers.*

Function of the Myelin Sheath — Saltatory Conduction. Myelin, which surrounds all large nerve fibers, is a lipid substance that will not conduct electric current. Therefore, it acts as an insulator around the nerve fiber. The myelin sheath is deposited around a nerve fiber by Schwann cells. The Schwann cell wraps itself around and around the fiber in the manner illustrated in Figure 9-11, leaving a portion of its own cell membrane each time around; this wrapping of membrane becomes the myelin sheath.

As illustrated schematically in Figure 9-12, at approximately every millimeter along the length of the fiber the myelin is broken by a *node of Ranvier,* which is the space between each pair of adjacent Schwann cells. At these nodes, typical membrane depo-

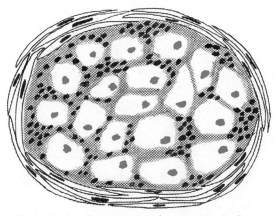

Figure 9-10. Cross-section of a nerve trunk, showing large, myelinated nerve fibers; small, unmyelinated nerve fibers; and the nerve sheath.

I mm. length

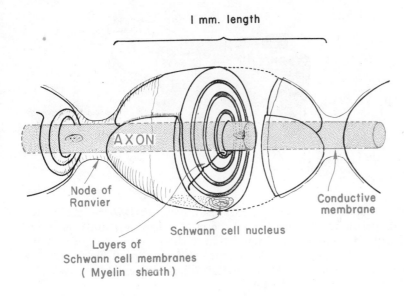

Figure 9–11. The myelin sheath and its formation by Schwann cells. (Modified from Elias and Pauley: Human Microanatomy. Davis Co., 1966.)

larization can occur, but beneath the myelin sheath, membrane depolarization does not take place because of the insulator properties of the myelin. Instead, impulses are transmitted along a myelinated nerve by a process called *saltatory conduction* which may be explained as follows: Referring once again to Figure 9–12 let us assume that the first node of Ranvier becomes depolarized. This causes electrical current to spread, as shown by the arrows, around the outside of the myelin sheath and also down the core of the fiber all the way to the next node of Ranvier, causing it to become depolarized also. Current generated by this node then causes the same effect at the next node; thus, the impulse "jumps" from node to node, which is the process of saltatory conduction.

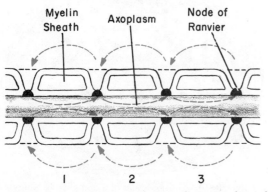

Figure 9–12. Saltatory conduction along a myelinated axon.

Saltatory conduction is valuable for two reasons:

1. By causing the depolarization process to jump long intervals along the nerve fiber, it increases the velocity of conduction along the fiber.

2. Perhaps even more important, saltatory conduction prevents the depolarization of large areas of the fiber and thereby prevents leakage of large amounts of sodium to the inside of the fiber as each nerve impulse is transmitted. This conserves the energy required by the sodium pump to expel the sodium. Therefore, the myelin sheath greatly decreases the amount of energy required by the nerve for impulse transmission.

Velocity of Conduction in Nerve Fibers. The larger the nerve fiber and the thicker the myelin sheath, the more rapidly can the nerve conduct an impulse. The largest nerve fibers are about 20 microns (micrometers) in diameter, and the smallest are about 0.5 micron. The very large fibers conduct impulses at a velocity as great as 100 meters — about the length of a football field — in 1 second, while the very small fibers conduct impulses at a velocity of only 0.5 meter per second, or approximately the distance from the foot to the knee. All sizes of nerve fibers exist between these smallest and largest sizes, so that a wide spectrum of impulse velocities occurs in the different nerves.

Number of Impulses That Can Be Transmitted Per Second. The number of impulses that can be transmitted by any one fiber per second is determined by the so-called "refractory period" of the fiber (which is the duration of time between the beginning of depolarization and the end of repolarization), and this depends also to a great extent on the size of the fiber. Large fibers (15 to 20 microns, or micrometers, in diameter) become repolarized in approximately $1/2500$ second. Therefore, a second nerve impulse can be transmitted $1/2500$ second after the first, or a total of up to 2500 impulses can be conducted each second. At the opposite extreme, the very smallest nerve fibers require as long as $1/250$ second to repolarize, which means that they can transmit no more than 250 impulses per second.

Functions of Myelinated and Unmyelinated Fibers. The myelinated fibers, which conduct impulses at high velocity and in great numbers, are used in the body to control the muscles, for these require extremely rapid signals from the brain if they are to function properly. They also transmit many of the sensory signals from the body to the brain, some of which likewise require rapid transmission if the person is to respond to his surroundings.

The unmyelinated fibers, which conduct impulses quite slowly, cannot cause rapid reactions. These control most of the subconscious activities of the body, such as the excitability of the heart, the contractions of the blood vessels, the gastrointestinal movements, and the emptying of the urinary bladder. Also, the small, unmyelinated fibers transmit those sensory signals that do not require immediate action, such as the aching type of pain sensations, crude touch sensations, and pressure sensations.

Further descriptions of the different types of fibers and their functions will be presented in connection with the discussions of sensory functions of the nervous system in Chapter 23 and of motor functions in Chapters 24, 25, and 26.

TRANSMISSION OF SIGNALS OF DIFFERENT STRENGTHS BY THE NERVE BUNDLE

The nerve bundle has two means by which it can transmit signals of different strengths — weak, strong, or intermediate. These are (1) to transmit impulses simultaneously over varying numbers of nerve fibers, which is called *spatial summation,* and (2) to transmit impulses at a slow or rapid frequency over the same fiber, which is called *temporal summation.*

As an example of spatial summation, if 100 nerve fibers are connected between the spinal cord and a foot muscle, stimulation of one of these fibers will cause only a weak response in the muscle, but simultaneous stimulation of all 100 fibers will cause a strong contraction. Obviously, any number of fibers between 1 and 100 can be stimulated at a time, giving any one of 100 different strengths of muscle contraction.

Temporal summation means changing the strength of a signal by sending a large or small number of impulses along the same fiber per second. If one impulse is transmitted each second, only a weak effect usually results, but if 5, 15, 25, 75 or more impulses are transmitted per second, the strength of the effect becomes progressively greater.

Ordinarily the nerve trunk transmits signals of different strengths by a combination of both the spatial and temporal methods. That is, when a strong signal is to be transmitted, large numbers of fibers are utilized and large numbers of impulses are also transmitted along each fiber. When a weak signal is to be transmitted, fewer fibers are used and fewer impulses are transmitted.

TRANSMISSION OF IMPULSES BY TISSUES OTHER THAN NERVE FIBERS

Transmission by Skeletal Muscle Fibers. Skeletal muscle fibers transmit impulses exactly as nerve fibers do. The normal velocity of transmission in skeletal muscle fibers is about 5 meters per second in contrast to 50 to 100 meters per second in the very large myelinated nerve fibers and 0.5 meter per second in the very small unmyelinated fibers. Because the nerve fibers that control the skeletal muscles are a large type, carrying impulses at about 60 meters per second, a signal travels from the brain to the muscle extremely rapidly but then decreases in velocity more than 10-fold as it goes into the muscle itself.

Transmission in Heart Muscle and Smooth Muscle. Transmission of impulses in heart muscle and smooth muscle also occurs like that in nerve and skeletal muscle, but at still lower velocities — about 0.3 meter per second in the heart and only a centimeter or so per second in smooth muscle. In heart muscle and in many smooth muscle masses the fibers interconnect with each other to form lattice-works so that stimulation of any one fiber always causes the impulse to travel over the entire muscle mass, resulting in complete contraction of the whole muscle rather than only part of it. This effect as it occurs in the heart is discussed in detail in Chapter 11.

Another difference between these two types of muscle and skeletal muscle is the duration of the action potential. In heart muscle it lasts about 0.3 second and in some smooth muscle as long as a second. So long as the membranes remain depolarized, the muscle fibers remain contracted. Therefore, the duration of contraction of both heart and smooth muscle is unusually long in comparison with that of skeletal muscle. This will be discussed further in the next two chapters.

THE NEUROMUSCULAR JUNCTION

The *neuromuscular junction* is the connection between the end of a large myelinated nerve fiber and a skeletal muscle fiber. In general, each skeletal muscle is supplied with only one neuromuscular junction but rarely more than one.

Figure 9–13A illustrates a neuromuscular junction, showing a nerve fiber attached to

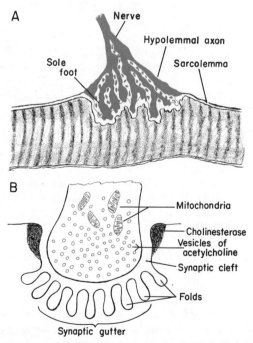

Figure 9–13. (A) The neuromuscular junction. (B) Invagination of a sole foot into the membrane of the muscle fiber.

the surface of the muscle fiber membrane and spreading to form many branching nerve terminals. These branches in turn end in club-like feet called *sole feet*. The entire nerve ending, including all the terminals, is called the *endplate*.

Figure 9–13B shows the invagination of a nerve terminal into the muscle fiber, forming a cavity called the *synaptic gutter*. Beneath the terminal is a small space called the *synaptic cleft,* and along the bottom of the gutter are many large folds of the muscle fiber membrane. Stored in the nerve terminals are many small vesicles containing the "transmitter" chemical *acetylcholine;* this substance is responsible for stimulating the muscle fiber membrane.

TRANSMISSION OF THE IMPULSE AT THE NEUROMUSCULAR JUNCTION

Secretion of Acetylcholine. When a nerve impulse reaches the neuromuscular junction, passage of the action potential over the membrane of the nerve terminal causes many of the small vesicles of acetylcholine stored in the terminal to rupture into the synaptic cleft between the terminal and the muscle fiber membrane. The acetylcholine then acts on the folded membrane to increase its permeability to sodium ions. The increased permeability in turn allows instantaneous leakage of sodium to the inside of the fiber, which carries positive charges to the inside and immediately depolarizes this local area of the muscle membrane. And this local depolarization sets off an action potential that travels in both directions along the fiber. In turn, the action potential traveling along the fiber causes muscle contraction, as will be explained in the following chapter.

Destruction of Acetylcholine by Cholinesterase. If the acetylcholine secreted by the nerve terminals should remain in contact with the muscle fiber membrane indefinitely, the fiber would transmit a continuous succession of impulses. However, on the surfaces of all the membrane folds in the synaptic gutter is a substance called *cholinesterase,* which enzymatically splits acetylcholine into acetic acid and choline in about $1/500$ second. Therefore, almost immediately after the acetylcholine has stimulated the muscle fiber, the acetylcholine itself is destroyed. This allows the membrane to repolarize and to become ready for stimulation again as soon as a new nerve impulse approaches.

The acetylcholine mechanism at the neuromuscular junction provides an amplifying system that allows a very weak nerve impulse to stimulate a very large muscle fiber. That is, the amount of electrical current generated by the nerve fiber is not enough by itself to elicit an impulse in the muscle fiber because the nerve fiber has a cross-sectional area only one-tenth or less that of the muscle fiber. Instead, the secreted acetylcholine causes the muscle fiber to generate its own impulse. Thus, each nerve impulse actually comes to a halt at the neuromuscular junction, and in its stead an entirely new impulse begins in the muscle.

PARALYSIS CAUSED BY MYASTHENIA GRAVIS

Sometimes a person has very poor transmission of impulses at the neuromuscular junction, an effect that obviously produces paralysis. One cause of this is the condition called *myasthenia gravis*. It is caused by an autoimmune response in which the immune system of the body has developed antibodies against the muscle membrane. Reaction of antibodies with the membrane in the synaptic gutter thickens the synaptic cleft and also destroys many of the membrane folds. These effects seriously depress the responsiveness of the muscle fiber to acetylcholine. Treatment with *neostigmine,* a drug that prevents the destruction of acetylcholine by cholinesterase, is often dramatic in overcoming the paralysis. This drug allows acetylcholine to accumulate in the neuromuscular junction from one nerve impulse to the next and, therefore, to exert a tremendous effect on the muscle fiber membrane. As a result, persons almost totally paralyzed by myasthenia gravis can sometimes be returned almost to complete normality within less than 1 minute after a single intravenous injection of neostigmine.

REFERENCES

Gage, P. W.: Generation of end-plate potentials. *Physiol. Rev., 56*:177, 1976.

Hille, B.: Gating in sodium channels of nerve. *Ann. Rev. Physiol., 38*:139, 1976.

Hodgkin, A. L.: The Conduction of the Nervous Impulse. Springfield, Ill., Charles C Thomas, Publisher, 1963.

Hodgkin, A. L., and Huxley, A. F.: Quantitative description of membrane current and its application to conduction and excitation in nerve. *J. Physiol. (Lond.), 117*:500, 1952.

Jack, J. J. B., Noble, D., and Tsien, R. W.: Electric Current Flow in Excitable Cells. New York, Oxford University Press, 1975.

Lester, H. A.: The response to acetylcholine. *Sci. Amer., 236(2)*:106, 1977.

Livett, B. G.: Axonal transport and neuronal dynamics: contribution to the study of neuronal connectivity. *Intern. Rev. Physiol., 10*:37, 1976.

Nachmansohn, D.: Chemical and Molecular Basis of Nerve Activity. New York, Academic Press, Inc., 1975.

Patrick, J., Heinemann, S., and Schubert, D.: Biology of cultured nerve and muscle. *Ann. Rev. Neurosci., 1*:417, 1978.

Pfenninger, K. H.: Organization of neuronal membrane. *Ann. Rev. Neurosci., 1*:445, 1978.

Ulbricht, W.: Ionic channels and gating currents in excitable membranes. *Ann. Rev. Biophys. Bioeng., 6*:7, 1977.

Varon, S. S., and Bunge, R. P.: Trophic mechanisms in the peripheral nervous system. *Ann. Rev. Neurosci., 1*:327, 1978.

Walker, J. L., and Brown, H. M.: Intracellular ionic activity measurements in nerve and muscle. *Physiol. Rev., 57*:729, 1977.

QUESTIONS

1. What is the role of the sodium pump in establishing ionic concentration differences and membrane potentials across the nerve membrane?
2. Describe the process of depolarization of the nerve membrane.
3. Describe the process of repolarization.
4. Summarize the steps in the action potential.
5. Describe the micropipet method for recording action potentials, as well as use of the oscilloscope for this purpose.
6. How can nerve fibers be excited?
7. What are the roles of spatial summation and temporal summation in transmission of signals of different strengths?
8. Describe the neuromuscular junction in skeletal muscle.
9. What are the roles of acetylcholine and cholinesterase in transmission of the signal through the neuromuscular junction?

10 MUSCLE PHYSIOLOGY

All physical functions of the body involve muscle activity. These functions include skeletal movements, contraction of the heart, contraction of the blood vessels, peristalsis in the gut, and many more. Three different types of muscle are responsible for these activities: skeletal muscle, cardiac muscle, and smooth muscle, all of which have some characteristics in common. For instance, the contractile process is the same or nearly the same in each, but, on the other hand, their strengths of contraction reactivities, durations of contraction, and other features differ greatly and are especially adapted in each type of muscle for the job to be performed.

Skeletal muscle and smooth muscle will be discussed in this chapter and cardiac muscle in the following chapter in relation to the pumping action of the heart.

SKELETAL MUSCLE

The skeletal muscles cause movements of the skeleton and, therefore, are responsible for movement of the different parts of the body.

PHYSIOLOGIC ANATOMY OF SKELETAL MUSCLE

Figure 10–1 illustrates a single muscle, showing schematically that the *muscle belly* is made up of thousands of individual *muscle fibers*. However, not shown in the figure is the fact that every fiber extends the entire length of the muscle and is attached at each end to *muscle tendons*. When stimulated, the muscle fibers contract, and the force of contraction is transmitted to the bones through the tendons.

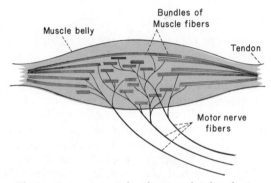

Figure 10–1. A muscle, showing the distribution of nerve fibers supplying three motor units.

Figure 10–2 illustrates cross-sectional and longitudinal views of a single muscle fiber. Each fiber is between 10 and 100 microns (micrometers) in diameter, and it varies from a few millimeters to 50 cm. in length, depending on the length of the muscle. The longitudinal view shows dark and light bands along the fiber, which are characteristic of skeletal and cardiac muscle but not of smooth muscle. The segment of the fiber between each two successive bands is called a *sarcomere*.

Myofibrils; Actin and Myosin Filaments. Each muscle fiber contains several hundred to several thousand myofibrils, which are illustrated by the small dots in the cross-sectional view of Figure 10–2. Each myofibril in turn has, lying side by side, about 1500 *myosin filaments* and two times this many *actin filaments,* which are elongated polymerized protein molecules that are responsible for muscle contraction. These can be seen in longitudinal view in the electron micrograph of Figure 10–3, and are represented diagrammatically in Figure 10–4. The thick filaments are *myosin* and the thin filaments

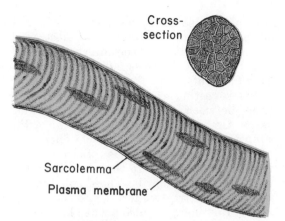

Cross-
section

Sarcolemma

Plasma membrane

Figure 10–2. Longitudinal and cross-sectional views of a skeletal muscle fiber.

are *actin*. Note that the myosin and actin filaments partially interdigitate and thus cause the myofibrils to have alternate light and dark bands. The light bands, which contain only actin filaments, are called *I bands*. The dark bands, which contain the myosin filaments as well as the ends of the actin filaments where they overlap the myosin, are called *A bands*. Note also the small projections from the sides of the myosin filaments. These are called *cross-bridges*. They protrude from the surfaces of the myosin filaments along the entire extent of the filament, except in the very center. It is interaction between these cross-bridges and the actin filaments that causes contraction.

Figure 10–4 also shows that the actin filaments are attached to the so-called *Z membrane* or *Z line,* and the filaments extend on either side of the Z membrane to interdigitate with the myosin filaments.

The portion of a myofibril (or of the whole muscle fiber) that lies between two successive Z membranes is the *sarcomere.* When the muscle fiber is at its normal fully stretched resting length, the length of the sarcomere is about 2 microns (micrometers).

The Sarcoplasm. The myofibrils are suspended inside the muscle fiber in a matrix called *sarcoplasm,* which is composed of usual intracellular constituents. Present are tremendous numbers of *mitochondria* that lie be-

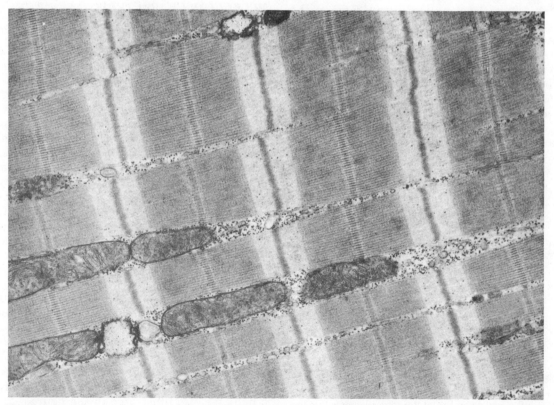

Figure 10–3. Electronmicrograph of a muscle fiber, showing the detailed organization of actin and myosin filaments in the myofibrils. Note the mitochondria lying between the myofibrils. (From Fawcett: The Cell. W. B. Saunders Co., 1966.)

tween and parallel to the myofibrils, a condition which is indicative of the great need of the contracting myofibrils for large amounts of ATP formed by the mitochondria.

Also in the sarcoplasm is an extensive endoplasmic reticulum, which in the muscle fiber is called the *sarcoplasmic reticulum*. This reticulum lies between all the myofibrils and has a special organization that is extremely important in the control of muscle contraction, which will be discussed in detail later in the chapter.

MOLECULAR MECHANISM OF MUSCLE CONTRACTION

Sliding Mechanism of Contraction. Figure 10–4 illustrates the basic mechanism of muscle contraction. It shows the relaxed state of three sarcomeres (above) and the contracted state (below). In the relaxed state, the ends of the actin filaments derived from two successive Z membranes barely overlap each other while at the same time completely overlapping the myosin filaments. On the other hand, in the contracted state these actin filaments have been pulled inward among the myosin filaments so that they now overlap each other to a major extent. Also, the Z membranes have been pulled by the actin filaments up to the ends of the myosin filaments. Thus, muscle contraction occurs by a *sliding filament mechanism*.

But what causes the actin filaments to slide inward among the myosin filaments? This is caused by attractive forces that develop between the actin and myosin filaments. Almost

certainly, these attractive forces are the result of mechanical, chemical, and electrostatic forces generated by the interaction of the cross-bridges of the myosin filaments with the actin filaments.

Under resting conditions, the attractive forces between the actin and myosin filaments are neutralized, but when an action potential travels over the muscle fiber membrane, this causes the release of large quantities of calcium ions into the sarcoplasm surrounding the myofibrils. These calcium ions activate the attractive forces and contraction begins. But energy is also needed for the contractile process to proceed. This energy is derived from the high energy bonds of adenosine triphosphate (ATP), which is degraded to adenosine diphosphate (ADP) to give the energy required.

In the next few sections we will describe what is known about the details of the molecular processes of contraction. To begin this discussion, however, we must first characterize the myosin and actin filaments themselves.

MOLECULAR CHARACTERISTICS OF THE CONTRACTILE FILAMENTS

The Myosin Filament. The myosin filament is composed of approximately 200 myosin molecules. Figure 10–5, section A, illustrates an individual molecule; section B illustrates the organization of the molecules to form a myosin filament as well as its interaction with two actin filaments. Note that the hinged portions of the myosin molecules protrude from all sides of the myosin filament, as illustrated in the figure. These protrusions constitute the *cross-bridges*. The heads of the cross-bridges lie in apposition to the actin filaments, while the rod portions of the cross-bridges act as hinged arms that allow the heads either to extend far outward from the body of the myosin filament or to lie close to the body.

The Actin Filament. The actin filament is also complex. It is composed of three different components: *actin, tropomycin,* and *troponin.*

The backbone of the actin filament is a double-stranded protein molecule, illustrated in Figure 10–6. The two strands are wound in a helix, as also illustrated in the figure.

Attached to the actin filament are also nu-

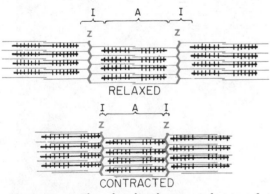

Figure 10–4. The relaxed and contracted states of a myofibril, showing sliding of the actin filaments into the channels between the myosin filaments.

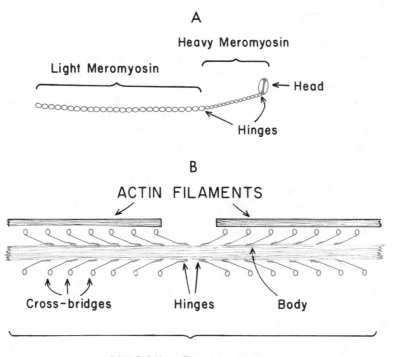

Figure 10–5. (A) The myosin molecule. (B) Combination of many myosin molecules to form a myosin filament. Also shown are the cross-bridges and the interaction between the heads of the cross-bridges with adjacent actin filaments.

merous molecules of ADP. It is believed that these ADP molecules are the active sites on the actin filaments with which the cross-bridges of the myosin filaments interact to cause muscle contraction.

The Troponin-Tropomyosin Complex. The actin filament also contains two additional strands of still another type of protein called *tropomyosin*. It is believed that these protein strands lie in the two grooves formed by the two protein strands of the actin helix, as illustrated for one of the grooves at the right-hand end of Figure 10–6.

Attached periodically along the tropomyosin strands are molecules of still a third protein, *troponin*. Troponin has a very high affinity for calcium ions. It is believed that the combination of calcium ions with troponin is the trigger that initiates muscle contraction, as will be discussed in the following section.

INTERACTION OF MYOSIN AND ACTIN FILAMENTS TO CAUSE CONTRACTION

Inhibition of the Actin Filament by the Troponin-Tropomyosin Complex During Relaxation; Activation by Calcium Ions During Contraction. A pure actin filament without the presence of the troponin-tropomyosin complex binds strongly with myosin molecules in the presence of magnesium ions and ATP, both of which are normally abundant in the myofibril. But, if the troponin-tropomyosin complex is added to the actin filament, this binding does not take place. Therefore, it is believed that the normal active sites on the actin filament of the relaxed

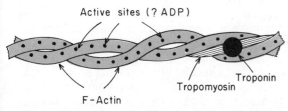

Figure 10–6. The actin filament, composed of two helical strands of F-actin. Also shown is a small portion of one of the two tropomyosin strands that lie in the grooves between the actin strands. On the surface of the tropomyosin one large molecule of troponin is shown schematically.

muscle are inhibited (or perhaps physically covered) by the troponin-tropomyosin complex. However, in the presence of calcium ions the inhibitory effect of the troponin-tropomyosin on the actin filaments is itself inhibited. When calcium ions combine with troponin, which has an extremely high affinity for calcium ions even when they are present in minute quantities, the troponin molecule supposedly undergoes a conformational change that in some way "uncovers" the active sites of the actin, thus allowing contraction to proceed. Therefore, the normal relationship between the tropomyosin-troponin complex and actin is altered by calcium ions — a condition which leads to contraction.

Interaction Between the "Activated" Actin Filament and the Myosin Molecule — the Ratchet Theory of Contraction. As soon as the actin filament becomes activated by the calcium ions, it is believed that the heads of the cross-bridges from the myosin filaments immediately become attracted to the active sites of the actin filament, and this then causes contraction to occur. Though the precise manner by which this interaction between the cross-bridges and the actin causes contraction is still unknown, a suggested hypothesis for which considerable circumstantial evidence exists is the so-called *ratchet theory of contraction.*

Figure 10–7 illustrates the postulated ratchet mechanism for contraction. This figure shows the heads of two cross-bridges attaching to and disengaging from the active sites of an actin filament. It is postulated that when the head attaches to an active site this attachment alters the bonding forces between the head and its arm, thus causing the head to tilt toward the center of the myosin filament, and to drag the actin filament along with it. This tilt of the head of the cross-bridge is called the *power stroke*. Then, immediately

after tilting, the head automatically splits away from the active site and returns to its normal perpendicular position. In this position it combines with an active site farther down along the actin filament; then, a similar tilt takes place again to cause a new power stroke. Thus the heads of the cross-bridges bend back and forth and step by step pull the actin filament toward the center of the myosin filament.

Each one of the cross-bridges is believed to operate independently of all others, each attaching and pulling in a continuous, alternating ratchet cycle. Therefore, the greater the number of cross-bridges in contact with the actin filament at any given time, the greater, theoretically, is the force of contraction.

ATP as the Source of Energy for Contraction. When a muscle contracts against a load, work is performed, and energy is required. ATP is cleaved to form ADP during the contraction process, and the greater the amount of work performed by the muscle, the greater is the amount cleaved. It is still not known exactly how ATP is used to provide the energy for contraction. However, it is believed that a molecule of ATP first combines with the head of each cross-bridge before the power stroke. Then after each power stroke of the head, the ATP is cleaved to form ADP, and the energy that is released is used to "recock" the head. Then the head is ready for another power stroke. Thus, during the process of muscle contraction, it is believed that one molecule of ATP is cleaved for each power stroke of the cross-bridge.

INITIATION OF MUSCLE CONTRACTION: EXCITATION–CONTRACTION COUPLING

We have now described the sliding mechanism by which muscle contracts. But we still must answer another question: How is the contraction initiated? This is achieved through a complex mechanism involving a system of tubules called the *sarcoplasmic reticulum.* These tubules are located throughout the skeletal muscle fiber, and their special function is to initiate and control the contraction process. When an action potential travels over the muscle fiber membrane, the tubules transmit the action potential deep into the muscle fiber to activate the contractile process itself. Therefore, let us first describe this intracellular tubular system.

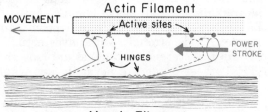

Figure 10–7. The ratchet mechanism for contraction of the muscle.

THE INTRACELLULAR TUBULAR SYSTEM — THE LONGITUDINAL AND TRANSVERSE TUBULES

The electronmicrograph of Figure 10–8 illustrates a system of tubules that lie in the spaces between myofibrils. Note the extensiveness of this tubular system and its close relationship with the contractile myofibrils. The tubular system is composed of two parts, the *longitudinal tubules* and the *transverse tubules* (also called *T tubules*).

In Figure 10–8 the longitudinal tubules lie parallel to the myofibrils. The transverse tubules are perpendicular to the myofibrils and in the figure can barely be seen only in cross-section. The two arrows illustrate two of these transverse tubules; many others also can be seen in the figure. Note that each longitudinal tubule terminates at each end in a bulbar structure called a *cisterna*. Also, each cisterna abuts against the side of a transverse tubule. This relationship of the transverse tubules to the cisternae is essential to the excitation process, as we shall see.

An especially important characteristic of the T tubules is that they pass all the way through the fiber from one side to the other; they open to the exterior rather than to the interior of the cell and contain *extracellular fluid* that is continuous with the fluid outside the cell rather than intracellular fluid. When an action potential spreads over the muscle fiber membrane, the action potential is also transmitted to the interior of the muscle fiber by way of the T tubule system, and it is this stimulus that causes the muscle to contract. Figure 10–9 illustrates this spread of the action potential, as illustrated by the dashed arrows, first along the cell fiber membrane, the sarcolemma, and then also into the T tubule system. Note that a local circuit of current also flows around each one of the T tubules.

RELEASE OF CALCIUM IONS BY THE CISTERNAE OF THE SARCOPLASMIC RETICULUM

Figure 10–9 also shows that the action potential of the T tubule causes current flow through the cisternae of the sarcoplasmic reticulum. This causes the cisternae to release calcium ions into the surrounding sarcoplasmic fluid. The calcium ions then diffuse to the adjacent myofibrils where they bind strongly with troponin, as discussed earlier, and this in turn elicits the muscle contraction, as has also been discussed.

The Calcium Pump for Removing Calcium Ions from the Sarcoplasmic Fluid. Once the calcium ions have been released from the cisternae and have diffused to the myofibrils, muscle contraction will continue as long as the calcium ions are still present in high concentration. However, a continually active calcium pump located in the walls of the longitudinal tubules of the sarcoplasmic reticulum pumps the calcium ions back into the longi-

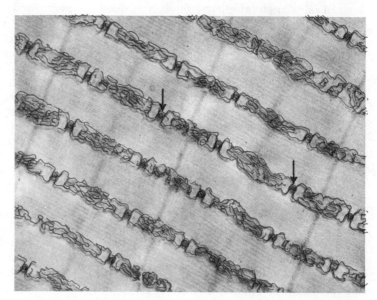

Figure 10–8. Sarcoplasmic reticulum surrounding the myofibril, showing the longitudinal system paralleling the myofibrils. Also shown in cross-section are the T tubules that lead to the exterior of the fiber membrane and that contain extracellular fluid (arrows). (From Fawcett: The Cell. W. B. Saunders Co., 1966.)

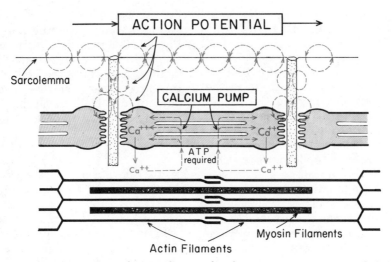

Figure 10–9. Excitation-contraction coupling in the muscle, showing an action potential that causes release of calcium ions from the sarcoplasmic reticulum and then re-uptake of the calcium ions by a calcium pump.

tudinal tubules. This pump can concentrate the calcium ions about 2000-fold inside the tubules, a condition which allows massive buildup of calcium in the sarcoplasmic reticulum and also causes almost total depletion of calcium ions in the fluid of the myofibrils except immediately after an action potential. Therefore, muscle contraction occurs after an action potential, but this removal of the calcium by the sarcoplasmic pump almost immediately thereafter causes relaxation of the muscle again.

The total duration of the calcium "pulse" in the usual skeletal muscle fiber is about $1/50$ second, though it may be several times as long in some skeletal muscle fibers and several times less in others. In heart muscle the calcium ions last for as long as 0.3 second. It is during this calcium pulse that muscle contraction occurs. If the contraction is to continue without interruption for longer intervals, a series of such pulses must be initiated by a continuous series of repetitive action potentials, as will be discussed in more detail later in the chapter.

CHARACTERISTICS OF WHOLE MUSCLE CONTRACTION

THE MOTOR UNIT

Figure 10–1 showed three separate nerve fibers entering a muscle. Actually, this was merely a figurative representation, because several hundred to several thousand nerve fibers enter most muscles. The figure also showed that each of the nerve fibers divides and spreads throughout the muscle belly to terminate on many different muscle fibers. On the average, a single motor nerve fiber innervates about 180 muscle fibers, which means that stimulation of 1 nerve fiber will cause contraction of 180 muscle fibers all at the same time. All the fibers innervated by the same nerve fiber are called a *motor unit* because they are always excited simultaneously and, therefore, always contract in unison.

The muscle fibers in a motor unit do not necessarily lie side by side but, instead, are divided into many bundles of only a few fibers, each spread throughout the muscle belly. Because of this, stimulation of the motor unit causes a weak contraction in a broad area of the muscle rather than a strong contraction at one specific point.

Those muscles that must control very fine movements usually have only a few muscle fibers in each motor unit, which means that the ratio of nerve fibers to muscle fibers is very high. For instance, the laryngeal muscles, which must control the extremely discrete movements necessary for speech have only two to three muscle fibers to each motor unit. On the other hand, some large muscles, which usually exhibit only very gross movements, may have as many as 1000 muscle fibers to the motor unit.

THE MUSCLE TWITCH

One of the laboratory methods for studying muscle contraction is to elicit a *muscle twitch*. To do this, a single instantaneous stimulus is applied to the nerve supplying the muscle. The duration of the resulting contraction, called a muscle "twitch," is between $1/5$ and $1/200$ second, depending on the type of muscle.

Isometric and Isotonic Contraction. Figure 10–10 shows the contraction of a muscle under two different conditions. To the left the muscle lifts weights in a pan, becoming shorter in the process. The total amount of weight applied to the muscle is always the same, for which reason the contraction is called *isotonic*, which means "same force." To the right the muscle is attached to an electronic transducer which will record the tension of muscle contraction on an electrical recorder even though the muscle contracts no more than $1/1000$ inch. Stimulation of the muscle under these conditions causes it to tighten but not to shorten significantly, and the contraction is called *isometric*, meaning "same length."

The characteristics of isometric and isotonic muscle contraction are somewhat different. The reasons for this are, first, the isometric system has no inertia while the isotonic system does, and, second, during isotonic contraction the shape of the muscle must change so that it can shorten, whereas during isometric concentration the muscle does not need to change shape but only to create force. Therefore, without inertia and without the necessity for changing shape, the isometric muscle twitch usually has a much shorter duration than the isotonic twitch. In expressing relative abilities of different muscles to contract, the isometric muscle twitch is the usual criterion employed, because its characteristics are dependent only on intrinsic characteristics of the muscle and not at all on extrinsic factors.

In the human body, muscle contraction is of both the isometric and isotonic types. When one is simply standing, he tenses his leg muscles to maintain a fixed position of the joints. This is isometric contraction. On the other hand, when he is walking and moving his legs, or when he is lifting his arms, the contraction is more of the isotonic type.

Duration of Contraction of Different Skeletal Muscles. Figure 10–11 shows recordings of isometric contractions by different muscles. The dashed curve of the figure shows the duration of depolarization of the fiber membrane caused by the action potential traveling over the muscle fiber. This is the period when calcium ions are being released into the fluids of the fiber. Immediately thereafter the contraction begins. The isometric contraction of an ocular muscle lasts for about $1/100$ second while that of a gastrocnemius muscle lasts about $3/100$ second and of a soleus muscle as long as $1/10$ second. It is evident, then, that different skeletal muscles have widely different durations of contraction. The ocular muscles, which must cause extremely rapid movement of the eyes from one position to another, contract more rapidly than almost any other muscle. The gastrocnemius muscle must contract moderately rapidly because it is used in jumping and in performing other rapid movements of the foot. The soleus usually does not need to contract rapidly at all, because it is used principally for support of the body against gravity.

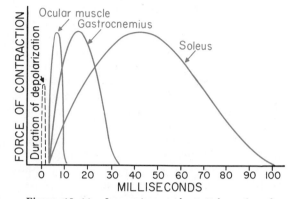

Figure 10–11. Isometric muscle twitches of ocular, gastrocnemius, and soleus muscles, illustrating the different durations of contraction.

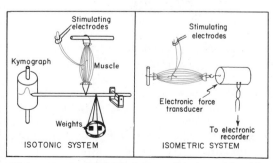

Figure 10–10. Methods for recording isotonic and isometric muscle twitches.

EFFECT OF INITIAL MUSCLE LENGTH ON THE FORCE OF CONTRACTION

The length to which a muscle is stretched before it contracts makes considerable difference in its force of contraction. When the length is much less than normal its force of contraction is greatly weakened, and, also, when it is stretched far beyond its normal limits, it fails to contract with as much force as would otherwise be possible. Figure 10–12 illustrates these effects, showing that a muscle in its normal stretched state will usually contract with the greatest possible force.

Fortunately, the normal length of a muscle in its most elongated position is almost exactly optimal for maximal strength of contraction. For instance, when the biceps is at its normal full length, it contracts with its greatest force, whereas, as it progressively shortens, its strength of contraction decreases.

The Lever Systems of the Body. Other factors that determine the force of a movement are (a) the manner in which the contracting muscles are attached to the skeletal system and (b) the structure of the joint at which movement will occur. Figure 10–13, as an example, illustrates movement of the forearm by biceps contraction. The fulcrum of the lever system is at the elbow, and the attachment of the biceps is approximately 2 inches in front of the fulcrum. If we assume that the total length of the forearm lever is about 14 inches, one immediately sees that the force of contraction of the biceps must be at least seven times as great as the force of movement of the hand. Thus, if the hand is to lift an object that weighs 50 pounds, the total force of

Figure 10–13. The lever system activated by the biceps muscle.

contraction of the biceps would have to be about 350 pounds. One can readily understand from this tremendous force why muscles sometimes actually pull their tendons out of the bone substance.

Every muscle of the body has its own peculiar shape and length that suits it to its particular function. For instance, the muscles of the buttocks are extremely broad but do not contract a long distance. They provide tremendous force for movement at the hip joint, and even a very slight distance of movement at this joint can cause tremendous movement of the foot. At the other extreme, some of the muscles of the anterior thigh are very long and can shorten as much as 6 inches, pulling the lower leg upward at the knee joint and flexing the upper leg at the hip joint at the same time.

The study of different types of muscle lever systems and their movements is called *kinesiology;* this is a very important phase of human physioanatomy.

CONTROL OF DIFFERENT DEGREES OF MUSCLE CONTRACTION – THE MECHANISM OF "SUMMATION"

In performing the different functions of the body, it is quite important that each muscle be able to contract with varying degrees of strength. This is accomplished by "summing" the contractions of varying numbers of muscle fibers at once. When a weak contraction is desired, only a few muscle fibers are contracted simultaneously. When a strong contraction is desired, a great number of fibers are contracted at the same time. In general, the dif-

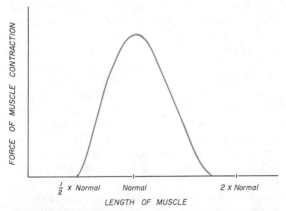

Figure 10–12. Effect of the initial length of a muscle on the contractile force developed following muscle excitation.

ferent "gradations" of muscle contraction are achieved by two different methods of summation called *multiple motor unit summation* and *wave summation.*

Multiple Motor Unit Summation. The force of muscle contraction increases progressively as the number of contracting motor units increases. In each muscle, the numbers of muscle fibers and their sizes in the different motor units vary tremendously, so that one motor unit may be as much as 50 times as strong as another. The smaller motor units are far more easily excited than are the larger ones because they are innervated by smaller nerve fibers whose cell bodies in the spinal cord have a naturally high level of excitability. This effect causes the gradations of muscle strength during weak muscle contraction to occur in very small steps, while the steps become progressively greater as the intensity of contraction increases.

Wave Summation. Figure 10–14 illustrates the principles of wave summation, showing in the lower left-hand corner a single muscle twitch followed by successive muscle twitches at various frequencies. When the frequency of twitches reaches 10 or more per second, the first muscle twitch is not completely over by the time the second one begins. Therefore, since the muscle is already in a partially contracted state when the second twitch begins, the degree of muscle shortening this time is slightly greater than that with the single muscle twitch.

At more rapid rates of contraction, the degree of summation of successive contractions becomes greater and greater, because the suc-cessive contractions appear at earlier times following the preceding contraction.

TETANIZATION. When a muscle is stimulated at progressively greater frequencies, a frequency is finally reached at which the successive contractions fuse together and cannot be distinguished one from the other. This state is called *tetanization,* and the lowest frequency at which it occurs is called the *critical frequency.* Once the critical frequency for tetanization is reached, further increase in rate of stimulation increases the force of contraction only a few more per cent, as shown in Figure 10–14.

Asynchronous Summation of Motor Units. Even when tetanization of individual motor units of a muscle is not occurring, the tension exerted by the whole muscle is still continuous and nonjerky because *the different motor units fire asynchronously;* that is, while one is contracting another is relaxing; then another fires, followed by still another, and so forth. Consequently, even when motor units fire as infrequently as five times per second, the muscle contraction, though weak, is nevertheless very smooth.

MUSCLE FATIGUE

Prolonged and strong contraction of a muscle leads to the well-known state of muscle fatigue. This results simply from inability of the contractile and metabolic processes of the muscle fibers to continue supplying the same work output. The nerve continues to function properly, the nerve impulses pass normally through the neuromuscular junction into the muscle fiber, and even normal action potentials spread over the muscle fibers, but the contraction becomes weaker and weaker because of depletion of ATP in the muscle fibers themselves.

Interruption of blood flow through a contracting muscle leads to almost complete muscle fatigue in a minute, even when the muscle is not very active, because of the obvious loss of nutrient supply.

EFFECT OF ACTIVITY ON MUSCULAR DEVELOPMENT

Exercise and Hypertrophy. The more a muscle is used, the greater becomes its size and strength, though the cause is yet unknown. Physical enlargement of the muscles

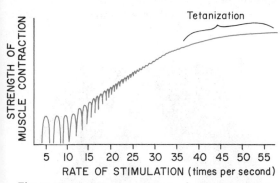

Figure 10–14. Wave summation, showing progressive summation of successive contractions as the rate of stimulation is increased. Tetanization occurs when the rate of stimulation reaches approximately 35 per second, and maximum force of contraction occurs at approximately 50 per second.

is called *hypertrophy:* examples are (1) the intense muscular development of weight lifters, (2) the great hypertrophy of the leg muscles in skaters, (3) the enlargement of the arm and hand muscles of carpenters, and (4) the enlargement of the thigh muscles of runners.

Associated with muscle hypertrophy is usually an increase in the efficiency of muscular contraction, for the hypertrophied muscle stores increased quantities of glycogen, fatty substances, and other nutrients, and the number of contractile myofibrils also increases. All these cause the efficiency of the contractile process to increase so that the percentage of energy lost as heat becomes considerably less in the athlete than in the non-athlete.

It should be noted that weak muscular activity, even when sustained over long periods of time, does not result in significant muscle hypertrophy. Instead, hypertrophy is mainly the result of forceful muscle activity even though the activity might occur for only a few minutes each day. For this reason, strength can be developed in muscles more rapidly by using "resistive exercises" or forceful "isometric exercises" than by prolonged mild exercise.

Muscle Denervation and Atrophy. When the nerve supply to a muscle is destroyed the muscle begins to *atrophy* — that is, the muscle fibers begin to degenerate. In about 6 months to 2 years the muscle will have atrophied to about one-fourth normal size, and the muscle fibers will have been replaced mainly by fibrous tissue.

For some reason, nerve stimulation of a muscle keeps the muscle tissue alive. Even when a person does not use his muscles to a great extent, the weak, intermittent tonic impulses are still sufficient to maintain a relatively normal muscle, but without these impulses the muscle fibers soon atrophy entirely. Perhaps this effect is caused by nutritional changes in the denervated muscle, for action potentials traveling down the fiber membrane alter its permeability markedly, which might be necessary for appropriate transfer of nutrients through the membrane.

The functional integrity of denervated muscle fibers can be maintained quite satisfactorily by daily electrical stimulation. The action potentials produced in this manner take the place of the nerve-induced potentials, and the muscle fibers do not atrophy.

SMOOTH MUSCLE

Most of the internal organs of the body contain *smooth muscle.* The name is derived from the fact that this muscle does not have microscopic striations similar to those in skeletal and cardiac muscle.

Smooth muscle is composed of fibers far smaller than skeletal muscle fibers — usually 2 to 5 microns (micrometers) in diameter and only 50 to 200 microns in length, in contrast to the skeletal muscle fibers that are as much as 20 times as large in diameter and thousands of times as long. Nevertheless, many of the same principles of contraction apply to both smooth muscle and skeletal muscle. Most important, the same chemical substances cause contraction in smooth muscle as in skeletal muscle, but the physical arrangement of smooth muscle fibers is entirely different, as we shall see.

TYPES OF SMOOTH MUSCLE

The smooth muscle of each organ is often distinctive from that of other organs in several different ways: physical dimensions, organization into bundles or sheets, response to different types of stimuli, characteristics of its innervation, and function. Yet, for the sake of simplicity, smooth muscle can generally be divided into two major types, which are illustrated in Figure 10–15: *multiunit smooth muscle* and *visceral smooth muscle.*

Visceral Smooth Muscle. By far most of the smooth muscle of the body is of the visceral type. As illustrated in Figure 10–15, this muscle usually forms large sheets of closely packed fibers. These sheets are then rolled into cylinders or spheres to form the various

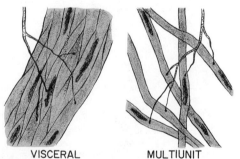

VISCERAL   MULTIUNIT

Figure 10–15. Visceral and multiunit smooth muscle fibers.

hollow organs such as the intestines, the urinary bladder, the gallbladder, the ureters, and so forth.

Another feature of visceral smooth muscle is that the individual muscle fibers abut against each other so tightly that many of their membranes actually fuse. And at the points where they do fuse, intracellular ions can pass with ease from one fiber to the next. Because of this, action potentials can travel freely from one fiber to the next. Therefore, when one muscle fiber is stimulated, the stimulus travels over the entire mass of muscle and causes it all to contract.

Multiunit Smooth Muscle. A small portion of the smooth muscle in the body is composed of so-called multiunit fibers, which are also shown in Figure 10–15. These fibers operate entirely independently of each other, and each one is often innervated by a single nerve ending as occurs in skeletal muscle fibers.

Some examples of multiunit smooth muscle are the smooth muscle fibers of the ciliary muscle of the eye, the iris of the eye, the nictitating membrane that covers the eyes in some lower animals, the piloerector muscles that cause erection of the hairs when stimulated by the sympathetic nervous system, and the smooth muscle of some of the larger blood vessels. Thus, in general, multiunit smooth muscle usually occurs in small muscle masses and performs discrete local functions, in contrast to visceral smooth muscle which is usually organized into large masses and performs a major function for an entire organ.

THE CONTRACTILE PROCESS IN SMOOTH MUSCLE

Smooth muscle contains exactly the same chemical substances as skeletal muscle, including both actin and myosin. However, the actin and myosin filaments do not have the same periodic spatial arrangement as that seen in the sarcomeres of skeletal muscle. Instead the filaments sometimes lie at odd angles to each other, though mostly oriented in the longitudinal direction of the muscle fibers.

The contractile process appears in many ways to be the same as in skeletal muscle: That is, it is activated by calcium ions; ATP is converted to ADP, just as occurs in skeletal muscle; and contraction occurs after depolari-

zation of the membrane and stops a fraction of a second to a few seconds following repolarization.

One significant difference between smooth muscle contraction and skeletal muscle contraction is the timing of the process; contraction develops only one-fourth to one-twentieth as rapidly in smooth muscle as in skeletal muscle. Also, relaxation at the end of the action potential is often very prolonged. This difference results at least partly from the slowness of the acion potential in smooth muscle, which will be discussed in a subsequent section.

The contractile filaments in smooth muscle seem to stick to each other so that they "drag" against each other during contraction and relaxation. This could explain a special ability of smooth muscle to maintain large amounts of tension over long periods of time without utilizing much ATP, an effect that is also quite different from that in skeletal muscle.

INITIATION OF CONTRACTION BY CALCIUM IONS

Contraction of smooth muscle is initiated by sudden increase of calcium ions in close association with the contractile fibrils, in the same way that the contractile process is initiated in skeletal muscle. However, there is one major difference: The source of the calcium ions differs between smooth muscle and skeletal muscle. In smooth muscle, the sarcoplasmic reticulum is poorly developed, indeed almost absent in some smooth muscle fibers. However, the smooth muscle fiber is so small that when the action potential spreads over its surface membrane, enough calcium ions diffuse into the cell during the positive phase of this action potential to initiate contraction.

After the calcium ions have caused the smooth muscle to contract, they must be removed from the fiber before relaxation can occur. This is accomplished by a powerful calcium pump in the smooth muscle fiber membrane that pumps the calcium back into the extracellular fluids. However, this pump is slow to act in comparison with the fast-acting sarcoplasmic reticulum pump in skeletal muscle. Therefore, the duration of smooth muscle contraction is often in the order of seconds rather than tens of milliseconds, as occurs for skeletal muscle.

ACTION POTENTIALS IN SMOOTH MUSCLE AND CONTROL OF CONTRACTION

The normal resting potential of smooth muscle is usually about −55 to −60 millivolts, or about 25 millivolts less negative than in skeletal muscle. Two different types of action potentials often occur and are associated with contraction. These are:

1. *Spike potentials.* Typical spikelike action potentials similar to those that occur in skeletal muscle fibers are seen in most types of visceral smooth muscle. Immediately following the action potential, a muscle contraction occurs. The duration of the muscle contraction depends on the speed with which the calcium pump removes calcium ions from the fiber; in many instances this contraction may last for several seconds.

Typical spike potentials are illustrated in Figure 10–16. Figure 10–16A shows a single spike potential elicited by some external stimulus. In Figure 10–16B the spike potentials are associated with slow undulations in the baseline membrane potential of the smooth muscle, called "slow waves," which will be discussed below.

2. *Action potentials with plateaus.* Another common type of action potential in visceral smooth muscle is that illustrated in Figure 10–17, having a plateau lasting 0.1 to 0.4 second. The cause of the plateau in the action

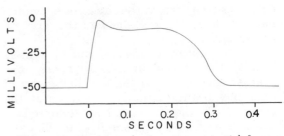
Figure 10–17. Monophasic action potential from a smooth muscle fiber.

potential is slowness of the membrane to repolarize at the end of the action potential. This same effect occurs in cardiac muscle, and its basic cause will be discussed in the following chapter in relation to cardiac contraction. So long as the plateau persists, the smooth muscle remains contracted. This type of action potential is frequently seen in the smooth muscle of the ureter and the uterus.

Rhythmicity of Smooth Muscle — Role of the "Slow Waves." Many types of smooth muscle exhibit rhythmicity, with contractions occurring periodically once every few seconds. Such rhythmic contractions are very important in many of the internal organs of the body because they are responsible for the peristaltic movements of the intestines, the ureters, the bile ducts, and so forth.

The cause of rhythmicity in smooth muscle is *slow waves*, which are illustrated in Figure 10–16B. These are not action potentials. Instead, they are slow, undulating changes in the membrane potential that continue indefinitely, sometimes almost imperceptible and at other times very intense. This slow wave rhythm is believed to result from waxing and waning of the activity of the sodium pump, the potential becoming more negative when the pump is very active and less negative when it is less active.

Even though the slow waves are not themselves action potentials, they do frequently elicit action potentials. Referring again to Figure 10–16B, note the spikes that occur at the top of each one of the slow waves. This is caused by the fact that the slow wave potential has risen high enough to reach the threshold at which the membrane spontaneously generates an action potential. Therefore, when the slow wave potentials are strong, one usually finds one or more action potentials occurring at their peaks, and it is these action potentials that cause the rhythmic contractions of the smooth muscle.

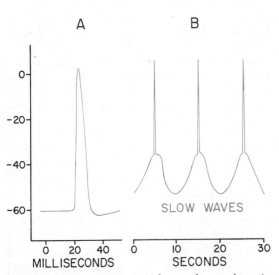

Figure 10–16. (A) A typical smooth muscle action potential (spike potential) elicited by an external stimulus. (B) A series of spike action potentials elicited by rhythmic, slow electrical waves occurring spontaneously in the smooth muscle wall of the intestine.

Spread of Action Potentials Through Visceral Smooth Muscle. Once an action potential begins in visceral smooth muscle, it spreads slowly through the entire muscle mass. For instance, if it begins at the upper end of the gastrointestinal tract, it often spreads downward along the intestinal wall, creating a constrictive ring that moves forward. The constrictive ring propels the intestinal contents forward. This process is called *peristalsis,* an important end result of smooth muscle function.

Excitation of Visceral Smooth Muscle by Stretch. When visceral smooth muscle is stretched sufficiently, spontaneous action potentials are usually generated. These result from a combination of the normal slow wave potentials plus a decrease in the membrane potential caused by the stretch itself. This response to stretch is especially important because it allows a hollow organ that is excessively stretched to contract automatically and therefore to resist the stretch. For instance, when the gut is overstretched by intestinal contents, a local automatic contraction sets up a peristaltic wave that moves the contents away from the superstretched intestine.

Control of Smooth Muscle by the Autonomic Nervous System. The autonomic nervous system, which will be discussed in Chapter 26, is composed of two separate parts, the *sympathetic* and the *parasympathetic systems.* Nerve fibers pass from each of these to most smooth muscle of the body. When stimulated, the sympathetics secrete a hormone called *norepinephrine,* and the parasympathetics secrete a hormone called *acetylcholine.* Norepinephrine stimulates the smooth muscle in some organs but in other organs inhibits it. Acetylcholine, on the other hand, usually inhibits all smooth muscle that norepinephrine stimulates and stimulates all that the norepinephrine inhibits. In other words, the hormones secreted by the two portions of the autonomic nervous system almost always have opposing effects. Some examples of this are the following: (1) The pupil of the eye is constricted by the parasympathetics but dilated by the sympathetics. (2) The coronary arteries are dilated by the sympathetics but constricted by the parasympathetics. (3) The intestines are constricted by the parasympathetics but dilated by the sympathetics. (4) The sphincters of the gut are constricted by the sympathetics but dilated by the parasympathetics. (5) The sphincters that regulate emptying of the urinary bladder are constricted by the sympathetics but dilated by the parasympathetics.

The reason why norepinephrine contracts some but relaxes other smooth muscle, while acetylcholine causes opposite effects, is not known, but this is believed to result from a difference in "receptor substances" in different types of smooth muscle cells. If the receptor substance is of an excitatory nature, norepinephrine will cause stimulation; if of an inhibitory nature, it will cause inhibition. It is presumed that the excitatory substance for norepinephrine is an inhibitory substance for acetylcholine, and the inhibitory substance for norepinephrine is an excitatory substance for acetylcholine. This would explain the antagonistic effects of the two hormones on smooth muscle.

SPECIAL CHARACTERISTICS OF SMOOTH MUSCLE CONTRACTION

Tone of Smooth Muscle — Summation of Individual Contractions. Smooth muscle can maintain a state of long-term, steady contraction that has been called either *tonus* contraction or simply *smooth muscle tone.* This is an important feature of smooth muscle contraction because it allows prolonged or even indefinite continuance of the smooth muscle function. For instance, the arterioles are maintained in a state of tonic contraction almost throughout the entire life of the person. Likewise, tonic contraction in the gut wall maintains steady pressure on the contents of the gut, and tonic contraction of the urinary bladder maintains a moderate amount of pressure on the urine in the bladder.

The tonic contraction of smooth muscle is caused by summation of individual contractile pulses. This effect is illustrated in Figure 10–18. Furthermore, rhythmic contractions can be superimposed on the tonic contraction, as also illustrated in the figure. The rhythmic contractions are caused by waxing and waning of the frequency of stimulation and, therefore, waxing and waning of the summation of the contractile process.

The Extreme Degree of Shortening of Smooth Muscle During Contraction. A special characteristic of smooth muscle—one that is also different from skeletal muscle—is its ability to shorten a far greater percentage of its length than can skeletal muscle. Skeletal

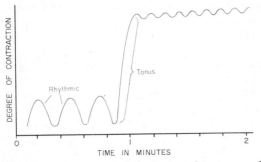

Figure 10–18. Record of rhythmic and tonic smooth muscle contraction.

REFERENCES

Basmajian, J. V.: Muscles Alive. 3rd ed. Baltimore, The Williams & Wilkins Company, 1974.

Buller, A. J.: The physiology of skeletal muscle. MTP International Review of Science: Physiology. Vol. 3. Baltimore, University Park Press, 1975, p. 279.

Clausen, J. P.: Effect of physical training on cardiovascular adjustments to exercise in man. *Physiol. Rev.,* 57:779, 1977.

Cohen, C.: The protein switch of muscle contraction. *Sci. Amer., 233(5):*36, 1975.

Ebashi, S.: Excitation-contraction coupling. *Ann. Rev. Physiol.,* 38:293, 1976.

Endo, M.: Calcium release from the sarcoplasmic reticulum. *Physiol. Rev.,* 57:71, 1977.

Fuchs, F.: Striated muscle. *Ann. Rev. Physiol.,* 36:161, 1974.

Holloszy, J. O., and Booth, F. W.: Biochemical adaptations to endurance exercise in muscle. *Ann. Rev. Physiol.,* 38:273, 1976.

Huxley, H. F.: Muscular contraction and cell motility. *Nature, 243:*445, 1973.

Murray, J. M., and Weber, A.: The cooperative action of muscle proteins. *Sci. Amer., 230:*58, 1974.

Taylor, C. R.: Exercise and environmental heat loads: different mechanisms for solving different problems? *Intern. Rev. Physiol., 15:*119, 1977.

Toida, N., Kuriyama, H., Tahsiro, N., and Ito, Y.: Obliquely striated muscle. *Physiol. Rev.,* 55:700, 1975.

Weber, A., and Murray, J. M.: Molecular control mechanisms in muscle contraction. *Physiol. Rev., 53:*612, 1973.

muscle has a useful distance of contraction equal to only 25 to 35 per cent of its length, while smooth muscle can contract quite effectively from a length of two times its normal length to as short as one-quarter to one-half its normal length, giving as much as a four- to eightfold distance of contraction. This allows smooth muscle to perform important functions in the hollow viscera — for instance, allowing the gut, the bladder, blood vessels, and other internal structures of the body to change their lumen diameters from almost zero up to very large values.

QUESTIONS

1. Describe the anatomy of skeletal muscle.
2. Explain the sliding mechanism of muscle contraction.
3. Describe the special characteristics of myosin and actin filaments.
4. Discuss the ratchet theory of contraction and explain the mechanism of the "power stroke."
5. What are the separate roles of the transverse tubules and the longitudinal tubules in the initiation of muscle contraction?
6. How do calcium ions react with the myofibrillar filaments to cause muscle contraction?
7. How many muscle fibers are, on the average, found in a motor unit?
8. Describe the effect of initial muscle length on force of muscle contraction.
9. What are the different types of summation by which the strength of muscle contraction can be changed?
10. What is the relationship between exercise and muscle hypertrophy?
11. How does smooth muscle differ anatomically from skeletal muscle?
12. How do "slow waves" cause rhythmic smooth muscle contraction?
13. What are the interrelationships between the sympathetic and parasympathetic systems for the control of smooth muscle contraction?
14. What is smooth muscle tone, its importance and cause?

THE CARDIOVASCULAR SYSTEM

IV

THE PUMPING ACTION OF THE HEART, AND ITS REGULATION

11

THE HEART AS A PUMP

The heart is actually two separate pumps, as shown in Figure 11–1: One pumps blood through the lungs, while the other pumps blood from the lungs through the remainder of the body. Thus, the blood flows around a continuous circuit.

Figure 11–2 shows the functional details of

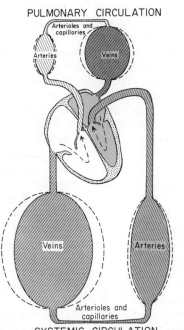

PULMONARY CIRCULATION

Arterioles and capillaries

Arteries

Veins

Veins

Arteries

Arterioles and capillaries

SYSTEMIC CIRCULATION

Figure 11–1. Schematic representation of the circulation, showing the two sides of the heart and the pulmonary and systemic circulations.

the heart as a pump. Blood entering the *right atrium* from the large veins is forced by atrial contraction through the *tricuspid valve* into the *right ventricle*. The right ventricle then pumps the blood through the *pulmonary valve* into the pulmonary artery, thence through the lungs, and finally through the pulmonary veins into the *left atrium*. Left atrial contraction then forces the blood through the *mitral valve* into the *left ventricle,* whence it is pumped through the *aortic valve* into the aorta and on through the systemic circulation.

The two atria are *primer pumps* that force extra blood into the respective ventricles immediately before ventricular contraction. This propulsion of extra blood into the ventricles makes the ventricles more efficient as pumps than they would be if they had no special filling mechanism. However, the ventricles are so powerful that they can still pump large quantities of blood even when the atria fail to function.

CARDIAC MUSCLE: ITS EXCITATION AND CONTRACTION

THE SYNCYTIAL CHARACTER OF CARDIAC MUSCLE

Figure 11–3 illustrates a microscopic section of cardiac muscle. Note that the fibers have the same striated appearance that is characteristic of skeletal muscle. However, note also that the cardiac muscle fibers interconnect with each other, forming a lattice-

123

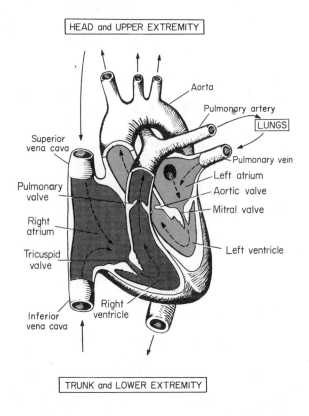

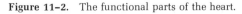

Figure 11–2. The functional parts of the heart.

work called a *syncytium*. This arrangement is similar to that which occurs in visceral smooth muscle where the smooth muscle fibers also fuse to form an interconnected mass of fibers called a syncytium. In the heart there are two separate muscle syncytia. One of these is the mass of cardiac muscle that wraps around the two atria, and the other is the mass of cardiac muscle that wraps around the two ventricles. These two muscle masses are separated from each other by fibrous rings surrounding the valves between the atria and the ventricles, the tricuspid and the mitral valves.

The importance of the two separate syncytial muscle masses is the following: When either one of the two muscle masses is stimulated, the action potential spreads over the entire syncytium and therefore causes the entire muscle mass to contract. Thus, when the atrial muscle mass is stimulated at any single point, the action potential spreads over both atria and this causes the whole complex of atrial walls to contract, thereby squeezing the atrial blood into the ventricles through the tricuspid and mitral valves. Then when the action potential spreads into the ventricles, here again it excites the entire ventricular muscle syncytium. Therefore, all of the ventricular walls now contract in unison, and the blood in their chambers is appropriately pumped through the aortic and pulmonary valves into the arteries. Later in the chapter we will discuss the manner in which the heart controls the contraction of both the atria and the ventricles.

Automatic Rhythmicity of Cardiac Muscle. Most cardiac fibers are capable of contracting rhythmically. This is especially true of a small group of cardiac fibers located on the posterior wall of the right atrium called

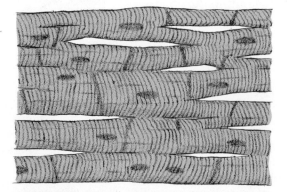

Figure 11–3. Cardiac muscle, showing the lattice arrangement of the fibers.

the *sinoatrial node* or, more simply, the S-A node. We will discuss the sinoatrial node in great detail later in this chapter because its strong propensity for rhythmic contraction makes it the normal controller of the heart beat. However, first, let us discuss the mechanism by which rhythmic contraction occurs.

Figure 11–4 illustrates rhythmic action potentials generated by an S-A nodal fiber. The cause of this rhythmicity is the following: The membranes of the S-A fibers, even in the resting state, are very permeable to sodium. Therefore, large numbers of sodium ions leak to the interior of the fiber, causing the resting membrane potential to drift continually toward a more positive value as illustrated in the figure. Just as soon as the membrane potential reaches a critical level, called the "threshold" level, an action potential suddenly occurs. At the end of the action potential, the membrane is highly permeable to potassium ions, and leakage of potassium ions out of the fiber carries positive charges to the outside. Therefore, the inside membrane potential now becomes more negative than ever, a state called *hyperpolarization,* because of loss of the extra positive charges. This condition persists for a fraction of a second and then disappears because the permeability to potassium ions returns to its normal state, whereupon the natural leakiness of the membrane to sodium ions then elicits another action potential. This process continues over and over throughout life, thereby providing rhythmic excitation of the S-A nodal fibers at a normal resting state of about 72 times per minute and for a total of about two billion heart beats during the life of the person.

Normally, action potentials originating in the S-A node spread from here through the entire heart and thereby elicit the rhythmic contractions of the heart. However, if the S-A node fails to generate rhythmic impulses an-

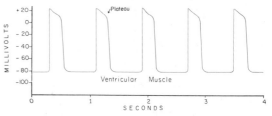

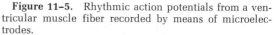

Figure 11–5. Rhythmic action potentials from a ventricular muscle fiber recorded by means of microelectrodes.

other area of the cardiac muscle will begin to generate impulses and will then take over control of the heart beat, as we shall discuss more fully later in the chapter.

Prolonged Duration of Cardiac Muscle Action Potential and of Cardiac Contraction. Another important difference between cardiac muscle and skeletal muscle is that cardiac muscle contraction lasts for a prolonged period of time in comparison with skeletal muscle, averaging a duration about 10 to 15 times as long as that of the average skeletal muscle.

The cause of the prolonged cardiac muscle contraction is that the cardiac muscle action potential also lasts for a long period of time. This is illustrated in Figure 11–5. Instead of the action potential being a sharp spike and then returning immediately to the baseline as occurs in large nerve fibers and in skeletal muscle fibers, this potential remains on a *plateau,* as illustrated in the figure, for almost 0.3 second before returning to the resting level. The cause of this plateau is slowness of the membrane to repolarize once it has become depolarized. Let us explain this more fully. It will be recalled from the discussion in Chapter 9 that the action potential is caused by two separate processes: first, the process of depolarization of the membrane resulting from rapid influx of sodium ions into the fiber, and, second, the repolarization process caused by rapid efflux of potassium ions from inside the fiber to the outside. The depolarization process causes the action potential to begin, and the repolarization process causes it to end. In large nerve fibers, skeletal nerve fibers, and some smooth muscle fibers, the repolarization process follows almost instantly after the depolarization process, thereby giving a very short spikelike action potential. In cardiac muscle and in some smooth muscle, the repolarization process is delayed. The cause of the delay is the following: In these types of muscle fibers the rapid leakage of

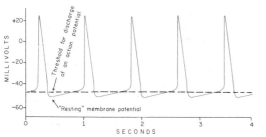

Figure 11–4. Rhythmic discharge of an S-A nodal fiber.

sodium ions to the interior of the fiber during depolarization has the peculiar effect of making the membrane temporarily very impermeable to potassium. Therefore, at the end of depolarization, failure of potassium ions to leak out of the fiber, because of this potassium impermeability, prevents immediate repolarization. However, slowly during the next few tenths of a second, the potassium permeability returns to normal, the repolarization process then takes place, and the action potential returns to the baseline.

In atrial muscle the total duration of the action potential as well as of the contraction is about 0.15 second, and in ventricular muscle about 0.3 second.

REGULATION OF CARDIAC RHYTHMICITY

The inherent rhythmicity of the heart beat is seen beautifully by simply observing an exposed heart. An especially instructive experiment is the one illustrated in Figure 11–6, which shows recordings of muscle contraction

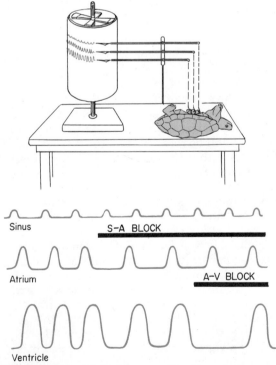

Sinus
S-A BLOCK
Atrium
A-V BLOCK
Ventricle

Figure 11–6. Method for recording the separate contractions of the sinus, atrium, and ventricle of the turtle heart. Below are illustrated the effects of S-A block and A-V block on impulse transmission.

in different portions of a turtle heart. The turtle heart has three chambers instead of the two that are present in each side of the human heart. The first chamber is called the *sinus* and the next two are the same as in the human heart, the *atrium* and the *ventricle*. The top record in Figure 11–6 shows contraction of the sinus, the middle record contraction of the atrium, and the bottom record contraction of the ventricle. No nerves are attached to the different portions of the heart in this preparation to make them contract, and no other signals arrive from outside the heart to cause the rhythmicity. In other words, the rhythmicity of the heart is vested in the heart itself, and if portions of the heart are removed from the body, they will continue to contract rhythmically as long as they are provided with sufficient nutrition.

The Sinoatrial Node as the Pacemaker of the Human Heart. Referring once again to Figure 11–6, it will be observed on the left-hand side of the figure that the sinus of the turtle heart contracts slightly ahead of the atrium, and the atrium slightly ahead of the ventricle. That is, every time the sinus contracts, an action potential spreads along the muscle fibers from the sinus to the atrium and then to the ventricle, making these parts of the heart contract in succession. The effects of blocking the action potential between the sinus and atrium (S-A block) and then between the atrium and ventricle (A-V block) are shown to the right in the figure. Note that the atrium and the ventricle then beat at their own natural rhythmic rates.

The human heart is different from that of the turtle, because no distinct sinus exists. However, as mentioned earlier, located in the posterior wall of the right atrium immediately beneath the point of entry of the superior vena cava is a small area known as the *sinoatrial node* (S-A node) which is the embryonic remnant of the sinus from lower animals. The rhythmic rate of contraction of muscle fibers excised from the S-A node is approximately 72 times per minute, while muscle excised from the atrium contracts about 60 times per minute and muscle from the ventricle about 20 times per minute. Because the S-A node has a faster rate of rhythm than other portions of the heart, impulses originating in the S-A node spread into the atria and ventricles, stimulating these areas so rapidly that they can never slow down to their natural rates of rhythm. As a result, the rhythm of the S-A node becomes the rhythm of the entire heart, for which reason the S-A node is called the *pacemaker* of the heart.

CONDUCTION OF THE IMPULSE THROUGH THE HEART

The Purkinje System. Even though the cardiac impulse can travel perfectly well along cardiac muscle fibers, the heart has a special conduction system called the Purkinje system, composed of specialized cardiac muscle fibers called *Purkinje fibers* that transmit impulses at a velocity approximately six times as rapidly as that in normal heart muscle, approximately 2 meters per second in contrast to 0.3 meter per second in cardiac muscle.

Figure 11–7 illustrates the organization of the Purkinje system. It originates in the *sinoatrial node* (S-A node), which has already been discussed. From here, several very small bundles of Purkinje fibers, called *internodal pathways,* pass in the wall of the atrium to a second node, the *atrioventricular node* (A-V node), that is located posteriorly and toward the center of the heart on the right atrial wall. From this node a large bundle of Purkinje fibers, called the *A-V bundle,* passes out of the atria and into the ventricles, entering first the ventricular septum. After traveling a short distance in the septum, the A-V bundle divides into two large bundles, a *left bundle branch* that continues into the wall of the left ventricle, and a *right bundle branch* that continues into the wall of the right ventricle. On reaching the walls of the ventricles, the bundles divide into many minute Purkinje fiber branches that make direct contact with the cardiac muscle. Therefore, an impulse traveling along the Purkinje fibers is conducted rapidly and directly into the cardiac muscle.

ROLE OF THE PURKINJE SYSTEM TO CAUSE COORDINATE CONTRACTION OF CARDIAC MUSCLE. The major function of the Purkinje system is to transmit the cardiac impulse rapidly throughout the atria and, after a short pause at the A-V node, then also rapidly throughout the ventricles. Rapid conduction of the impulse in this manner will cause all portions of each cardiac muscle syncytium — the atrial syncytium and then the ventricular syncytium — to contract in unison so that it will exert a coordinated pumping effort. Were it not for the Purkinje system, the impulse would travel much more slowly through the muscle, allowing some of the muscle fibers to contract long before others and then also to relax before the others. Obviously, this would result in decreased compression of the blood and, therefore, decreased pumping power.

SEQUENCE OF IMPULSE TRANSMISSION THROUGH THE HEART — IMPULSE DELAY AT THE A-V NODE. After the cardiac impulse originates in the S-A node, it travels first throughout the atria, causing the atrial muscle to contract. A few hundredths of a second after leaving the S-A node the impulse reaches the A-V node. However, the A-V node delays the impulse another few hundredths of a second before allowing it to pass on into the ventricles. This delay allows time for the atria to force blood into the ventricles prior to ventricular contraction. After the delay, the impulse then spreads very rapidly through the Purkinje system of the ventricles, causing both ventricles to contract with full force within the next few hundredths of a second.

The A-V node delays the cardiac impulse by the following mechanism: The fibers in this node are extremely small. This is very different from the remainder of the Purkinje system, and it allows these special fibers to conduct the cardiac impulse very slowly, at a velocity only one-tenth that of the cardiac muscle fibers and only one-sixtieth that of the large Purkinje fibers. Therefore, the cardiac impulse travels at a snail's pace through this node, causing a delay of more than 0.1 second between contraction of the atria and the ventricles.

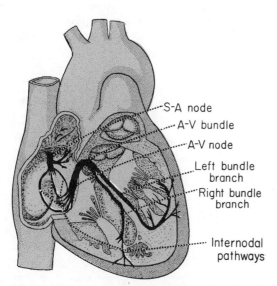

S-A node

A-V bundle

A-V node

Left bundle branch

Right bundle branch

Internodal pathways

Figure 11–7. Transmission of the cardiac impulse from the S-A node into the atria via internodal pathways, then into the A-V node, and finally through the A-V bundles to all parts of the ventricles.

Block of Impulse Conduction in Damaged Hearts. Occasionally the cardiac impulse is blocked at some point in its pathway because of damage to the heart. For instance, a portion of heart muscle or of the Purkinje system may be destroyed and replaced by fibrous tissue that cannot transmit the impulse. Returning to Figure 11-6, note the effect of artificially blocking the impulse at two critical points in the turtle heart. First, a *sinoatrial block* has been effected by tying a string tightly around the heart between the sinus and the atrium. This stops impulse conduction from the sinus into the atrium, and, as is evident from the illustration, the sinus continues to beat at its own natural rate while the atrium and ventricle assume a rate equal to that of the natural rate of the atrium. In other words, the atrium becomes the pacemaker for the ventricle because the ventricle's natural rate of rhythmicity is much slower than that of the atrium. Another ligature is then tied tightly around the heart, this time between the atrium and the ventricle to cause *atrioventricular block*. After this, the sinus, the atrium, and the ventricle all beat at their own respective natural rates of rhythm. This experiment shows that the portion of the heart that beats most rapidly controls the rate of rhythm of the remainder of the heart only so long as functioning conductive fibers exist between the different areas.

In the human heart, block rarely occurs between the S-A node and the atrial muscle, but very frequently conduction from the atria into the ventricles through the A-V bundle is blocked. It is only through this bundle that the normal impulse can pass from the atria into the ventricles, because elsewhere the atria are connected to the ventricles not by conductive fibers, but instead by fibrous tissue that cannot conduct impulses. Therefore, whenever the A-V bundle is blocked, the atria will beat at the rhythm of the S-A node, and the ventricles at their own natural rate. In other words, the rate of the atria will remain at approximately 72 beats per minute, while that of the ventricles will decrease to 15 to 45 beats per minute. Despite this asynchrony of the atria and ventricles, the heart still operates reasonably satisfactorily as a pump, though its pumping ability may be decreased as much as 50 per cent. Nevertheless, it is evident that the atria are not absolutely essential for the heart to pump blood through the circulatory system.

Cessation of the Impulse at the End of Each Heart Beat — the Refractory Period. Normally, when an impulse spreads along the membranes of the heart muscle fibers, a second impulse cannot spread along these same membranes until approximately 0.3 second later because this is the duration of the cardiac muscle action potential as explained earlier in the chapter. During this time the heart is said to be *refractory* to new impulses.

After an impulse enters the ventricles from the atria, it spreads all the way around the ventricles in about 0.06 second. Since the ventricular fibers cannot conduct again for 0.3 second, the impulse completely stops. The upper part of Figure 11-8 illustrates this principle diagrammatically. It shows a circular strip of cardiac muscle in which an impulse starts at the 12 o'clock point, then travels around the heart and finally returns to the 12 o'clock point. When the impulse reaches the starting point the entire heart is still refractory, which causes the impulse to die.

The Circus Movement. Occasionally conditions become sufficiently abnormal that a cardiac impulse does continue on and on around the heart, never stopping. For instance, in the lower part of Figure 11-8, the length of the pathway has been increased so greatly by the larger circle (which represents an enlarged heart) that after traveling all the way around the heart the impulse now returns to the 12 o'clock position more than 0.3 second after it starts. By this time, the originally stimulated portion of the muscle is no longer refractory, which allows the impulse to travel around again. As the impulse proceeds, the refractory state recedes ahead of it, allow-

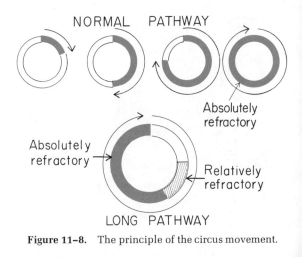

Figure 11-8.   The principle of the circus movement.

ing it to continue indefinitely. This effect is known as a *circus movement,* and it is often lethal as explained below.

CAUSES OF CIRCUS MOVEMENTS. Circus movements can result from any of four different abnormalities of the heart. First, as explained in the preceding example, a circus movement is likely to occur when the heart becomes greatly enlarged, thus creating a long pathway. A second cause is slow conduction of the impulse through the heart. For instance, failure of the Purkinje system causes the impulse then to be transmitted by the cardiac muscle itself. This slows impulse transmission about sixfold, often causing the impulse to return to the starting point after the originally excited muscle is no longer refractory. A third cause of a circus movement may be decreased refractory period of the heart muscle. This sometimes results from increased cardiac excitability caused by epinephrine, sympathetic stimulation, or irritation of the heart as a result of disease. Fourth, a very common cause is transmission of impulses in figure 8's, in zigzags, or in any other odd pattern, the impulses sometimes traveling deep in the muscle and then later traveling at shallow levels, recrossing the same area that had already been stimulated at a deeper level. In this way the pathway becomes extremely lengthened, thus allowing development of an odd-shaped circus movement. Such a pathway, as illustrated in Figure 11–9B, causes ventricular fibrillation, which is discussed below.

EFFECT ON CARDIAC PUMPING. A circus movement is disastrous to the pumping action of the heart because normal pumping requires that the muscle relax as well as contract. A period of total ventricular relaxation is not possible if the impulse continually travels through the muscular mass. Therefore, the entire muscle mass never relaxes or contracts simultaneously, never allowing an alternating filling and squeezing action.

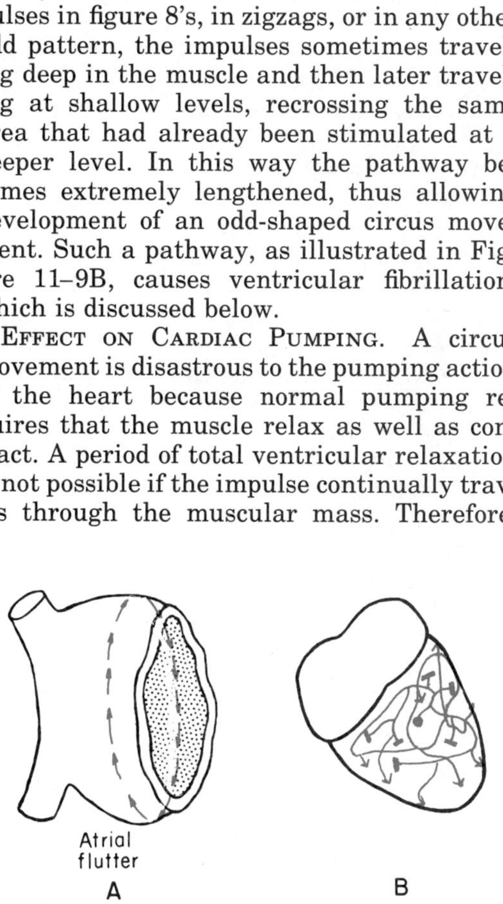

ATRIAL FLUTTER AND ATRIAL FIBRILLATION. Occasionally a circus movement occurs around and around the two atria, as shown in Figure 11–9A, at a rate of 200 to 400 times per minute, but not involving the ventricles. This causes *atrial flutter,* during which the atria are "fluttering" rapidly but pumping almost no blood. At other times, an odd-shaped, zigzag, or figure 8 pattern of circus movement occurs in the atria at such a rapid rate that one can see only minute fibrillatory movements of the muscle. This is called *atrial fibrillation,* and when it occurs the atria are of no use whatsoever as primer pumps for the ventricles.

VENTRICULAR FIBRILLATION. A circus movement having the very odd pattern illustrated in Figure 11–9B frequently develops in the ventricles, in which case impulses travel in all directions and cause *ventricular fibrillation.* The impulses go around refractory areas of muscle, dividing into multiple wave fronts in doing so. Some of the impulses die, but for every impulse that does die, another impulse divides to form two impulses going in separate directions in the heart. This type of circus movement has been called the *chain-reaction movement,* for it is similar to the chain reaction that occurs in nuclear bomb explosions.

Ventricular fibrillation causes the ventricles to contract continuously in very fine, rippling fibrillatory movements. The contracting areas are widespread in the ventricles, interspersed with relaxing areas. As a consequence, the ventricles are incapable of pumping any blood, and the person dies in the next few seconds.

Ventricular fibrillation is frequently initiated by electric shock, particularly by 60 cycle alternating current; this causes impulses to go in many different directions at once in the heart, setting up the odd-shaped pattern of impulse transmission illustrated in Figure 11–9B. Ventricular fibrillation also occurs in diseased ventricles that (a) become overly excitable, (b) develop a damaged Purkinje system, or (c) become greatly enlarged.

THE ELECTROCARDIOGRAM

Figure 11–9. (A) Impulse transmission in *atrial flutter.* (B) Impulse transmission in *ventricular fibrillation.*

The electrocardiogram is a very important tool for assessing the ability of the heart to

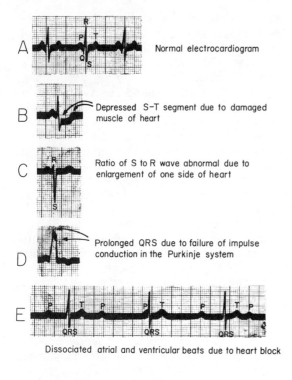

A — Normal electrocardiogram

B — Depressed S-T segment due to damaged muscle of heart

C — Ratio of S to R wave abnormal due to enlargement of one side of heart

D — Prolonged QRS due to failure of impulse conduction in the Purkinje system

E — Dissociated atrial and ventricular beats due to heart block

Early beat due to irritable focus in heart

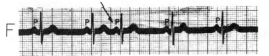

F

Figure 11-10. The normal and several abnormal electrocardiograms.

transmit the cardiac impulse. When the impulse travels through the heart, electrical current generated by the action potential on the surface of the heart muscle spreads into the fluids surrounding the heart, and a minute portion of the current actually flows as far as the surface of the body. By placing electrodes on the skin over the heart or on any two sides of the heart and connecting these to an appropriate recording instrument, the impulse generated during each heart beat can be recorded.

In the normal electrocardiogram illustrated in Figure 11-10A, the small hump in the recording labeled "P" is caused by electrical current generated by passage of the impulse through the atria. The spikes marked "Q," "R," and "S" are caused by passage of the impulse through the ventricles, and the hump "T" is caused by return of the membrane potential in the ventricular muscle fibers to its normal resting level at the end of contraction.

When cardiac abnormalities are caused by various diseases, the electrocardiogram often becomes changed from the normal. Figure 11-10B shows the effect when some of the ventricular muscle has been damaged. In this record the portion of the electrocardiogram between the S and T waves is depressed. This results from abnormal leakage of electrical current from the heart between heart beats. It indicates damage to the membranes of the ventricular muscle fibers, which often occurs when one has had an acute heart attack.

Figure 11-10C illustrates the effect seen when one side of the heart is enlarged more than the other. The record shows abnormal enlargement of the S wave and diminishment of the R wave, indicating more current flow from the left side of the heart than from the right. High blood pressure very frequently causes this type of electrocardiogram because of excessive pressure load on the left ventricle.

Figure 11-10D shows the electrocardiogram in a person who has a partially blocked Purkinje system. In this instance, the impulse is transmitted through much of the ventricles by way of slowly conducting cardiac muscle fibers rather than rapidly conducting Purkinje fibers, so that the QRS complex lasts for a prolonged period of time and also develops an abnormal shape.

Figure 11-10E illustrates the effect of blocking the impulse at the A-V bundle. The P waves occur regularly, and the QRST waves also occur regularly but in no definite relationship to the P waves. The atria are beating at the natural rate of rhythm of the S-A node, at 72 beats per minute, while the ventricles have assumed their own rate of rhythm at 38 beats per minute.

Finally, Figure 11-10F illustrates a record, indicated by the arrow, of a *premature beat* of the heart. The only abnormality here is that the impulse occurs too soon after the previous heart beat. This is usually caused by an irritable heart resulting from such factors as too much smoking, too much coffee drinking, or lack of sleep.

A study of Figure 11-10 shows how various abnormalities of heart function can be discovered from electrocardiographic recordings, and why this diagnostic procedure is used in all persons with heart disease.

FUNCTION OF THE HEART VALVES

Referring once again to Figure 11-2, one can see that the four valves of the heart are

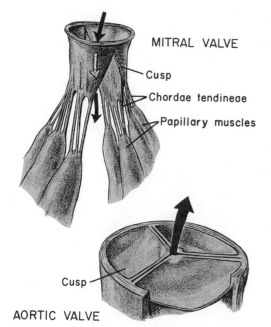

MITRAL VALVE

— Cusp

— Chordae tendineae

— Papillary muscles

Cusp —

AORTIC VALVE

Figure 11–11. Structure of the mitral and aortic valves.

all oriented so that blood can never flow backward but only forward when the heart contracts. The tricuspid valve prevents backflow from the right ventricle into the right atrium, the mitral valve prevents backflow from the left ventricle into the left atrium, and the pulmonary and aortic valves prevent backflow into the right and left ventricles from the pulmonary and systemic arterial systems. These valves have the same functions as valves in any compression pump, for no pump of this type can possibly operate if fluid is allowed to flow backward as well as forward.

Figure 11–11 illustrates in more detail the structures of the valves. The tricuspid and mitral valves (the atrioventricular valves) are similar to each other, having rather expansive filmlike vanes, called "cusps," held in place by special ligaments, the *chordae tendineae,* that extend from the papillary muscles, as illustrated in the upper panel of the figure. The papillary muscles contract at the same time that the ventricles contract, which keeps these valves from bulging backward through the mitral opening when the ventricles pump blood.

The *pulmonary* and *aortic valves* also are similar to each other but quite different from the mitral and tricuspid valves. They have no chordae tendineae and no papillary muscles but instead have very strong cup-shaped cusps that open for forward flow of blood and

close for backflow. The probable reason for the differences between these valves and the atrioventricular valves is that blood must flow with great ease from the atria into the ventricles because the atria do not pump with much force. This requires very easily movable, filmy-type valves. The aortic and pulmonary valves do not have to function with such extreme ease because of the great force of contraction of the ventricles; this allows the valves to be of simpler but of stronger construction than the A-V valves.

THE HEART SOUNDS

When one listens with a stethoscope to the beating heart he normally hears two sounds that are aptly described as "lub, dub; lub, dub; lub, dub." The "lub" is called the *first heart sound* and the "dub" the *second heart sound.* The first sound is caused by closure of the A-V valves when the ventricles contract, and the "dub" sound is caused by closure of the aortic and pulmonary valves at the end of contraction. This relationship to the cycle of heart beat is illustrated in Figure 11–12 by a *phonocardiogram,* which is a graphic representation of the heart sounds. When the ventricles contract, the increasing pressures in the two ventricles force the vanes of the A-V valves closed. The sudden stoppage of backflow from the ventricles into the atria creates vibration of the blood and of the heart walls; these vibrations are transmitted through the chest to be heard as the first sound, the "lub" sound. Immediately after the ventricles have discharged their blood into the arterial systems, the subsequent ventricular relaxation allows blood to begin flowing backward from the arteries toward the ventricles, thereby causing the aortic and pulmonary valves to close suddenly. This also sets up vibrations, this time in the blood and walls of the arteries as well as the ventricles. These vibrations are transmitted to the chest wall, causing the "dub" sound.

VALVULAR HEART DISEASE

The most frequent cause of valvular heart disease is rheumatic fever, a disease that results from an immune reaction to toxin secreted by streptococcic bacteria as follows: The acute phase of the rheumatic fever usually occurs 2 to 4 weeks after a person has had a streptococcic sore throat, scarlet fever, a

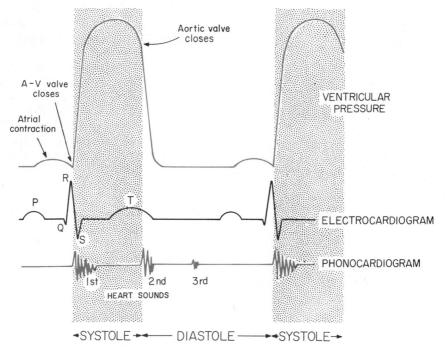

Figure 11–12. Relationship of ventricular pressure to the electrocardiogram and phonocardiogram during the cardiac cycle, showing also the periods of systole and diastole.

streptococcic ear infection, or some other streptococcic infection. Antibodies formed against the streptococcus toxin attack the valves, causing small, cauliflowerlike growths on their edges, and also erode the valves as well as cause ingrowth of fibrous tissue. The valve sometimes is eaten away completely or becomes so constricted and hardened by the fibrous tissue that it cannot close. Obviously, such a valve is then merely a constricted hole rather than a valve at all.

VALVULAR STENOSIS. Sometimes in rheumatic fever the valve openings are so greatly narrowed by fibrous scar tissue that blood flows through the opening only with much difficulty. This is called *stenosis*. If the aortic valve becomes stenosed, blood dams up in the left ventricle. If the mitral valve becomes stenosed, blood dams up in the left atrium and lungs. Likewise, stenosis of the pulmonary or tricuspid valves causes blood to dam up in the right ventricle or systemic circulation respectively.

VALVULAR REGURGITATION. Often the valves do not become stenotic but instead are eroded so much that they cannot close. As a result, blood leaks backward through the valve that should be stopping this backward flow; this effect is called *regurgitation*. For instance, in aortic regurgitation much of the blood pumped into the aorta by the left ventricle returns to the ventricle at the end of heart contraction rather than flowing on through the systemic circulation. Likewise, failure of the mitral valve to close when the left ventricle contracts allows blood to flow backward into the left atrium as well as forward into the aorta. Thus, leaking valves are as disastrous to cardiac function as are narrowed valvular openings. Often a valve is both leaky and narrowed, decreasing the effectiveness of cardiac pumping in two ways.

PULMONARY EDEMA. The valve most frequently affected by rheumatic fever is the mitral valve, though the aortic valve is affected almost as often. On the other hand, the valves of the right heart are rarely damaged severely. Severe damage to the mitral valve leads to excessive damming of blood in the left atrium and lungs. Also, aortic valvular disease causes blood to dam up in the left ventricle, the left atrium, and the lungs. Therefore, following damage to either of these valves, blood often engorges the lungs, causing fluid to leak from the pulmonary capillaries into the pulmonary tissues and alveoli, resulting in severe *pulmonary edema* and often drowning the patient in his own fluid. This will be explained in detail in Chapter 15 in the discussion of heart failure.

THE CARDIAC CYCLE

Now that the pumping action of the heart, the rhythmicity of the heart, and the function of the heart valves have been described, it is possible to synthesize this information into a sequence of events called the *cardiac cycle*. Figure 11–12 illustrates the major events of this cycle, showing curves from above downward for (1) pressure changes in a ventricle, (2) the electrocardiogram, and (3) the phonocardiogram. The cardiac cycle rightfully begins with initiation of the rhythmic impulse in the S-A node. Then transmission of the impulse through the heart causes the muscle fibers to contract. Thus, as shown in Figure 11–12, the P wave of the electrocardiogram occurs immediately before onset of the atrial pressure wave caused by atrial contraction. Approximately 0.16 second after the P wave begins, the electrical impulse has completed its passage through the atria, the A-V node, and the A-V bundle. Then it begins to spread rapidly over the ventricles, causing the QRS wave of the electrocardiogram and stimulating the ventricular muscle to contract. The rising ventricular pressure closes the mitral and tricuspid valves, thereby generating the first heart sound, and it opens the aortic and pulmonary valves. The ventricles remain contracted approximately 0.3 second and then relax. During the process of relaxation, ions retransfer through the fiber membranes to reestablish the normal negative electrical charge inside the cardiac muscle fibers. This causes the T wave of the electro-cardiogram. Immediately after the ventricular muscle relaxes, a small amount of blood flows backward from the arteries toward the ventricles, closing the aortic and pulmonary valves, which elicits the second heart sound. Following ventricular relaxation, no further contraction occurs until a new electrical impulse is initiated in the S-A node.

SYSTOLE AND DIASTOLE

The period during the cardiac cycle when the ventricles are contracting is called *systole*, and the period of relaxation is called *diastole*. A clinician examining the heart can note the periods of systole and diastole either from the electrocardiogram or from the heart sounds; systole begins with the QRS wave and ends with the T wave, or it begins with the first heart sound and ends with the second sound. Diastole, on the other hand, begins with the T wave and ends with the QRS wave, or it begins with the second heart sound and ends with the first heart sound.

Sometimes it is quite important to distinguish between systole and diastole. This is particularly true when one is studying valvular disorders or abnormal openings between the two sides of the heart. For instance, leakage of the aortic or the pulmonary valve causes a "swishing" sound (a *murmur*) during diastole. On the other hand, a murmur caused by leakage of an A-V valve occurs during systole, because that is the period when these valves leak if they are abnormal.

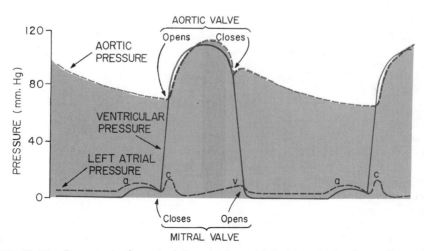

Figure 11–13.   Pressures in the aorta, left ventricle, and left atrium during the cardiac cycle.

PRESSURE CHANGES DURING THE CARDIAC CYCLE

Figure 11–13 shows the changes in pressure in the left atrium, left ventricle, and aorta during a typical cardiac cycle. During diastole, the left atrial pressure is a little higher than that of the left ventricle. This obviously causes blood to flow at this time from the left atrium into the left ventricle. Toward the end of diastole, contraction of the left atrium elevates the left atrial pressure to an even higher level and forces an extra quantity of blood into the ventricle. Then, suddenly, the left ventricle contracts, the mitral valve closes, and the ventricular pressure rises rapidly. When this pressure rises higher than that in the aorta, the aortic valve opens and blood flows into the aorta during the entire remainder of systole. When the ventricle relaxes, the ventricular pressure falls precipitously, allowing a slight backflow of blood that immediately closes the aortic valve. Throughout diastole the aortic pressure remains high because a large quantity of blood has been stored in the very distensible arteries during systole. This blood runs off slowly through the capillaries back to the right atrium, allowing the aortic pressure to fall from a peak during systole of approximately 120 mm. Hg down to a minimum of approximately 80 mm. Hg by the end of diastole. Therefore, the normal systemic arterial blood pressure is said to be 120/80, meaning by this 120 mm. Hg *systolic pressure* and 80 mm. Hg *diastolic pressure*.

THE LAW OF THE HEART

The amount of blood pumped by the heart is normally determined by the amount of blood flowing from the veins into the right atrium. This principle is frequently called the "law of the heart." That is, the heart is simply an automaton that continues to pump all of the time, and whenever blood enters the right atrium it is pumped on through the heart. Of course, there is a maximum rate at which the heart can pump, for which reason, to be completely accurate, the law of the heart is more correctly stated as follows: *Within physiologic limits, the heart pumps all of the blood that flows into it, and it does so without significant damming of blood in the veins.* Thus, the heart is analogous to a sump pump, because any time fluid enters the chamber of a sump pump, it is pumped out of the chamber immediately. In the case of the heart, whenever blood enters one of the atria, it is immediately pumped on into a ventricle and then on into the arteries.

Cardiac muscle has a special characteristic that gives the heart this ability to pump varying amounts of blood in response to changing rates of venous inflow. When cardiac muscle is stretched beyond its normal length, it contracts with greater force than when it is not stretched. Therefore, when only a small quantity of blood enters the heart, the muscle fibers are not stretched greatly, and the force of contraction is weak. On the other hand, if large quantities of blood enter, the heart chambers dilate greatly, the muscle fibers stretch, and the force of contraction becomes very great. As a result, the increased quantity of blood returning to the heart is pumped on through.

The law of the heart often fails when the heart has been damaged, for then even normal quantities of blood returning to the heart are more than the heart can cope with. As a consequence, blood begins to dam up in the veins of either the lungs or the systemic circulatory system. In this case the heart is said to be *failing*. The subject of heart failure is so very important that it will be discussed in detail in Chapter 15.

The Heart-Lung Preparation. One method frequently used in the physiology laboratory to demonstrate the automatic ability of the heart to pump blood is the heart-lung preparation illustrated in Figure 11–14. In this preparation, blood flow from the heart is channeled from the aorta, through an external system and then back into the right atrium, rather than through the animal's body. Artificial respiration is supplied to the lungs to keep the blood appropriately oxygenated, and nutrient in the form of glucose is added to the blood. The heart will continue to beat for many hours, with the external flow system taking the place of the systemic circulation. One can vary the resistance of the external circuit by tightening or loosening a screw clamp. He can measure the blood flow with a flowmeter, and he can change the amount of blood flowing into the left atrium by raising or lowering the venous reservoir.

Several principles that demonstrate the independent function of the heart can be illustrated beautifully with the heart-lung preparation. First, it can be shown that, within physiologic limits, raising or lowering the ve-

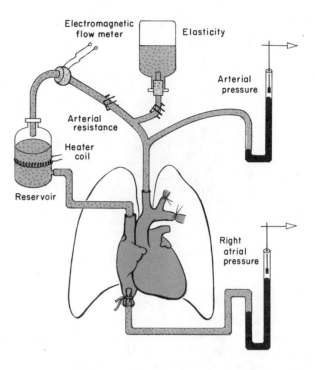

Figure 11–14.   The heart-lung preparation.

nous reservoir is the main factor that determines how much blood will be pumped. In other words, *the greater the input pressure forcing blood from the veins into the heart, the greater the volume of blood pumped,* which is another expression of the "law of the heart."

Second, it can be demonstrated that changing the resistance of the external circuit within normal limits *does not* greatly affect the amount of blood pumped. However, the pressure in the aorta increases approximately in proportion to the increase in resistance. That is, *the amount of resistance to blood flow through the circulatory system changes arterial pressure greatly, but hardly affects the amount of blood pumped by the heart* unless the resistance becomes so great that the heart simply cannot pump with enough force to overcome it.

A third interesting effect that can be demonstrated is that caused by changing the distensibility of the arterial system. This can be done by connecting an air bottle to the arterial system. Blood flows into the bottle and compresses the air when the pressure rises during systole, but during diastole the compressed air forces blood back into the arteries, thereby helping to maintain a relatively high level of arterial pressure even between heart beats. Therefore, with the bottle in the system, the arterial pressure pulsates far less than when it is out of the system. This effect illustrates

that *distensibility of the arterial walls is very important to smooth out pulsations in the arterial pressure.*

Finally, the heart-lung preparation can be used to demonstrate the effect of temperature, drugs, or abnormal blood constituents on the heart. For instance, *increasing the temperature of the blood 10° F. increases the heart rate approximately 100 per cent.* Also, increasing the amount of *calcium* in the blood makes the heart contract with increased vigor, while increasing the amount of *potassium* decreases the vigor of heart contraction. Finally, when the nutrients of the blood fall to low values, addition of glucose or other nutrients can greatly enhance the function of the heart.

NERVOUS CONTROL OF THE HEART

Though the heart has its own intrinsic control systems and can continue to operate without any nervous influences, the efficacy of heart action can be changed greatly by regulatory impulses from the central nervous system. The nervous system is connected with the heart through two different sets of nerves, the *parasympathetic nerves* and the *sympathetic nerves.* These will be discussed in detail in Chapter 26 in connection with function of the nervous system. The connections of the parasympathetic (vagi) and sympathetic

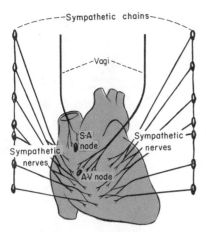

Figure 11–15.   Innervation of the heart.

nutrition to the heart muscle. These effects can be summarized by saying that *sympathetic stimulation increases the activity of the heart as a pump,* sometimes increasing the ability to pump blood as much as 100 per cent. This effect is necessary when a person is subjected to stressful situations such as exercise, disease, excessive heat, and other conditions that demand rapid blood flow through the circulatory system. Therefore, the sympathetic effects on the heart are a standby mechanism held in readiness to make the heart beat with extreme vigor when necessary.

nerves with the heart are shown in Figure 11–15.

Parasympathetic Stimulation. Stimulation of the parasympathetic nerves causes the following four effects on the heart: (1) decreased rate of heart beat, (2) decreased force of contraction of the atrial muscle, (3) delayed conduction of impulses through the A-V node, which lengthens the delay period between atrial and ventricular contraction, and (4) decreased blood flow through the coronary blood vessels, which are the vessels that supply nutrition to the heart muscle itself. All of these effects may be summarized by saying that *parasympathetic stimulation decreases all activities of the heart.* Usually, heart function is reduced by the parasympathetics during periods of rest; this allows the heart to rest at the same time that the remainder of the body is resting. This preserves the resources of the heart; without such periods of rest the heart undoubtedly would wear out at a much earlier age than it normally does.

Sympathetic Stimulation. Stimulation of the sympathetic nerves has essentially the opposite effects on the heart: (1) increased heart rate, (2) increased vigor of cardiac contraction, and (3) increased blood flow through the coronary blood vessels to supply increased

REFERENCES

Bishop, V. S., Peterson, D. F., and Horowitz, L. D.: Factors influencing cardiac performance. *Intern. Rev. Physiol., 9*:239, 1976.

Brutsaert, D. L., and Sonnenblick, E. H.: Cardiac muscle mechanics in the evaluation of myocardial contractility and pump function: problems, concepts, and directions. *Prog. Cardiovasc. Dis., 16*:337, 1973.

Burch, G. E., and Winsor, T.: A Primer of Electrocardiography. 6th ed. Philadelphia, Lea & Febiger, 1972.

Cranefield, P. F.: The Conduction of the Cardiac Impulse. New York, Futura Publishing Co., Inc., 1975.

Fozzard, H. A.: Heart: excitation-contraction coupling. *Ann. Rev. Physiol., 39*:201, 1977.

Gibbs, C. L.: Cardiac energetics. *Physiol. Rev., 58*:174, 1978.

Guyton, A. C.: Determination of cardiac output by equating venous return curves with cardiac response curves. *Physiol. Rev., 35*:123, 1955.

Guyton, A. C., Jones, C. E., and Coleman, T. G.: Circulatory Physiology: Cardiac Output and Its Regulation. 2nd ed. Philadelphia, W. B. Saunders Company, 1973.

Guyton, R. A., and Daggett, W. M.: The evolution of myocardial infarction: physiological basis for clinical intervention. *Intern. Rev. Physiol., 9*:305, 1976.

Irisawa, H.: Comparative physiology of the cardiac pacemaker mechanism. *Physiol. Rev., 58*:461, 1978.

Langer, G. A., and Brady, A. J.: The Mammalian Myocardium. New York, John Wiley & Sons, Inc., 1974.

Langer, G. A., Frank, J. S., and Brady, A. J.: The myocardium. *Intern. Rev. Physiol., 9*:191, 1976.

Starling, E. H.: The Linacre Lecture on the Law of the Heart. London, Longmans Green & Company, 1918.

Sugimoto, T., Allison, J. L., and Guyton, A. C.: Effect of maximal work load on cardiac function. *Jap. Heart J., 14*:146, 1973.

QUESTIONS

1. Why is it important that cardiac muscle have a syncytial structure?
2. What causes automatic rhythmicity of cardiac muscle?
3. What are the differences between cardiac muscle contraction and skeletal muscle contraction?

4. Explain why the sinoatrial node is normally the pacemaker of the heart, while at times some other area of the heart might become the pacemaker.
5. How does the Purkinje system increase the effectiveness of the heart as a pump?
6. Explain the circus movement and how it can cause flutter or fibrillation of the heart.
7. In the electrocardiogram, what does the "P" wave represent? The "QRS" wave? The "T" wave?
8. What is the structural difference between the A-V valves and the pulmonary and aortic valves, and what is an explanation of this difference?
9. What causes each of the heart sounds?
10. In the cardiac cycle, what is the relationship between left ventricular pressure and aortic pressure?
11. Explain the "law of the heart" and its significance.
12. What are the effects of sympathetic and parasympathetic stimulation of the heart?

12 BLOOD FLOW THROUGH THE SYSTEMIC CIRCULATION AND ITS REGULATION

All of the circulation besides the heart and the pulmonary circulation is called the *systemic circulation*. The blood flowing through this part of the circulation provides nutrition to the tissues, transport of excreta away from the tissues, cleansing of the blood as it passes through the kidneys, absorption of nutrients from the gastrointestinal tract, and mixing of all the fluids of the body, as explained in Chapter 1. The rate of blood flow to each respective tissue is almost exactly the amount required to provide adequate function, no more, no less. The purpose of the present chapter, therefore, is to describe, first, the basic principles of blood flow through the circulation and, second, the mechanisms that control the blood flow to each respective tissue in proportion to its needs.

HEMODYNAMICS

The study of the physical principles that govern blood flow through the vessels and the heart is known as *hemodynamics*. The heart forces blood into the aorta, distending it and creating pressure within it. This pressure then pushes the blood through the arteries, arterioles, capillaries, venules, and veins, finally back to the heart. As long as the animal remains alive, this flow of blood around the continuous circuit never ceases.

The small arteries, arterioles, capillaries, venules, and small veins have such small diameters that blood flows through them with considerable difficulty. In other words, the vessels are said to offer *resistance* to blood flow. Obviously, the smaller the vessel the greater is the resistance, and the larger the vessel the less the resistance.

In essence, the discussion in the present chapter centers on the effects of pressure and resistance on blood flow.

BLOOD FLOW AND CARDIAC OUTPUT

The amount of blood pumped by the heart when a person is at rest is approximately 5 liters per minute. This is called the *cardiac output,* and it can increase to as much as 20 to 30 liters per minute during the most extreme exercise, or it can decrease following severe hemorrhage to as low as 1.5 liters per minute without causing immediate death.

Blood flow to the different parts of the body during rest is given in Table 12–1. One will note that the brain receives approximately 14 per cent of the total blood flow, the kidneys 22 per cent, the liver 27 per cent, and the muscles, which comprise almost half of the body, only 15 per cent. However, during exercise quite a different picture develops, for then essentially all of the increase in blood flow occurs in the muscles, this increasing as much as 15-fold during very intense exercise and then representing as much as 75 per cent of the total blood flow.

Methods for Measuring Blood Flow. Many different methods have been used for measuring blood flow, most of which require cutting the blood vessel and then allowing the blood to flow through some physical device that measures the rate of flow.

Table 12-1. **Blood Flow to Different Organs and Tissues Under Basal Conditions***

	PER CENT	ML./MIN.
Brain	14	700
Heart	4	200
Bronchi	2	100
Kidneys	22	1100
Liver	27	1350
Portal	(21)	(1050)
Arterial	(6)	(300)
Muscle	15	750
Bone	5	250
Skin (cool weather)	6	300
Thyroid gland	1	50
Adrenal glands	0.5	25
Other tissues	3.5	175
Total	100.0	5000

*Based mainly on data compiled by Dr. L. A. Sapirstein.

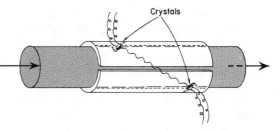

Figure 12-2. Basic construction of the ultrasonic flowmeter probe.

However, there is much advantage in using types of flowmeters that do not require opening the vessels. Two such flowmeters are the *electromagnetic flowmeter* and the *ultrasonic flowmeter*.

THE ELECTROMAGNETIC FLOWMETER. Figure 12-1 illustrates the electromagnetic flowmeter. The electromagnet creates a very strong magnetic field that passes through the blood vessel. As the blood flows through this field, it "cuts" the magnetic lines of force and develops an electrical potential at right angles to these lines of force, and this potential is proportional to the rate of flow. Two electrodes touching the surface of the vessel at right angles to the magnet, therefore, can pick up this potential and transmit it to an appropriate electronic recording apparatus. Obviously, the blood vessel does not have to be opened when this type of flowmeter is used. Also, the response of the instrument is so rapid that it can measure even very transient changes in flow that take place in as little as

0.01 second. Because of these advantages, this type of flowmeter has become very widely used in physiologic studies of the circulation.

THE ULTRASONIC FLOWMETER. Another type of flowmeter that can be applied to the outside of a vessel, and that has many of the same advantages as the electromagnetic flowmeter, is the ultrasonic flowmeter illustrated in Figure 12-2. A minute piezoelectric crystal is mounted in the wall of each half of the device; one of these crystals transmits sound diagonally along the vessel and the other receives the sound. An electronic apparatus alternates the direction of sound transmission several hundred times per second, transmitting first downstream and then upstream. Sound waves travel downstream at greater velocity than upstream. An appropriate electronic apparatus measures the difference between these two velocities, which is a measure of blood flow. This type of flowmeter is called *ultrasonic* because the sound frequency that is used is usually somewhere between 100,000 and 4,000,000 cycles per second, which is far beyond the sound range that the human ear can hear.

Velocity of Blood Flow in Different Portions of the Circulation. The term *blood flow* means the actual *quantity* of blood flowing through a vessel or group of vessels in a given period of time. In contradistinction to this, the *velocity* of blood flow means the distance that the blood travels along a vessel in a given period of time.

If the quantity of blood flowing through a vessel remains constant, the velocity of blood flow obviously decreases as the size of the vessel increases. The aorta as it leaves the heart has a cross-sectional area of approximately 2.5 sq. cm. Then it branches into the large arteries, the small arteries, and the capillaries with a portion of the aortic blood flowing into each of these vessels. The total

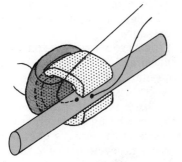

Figure 12-1. The electromagnetic flowmeter.

cross-sectional area of the branching vessels is considerably greater than that of the aorta; in the capillaries, for instance, it is 1000 times that in the aorta. As a consequence, the velocity of blood flow is greatest in the aorta and least in the capillaries, where it is only 1/1000 that in the aorta. Numerically, the velocities are approximately the following: aorta, 30 cm. per second; arterioles, 1.5 cm. per second; capillaries, 0.3 mm. per second; venules, 3 mm. per second; and venae cavae, 8 cm. per second.

TRANSIT TIME FOR BLOOD IN THE CAPILLARIES. The velocity of blood flow in the capillaries is particularly significant because it is here that oxygen, other nutrients, and excreta pass back and forth between the blood and the tissue spaces. The length of the average capillary is about 0.5 to 1 mm. Therefore, at a velocity of blood flow of 0.3 mm. per second, the length of time that blood remains in each capillary averages 1 to 3 seconds. Despite this short time, an extremely large proportion of the substances in the blood can transfer through the capillary membrane into or out of the tissues. This signifies the extreme capability of the capillaries as an exchange site between the blood and the tissue fluids. Without this exchange in the capillaries, all the other hemodynamic functions of the circulation would be useless.

BLOOD PRESSURE

The pressure in a blood vessel is the force that the blood exerts against the walls of the vessel. This force distends the vessel because all blood vessels are distensible, the veins eight times as much as the arteries. Pressure also makes blood attempt to leave a vessel by any available opening, which means that the normally high pressure in the arteries forces blood through the small arteries, then through the capillaries, and finally into the veins. The importance of blood pressure, then, is that it is the force that makes the blood flow through the circulation.

Measurement of Blood Pressure — the Mercury Manometer. Historically, the standard device for measuring blood pressure has been the mercury manometer shown in Figure 12–3. The tube is attached to a blood vessel by means of a cannula that enters the vessel. The blood pressure in the vessel pushes on the fluid in the tube, and this presses downward on the left column of mer-

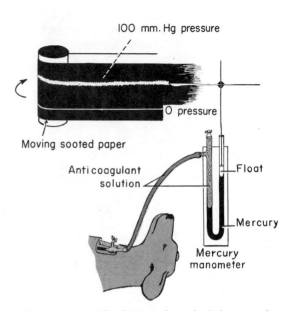

Figure 12-3. The historical method for recording arterial pressure using a mercury manometer.

cury, which then forces the right column upward. A float resting on the mercury rises and falls with the level of the mercury, and a recording arm connected to the float records the pressure on a moving paper. If the level of mercury in the right column is 100 millimeters above the level in the left column, the blood pressure is said to be 100 millimeters of mercury (mm. Hg). If the level to the right is 200 millimeters above that on the left, then the pressure is 200 mm. Hg.

Because of the convenience of the mercury manometer for measuring pressure, almost all pressures of the circulatory system are expressed in millimeters of mercury. However, pressure can also be expressed as water pressure because one can equally well connect an artery to a column of water and determine how high the water rises. Mercury weighs 13.6 times as much as water, which means that the level of water in a manometer will rise this many times as high as the level of mercury in a manometer. Therefore, 13.6 mm. or 1.36 cm. of water pressure equals 1 mm. Hg pressure.

HIGH FIDELITY METHODS FOR RECORDING BLOOD PRESSURE. Unfortunately, the mercury in the mercury manometer has so much *inertia* that it cannot rise and fall rapidly. For this reason, this type of manometer, though excellent for recording *mean* pressure, cannot respond to pressure *changes* that occur more rapidly than one cycle every 2 to 3 seconds. Also, it is a cumbersome apparatus to use.

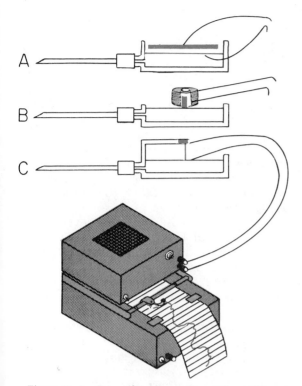

Figure 12–4.   Principles of operation of three different electronic transducers for recording rapidly changing blood pressure.

Therefore, in modern physiology a type of pressure recorder such as one of those illustrated in Figure 12–4 is used. Each of the three "transducers" illustrated in this figure converts pressure into an electrical signal that is then recorded on a high-speed recorder. A syringe needle is connected to each transducer; this needle is inserted directly into the vessel whose pressure is to be measured, and the pressure is transmitted into a chamber bounded on one side by a thin membrane. In transducer A this membrane forms one plate of a variable capacitor. As the pressure rises, the membrane moves closer to the other plate and changes the electrical *capacitance,* which is detected and recorded by an electronic instrument. In transducer B a small iron slug on the membrane moves upward into a coil and changes the coil's *inductance,* which likewise is detected and recorded electronically. In transducer C the membrane is connected to a thin, stretched wire. As the membrane moves, the length of the wire changes, which changes its *resistance,* once again allowing the pressure to be recorded by an electronic recorder.

Relationship of Pressure to Blood Flow.   When the blood pressure is high at one end of a vessel and low at the other end, blood will attempt to flow from the high toward the low pressure area because the two forces are unequal. *The rate of blood flow is directly proportional to the difference between the pressures.* Figure 12–5 illustrates this principle, showing that the flow from the spout becomes greater in proportion to the water level in the chamber.

It should be noted that it is not the actual pressure in a vessel that determines how rapidly blood will flow, but the *difference* between the pressures at the two ends of the vessel. For instance, if the pressure at one end of the vessel is 100 mm. Hg and 0 mm. Hg at the other end, the pressure difference is 100 mm. Hg. If the pressure at the two ends of this same vessel are suddenly changed to 5000 mm. Hg at one end and 4950 mm. Hg at the other end, the pressure difference now is only 50 mm. Hg, and the flow will be reduced to one-half as much (provided the vessel's diameter does not become changed by the pressure).

RESISTANCE TO BLOOD FLOW

Resistance is essentially the same as friction, for it is friction between the blood and the vessel walls that creates impediment to flow. The amount of resistance is dependent on the length of the vessel, the diameter of the vessel, and the viscosity of the blood.

Effect of Vessel Length on Resistance to Blood Flow.   The longer a vessel, the greater

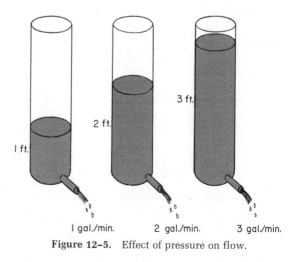

Figure 12–5.   Effect of pressure on flow.

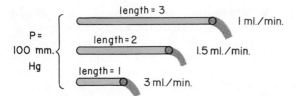

Figure 12–6. Effect of vessel length on flow.

the vascular surface along which the blood must flow, and consequently the greater the friction between the blood and vessel wall. For this reason, the *resistance to blood flow is directly proportional to the length of the vessel.*

This principle is illustrated by the three vessels in Figure 12–6, each of which has a pressure head of 100 mm. Hg at its inlet. The vessel with a length of 1 has an output of 3 ml. per minute while the one with a length of 3 but of exactly the same diameter has only one-third as much output, illustrating the inverse relationship between the length of the vessel and the amount of fluid that will flow when all other factors remain constant.

Effect of Vessel Diameter on Resistance to Blood Flow. Fluid flowing through a vessel is retarded mainly along the walls. For this reason, the velocity of blood flow in the middle of a vessel is very great while the velocity along the surface is very low, and the larger the vessel the more rapidly can the central portion of blood flow. Because of this effect, the blood flow through a vessel — that is, the total quantity of blood that passes through the vessel each minute — increases very markedly as the diameter of the vessel increases. In fact, if all other factors stay constant, the blood flow through the vessel is directly proportional to the *fourth power* of the diameter. Thus, in Figure 12–7 three vessels of the same length but with diameters of 1, 2, and 4 are shown. Note that even these small changes in diameter increase the flow 256-fold. Because of this extreme dependence of flow on diameter, very slight changes in vascular diameter can affect tremendously blood flow to different vascular regions. The diameter of most vessels can change approxi-

mately fourfold, and can therefore change the blood flow as much as 256-fold.

Effect of Viscosity on Resistance to Blood Flow. The more viscous the fluid attempting to flow through a vessel, the greater the friction with the wall, and, consequently, the greater the resistance. This effect is illustrated in Figure 12–8, which shows water, plasma, and normal blood attempting to go through spouts of equal dimensions. In each instance the amount of fluid flowing through the spout is 100 ml. per minute, but, because of different viscosities of the three fluids, the pressure required to force each through the spout is also different. The level of fluid in the vertical tube, which is an indicator of the pressure required to cause the flow, is 1 cm. when water is flowing, 1.5 cm. for plasma, and 3.5 cm. for blood. In other words, the *relative* viscosoties of water, plasma, and blood are 1, 1.5, and 3.5.

The most important factor governing blood viscosity is the concentration of red blood cells. The viscosity of normal blood is about 3.5 times that of water. However, when the red blood cell concentration falls to one-half normal, the viscosity becomes only 2 times that of water, and when the concentration rises to approximately 2 times normal, the viscosity can increase to as high as 20 times that of water. For these reasons, the flow of blood through the blood vessels of anemic persons — that is, persons who have low concentrations of red blood cells — is extremely rapid, while blood flow is very sluggish in persons who have excess red blood cells (polycythemia).

INTERRELATIONSHIPS BETWEEN PRESSURE, FLOW, AND RESISTANCE

It is obvious from the preceding discussions that pressure and resistance oppose each other in affecting blood flow, pressure tending to increase flow and resistance tending to decrease flow. One can state this relationship mathematically by the following formula:

$$\text{Blood flow} = \frac{\text{Pressure}}{\text{Resistance}}$$

The same formula can be expressed in two other algebraic forms:

$$\text{Pressure} = \text{Blood flow} \times \text{Resistance}$$

$$\text{Resistance} = \frac{\text{Pressure}}{\text{Blood flow}}$$

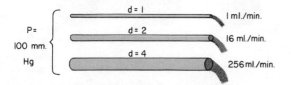

Figure 12–7. Effect of vessel diameter on flow.

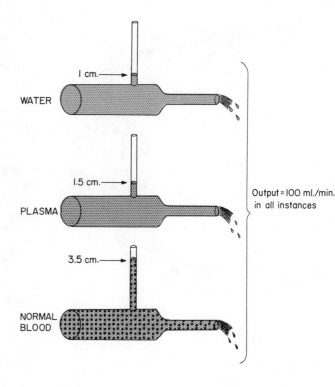

Figure 12–8. Effect of viscosity on flow.

WATER — 1 cm.

PLASMA — 1.5 cm.

NORMAL BLOOD — 3.5 cm.

Output = 100 ml./min. in all instances

These formulas are basic in almost all hemodynamic studies of the circulatory system, and therefore must be thoroughly understood prior to any attempt to analyze the operation of the circulation.

Poiseuille's Law.   By inserting the various factors that affect resistance into the preceding formulas, one can derive still another formula known as Poiseuille's law:

$$\text{Blood flow} = \frac{\text{Pressure} \times (\text{Diameter})^4}{\text{Length} \times \text{Viscosity}}$$

This formula expresses the ability of blood to flow through any given vessel, showing that the rate of blood flow is directly proportional to the pressure difference between the two ends of the vessel, directly proportional to the fourth power of the vessel diameter, and inversely proportional to the vessel length and blood viscosity.

REGULATION OF BLOOD FLOW THROUGH THE TISSUES

THE ARTERIOLES AS THE MAJOR REGULATORS OF BLOOD FLOW

The rate of flow through each respective tissue is controlled mainly by the arterioles.

There are three separate reasons for this:

First, approximately one-half of all the resistance to blood flow in the systemic circulation is in the arterioles, which can be seen by carefully studying figure 12–9. This figure shows the minuteness of the flow channel through the arteriole, which gives this vessel its high resistance. Therefore, a slight change in arteriolar diameter can change the total resistance to flow through any given tissue far more than can a similar change in diameter in any other vessel.

Second, the arterioles have a very strong muscular wall constructed in such a manner that the diameter of an arteriole can change as much as three- to fivefold. Remembering once again that resistance is inversely proportional to the *fourth power* of a vessel's diameter, it immediately becomes obvious that the resistance to blood flow through the arterioles can be changed as much as several hundred- or even a thousandfold by simply relaxing or contracting the smooth muscle walls.

Third, the smooth muscle walls of the arterioles respond to two different types of stimuli that regulate blood flow. First, they respond to the local needs of the tissues, increasing blood flow when the supply of nutrients to the tissues falls too low and decreasing the flow when the nutrient supply becomes too great. This mechanism is called

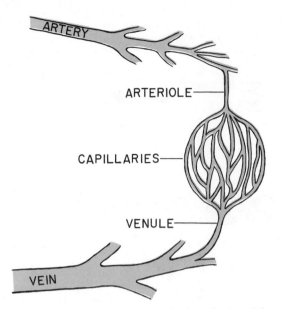

Figure 12–9. Demonstration of the minute size of the arteriole relative to the remainder of the vascular system, which explains its tremendous resistance.

autoregulation. Second, autonomic nerve signals, particularly *sympathetic signals,* have a profound effect on the degree of contraction of the arterioles.

AUTOREGULATION OF BLOOD FLOW – ROLE OF OXYGEN

The term autoregulation means automatic adjustment of blood flow in each tissue to the need of that tissue. In most instances the need of the tissue is nutrition, but in a few instances other factors that depend on blood flow are needed even more than nutrition. For instance, in the kidneys, the need to excrete end-products of metabolism and electrolytes is prepotent, and the concentrations of these substances in the blood play a major role in controlling renal blood flow. In the brain it is essential that carbon dioxide concentration remain very constant because reactivity of the brain cells increases and decreases with the amount of carbon dioxide; in this tissue it is primarily the need to remove carbon dioxide that determines the rate of blood flow.

However, in most tissues it is the need of the tissue for the nutrient *oxygen* that seems to be the most powerful stimulus for autoregulation. For instance, if a tissue becomes more active than usual, the need for oxygen might increase as much as five- to tenfold,

and the blood flow automatically increases to help supply this extra oxygen.

Mechanism by Which Oxygen Deficiency Causes Arteriolar Dilatation. Even though we know that blood flow through most tissues is controlled in proportion to their need for oxygen, we still do not know the exact mechanism by which oxygen need can affect the degree of vascular constriction. There are two basic theories for arteriolar vasodilatation in response to oxygen need in the tissues. One of these, the "oxygen demand" theory, is illustrated in Figure 12–10. This shows that the tissue cells lie in close proximity to the arterioles and that they are in competition with the arterioles for the oxygen, which is represented by the small dots. It is assumed that the smooth muscle of the arteriolar wall requires oxygen to contract. Should the oxygen supply diminish, the tissue cells would deplete the oxygen; therefore, the vessels would dilate simply because of failure to receive adequate oxygen to maintain contraction. And if the tissue should utilize an increased quantity of oxygen, the amount of oxygen available to the arterioles would again decrease, this too causing increased blood flow. The resulting increase in blood flow would then increase the tissue oxygen concentration back toward a normal level.

Another theory is that oxygen lack causes *vasodilator substances* to be formed in the tissue. That is, decreased availability of oxygen

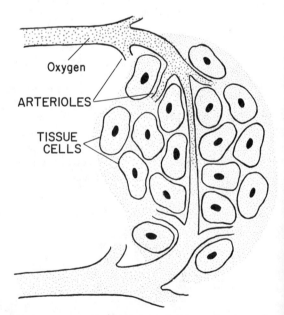

Figure 12–10. Postulated mechanism of autoregulation.

to the metabolic systems of the cells supposedly causes them to release a chemical substance or substances that then act directly on the blood vessels to cause active vasodilatation. The most likely vasodilator is the substance adenosine, which is always released from very active or hypoxic cells. However, some other suggested vasodilator substances have been histamine, lactic acid, carbon dioxide, and potassium ions, all of which are known to be released from hyperactive tissues and all of which are known to cause at least some dilatation of the arterioles.

Long-term Autoregulation. If the needs of the tissues for oxygen and other nutrients are more than the acute autoregulatory mechanism can supply, a long-term autoregulatory mechanism, requiring several weeks to develop, can cause still more increase in blood flow. This mechanism causes the actual number of blood vessels to increase and also causes enlargement of the existing vessels. For instance, when a blood vessel such as a coronary vessel of the heart has become partially occluded, new vessels grow into the area of poor blood supply, returning the blood supply toward normal after a few weeks. Also, if a person ascends to a high altitude and remains there for many weeks, he gradually develops enlarged vessels as well as an increased number of blood vessels in all his tissues, thereby supplying the tissues with increased amounts of oxygen. Conversely, when he comes back to a lower altitude, some of the vessels actually disappear after several more weeks, thus illustrating that this long-term vascular phenomenon helps to autoregulate blood flow in response to the tissues' needs.

To state this mechanism another way, in the long run, most tissues of the body try to adjust their vascular supplies — their actual numbers of vessels — to their needs for nutrients.

NERVOUS CONTROL OF BLOOD FLOW

All the arteries, arterioles, and veins of the systemic circulation are supplied by nerves from the sympathetic nervous system, as illustrated in Figure 12–11. Stimulation of the sympathetic nerves causes especially the arterioles and veins, and the large arteries to a lesser extent, to constrict.

Vasoconstrictor Tone of the Systemic Vessels. The sympathetic vasoconstrictor nerves normally transmit a continual stream

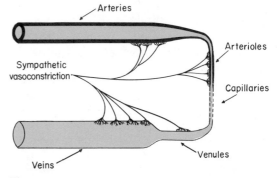

Figure 12–11. Innervation of the peripheral vascular system, showing especially the abundant innervation of the arterioles.

of impulses to the blood vessels, maintaining the vessels in a moderate state of constriction all of the time. This is called *vasomotor tone.* Then, when the sympathetic nervous system is required to constrict vessels more than their normal state of constriction, it does so by increasing the number of sympathetic impulses above normal. And the mechanism by which the sympathetic nervous system dilates the vessels is to decrease the impulses to less than normal. In this manner the sympathetic system can cause both vasoconstriction and vasodilatation. Therefore, special nerves are not required to cause vasodilatation in most parts of the circulation.

Importance of Nervous Control of the Blood Vessels. In contrast to the role of autoregulation, nervous control of the blood vessels is not generally concerned with regulating the delivery of nutrients to the tissues. Instead, nervous control is concerned with overall distribution of blood flow to major sections of the body. For instance, when one exercises, the muscles require such an extreme increase in blood flow that the heart often cannot pump an adequate quantity of blood both through the dilated muscle vessels and at the same time through such areas as the skin, the kidneys, and the gastrointestinal tract, areas which normally receive three-fourths of the total cardiac output. Therefore, during exercise, sympathetic impulses cause vasoconstriction in these areas while the muscle vessels become dilated mainly because of the local muscle autoregulation vasodilator mechanism. As a result, a major portion of the total blood flow shifts to the muscles, thereby allowing much more muscular activity than could otherwise occur.

The sympathetic nervous system can also shift large amounts of blood flow to help regulate body temperature. When the body temperature rises too high, the sympathetic nervous system dilates the arterioles of the skin, increasing the flow of warm blood to the skin; this promotes heat loss until the temperature returns to normal. On the other hand, when the body temperature falls below normal, sympathetic-induced vasoconstriction occurs in the skin vessels, skin blood flow decreases, heat loss becomes less, and body temperature rises back toward normal.

A third instance of important nervous regulation of blood flow occurs when the circulatory system has been damaged so greatly that the heart pumps an insufficient quantity of blood to supply all portions of the body. Under these conditions sympathetic stimulation causes vasoconstriction in the less vital areas such as the skin and gut, while organs such as the brain and heart continue to receive adequate flow.

And, finally, nervous control of the arterioles plays a very important role in reflex regulation of arterial pressure. This effect is so important that it will be discussed in detail in Chapter 14. Briefly, when the arterial pressure begins to fall as a result of hemorrhage or cardiac damage or for any other cause, pressure sensitive detectors in the walls of several large arteries, called *baroreceptors,* detect the falling pressure and transmit signals to the brain. The brain in turn reflexly transmits nerve signals back to the heart to increase heart activity and to the arterioles throughout the body to cause vasoconstriction, thereby elevating the arterial pressure back toward normal.

DISTRIBUTION OF BLOOD VOLUME IN THE BODY

Figure 12–12 shows the approximate percentages of the total blood volume in the different portions of the circulatory system. It is evident from this figure that about three-fourths of the blood is in the systemic circulation and one-fourth in the heart and pulmonary circulation. Also, a far higher proportion is in the veins than in the arteries. For instance, the arteries of the systemic circulation contain only 13 per cent of the total blood, while the veins, venules, and venous sinuses contain approximately 50 per cent. The arterioles and capillaries, despite their extreme importance to the cirulation, contain only a few per cent of the blood.

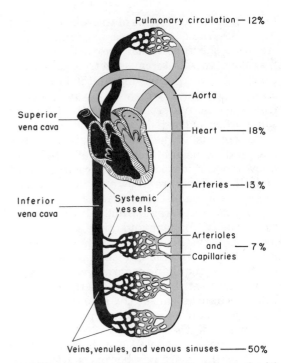

Figure 12–12. Distribution of blood volume in the different portions of the circulatory system.

THE BLOOD PRESSURE – THE RESERVE SUPPLY OF BLOOD

When the total blood volume falls low enough that the vessels are no longer adequately filled, blood cannot circulate normally through the tissues. For this reason, it is important to have an extra supply of blood. The entire venous system acts as a *blood reservoir,* for the veins exhibit a plastic quality whereby their walls can distend and contract in response to the amount of blood available in the circulation. Also, the veins are supplied with nerves from the sympathetic nervous system so that any time the tissues begin to suffer for lack of blood flow, nervous reflexes cause a large number of sympathetic impulses to pass to the veins, constricting them and translocating blood into the heart and other vessels. It is mainly this contractile and expansile quality of the venous system that protects the circulation against the disastrous effects of blood loss.

Certain portions of the venous system are especially important for storing blood:

First, the *large veins of the abdominal region* are particularly distensible, and, therefore, normally hold a tremendous amount of extra blood. Yet they can contract when the blood is needed elsewhere in the circulation.

Second, the *venous sinuses of the liver* can expand and contract many times over, so that the liver under certain circumstances may hold as much as a liter and a half of blood, but at other times only a few hundred milliliters.

Third, the *spleen* normally contains approximately 200 ml. of blood, but can expand to hold as much as 1 liter, or can contract to hold as little as 50 ml.

The *venous plexuses of the skin* are a fourth important blood reservoir. Normally the blood in these plexuses is used to regulate the heat of the body—the more rapidly blood flows through them, the greater is the loss of heat. However, when the vital organs need extra blood flow, the sympathetic nervous system can markedly contract the skin's venous plexuses, transferring the stored blood into the main stream of flow.

A fifth blood reservoir is the *pulmonary vessels*. Approximately 12 per cent of the blood is normally in the pulmonary circulation, but much of this can be displaced into other portions of the circulation without impairing the function of the lungs. Therefore, the lungs also act as a major source of blood in times of need.

REFERENCES

Cooper, M. D., and Lawton, A. R., III: The physiology of the giraffe. *Sci. Amer., 231(5)*:96, 1974.

Dobrin, P. B.: Mechanical properties of arteries. *Physiol. Rev., 58*:397, 1978.

Green, H. D.: Circulation: physical principles. *In* Glasser, O. (ed.): Medical Physics. Chicago, Year Book Medical Publishers, 1944.

Guyton, A. C., and Jones, C. E.: Central venous pressure: physiological significance and clinical implications. *Am. Heart J., 86*:434, 1973.

Guyton, A. C., Ross, J. M., Carrier, O., Jr., and Walker, J. R.: Evidence for tissue oxygen demand as the major factor causing autoregulation. *Circ. Res., 14*:60, 1964.

Guyton, A. C., Cowley, A. W., Jr., Young, D. B., Coleman, T. G., Hall, J. E., and DeClue, J. W.: Integration and control of circulatory function. *Intern. Rev. Physiol., 9*:341, 1976.

Intaglietta, M., Tompkins, W. R., and Richardson, D. R.: Velocity pressure, flow, and elastic properties in microvessels of cat omentum. *Am. J. Physiol., 221*:922, 1971.

Lundgren, O., and Jodal, M.: Regional blood flow. *Ann. Rev. Physiol., 37*:395, 1975.

Mancia, G., Lorenz, R. R., and Shepherd, J. T.: Reflex control of circulation by heart and lungs. *Intern. Rev. Physiol., 9*:111, 1976.

McDonald, D. A.: Blood Flow in Arteries, 2nd ed. Baltimore, The Williams & Wilkins Company, 1974.

Patel, D. J., Vaishnav, R. N. Gow, B. S., and Kot, P. A.: Hemodynamics. *Ann. Rev. Physiol., 36*:125, 1974.

Roach, M. R.: Biophysical analyses of blood vessel walls and blood flow. *Ann. Rev. Physiol., 39*:51, 1977.

Talbot, L., and Berger, S. A.: Fluid-mechanical aspects of the human circulation. *Am. Sci., 62*:671, 1974.

QUESTIONS

1. Describe several methods for measuring blood flow.
2. Describe the methods for measuring blood pressure.
3. How is resistance to blood flow calculated?
4. Why does the diameter of the blood vessel play such an important role in determining vascular resistance?
5. If the blood pressure is decreased to one-half normal but the diameter of the blood vessel is increased at the same time to two times normal, what happens to the blood flow through the vessel?
6. Explain why the arterioles are the major regulators of blood flow.
7. What is meant by autoregulation of blood flow? Explain its significance as well as the role of oxygen in its mechanism.
8. Discuss the role of the sympathetic nervous system in the control of local blood flow throughout the body.
9. Name the major blood reservoirs in the body.

13 SPECIAL AREAS OF THE CIRCULATORY SYSTEM

✗ THE PULMONARY CIRCULATION

The pulmonary circulation is the vascular system of the lungs. Its function is to transport blood through the pulmonary capillaries where oxygen is absorbed into the blood from the alveolar air and where carbon dioxide is excreted from the blood into the alveoli.

The physiologic anatomy of the pulmonary circulation, illustrated in Figure 13–1, is very simple. The right ventricle pumps blood into the pulmonary artery. From there, the blood flows through the pulmonary capillaries into the pulmonary veins and finally into the left atrium. Because all portions of the lungs have

the same function — that is, to aerate the blood — the arrangement of the vessels is essentially the same in all areas of the pulmonary circulation.

The pulmonary capillaries abut against the epithelial linings of the alveoli, and the membrane between the blood in the capillaries and the air in the alveoli is called the *pulmonary membrane,* the total thickness of which is only 0.4 micron (micrometer). The pores of this membrane are large enough for oxygen and carbon dioxide to diffuse through them with ease. Yet they are small enough that fluid does not leak from the blood into the alveoli.

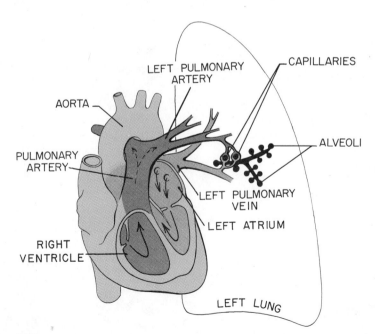

Figure 13–1. The pulmonary circulation and its relation to the heart.

FLOW OF BLOOD THROUGH THE LUNGS

Because blood flows around a continuous circuit in the body, the same amount of blood must flow through the lungs as through the systemic circulation. The vessels of the lungs are very expansile so that whenever the amount of blood entering the lungs increases, the pulmonary vessels are automatically stretched to allow the more rapid flow demanded in the pulmonary circulation. This allows facile transport of blood through the lungs under widely varying conditions.

Regulation of Blood Flow Through the Lungs. Since all portions of the lungs perform essentially the same function, there is no need for extensive regulation of the pulmonary blood vessels. However, one feature of blood flow regulation in the lungs is different from that in the remainder of the body. It is quite important for blood to flow only through those portions of the lungs that are adequately aerated and not through portions not aerated. Research studies in the past few years have shown that a low concentration of oxygen lasting longer than a minute in any area of the lungs automatically causes the vessels there to constrict. Therefore, when the bronchi to an area of the lungs become blocked, the alveolar oxygen is rapidly used up, and as a consequence, the vessels constrict, thereby forcing the blood to flow instead through other areas of the lungs that are still aerated.

PULMONARY VASCULAR PRESSURES

The resistance to blood flow in the pulmonary circulation is so little that the mean pulmonary arterial pressure averages only 13 mm. Hg, and the systolic pressure (the highest pressure during contraction of the ventricle) is only 22 mm. Hg, while the diastolic (the lowest pressure between contractions) is 8 mm. Hg. The mean pulmonary arterial pressure is approximately one-seventh the mean systemic arterial pressure, which is about 100 mm. Hg, as is discussed in the following chapter.

The pulmonary venous pressure is about 2 mm. Hg. This is also the pressure in the left atrium. The pulmonary capillary pressure has never been measured, but it must be greater than the pulmonary venous pressure and less than the pulmonary arterial pressure — that is, somewhere between 2 and 13 mm. Hg. Indirect measurements of pulmonary capillary pressure indicate that it is probably about 7 mm. Hg.

The mean pressure gradient from the pulmonary artery to the pulmonary vein is 13 mm. Hg minus 2 mm. Hg, or 11 mm. Hg. This compares with a pressure gradient in the systemic circulation of approximately 100 mm. Hg. To express this differently, the same amount of blood flows through the lungs as through the systemic circulation with about one-ninth the force propelling it. This means also that the total resistance of the pulmonary circulation is normally only one-ninth the total resistance of the systemic circulation.

Pressure in the Pulmonary Artery During Exercise. When one exercises, the amount of blood flowing through the systemic circulation sometimes increases to as high as six times normal, which means that the amount flowing through the lungs must also increase this much. Still, the pulmonary arterial pressure does not increase more than 50 per cent because the vessels of the lungs passively enlarge to accommodate whatever amount of blood needs to flow. This obviously keeps the right heart from becoming overloaded during exercise.

Pulmonary Capillary Dynamics. Only about one-tenth of the blood of the lungs is actually in the capillaries at any one time, and the blood flows through them so rapidly that it remains in the capillaries at most only 1 to 2 seconds. During exercise, when blood flow is very rapid, the blood may remain in the pulmonary capillaries only one-quarter to one-half second. Yet, because the pulmonary membrane is extremely permeable to oxygen and carbon dioxide, even during this very minute period of time the blood can become almost completely oxygenated and can excrete the necessary amount of carbon dioxide.

CAPILLARY MECHANISM FOR MAINTAINING DRY LUNGS. Another extremely important feature of pulmonary capillary dynamics is the mechanism for keeping the air spaces of the lungs dry. The plasma of the blood has a colloid osmotic pressure of 28 mm. Hg, which causes a continual tendency for fluid to be absorbed into the capillaries. The mean pulmonary capillary pressure, which tends to filter fluid out of the capillaries, is only 7 mm. Hg. Therefore, the tendency for fluid to be absorbed into the capillaries is about 21 mm. Hg greater than for fluid to filter out of the capillaries. Thus, the plasma colloid osmotic pressure continually promotes fluid absorption from the tissues and alveoli of the lungs, and if a small amount of fluid does enter the alveoli, it is usually absorbed within a few minutes.

ABNORMALITIES OF THE PULMONARY CIRCULATION

Pulmonary congestion means too much blood and fluid in the lungs. It results from too high a pressure in the pulmonary circulation, which in turn is caused most frequently by failure of the left heart to pump blood adequately from the lungs into the systemic circulation. Once the pulmonary capillary pressure rises above the plasma colloid osmotic pressure, about 28 mm. Hg, fluid leaks out of the capillaries extremely rapidly, and the alveoli no longer remain dry. Then, severe *pulmonary edema* (excess fluid in the lungs) develops, sometimes so rapidly that it causes death in 30 minutes to 2 hours.

Atelectasis means collapse of a lung or part of a lung. Occasionally some abnormality blocks one or more of the bronchi, cutting off air flow. When this happens, the air in the blocked alveoli is absorbed into the blood within a few hours, causing the lung to collapse. Simultaneously, the lack of oxygen in the collapsed alveoli causes vascular constriction, a mechanism that was discussed earlier in the chapter, and, in addition, the collapse of the lung tissue itself causes external compression and kinking of the vessels. As a result of the combined effects, the blood flow through the collapsed lung decreases to about one-fifth its previous value, which causes most of the blood to pass through the uncollapsed areas of the lungs where it can receive oxygen.

Surgical removal of large portions of the lungs causes excessive blood flow through the remaining lung tissue. Ordinarily blood can flow through each lung four times as rapidly as normal before the pulmonary arterial pressure begins to rise significantly. Therefore, a person can lose one whole lung, which increases the blood flow through the opposite lung about twofold, without causing any serious disability. Yet, if this person attempts too heavy exercise, he rapidly reaches the upper limit that the blood can flow before pulmonary hypertension and failure of the right heart develop. These same effects also occur in *emphysema*, a disease of the lungs in which large portions of the lung have been destroyed, most often the result of smoking.

EFFECT OF CONGENITAL HEART DISEASE ON THE PULMONARY CIRCULATION

Congenital heart disease means an abnormality of the heart or of closely allied blood vessels that is present at birth. Many types of congenital heart disease affect the pulmonary

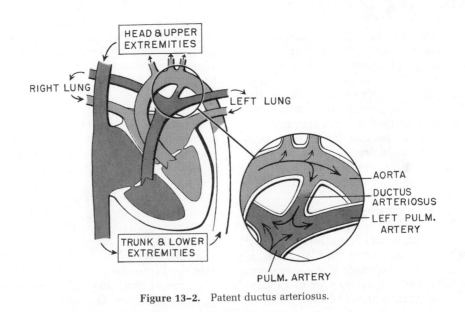

HEAD & UPPER EXTREMITIES

RIGHT LUNG

LEFT LUNG

AORTA

DUCTUS ARTERIOSUS

LEFT PULM. ARTERY

TRUNK & LOWER EXTREMITIES

PULM. ARTERY

Figure 13–2. Patent ductus arteriosus.

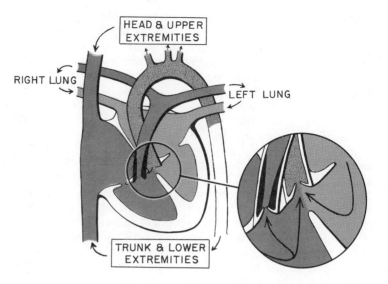

Figure 13-3.   Tetralogy of Fallot.

circulation; two of these, called *patent ductus arteriosus* and *tetralogy of Fallot,* are illustrated in Figures 13–2 and 13–3.

Patent Ductus Arteriosus.   During normal fetal life, blood by-passes the lungs, flowing from the pulmonary artery through the *ductus arteriosus* into the aorta. Immediately after birth, expansion of the lungs dilates the pulmonary vessels so that blood can then flow through them with ease. As a result, pulmonary arterial pressure falls, causing blood now to reverse its flow in the ductus, and to flow backward from the aorta into the pulmonary artery. Also, the blood now flowing through the lungs becomes oxygenated before it enters the aorta. The backward flow of *oxygenated* blood through the ductus arteriosus causes contraction of the walls of the ductus and causes it to become occluded during the first few days or weeks of life in almost all babies.

However, in about 1 out of every 2000 babies the ductus never closes — that is, a *patent ductus arteriosus,* an open vessel connecting the aorta and the pulmonary artery, remains present throughout life (see Fig. 13–2). As the child grows older and his heart becomes stronger, the pressure in the aorta becomes much greater than the pressure in the pulmonary artery, forcing a tremendous quantity of blood from the aorta into the pulmonary artery. This blood flows a second time through the lungs, then through the left heart and through the lungs again, and around and around this circuit several times before finally passing into the systemic circulation. In other words, greatly excessive amounts of blood are pumped through the lungs. In early life this is a great strain on the heart but does not cause extensive damage to the lungs. However, in early adulthood the very high pressure in the pulmonary vessels eventually promotes fibrosis of the vessels and congestion of the lungs. Finally, the person has such difficulty aerating his blood that this alone may cause his death.

Tetralogy of Fallot — "Blue Baby."   Figure 13–3 shows the congenital abnormality called *tetralogy of Fallot.* This disease presents four abnormalities of the heart, including (1) a hole in the septum (partition) between the right and left ventricles, (2) a greatly constricted pulmonary artery, (3) displacement of the aorta to the right, and (4) a hypertrophied right ventricular muscle. Very little blood passes from the right ventricle through the constricted pulmonary artery. Instead, most of it is forced through the hole in the septum and then into the aorta. Thus, most of the blood pumped by the right ventricle by-passes the lungs and is not aerated. Because nonaerated blood is bluish, the whole body assumes this hue. Therefore, tetralogy of Fallot is one of the congenital anomalies of the heart that cause the so-called *blue baby.*

Other Congenital Heart Abnormalities.   Other often-encountered congenital abnormalities of the heart are: (1) an abnormal opening through the ventricular septum, (2) an abnormal opening through the atrial septum, (3) constriction or occlusion of the descending aorta, so that blood must flow to the lower part of the body through many small vessels in the body wall rather than through the aorta, (4) abnormal connection of one of the pulmonary veins to the right atrium instead of to the left atrium, and (5) partial constriction of the pulmonary artery, resulting in high right ventricular pressure and overloading of the right ventricle.

Surgical Treatment.   In the past few years many advances have been made in the surgical repair of cardiac abnormalities. For instance, any patient with a patent ductus arteriosus can be treated very simply by placing a tight ligature around the ductus. Treatment of tetralogy of Fallot is more complicated but is usually accomplished reasonably satisfactorily by open heart surgery in which extensive repair of the heart defects is performed inside the heart itself. Surgical procedures have also been devised to transpose blood vessels when necessary and to close abnormal openings in the ventricular or atrial septum.

✗ THE CORONARY CIRCULATION

Contrary to what might be expected, blood does not pass directly from the chambers of the heart into the heart muscle. Instead, the heart has its own special blood supply, which is shown in Figure 13–4. Two small arteries called the *coronary arteries* originate from the aorta immediately above the aortic valve. Blood flows from these vessels through branches over the *outer surface* of the heart, then into smaller arteries and capillaries in the cardiac muscle, and finally to the right atrium mainly through a very large vein called the *coronary sinus*.

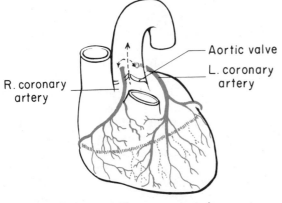

Figure 13–4.   The coronary arteries.

R. coronary artery

Aortic valve

L. coronary artery

FLOW OF BLOOD THROUGH THE CORONARY VESSELS

The amount of blood flowing through the coronaries each minute is approximately 225 ml., which amounts to about 4 to 5 per cent of all the blood pumped by the heart.

In contrast to the flow of blood in other portions of the circulation, flow in the coronaries is greater during diastole than during systole. The reason for this is that systolic contraction of the heart muscle compresses the vessels in the wall of the heart. This momentarily partially or totally occludes these vessels during systole, though this occlusion is not present during diastole. Therefore, the rate of coronary flow is determined more by the diastolic level of arterial pressure than by either the mean pressure or the systolic pressure. This has importance in some circulatory diseases in which diastolic pressure is especially low, such as in aortic regurgitation in which the aortic valve is destroyed, thus allowing blood to empty during diastole from the aorta backward into the left ventricle.

Regulation of Coronary Blood Flow. Blood flow through the coronaries is regulated mainly in proportion to the need of the heart for nutrition. The most important method of regulation is the *autoregulation mechanism,* which was discussed in the previous chapter. That is, when the metabolic need for nutrients — especially for oxygen — becomes greater than the supply, the arterioles automatically dilate. As a result, blood flow through the coronaries increases until the level of nutrition matches the degree of cardiac activity.

The sympathetic nervous system also helps slightly to control blood flow through the coronaries. Increased sympathetic stimulation increases coronary flow and decreased sympathetic stimulation decreases flow. Physiologists are not sure, however, how much of this is a direct effect of sympathetic stimulation and how much is an indirect effect as follows: Sympathetic stimulation tremendously increases the degree of activity of the cardiac muscle, and this in turn increases coronary blood flow by the autoregulation mechanism just discussed. Therefore, some physiologists believe that most of the increased coronary flow following sympathetic stimulation is simply another manifestation of autoregulation.

At any rate, nervous control of coronary blood flow is relatively unimportant in comparison with the autoregulatory control of flow.

CORONARY OCCLUSION AND THE HEART ATTACK

Coronary occlusion means blockage of one of the coronary vessels. When this occurs, that portion of the heart formerly supplied by the vessel loses its nutrition and stops contracting. And the person is said to have had a *heart attack*. Obviously, when a major portion of the heart is affected, the heart's pumping capability becomes so depressed that it can no longer pump sufficient blood to maintain life.

Coronary occlusion is the cause of approximately one-third of all deaths. Occlusion may occur rapidly or slowly, and in almost all instances the cause is *atherosclerosis*. Therefore, to explain why people have heart attacks, it is first necessary to discuss the cause and the effects of atherosclerosis.

Atherosclerosis. Atherosclerosis is a disease in which fatty deposits containing mainly cholesterol but also smaller amounts of phospholipids and neutral fat appear in the walls of the arteries. Gradually, fibrous tissue grows around or into the fatty deposits, and calcium from the body fluids frequently combines with the fat to form solid calcium compounds, eventually evolving into hard plates similar to bone. Thus, in the early stage of atherosclerosis, only fatty deposits occur in the walls of the vessels, but in the late stage the vessels may become extremely fibrotic and constricted, or even bony hard in consistency, a condition called *arteriosclerosis* or "hardening of the arteries."

Acute Coronary Occlusion. Atherosclerosis often causes acute occlusion of the coronary vessels in one of two ways: First, the fatty deposits may break through the inside surface of the vessel and cause clotting of the blood, which in turn plugs the vessel. Second, the protruding fat may actually break away from its original deposit and flow into and occlude a smaller vessel. In either instance, the resulting coronary occlusion can cause sudden diminishment of cardiac function, which is the familiar condition known as the "heart attack." If the occlusion is extremely severe, it will cause immediate death; if it affects only a small vessel, the heart is often only temporarily weakened for 1 to 3 months until some of the vascular connections with neighboring blood vessels enlarge to supply

new blood flow. After a heart attack, the heart sometimes returns to full strength, though more often it remains weakened for life.

Slowly Developing Occlusion. A slowly developing type of coronary occlusion results from fibrous tissue invading the atherosclerotic fatty deposits. The fibrous tissue contracts slowly and continually, gradually narrowing the vessel. This occurs to some extent in all people as they become old, though it occurs in some at much earlier ages and much more severely than in others. Actually, coronary occlusion eventually occurs in everyone who lives long enough. Indeed, almost 100 per cent of hearts examined after death in persons beyond the age of 60, whether the persons had ever had obvious heart attacks or not, show at least one fairly large coronary vessel totally occluded. Ordinarily, accessory blood vessels will have grown into the area previously supplied by the occluded vessel so that the person may never have known of his coronary disease. The point of this discussion is that all older people develop at least some degree of coronary heart disease, which means that it is important for each person as he becomes older to prevent major stresses on his heart.

Relationship of Diet and Smoking to Atherosclerosis and Heart Attacks. Recent studies have shown that atherosclerosis, and consequently heart attacks as well, can be minimized by two dietary measures: First, the total intake of food should be low enough for the person's weight to remain either normal or even slightly subnormal. Second, the amounts of saturated fat and cholesterol in the diet should be as little as possible. The average American and northwestern European diets contain far more of both saturated fat and cholesterol than those of other sections of the world, and as a result the incidence of heart attacks is far higher than among such people as the Italians, Japanese, and Chinese.

Statistics have also shown that smoking *doubles* the number of deaths from heart disease, but why this occurs is not yet clear — probably because smoking both constricts the coronary arteries and excessively excites the heart at the same time.

THE CEREBRAL CIRCULATION

Figure 13–5 illustrates the arterial system on the base of the brain, showing the two *vertebral arteries* passing upward along the

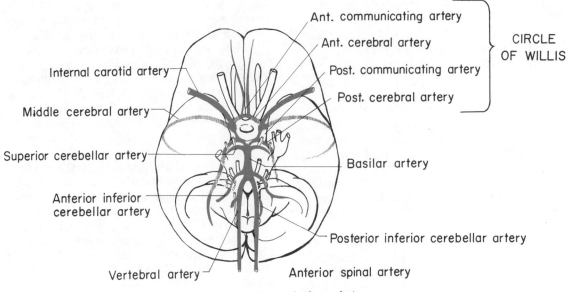

Figure 13–5. The cerebral circulation.

medulla and the two *internal carotid arteries* approaching the brain on its ventral surface. Communicating branches between these four major arteries protect the brain against damage should any one or even two of them become occluded. However, the smaller arteries that spread over the surface of the brain do not have many communications from one to another. Therefore, occlusion of any one of them usually causes destruction of the respective brain tissue.

CEREBRAL BLOOD FLOW AND ITS REGULATION

The total blood flow to the brain averages 650 to 700 ml. per minute, which is about 13 per cent of the blood pumped by the heart.

Autoregulation of Cerebral Blood Flow. Cerebral blood flow is autoregulated better than that in almost any other area of the body except in the kidneys. Even though the arterial pressure might fall to as low as 40 mm. Hg or rise to as high as 200 mm. Hg, the blood flow through the brain still varies no more than a few per cent.

Autoregulation of blood flow in the brain results mainly from a carbon dioxide autoregulation mechanism. The cerebral circulation is different from that of other tissues in that it responds very markedly to carbon dioxide concentration as well as to oxygen concentration. When the carbon dioxide concentration rises above normal, the cerebral vessels dilate and cerebral blood flow automatically increases. This then washes carbon dioxide out of the tissues, bringing the carbon dioxide concentration back toward normal. Conversely, decreased carbon dioxide in the brain decreases the blood flow, which allows the carbon dioxide being formed in the tissues to accumulate and therefore to increase the carbon dioxide concentration toward normal.

The carbon dioxide mechanism is much more powerful in the cerebral circulation than is the oxygen deficiency mechanism of autoregulation (which is far more powerful in most other tissues), and it is mainly because of this carbon dioxide mechanism that cerebral blood flow is autoregulated to an almost exactly constant level all the time. The only exception to this constancy of blood flow occurs when the brain itself becomes overly active. Under these conditions cerebral blood flow can on rare occasions rise to as high as 30 to 50 per cent above normal, but even this increase in flow is a manifestation of autoregulation, the flow increasing in proportion to the increase in carbon dioxide formation in the brain.

The unvarying rate of cerebral blood flow is very advantageous to cerebral function, for neurons must never become excessively excitable nor excessively inhibited if the brain is to function properly, and, obviously, changes in the nutritional status of the neurons could quite easily change their degree of excitability. For instance, if the carbon dioxide concentration in the tissues should rise too high, the tissue fluids would become excessively acidic, which is known to depress neuronal function severely.

VENOUS DYNAMICS IN THE CEREBRAL CIRCULATION

Function of the veins in the cerebral circulation is somewhat different from that of veins in the remainder of the body because the cerebral veins are noncollapsible. They are formed as *sinuses* in the fibrous lining, the *dura mater,* of the cerebral vault, and the sides of the sinuses are so firmly attached to the skull and other tissues that they cannot collapse completely. Also, because the head is in an upright position most of the time, the weight of the venous blood makes it run rapidly downward toward the heart, creating a semivacuum in the cerebral veins. If one of these veins is inadvertently punctured while a person is in the upright position, air is often sucked into the venous system and occasionally may even flow so rapidly to the heart that it blocks the action of the heart valves and kills the person. Because the veins in all other parts of the body are flimsy and collapse very easily whenever suction is applied to them, any sucking action that might occur simply closes the veins rather than sucking air.

CEREBRAL VASCULAR ACCIDENTS – THE "STROKE"

When a blood vessel supplying blood to an important part of the brain suddenly becomes blocked or when a cerebral vessel ruptures, the person is said to have had a *cerebral vascular accident* or a "stroke." This is the cause of death in perhaps 5 to 10 per cent of all people. In about one-fourth of all cases the damage is caused by a blood clot developing on an atherosclerotic fatty deposit in a major cerebral artery. In the other three-fourths, an artery ruptures because of excessively high arterial pressure or because the vessel itself has been weakened by the atherosclerosis disease process. Such hemorrhages into the tissues of the brain often compress the neuronal cells enough to destroy them or at least to stop their functioning. Also, the blood clot developing in the hemorrhagic area extends backward into the artery, blocking further flow

inside the vessel. Therefore, still more brain tissue dies for lack of nutrition. Thus, regardless of whether the cerebral vascular accident is initially caused by simple vascular occlusion or by hemorrhage, the results are essentially the same, that is, destruction of a major part of the brain itself.

The effects of a stroke depend on which vessel is destroyed. Frequently it is the middle cerebral artery supplying the area of the brain that controls muscular function, in which case the opposite side of the body becomes paralyzed. Blockage of the posterior cerebral artery, on the other hand, causes partial blindness. Blockage of arteries in the hindbrain region is likely to destroy nerve tracts connecting the brain with the spinal cord, causing any number of abnormalities such as paralysis, loss of sensation, loss of equilibrium, and so forth.

THE PORTAL CIRCULATORY SYSTEM

The portal circulatory system, shown in Figure 13–6, is composed of the veins leading from the intestines and spleen to the liver. The blood then flows through the liver into the vena cava. The function of this system is to pass the blood from the intestines through the liver prior to its entering the general circulation. On flowing through the minute sinuses of the liver the blood is "processed" by the liver in two ways: First, the blood comes into contact with special reticuloendothelial phagocytic cells that line the sinuses, the *Kupffer cells,* that are capable of removing abnormal debris. These cells, therefore, cleanse the intestinal blood before it reenters the general circulation. This is quite important because a few of the literally trillions of bacteria in the gastrointestinal tract make their way into the portal blood each minute. The Kupffer cells are so effective in removing the bacteria that probably not one in a thousand escapes through the liver into the general circulation.

Second, the liver removes various absorbed nutrients from the circulation. For instance, the liver cells normally remove two-thirds of the glucose absorbed into the portal blood from the intestines and as much as one-half of the proteins before the blood ever reaches the general circulation. These nutrients are stored in the liver until they are needed later by the remainder of the body and then are released back into the blood for use else-

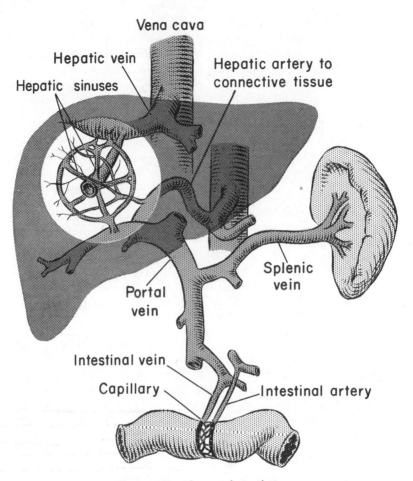

Figure 13–6.   The portal circulation.

where. The storage of glucose and proteins "buffers" the blood concentrations of these nutrients, preventing sudden increases in their concentrations after a meal and also preventing low concentrations during the long intervals between meals.

VASCULAR DYNAMICS OF THE PORTAL SYSTEM

Portal Vascular Pressure, and Ascites. The small vessels in the liver offer a small amount of resistance to blood flow, and this causes the portal venous pressure normally to be about 7 mm. Hg greater than the zero pressure in the vena cava. Ordinarily this elevated portal pressure is of no major concern, but when some abnormality causes the systemic venous pressure to rise, the portal pressure increases a corresponding amount, always remaining about 7 mm. Hg

greater than the systemic venous pressure. Consequently, the portal capillary bed is one of the first to suffer high capillary pressure as a result of excessively high systemic venous pressures, a condition that occurs most frequently when the heart fails and this in turn causes blood to dam up in the veins. The high capillary pressure forces fluid to leak through the capillary walls into the intestinal tissues and peritoneal cavity. In addition, high pressure in the small liver sinuses causes fluid also to leak profusely from the surface of the liver. Therefore, instead of the norm of only a few milliliters of fluid in the abdomen, many liters may accumulate. This accumulation of intra-abdominal fluid is called *ascites*.

Portal Obstruction.   Occasionally the portal vein leading from the intestines and spleen to the liver becomes totally occluded by a large blood clot. And even more frequently the small vessels in the liver become blocked or severely constricted because of alcoholic liver disease called *liver cirrhosis*. In either event, the portal venous pressure rises, which in turn causes high portal capillary pressure, edema of the intestinal walls, enlargement of the spleen, and ascites (free fluid in the abdomen, sometimes as much as 5 gallons!).

FUNCTION OF THE SPLEEN

Reservoir Function. As was pointed out in Chapter 12, the spleen, which is part of the portal system, is a blood reservoir. It is capable of enlarging in some pathologic conditions up to a total volume of more than 1000 ml. or of contracting down to a minimal volume of less than 50 ml. The internal structure of a small segment of the spleen is illustrated in Figure 13–7, which shows several small arteries leading into capillaries in the *pulp* of the spleen and finally connecting with large venous sinuses. The blood flows from the venous sinuses into the large circumferential veins and then back into the general circulation. Much of the blood, including the red blood cells, leaks from the capillaries into the pulp and then squeezes through this area, finally reentering the general circulation through the walls of the venous sinuses. Large numbers of cells are often trapped in the pulp in this manner while the plasma is returned to the circulation. In this way the spleen stores a much higher percentage of red blood cells than plasma, but when it contracts, the cells are forced from the pulp back into the circulation. Therefore, the spleen is

often said to be a *red blood cell reservoir* rather than a general blood reservoir.

In lower animals, sympathetic stimulation causes smooth muscle in the capsule of the spleen to contract, which expels blood both from the splenic pulp and splenic sinuses into the general circulation. In the human being there is very little smooth muscle in the capsule. Therefore, it is believed that sympathetic stimulation causes the spleen of the human being to empty its blood mainly by direct constrictive effects on the vascular walls inside the spleen itself.

Marked splenic contraction occurs during muscular exercise. The extra red blood cells expressed from the spleen into the circulation aid in the transport of oxygen to the active muscles.

Phagocytic Function of the Spleen. Another major function of the spleen is to cleanse the blood. This results from phagocytosis by reticuloendothelial cells that line the splenic sinuses and that are located throughout the splenic pulp.

MUSCLE BLOOD FLOW

Under resting conditions the blood flow through the muscles is only 15 to 20 per cent of the total flow in the body. However, during exercise, muscle blood flow increases markedly in response to the increase in muscle metabolism — that is, in response to the increase in use of nutrients by the muscle cells. During extreme exercise, muscle blood flow often increases 20-fold.

MECHANISM OF BLOOD FLOW REGULATION IN THE MUSCLES

Autoregulation. The principal mechanism by which blood flow is regulated in the muscles is the autoregulation mechanism which was described in the previous chapter. That is, increased muscle metabolism increases the utilization of nutrients from the blood, including oxygen, which in turn has a vasodilator effect on the blood vessels to increase the rate of blood flow. It is possible that the diminished oxygen causes vasodilator substances to be formed in the muscle, but another likely effect is simply that lack of oxygen directly causes the blood vessels to dilate.

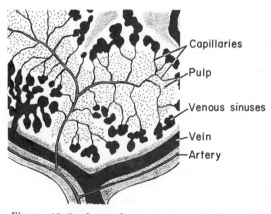

Capillaries

Pulp

Venous sinuses

Vein

Artery

Figure 13–7. Internal organization of the spleen. (Modified from Bloom and Fawcett: A Textbook of Histology, 8th ed.)

Nervous Control of Muscle Blood Flow. In addition to the autoregulation mechanism for blood flow control, the blood vessels of the skeletal muscles are also provided with vascular control by the sympathetic nerves. However, this is a weak mechanism. It does cause constriction of the blood vessels of muscles following severe hemorrhage and thus helps to maintain the arterial pressure. Otherwise, nervous control of the blood vessels is far less important in the muscles than in most other tissues of the body and is of little significance during exercise.

CIRCULATORY READJUSTMENTS DURING EXERCISE

Very strenuous exercise is the most stressful condition encountered by the normal circulatory system. This is true because exercise of large muscle areas of the body at the same time can increase the cardiac output up to a level as great as the heart can pump, sometimes to as high as five to seven times the normal cardiac output level.

Aside from the large decrease in vascular resistance that occurs in exercising muscles, three other major factors are essential for the circulatory system to supply the tremendous blood flow required by the muscles during exercise. These are (1) mass discharge of the sympathetic nervous system throughout the body, with consequent stimulatory effects on the circulation, (2) increase in pumping activity of the heart, and (3) increase in arterial pressure.

Mass Sympathetic Discharge. At the onset of exercise, signals are transmitted not only from the motor area of the brain to the muscles but also simultaneously from the vasomotor center of the brain to initiate mass sympathetic discharge. At the same time, the parasympathetic signals to the heart, which keep the heart rate decreased under normal conditions, are considerably attenuated. Therefore, two major circulatory effects result. First, the heart is stimulated to greatly increased activity. Second, all the blood vessels of the peripheral circulation are strongly contracted except the vessels in the active muscle — these vessels are strongly vasodilated by the local vasodilator effects in the muscles themselves. Thus, the heart is stimulated to provide the increased blood flow required by the muscles, and the blood flow through most nonmuscular areas of the body is temporarily reduced, thereby "lending" this blood supply to the muscles. This extra blood flow to the muscles often makes the difference between life and death to a wild animal that is running for its life.

Fortunately, when the mass sympathetic discharge constricts most of the blood vessels of nonmuscular areas during exercise, two of the organ circulatory systems, the coronary and cerebral systems, are spared this vasoconstrictor effect. This occurs because both of these circulatory areas have very poor vasoconstrictor innervation, fortunately so because both the heart and the brain are just as essential to exercise as are the skeletal muscles themselves.

Increase in Pumping Activity by the Heart. The pumping activity of the heart is heightened in two ways. First, the sympathetic signals to the heart stimulate the heart muscle to contract with increased force and heart rate. Thus, the heart actually becomes a much more effective pump during exercise than during periods of rest. However, greater activity of the heart will not by itself boost the cardiac output. An increased supply of blood entering the heart by way of the veins is also necessary. This is achieved in exercise in the following two ways: (a) The mass sympathetic discharge also constricts most of the veins of the body, which pushes extra amounts of blood toward the heart, increasing the so-called *venous return*. (b) When the muscles contract they squeeze the vessels within the muscle mass, and this too augments the venous return. It will be recalled from the discussion in Chapter 11 that whenever greater amounts of blood flow into the heart, the heart automatically pumps this on into the arteries, a mechanism called the "law of the heart."

Thus, by a combination of nervous stimulation, which heightens the pumping effectiveness of the heart itself, plus a greatly increased venous return, the cardiac output is vastly augmented during heavy exercise. It is this increased cardiac output that supplies the nutrients necessary for continued muscle activity. Indeed, during prolonged exercise, such as during marathon races, the level of exercise that can be maintained (the speed at which the runner can run, for instance) is determined to a great extent by the level of cardiac output that can be achieved by the heart.

Increase in Arterial Pressure. The mass sympathetic discharge that occurs during exercise not only boosts the pumping activity of the heart but also increases the arterial pressure at the same time. This results from a combination of vasoconstriction of the vessels in nonmuscular areas and increased cardiac output by the heart. The rise in pressure can be as little as 20 mm. Hg or as great as 80 mm. Hg, depending on the conditions under which the exercise is performed. The value of the increased pressure is that it further augments the amount of blood that flows through the muscles, partly by forcing more blood through the vessels but chiefly by distending the vessels as well, thus further increasing the flow.

When the arterial pressure fails to rise in exercising animals, the amount of work output that can be achieved may be decreased to as little as two-thirds that which can be achieved when the arterial pressure does rise.

BLOOD FLOW THROUGH THE SKIN

Blood flow through the skin has two functions: first, to regulate the temperature of the body and, second, to supply nutrition to the skin itself.

Relationship of Skin Blood Flow to Body Temperature Regulation. Figure 13–8 illustrates the vascular architecture of the skin, showing especially an extensive venous plexus lying a few millimeters beneath the surface of the skin. The rate of blood flow through this plexus can be altered up or down as much as 50- to 100-fold. When the arteries that supply blood to the skin venous plexus are constricted, the blood flow may be as little as 20 to 50 ml. per minute to the skin of the entire body, whereas, when the arteries are completely dilated, the blood flow to the skin may be as great as 2 to 3 liters per minute.

When the blood flow to the skin is very rapid, the amount of heat transmitted by the blood from the internal structures to the surface of the body is very great, and large amounts of heat are then lost from the skin. Conversely, when the quantity of blood flowing to the skin is slight, the amount of heat lost from the body will also be slight. Special temperature regulatory mechanisms, controlled by the hypothalamus in the brain and acting through the sympathetic nerves, can either vasoconstrict or vasodilate the skin vessels, thereby helping to regulate body temperature. This will be discussed in much more detail in Chapter 33.

Certain areas of the body, such as the hands, feet, and ears, have special muscular walled vessels called *arteriovenous anastomoses* that connect the arteries directly with the venous plexus. When these anastomoses are dilated, blood bypasses the capillary system and flushes extremely rapidly into the plexus. In this way the hands, feet, and ears receive tremendous amounts of blood flow when they are exposed to excessive cold, and the increased blood flow obviously acts to protect these tissues from freezing.

The Nutrient Blood Flow. Ordinarily the blood flow through the skin is some 20 to 30 times that required to supply the necessary amount of nutrition to the skin tissues. Yet, when the skin becomes extremely cold the body temperature mechanisms constrict the skin blood flow so greatly that nutrition can then become impaired. And this also occurs occasionally in constrictive diseases of the vasculature supplying the skin—for instance, in atherosclerosis of arteries leading to the skin, especially in the hands and feet. It is only under such conditions as these that one need be concerned with the skin nutrient vascular supply.

Figure 13–8 illustrates capillary loops that carry blood into all the tufts beneath the skin. The nutritive blood flow is subject to the same local autoregulatory mechanism for blood flow as that found elsewhere in the body. For instance, when a person compresses an area of the skin so hard that the blood flow is completely blocked for a few minutes, the skin flow becomes far greater than normal immediately after release of the pressure, and this automatically makes up for the deficiency in nutrients. Fortunately, the skin can be without blood flow for many hours before serious damage occurs.

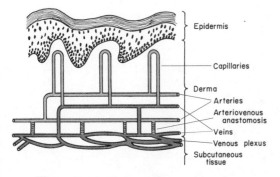

Figure 13–8. The circulation in the skin.

REFERENCES

Bakhle, Y. S., and Vane, J. R.: Pharmacokinetic function of the pulmonary circulation. *Physiol. Rev., 54*:1007, 1974.

Bergofsky, E. H.: Mechanisms underlying vasomotor regulation of regional pulmonary blood flow in normal and disease states. *Am. J. Med., 57*:378, 1974.

Fein, J. M.: Microvascular surgery for stroke. *Sci. Amer., 238(4)*:58, 1978.

Grayson, J.: The gastrointestinal circulation. *In* MTP International Review of Science: Physiology. Vol. 4. Baltimore, University Park Press, 1974, p. 105.

Greenway, C. V., and Stark, R. D.: Hepatic vascular bed. *Physiol. Rev., 51*:23, 1971.

Hughes, J. M. B.: Pulmonary circulatory and fluid balance. *Intern. Rev. Physiol., 14*:135, 1977.

Lassen, N. A.: Control of cerebral circulation in health and disease. *Circ. Res., 34*:749, 1974.

Morse, R. L. (ed.): Exercise and the Heart. Springfield, Ill., Charles C Thomas, Publisher, 1974.

Sparks, H. V., Jr., and Belloni, F. L.: The peripheral circulation: local regulation. *Ann. Rev. Physiol., 40*:67, 1978.

Staub, N. C.: Pulmonary edema. *Physiol. Rev., 54*:678, 1974.

Svanik, J., and Lundgren, O.: Gastrointestinal circulation. *Intern. Rev. Physiol., 12*:1, 1977.

West, J. B.: Blood flow to the lung and gas exchange. *Anesthesiology, 41*:124, 1974.

QUESTIONS

1. Compare the pressures in the pulmonary circulation with those in the systemic circulation.
2. What are the effects of exercise on blood flow through the lungs and on pulmonary artery pressures?
3. Discuss the transfer of oxygen and carbon dioxide through the capillary membrane as well as the capillary mechanism for maintaining dry lungs.
4. What is the major mechanism for regulation of coronary blood flow?
5. What is the cause of coronary occlusion, and what are its effects?
6. What is the major mechanism for regulation of cerebral blood flow, and what is the role of carbon dioxide in this?
7. Discuss the causes of "stroke."
8. Describe the portal circulatory system and explain its relationship to the liver.
9. Discuss the relative roles of autoregulation and nervous regulation of blood flow in muscles.
10. How does the heart increase the cardiac output during exercise?
11. Distinguish between blood flow through the skin for the purpose of controlling body temperature and blood flow through the skin to provide nutrition.

SYSTEMIC ARTERIAL PRESSURE AND HYPERTENSION

14

When the left ventricle contracts, it forces blood into the systemic arteries, thereby creating pressure that drives the blood through the systemic circulation. The *regulation of arterial pressure* is among the most important subjects of circulatory physiology, and the disease *hypertension,* which means high arterial pressure and is commonly called simply "high blood pressure," is one of the most common of all disorders of the body. These two topics receive special emphasis in the present chapter.

PULSATILE ARTERIAL PRESSURE

Instead of pumping a continuous stream, the heart pumps a small quantity of blood with each beat. As a result, the arterial pressure rises during systole but falls during diastole. Figure 14–1 shows graphically this pulsatile pressure under both normal and abnormal conditions.

SYSTOLIC AND DIASTOLIC PRESSURES

The pressure at its highest point during the pressure cycle is called the *systolic pressure,* and the pressure at its lowest point is called the *diastolic pressure.* As shown by the top curve of Figure 14–1, systolic pressure in a normal young adult is approximately 120 mm. Hg. while the diastolic pressure is approximately 80 mm. Hg. The usual method for writing these pressures is 120/80. As another example, if a person's blood pressure is stated to be 210/125 this means that the systolic pressure is 210 mm. Hg and the diastolic pressure 125 mm. Hg.

The systolic and diastolic pressures change with age, as demonstrated graphically in Figure 14–2. In a newborn baby the systolic pressure is about 90 mm. Hg and the diastolic pressure about 55 mm. Hg. The pressure usually reaches 120/80 by young adulthood, and in old age the average is about 150/90.

Measurement of Systolic and Diastolic Pressures. RAPID-RESPONSE DIRECT RECORDING METHODS. The arterial pressure varies so rapidly during the cardiac cycle that special high-speed recorders designed to follow the rapid changes in pressure must be used to measure it. Several electronic transducers, used with appropriate electronic recorders, were described in Chapter 12 and illustrated in Figure 12–4. The basic principle of all these transducers is the following: The vessel from

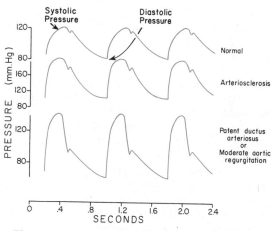

Figure 14–1. Pressure pulse contours in the normal circulation, in arteriosclerosis, in patent ductus arteriosus, and in moderate aortic regurgitation.

161

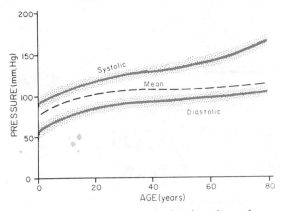

Figure 14–2. Changes in systolic, diastolic, and mean pressures with age. The shaded areas show the normal range.

which the pressure is to be measured is connected by way of a nondistensible tube with a small chamber bounded on one side by a thin elastic membrane. When the pressure rises, the membrane bulges outward, and when the pressure falls, the bulge decreases. The bulge in turn activates an electric sensor, the voltage output from which is amplified and recorded on a moving strip of paper.

Appropriately designed membrane transducers can record pressure changes that occur in less than 0.01 second, which is more than adequate for recording any significant pressure change in the circulatory system. The arterial pressure pulse curves illustrated in Figure 14–1 were recorded using a membrane type of electronic transducer.

INDIRECT MEASUREMENT OF SYSTOLIC AND DIASTOLIC PRESSURES BY THE AUSCULTATORY METHOD. Figure 14–3 illustrates the method usually employed in the doctor's office for measuring systolic and diatolic pressures. An inflatable blood pressure cuff is placed around the upper arm and is connected by a tube to a mercury manometer. When the pressure in the cuff is elevated above that in the brachial artery, the wall of the artery collapses because more pressure is then exerted against the outside of the arterial wall than against the inside by the blood. If the pressure in the cuff is gradually decreased until it falls below systolic pressure, small spurts of blood then begin to jet intermittently through the artery each time the arterial pressure rises to its systolic value, a value now great enough to oppose the outside pressure and to open the vessel intermittently at the peaks of the systolic pressure waves. However, during diastole blood still does not flow through the artery. This intermittent flow of blood causes

vibrations in the arteries of the lower arm that can be heard with a stethoscope as shown in the figure. Therefore, to determine systolic pressure, one simply inflates the cuff until the pressure rises to a high value and then deflates it gradually until the intermittent sounds can be heard. At this instant the pressure recorded by the mercury manometer is a reasonably accurate measure of systolic pressure.

To determine diastolic pressure, the cuff pressure is reduced still more. When the pressure falls below the diastolic value, blood then flows through the artery all of the time and no longer in jetlike bursts. As a result, the vibrations caused by the jetting blood disappear, and sounds can no longer be heard with the stethoscope. The pressure recorded at this instant is a close estimate of the diastolic pressure. The intensity of the sounds heard through the stethoscope at different cuff pressure levels is shown graphically at the top of Figure 14–3.

FACTORS THAT AFFECT PULSE PRESSURE

The arterial *pulse pressure* is the pressure difference between the systolic and diastolic pressures. The higher the systolic pressure and the lower the diastolic pressure, the greater is the pulse pressure, but the more closely these two pressures approach each other, the less becomes the pulse pressure. Thus, when the arterial pressure is normal, at a level of 120/80, the pulse pressure is 40 mm. Hg. Two important factors affect the pulse pressure in the arterial system: These are (1)

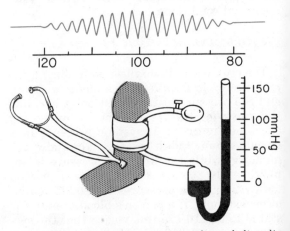

Figure 14–3. Measurement of systolic and diastolic pressures by the auscultatory method.

stroke volume output of the heart and (2) distensibility of the arterial system.

Stroke Volume Output   The stroke volume output is the amount of blood pumped by the heart with each beat. Normally, it is approximately 70 ml., but it can on occasion fall to as low as 10 to 20 ml. or can rise to as high as 200 ml. Obviously, if only a small quantity of blood is pumped into the arterial system with each beat, the pressure will not rise and fall greatly during each cardiac cycle. But if the stroke volume output is great, the pressure will rise and fall tremendously, thereby giving an extremely high pulse pressure. In other words, the pulse pressure is roughly proportional to the stroke volume output.

The slower the rate of the heart, the greater must be the stroke volume output to maintain adequate blood flow through the body. Very slow heart beats are quite often found in well-trained athletes, the slowness actually indicating efficient hearts. The large stroke volume output is accompanied by a high pulse pressure. On the other hand, persons with fever, toxicity, or weakened hearts very frequently have fast heart beats with accompanying small stroke volume outputs. The total amount of blood pumped per minute may be the same as that pumped by the athlete's heart, but because of the small stroke volume output the pulse pressure is very slight.

Distensibility of the Arterial System.   The more distensible the arteries, the greater the quantity of blood that can be compressed into the arterial system by a given amount of pressure. Therefore, each beat of the heart causes far less pressure rise in very distensible arteries than in nondistensible arteries.

Conditions That Cause Abnormal Pulse Pressure.   In old age, the distensibility of the arterial system decreases tremendously because of arteriosclerotic changes in the walls of the vessels. *Arteriosclerosis* is usually the end result of atherosclerosis, which was described in the previous chapter, and arteriosclerosis actually means "hardening of the arterial walls." In many older persons who have arteriosclerosis, the walls even become calcified, bone-hard tubes. As a result, the arterial system then cannot stretch adequately during systole and cannot recoil to a smaller size during diastole. Therefore, pulses of blood entering the arterial tree cause tremendous changes in pressure, as illustrated by the middle curve of Figure 14–1, sometimes causing a pulse pressure of 100 mm. Hg or more.

Aortic regurgitation occurs when the aortic valve has been partially destroyed. After blood has been pumped by the heart into the arterial system during systole, the valve fails to close and as a result much of the blood flows backward into the relaxed left ventricle during diastole, thus causing the diastolic pressure to fall very low — often to zero, as shown by the third curve in Figure 14–1. Consequently, the pulse pressure becomes the greatest possible, sometimes as great as 160 mm. Hg.

TRANSMISSION OF PRESSURE PULSES TO THE SMALLER VESSELS

When the blood pumped by each beat of the heart is suddenly thrust into the aorta, the resulting increase in pressure at first distends only the initial portion of the aorta adjacent to the heart, as shown in Figure 14–4. A short period of time is required for some of this blood to push forward along the arterial tree and build up the pressure further peripherally. This movement of the pressure along the arteries is called *transmission of the pressure pulse.*

Damping of the Pressure Pulses.   The pressure pulse becomes less and less intense as the wave spreads toward the smaller vessels.

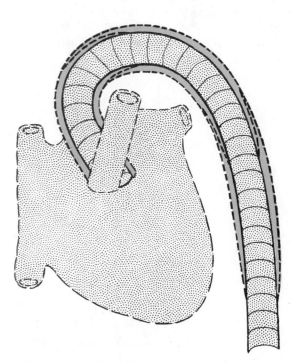

Figure 14–4. Progressive movement of the pressure pulse wave along the aorta.

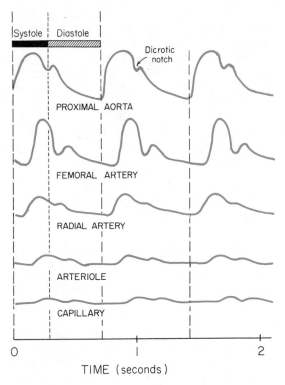

Figure 14–5. Changes in the pulse pressure contour as the pulse wave travels toward the smaller vessels.

This is called *damping* of the pulsations, and it results mainly from (1) the resistance to blood flow in the vessels and (2) the distensibility of the vessels. Figure 14–5 illustrates

this damping of the pulse wave as it spreads peripherally, showing especially the almost complete absence of pulsations in the capillary.

PRESSURES AT DIFFERENT POINTS IN THE SYSTEMIC CIRCULATION

Figure 14–6 shows the pressures in the different vessels from the aorta to the venae cavae. From this chart it is obvious that about one-half of the entire fall in pressure is in the arterioles and about one-fourth in the capillaries. The reason for this is that about one-half of the resistance to blood flow is in the arterioles and about one-fourth in the capillaries.

In the normal person, the pressure in the aorta, though highly pulsatile, averages approximately 100 mm. Hg, and it falls to about 85 mm. Hg at the end of the small arteries. The pressure then falls another 55 mm. Hg in the arterioles down to 30 mm. Hg at the beginning of the capillaries. At the end of the capillaries the pressure normally is about 10 mm. Hg, and it gradually decreases in the venules, small veins, and large veins until it reaches 0 mm. Hg in the right atrium.

MEAN ARTERIAL PRESSURE

The *mean arterial pressure* is the arterial pressure averaged during a complete pressure

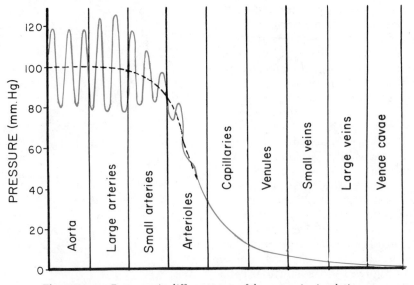

Figure 14–6.   Pressures in different parts of the systemic circulation.

pulse cycle. The mean pressure is not always equal to the average of systolic and diastolic pressures, for during each pressure cycle the pressure usually remains at systolic levels for a shorter time than at diastolic levels. Therefore, when the pressures at all stages of the pressure cycle are averaged, the resulting mean pressure is usually slightly nearer diastolic pressure than systolic pressure.

So far as flow of blood in the circulatory system is concerned, *the mean arterial pressure is much more important than is either systolic or diastolic pressure,* because it is the mean pressure that determines the average rate at which blood will flow through the systemic vessels. For this reason, in most physiologic studies it is not necessary to record the pressure changes throughout the pressure cycle but instead simply to record the mean arterial pressure.

REGULATION OF ARTERIAL PRESSURE

RELATIONSHIP OF MEAN ARTERIAL PRESSURE TO TOTAL PERIPHERAL RESISTANCE AND CARDIAC OUTPUT

Recalling once again the basic formula given in Chapter 12 for the relationship of blood flow to pressure and resistance, one can derive the following formula that is basic to understanding arterial pressure regulation:

Arterial pressure = Cardiac output
× Total peripheral resistance

The *cardiac output* is the rate at which blood is pumped by the heart, and the *total peripheral resistance* is the total resistance of all the vessels in the systemic circulation, from the origin of the aorta back to the veins entering the right atrium. Applying this formula, one can see that either generalized constriction of all the blood vessels or an increase in cardiac output will raise the arterial pressure. Conversely, dilatation of the vessels or a decrease in cardiac output will lower the pressure.

BASIC MECHANISMS OF PRESSURE REGULATION

Under resting conditions, the mean arterial pressure is normally regulated at a level of almost exactly 100 mm. Hg. The body has at least four separate types of arterial pressure regulatory systems which are responsible for maintaining the normal pressure. These are: (1) nervous mechanisms that regulate arterial pressure by controlling the degree of arteriolar constriction, (2) a capillary fluid shift mechanism which regulates arterial pressure by altering the blood volume, (3) a kidney excretory mechanism which also regulates arterial pressure by altering the blood volume, and (4) hormonal mechanisms that regulate either the blood volume or the degree of arteriolar constriction.

Nervous Regulation of Arterial Pressure. In Chapter 13, it was pointed out that almost all of the blood vessels of the body are supplied with sympathetic nerve fibers, which, when stimulated, cause most of the blood vessels to constrict. Constriction of the vessels obviously impedes blood flow and thereby increases the arterial pressure.

The anatomy of the nervous system for control of the circulation is illustrated in Figure 14–7. The insert to the left shows the medullary portion of the lower brain stem in which is located the so-called *vasomotor center* for control of (a) the degree of vasoconstriction of the blood vessels, (b) the degree of vasodilatation, and (c) the heart rate (cardioacceleration and cardioinhibition).

SYMPATHETIC EFFECTS ON THE CIRCULATION. The vasomotor center controls the circulation mainly through the sympathetic nervous system, which is illustrated to the right in Figure 14–7. Nerve impulses are transmitted down the spinal cord into the sympathetic chains and then to the heart and blood vessels. These impulses increase the *rate of the heart* and its *strength of contraction,* both of which tend to increase the arterial pressure. They also cause *vasoconstriction* of the blood vessels in most parts of the body, which increases the total peripheral resistance and, therefore, increases the arterial pressure.

PARASYMPATHETIC EFFECTS ON THE HEART. The vasomotor system also helps to control the circulation through the vagus nerves which carry so-called *parasympathetic* fibers to the heart. When the sympathetics are stimulated the parasympathetic fibers in the vagi are inhibited. Since parasympathetic stimulation has opposite effects on the heart to those of sympathetic stimulation, that is, it decreases the heart's activity, parasympathetic inhibition has the same effect on heart pumping as does sympathetic stimulation.

Except for this effect on the heart, the para-

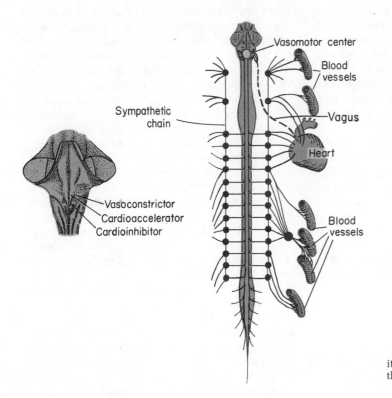

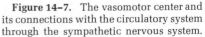

Figure 14–7. The vasomotor center and its connections with the circulatory system through the sympathetic nervous system.

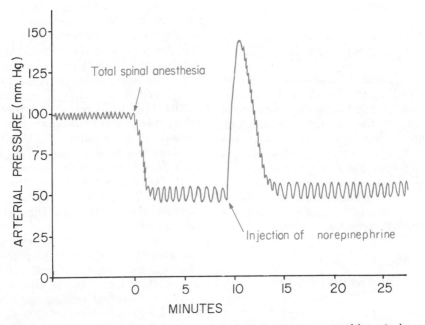

Figure 14–8. Effect on arterial pressure of greatly decreased vasomotor tone caused by spinal anesthesia, and of greatly increased vasomotor tone caused by injection of norepinephrine.

sympathetics play very little role in control of the circulation.

VASOMOTOR TONE. Even normally, the sympathetic nervous system transmits impulses at a low but continuous rate to the heart and vascular system, thus maintaining a moderate degree of vasoconstriction in the blood vessels. The sympathetic system decreases the arterial pressure by simply decreasing the number of impulses transmitted to the blood vessels, thus allowing the arterioles to dilate and reduce the pressure. Conversely, it increases the arterial pressure by increasing the number of impulses above those normally transmitted, thus further constricting the arterioles.

Figure 14–8 demonstrates the importance of vasomotor tone in the regulation of arterial pressure. At the beginning of this record, the mean arterial pressure was 100 mm. Hg; then total spinal anesthesia was instituted by injecting a local anesthetic all the way up the spinal canal. This blocked all nerves coming from the spinal cord, including the sympathetic nerves. Note that the arterial pressure immediately fell to 55 mm. Hg, which is the pressure in the normal circulatory system when there is no vasomotor tone. After a few minutes, a small amount of norepinephrine (the substance normally secreted at the sympathetic nerve endings) was injected into the blood. This caused vasoconstriction throughout the body and thereby immediately elevated the arterial pressure up to 150 mm. Hg. Yet, when the effect of the norepinephrine wore off after another minute, the arterial pressure returned to its basal level. Thus, we see from this experiment that arterial pressure can be reduced far below normal by decreasing the degree of vasomotor tone, and can be elevated far above normal by increasing the tone.

EFFECT OF ISCHEMIA OF THE VASOMOTOR CENTER ON ARTERIAL PRESSURE

The term *ischemia* means insufficient blood flow to supply the normal needs of a tissue. When the vasomotor center becomes ischemic it becomes greatly excited, thereby raising the arterial pressure. This effect is illustrated by the experiment of Figure 14–9, which shows fluid being injected under pressure into the space around the brain. Compression of the brain by the fluid also compresses the arteries that supply the vasomotor center in the medulla, and the resultant ischemia excites the center, thereby elevating the arterial pressure.

The medullary ischemia mechanism normally protects the brain from being injured because of insufficient pressure to supply adequate blood flow to the brain tissue. That is, the resulting rise in pressure increases the brain blood flow back toward normal.

This vasomotor ischemia mechanism for raising the arterial pressure is by far the most powerful of all the nervous controls of pressure. Indeed, severe ischemia of the brain can increase the mean arterial pressure to as high as 260 mm. Hg, the highest pressure that the normal heart can achieve.

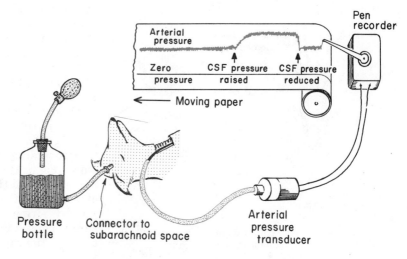

Figure 14–9. Elevation of arterial pressure caused by increased cerebrospinal fluid pressure.

THE BARORECEPTOR PRESSURE CONTROL SYSTEM

The pressure regulatory system that has been studied more than any other is the *baroreceptor system* illustrated in Figure 14-10. In the arch of the aorta and in the internal carotid arteries near their origins in the neck (areas called the *carotid sinuses*) are many small nerve receptors that detect the degree of stretch caused in the walls of the arteries by the pressure. These receptors are called *baroreceptors* or *pressoreceptors*. The number of impulses transmitted from the baroreceptors increases as the arterial pressure rises. On passing to the brain, the impulses *inhibit* the vasomotor center and thereby decrease the arterial pressure back toward normal. On the other hand, when the pressure in the arteries falls too low, the baroreceptors lose their stimulation so that the vasomotor center now becomes excessively excited, elevating the arterial pressure again back toward normal.

One can readily see that the barorecptor system opposes either a rise or a fall in pressure. For this reason, it is sometimes called a *moderator system* or a *buffer system*. As an example of this system's effectiveness, if 500 ml. of blood is injected into a person's circulation over a period of about 30 seconds, the arterial pressure will normally rise no more

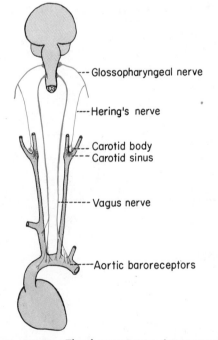

Figure 14-10. The baroreceptors (pressoreceptors) and their connections with the vasomotor center.

than 15 mm. Hg. However, injection of the same quantity of blood into the circulation of someone whose baroreceptors have been inactivated will cause the pressure to rise almost seven times this much, or 100 mm. Hg.

Function of the Baroreceptors When a Person Stands. The pressure in the arteries of the upper body tends to fall when one stands because the blood vessels in the lower body stretch when exposed to the weight of the vertical columns of blood, this in turn causing loss of blood from the heart while *pooling* the blood in the lower vessels. But the falling pressure automatically excites the baroreceptor reflex, returning the pressure in the upper body almost to normal. One of the values of this constancy of pressure in the upper part of the body is that it helps to maintain an adequate blood supply to the brain regardless of the position of the body.

THE CAPILLARY FLUID SHIFT MECHANISM FOR PRESSURE REGULATION

Because of the rapidity with which the various nervous control systems can return arterial pressure toward normal when it becomes abnormal, it is very tempting to explain arterial pressure regulation entirely on the basis of nervous mechanisms alone. However, the nervous mechanisms have a major failing that prevents their continuing to control arterial pressure over a long period of time. This failing is that the baroreceptors in the walls of the vessels *adapt* after a while to the new level of pressure and soon stop sending their signals indicating abnormality. Therefore, after a few days, most nervous regulatory mechanisms become ineffective. By that time, nonnervous mechanisms take over the pressure regulation. One of these is the *capillary fluid shift mechanism,* which is particularly important in helping to regulate arterial pressure when the blood volume tends to become either too little or too great.

An increase in blood volume, such as might occur following a transfusion, increases the pressures in all parts of the systemic circulation, including the pressure in the capillaries. However, in the next few minutes to several hours, even in the absence of nervous control of pressure, the pressures still return back toward normal. Much of the decline in pressure is caused by shift of fluid out of the blood through the capillary membranes into the interstitial spaces, thereby decreasing the blood volume back toward normal.

The mechanism of the capillary fluid shift is simply the following: Too much blood volume increases the pressures in the capillaries as well as in the arteries. The increased capillary pressure causes fluid to leak out of the circulation into the interstitial spaces. Conversely, when the blood volume becomes too low, the capillary pressure falls, and now fluid is pulled by osmosis from the interstitial spaces into the circulation because of the osmotic pressure effect of the proteins in the plasma.

The capillary fluid shift mechanism is much slower to regulate arterial pressure than are the nervous mechanisms, for it requires from 10 minutes to several hours to readjust the arterial pressure back toward normal.

REGULATION OF ARTERIAL PRESSURE BY THE KIDNEYS

Kidney Control of Pressure by Controlling Blood Volume. The kidneys, like the capillaries, can regulate arterial pressure by increasing or decreasing the blood volume. After even a slight decrease in arterial pressure, the kidneys often stop or almost stop forming urine because the glomerular pressure falls too low for normal glomerular filtration to take place, and also because the low pressure in the capillaries in the kidneys causes almost total reabsorption by the kidney tubules of the filtrate that is formed. Therefore, the fluid and electrolytes taken in by mouth gradually accumulate in the body until the blood volume rises enough to reestablish normal arterial pressure.

Conversely, when the arterial pressure rises too high, the urinary output increases. Gradually, over a period of hours, the blood volume decreases, causing the arterial pressure again to return toward normal.

Function of "Renin" and Angiotensin in Arterial Pressure Regulation. The kidneys also have a hormonal effect in regulating arterial pressure. When the arterial pressure falls to a low value, diminished blood flow through the kidneys causes them to secrete a special hormone called *renin* into the blood. Renin in turn acts as an enzyme to convert one of the plasma proteins of the plasma to a substance called *angiotensin*. Angiotensin then acts directly on the blood vessels to cause

vasoconstriction, which increases the arterial pressure back toward normal.

Importance of Kidney Regulation of Arterial Pressure. *The kidneys are by far the most important of all parts of the body for long-term regulation of arterial pressure.* This is shown very dramatically by the experiment of Figure 14–11. If clamps are placed on the arteries of both kidneys so that the renal blood supply is diminished, the arterial pressure begins to rise. In several weeks, the mean pressure may rise to twice its original value. The elevated pressure will then force normal blood flow through the kidneys despite the continued constriction of the arteries, and the kidneys will function normally again. When the constrictions are removed, the arterial pressure falls back to normal within another few days. Therefore, it is obvious that the kidneys are capable of protecting themselves against diminished blood flow; that is, *diminished flow through the kidneys causes a general arterial pressure rise until the kidneys again receive an adequate blood supply.* On the other hand, if the blood flow through the kidneys becomes excessive, the arterial pressure falls until the renal flow again returns to normal.

HORMONAL REGULATION OF ARTERIAL PRESSURE – THE EFFECT OF ALDOSTERONE

At least one major hormonal system is also involved in the regulation of arterial pressure. This is the secretion of aldosterone by the *adrenal cortex*. The adrenal cortex is the outer shell of each of two small endocrine glands, the adrenals, one located immediately above each of the two kidneys. The cortex secretes adrenocortical hormones, one of which, *aldosterone,* decreases the kidney output of salt and water. The amount of salt and water in the body in turn helps to determine the volumes of blood and interstital fluids in the body.

The manner in which the aldosterone mechanism enters into blood pressure regulation is the following: When the arterial pressure falls very low, lack of adequate blood flow through the body's tissues causes the adrenal cortices to secrete aldosterone. One of the probable causes of this effect is stimulation of the adrenal glands by angiotensin that is formed when the arterial pressure falls. The aldosterone acts to build up the blood volume and return the arterial pressure to normal. Conversely, an elevated arterial pressure reverses this mechanism so that fluid volumes, and consequently the arterial pressure, decrease.

REGULATION OF MEAN ARTERIAL PRESSURE DURING EXERCISE

One of the most severe types of stress to which the circulation can be subjected is strenuous exercise, for rapidly acting muscles need tremendous blood supply. One of the means for increasing blood flow to the muscles during exercise is to elevate the arterial pressure. The mechanisms causing this are: First, increased muscular metabolism during exercise increases the concentrations of carbon dioxide, lactic acid, and other metabolites in the blood, and these circulate to the vasomotor center and excite it, elevating the pressure. Second, and probably even more important, the vasomotor center is stimulated by nerve impulses generated in the muscle-controlling "motor" area of the brain. Therefore, at the same time that the motor area causes the muscles to contract, it also excites the vasomotor center and thereby elevates the arterial pressure.

HYPERTENSION

Hypertension means high arterial pressure, or simply "high blood pressure," and it occurs in approximately one out of every five persons before the end of life, usually in middle or old age. The excessive arterial pressure of hypertension can rupture blood vessels in the brain to cause "strokes"; in the kidney to cause "renal failure"; or in other vital organs to cause blindness, deafness, heart attacks, and so forth. Also, it can place excessive strain on the heart and cause it to fail. For these reasons one of the most important research problems in physiology is to determine the causes of hypertension. Some of the known causes are the following:

HYPERTENSION CAUSED BY ABNORMAL FUNCTION OF THE PRESSURE REGULATORY MECHANISMS

Renal Hypertension. Many conditions that damage the kidneys can cause *renal hypertension.* For instance, constriction of the renal arteries, as illustrated in Figure 14–11, causes the arterial pressure to rise; and the greater the degree of constriction, the greater the elevation of pressure. Diseases of the kidneys such as kidney infections, sclerosis of the renal arterioles, kidney inflammation, and so forth can all elevate the pressure.

A particularly interesting type of renal hypertension occurs when the aorta is occluded above the kidneys, which occurs frequently before birth, in which case the person goes through life with an occluded aorta, a condition called *coarctation of the aorta.* Blood finds

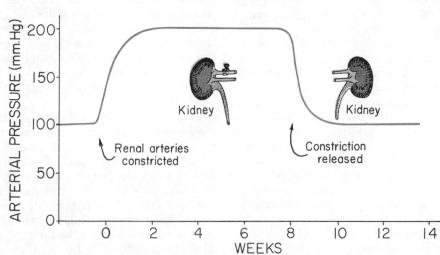

HYPERTENSION

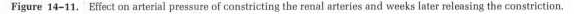

Figure 14–11.   Effect on arterial pressure of constricting the renal arteries and weeks later releasing the constriction.

its way through many smaller arteries of the body wall from the upper segment of the aorta into the lower segment, but the resistance to flow through these small arteries is so great that the arterial pressure in the lower aorta is much lower than that in the upper aorta. The insufficient blood flow to the kidneys from the lower aorta causes the pressures throughout the body to rise until the pressure in the lower aorta returns to normal. Obviously, by this time, the pressure in the upper aorta will have risen to a very high level. Thus, the kidneys automatically maintain a normal arterial pressure at the kidney level even at the expense of the arterial pressure in the upper part of the body.

EXCESS BLOOD VOLUME AS ONE OF THE MAJOR MECHANISMS OF RENAL HYPERTENSION. One mechanism of renal hypertension is development of too much blood volume when renal function becomes abnormal. Obviously, if renal output becomes reduced as a result of renal disease or poor blood supply, the retention of salt and water would be expected to increase both the interstitial fluid and blood volumes. Experiments have recently demonstrated that removal of large portions of the kidney plus a high daily intake of water and salt does indeed increase both of these volumes considerably. These greater volumes elevate the arterial pressure in the following ways: (1) The heightened blood volume increases the blood flow into the heart, which obviously raises the cardiac output above normal. (2) The excess cardiac output causes hypertension. And it also causes too much blood to flow through the tissues. (3) The excess tissue blood flow causes the tissue vessels to constrict during the next few days because of the long-term tendency of each tissue to reset its own blood flow back to normal. (4) This constriction of the tissue vessels increases the total peripheral resistance far above normal, which automatically returns both the blood volume and the cardiac output back toward normal even though the arterial pressure remains elevated.

Thus, the initial increase in blood volume leads to hypertension but also to a further sequence of events that returns both the blood volume and the cardiac output almost back to normal even though the total peripheral resistance remains very high. In fact, except during the first few days, both the blood volume and the cardiac output are usually only 5 to 8 per cent above normal even in pure volume-loading hypertension. And these small increases can rarely be measured, so that most persons with volume-loading hypertension are said to have normal blood volume and cardiac output but high total peripheral resistance and arterial pressure.

ROLE OF THE RENIN-ANGIOTENSIN SYSTEM IN RENAL HYPERTENSION. Sometimes, damaged kidneys secrete large amounts of renin, which in turn leads to formation of angiotensin and this to vasoconstriction and increased total peripheral resistance. This sequence occurs especially in the condition called *malignant hypertension.* The vasoconstriction obviously can act in concert with the excess blood volume effect that occurs in kidney disease to cause very severe hypertension.

Hormonal Hypertension. Occasionally the adrenal *cortices* secrete excessive quantities of aldosterone, either because of an aldosterone-secreting tumor in one of the adrenal glands or because of excessive stimulation of the adrenals by the anterior pituitary gland. In any event, the increased production of aldosterone causes the kidneys to retain excessive quantities of water and salt. The fluid volumes throughout the body increase, and the arterial pressure often rises considerably above normal.

A second type of hormonal hypertension is that caused by a *pheochromocytoma.* This is a tumor of the adrenal *medulla,* which is the central portion of the adrenal gland, entirely distinct from the adrenal cortex. Because the medulla is part of the sympathetic nervous system, the pheochromocytoma secretes large quantities of epinephrine and norepinephrine, the hormones normally secreted by sympathetic nerve endings. These hormones circulate in the blood to all the vessels and promote intense vasoconstriction. As a result, the mean arterial pressure may rise to as high as 200 mm. Hg or higher. The pheochromocytoma usually secretes its hormones only when stimulated by the sympathetic nervous system. This means that when a person who has this type of tumor becomes excessively excited, the secretion of epinephrine and norepinephrine may be tremendous, causing the blood pressure to become very high. Between periods of excitement the pressure may be essentially normal.

Neurogenic Hypertension. Hypertension often results from abnormalities in the brain. For instance, sudden blockage of one of the arteries to vital portions of the vasomotor center sometimes causes hypertension that lasts for life. Also, experiments in animals have indicated that destruction of certain areas in the hypothalamus can cause prolonged hypertension. These abnormalities cause hypertension by continually exciting the vasomotor center; very recent studies show that the main method by which vasomotor activity causes hypertension is to constrict the arteries of the kidneys, thus leading to kidney retention of fluid, increased fluid volumes, and hypertension.

Many psychiatrists believe, though it has not yet been proved, that overexcitation of certain of the conscious regions of the brain in neurotic states such as anxiety, worry, and so forth can excite the vasomotor center and permanently elevate the arterial pressure.

ESSENTIAL HYPERTENSION

The types of hypertension described above all have known causes. Yet in approximately 90 per cent of all hypertensive persons the cause is unknown, and they are said to have *essential hypertension,* the word essential meaning simply "of unknown cause." Many efforts have been made to prove that essential hypertension is caused by a kidney abnormality, a glandular abnormality, or excessive activity of the vasomotor center. Authentic proof that any of these factors is involved in essential hypertension is yet lacking, but most of the recent focus has been on depressed kidney excretion of water and salt caused either by an abnormality of the kidneys themselves or by nervous or hormonal influences on the kidneys. A possible cause of essential hypertension might be a genetic abnormality of the arterioles throughout the body, including those in the kidneys, causing them always to constrict to excessive degrees, Indeed, it is

a known fact that essential hypertension exhibits a very strong hereditary tendency. If one person should inherit excessivley active arterioles and another more relaxed arterioles, the first would have hypertension and the second would be normal. Also, it has been suggested that the person with essential hypertension might have a genetic deficiency in the number of arterioles, which obviously would impede the flow of blood and cause hypertension. Thus far, researchers have not found means of exploring these possibilities, though they must be included among the others as possible causes of essential hypertension.

REFERENCES

Cowley, A. W., Jr., and Guyton. A. C.: Baroreceptor reflex contribution in angiotensin-II-induced hypertension. *Circulation, 50*:61, 1974.

Freis, E. D.: Hemodynamics of hypertension. *Physiol. Rev., 40*:27, 1960.

Guyton, A. C.: Acute hypertension in dogs with cerebral ischemia. *Am. J. Physiol., 154*:45, 1948.

Guyton, A. C.: Integrative hemodynamics. *In* Sodeman, W. A., Jr., and Sodeman, W. A. (eds.): Patholgic Physiology: Mechanisms of Disease. 5th ed. Philadelphia. W. B. Saunders Company, 1974, p. 149.

Guyton, A. C., Coleman, T. G., and Granger, H. J.: Circulation: overall regulation *Ann. Rev. Physiol., 34*:13, 1972.

Kirchheim, H. R.: Systemic arterial baroreceptor reflexes. *Physiol. Rev., 56*:100, 1976.

Kovach, A. G. B., and Sandor, P.: Cerebral blood flow and brain function during hypotension. *Ann. Rev. Physiol., 38*:571, 1976.

Oberg, B.: Overall cardiovascular regulation. *Ann. Rev. Physiol., 38*:537, 1976.

Peach, M. J.: Renin-angiotensin system: biochemistry and mechanisms of action. *Physiol. Rev., 57*:313, 1977.

Pickering, G.: Hypertension. 2nd ed. New York, Churchill Livingstone. Div. of Longman, Inc., 1974.

Reid, I. A., Morris, B. J., and Ganong W. F.: The renin-angiotensin system. *Ann. Rev. Physiol., 40*:377, 1978.

QUESTIONS

1. What is meant by systolic pressure, diastolic pressure, and pulse pressure?
2. How do the stroke volume output of the heart and the distensibility of the arterial system affect the pulse pressure?
3. In the systemic circulation, what are the approximate normal pressures in the aorta, the arterioles, the capillaries, the venules, and the venae cave?
4. If the arterial pressure increases 50 per cent while the cardiac output increases twofold, how much does the total peripheral resistance change?
5. Explain the nervous mechanisms for regulation of arterial pressure, including the baroreceptor control system.
6. How does shift of fluid into or out of the capillaries act as a mechanism for pressure regulation?

7. Explain how the kidneys help to control arterial pressure by increasing or decreasing blood volume.
8. Explain how hypertension can result from abnormal kidney function.
9. What is the role of the renin-angiotensin system in the control of arterial pressure and also in hypertension?
10. What is meant by "essential hypertension"?

15 CARDIAC OUTPUT, VENOUS PRESSURE, CARDIAC FAILURE, AND SHOCK

CARDIAC OUTPUT

Cardiac output is the rate at which the heart pumps blood. Ordinarily it is expressed in liters of flow per minute. Because the function of the circulatory system is to supply adequate nutrition to the tissues, the subject of cardiac output is one of the most important of all phases of physiology.

NORMAL VALUES OF CARDIAC OUTPUT

The average cardiac output in an adult lying down and in a state of complete rest is about 5 liters per minute. If he walks, it rises to perhaps 7.5 liters per minute. If he performs strenuous exercise, it might rise to as high as 25 liters per minute in the normal person or to as high as 35 liters per minute in the well-trained athlete.

One can see, therefore, that the cardiac output varies in proportion to the degree of activity of the person. The ability to regulate the cardiac output in accord with the needs of the body is one of the most remarkable feats of the circulation. Much of the present chapter is devoted to discussing this regulation.

REGULATION OF CARDIAC OUTPUT

The amount of blood that the heart can pump each minute is determined by two major factors: (1) the pumping effectiveness of the heart itself and (2) the ease with which blood can flow through the body and return to the heart from the systemic circulation after it is pumped by the heart. To understand the regulation of cardiac output it is essential that one be familiar with the way in which each of these factors affects the circulation.

The Pumping Effectiveness of the Heart. Normally the heart is an automatic pump that pumps any blood that flows into it from the veins. If the amount flowing from the veins increases, the heart becomes stretched and automatically adjusts to accommodate the extra blood. This ability of the heart to pump any amount of blood, within limits, is called the *law of the heart,* which was explained in detail in Chapter 10. Figure 15–1 illustrates this principle pictorially and graphically, showing that as the heart becomes progressively stretched by greater and greater inflow of blood, it automatically pumps increasing amounts of blood until the heart muscle finally becomes stretched beyond its physiologic limits.

EFFECT OF NERVOUS STIMULATION ON THE EFFECTIVENESS OF THE HEART AS A PUMP. Stimulation of the sympathetic nerves to the heart greatly enhances the effectiveness of the heart as a pump, while stimulation of the parasympathetic nerves (the vagi) greatly decreases the pumping effectiveness. Under normal resting conditions parasympathetic stimulation keeps the heart activity depressed, which probably prolongs the life

174

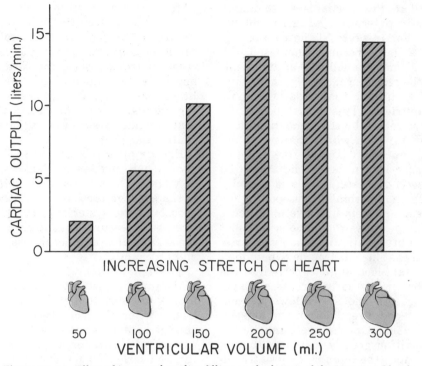

Figure 15–1.   Effect of increased cardiac filling on the heart's ability to pump blood.

of the heart. But during exercise and other circulatory states in which the cardiac output needs to be enhanced very greatly, sympathetic stimulation takes over and makes the heart a very effective pump, about two times as effective as the normal unstimulated heart.

Therefore, the pumping effectiveness of the heart can be increased in two separate ways: (1) by local adaptation of the heart caused by increased inflow of blood, the mechanism called the law of the heart, and (2) by a shift from parasympathetic to sympathetic stimulation.

Flow of Blood into the Heart (Venous Return).   Because the heart pumps whatever amount of blood enters its chambers (up to its limit to pump), cardiac output normally is regulated mainly by the amount of blood that returns to the heart from the peripheral vessels. This flow of blood into the heart is called *venous return.* When a person is at rest, venous return averages 5 liters per minute. Therefore, this is also the amount of blood that is pumped, and it is also the amount of cardiac output. However, if 40 liters per minute, a very abnormal amount, should attempt to return to the heart, even a heart strongly stimulated by the sympathetics would be able

to pump only about 25 liters of this because this is its limit. Therefore, blood would dam up in the veins rather than being pumped into the arteries.

The three important factors that determine the amount of venous return per minute are (1) the resistance of blood flow through the systemic vessels, (2) the average pressure of the blood in all parts of the systemic circulation, which is called the *mean systemic filling pressure,* and (3) the right atrial pressure.

EFFECT OF VASCULAR RESISTANCE ON VENOUS RETURN.   The factor that most often controls venous return is the vascular resistance. That is, if a large share of the peripheral circulatory vessels become widely opened so that the vascular resistance decreases, large amounts of extra blood will flow from the arteries into the veins. This obviously increases the rate of venous return and also augments the cardiac output. Venous return is actually the sum of all the blood flows through the different peripheral tissues. Therefore, every time the blood flow in any single peripheral tissue increases, the venous return likewise increases.

The phenomenon of autoregulation, which has already been discussed, plays a major role in determining blood flow in each respective tissue. Therefore, this mechanism also figures significantly in determining the rate of venous return. For instance, if the rate of metabolism in a tissue becomes increased, the autoregulatory mechanism will automatically decrease the resistance, thereby increasing the blood flow through that tissue to supply the needed nutrients. This in turn boosts the rate of venous return by the same amount.

Another instance in which changes in peripheral resistance play a major role in determining venous return is in the condition called *arteriovenous fistula.* This is a condition in which an abnormal direct opening develops between an artery and a vein. It occurs frequently in persons who have a gunshot wound that ruptures simultaneously an artery and a vein that lie side by side. Upon healing, the arterial blood flows directly into the vein. This obviously greatly decreases the peripheral resistance and also allows tremendous flow of blood directly from the arterial tree into the veins. One can well understand that this condition greatly increases the venous return and, therefore, also greatly increases cardiac output.

Figure 15–2 illustrates the effect on venous return and cardiac output of changing the systemic vascular resistance. Note in Figure 15–2A that when all the peripheral resistances are decreased to one-half normal, the cardiac output rises from its normal value of 5

liters per minute up to double this value — that is, to 10 liters per minute. Conversely, Figure 15–2B shows that a twofold increase in resistances everywhere in the peripheral circulation will decrease the cardiac output to approximately one-half normal, or to 2.5 liters per minute. Thus, it is clear that changing the resistance to blood flow in the peripheral circulatory system is a potent means for controlling cardiac output.

EFFECT OF MEAN SYSTEMIC FILLING PRESSURE AND RIGHT ATRIAL PRESSURE ON VENOUS RETURN. The *mean systemic filling pressure* is the average pressure in all the distensible vessels of the systemic circulation. It is this average pressure that continually pushes blood from the peripheral vessels toward the right atrium.

The *right atrial pressure,* on the other hand, is the input pressure to the heart; this acts as a back pressure to impede blood flow into the heart. When the heart becomes weakened so that it cannot pump blood out of the right atrium with ease, or whenever the heart becomes overloaded for any reason, the right atrial pressure rises and tends to keep blood from entering the heart from the systemic circulation.

Thus, the mean systemic filling pressure is the average pressure in the periphery tending to push blood toward the heart, while the right atrial pressure is the input pressure against which the blood must flow. Therefore, venous return is determined to a great extent by the difference between these two pressures, *mean systemic filling pressure* minus *right atrial pressure.* This difference is called the *pressure gradient for venous return.*

Figure 15–3 illustrates the effect of both mean systemic filling pressure and right atrial pressure on the control of venous return and cardiac output. In Figure 15–3A, the mean systemic pressure is 7 mm. Hg and the right atrial pressure 0 mm. Hg. The pressure gradient for venous return in this case is 7 – 0, or 7 mm. Hg, and the venous return and cardiac output are 5 liters per minute. In Figure 15–3B, the mean systemic pressure is 18 mm Hg, and the right atrial pressure, because the heart is now beginning to be overloaded, has risen to 4 mm. Hg. In this case, the pressure gradient for venous return is 18 – 4 mm. Hg, or 14 mm. This pressure gradient is twice that in Figure 15–3A. Likewise, the cardiac output and venous return are now twice that in Figure 15–3A, or 10 liters per minute, showing the importance of the pressure gra-

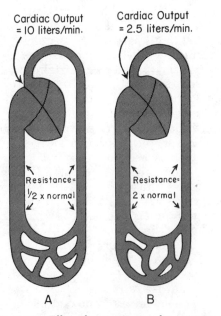

Figure 15–2. Effect of systemic vascular resistance on venous return and cardiac output.

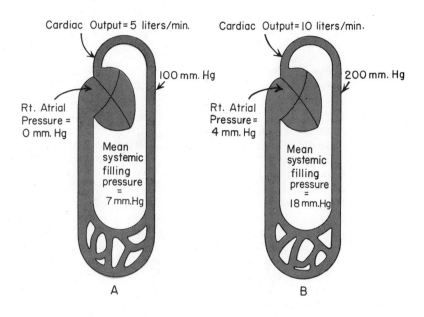

Cardiac Output = 5 liters/min.   Cardiac Output = 10 liters/min.

100 mm. Hg   200 mm. Hg

Rt. Atrial   Rt. Atrial
Pressure =   Pressure =
0 mm. Hg   4 mm. Hg

Mean   Mean
systemic   systemic
filling   filling
pressure   pressure
=   =
7 mm.Hg   18 mm.Hg

A   B

Figure 15–3. Effect of mean systemic filling pressure and right atrial pressure on venous return.

dient for venous return in determining cardiac output.

Effect of Exercise on Venous Return and Cardiac Output. Exercise greatly increases the venous return and cardiac output for several reasons. First, exercise is one of the most potent stimulators of metabolism that is known; indeed, intense muscle activity occasionally causes as much as 50-fold increase in total body metabolism for short periods of time and 20-fold for long periods of time. This obviously increases the rate of utilization of nutrients, including greatly enhanced use of oxygen. Almost instantly, the autoregulation mechanism is set into motion, and the blood vessels of the muscles open widely to supply the tissue with needed nutrients. And this obviously greatly increases venous return and cardiac output.

However, in addition to the vasodilatation that occurs in exercise, the mean systemic filling pressure also increases for two reasons. First, sympathetic activity is greatly heightened during exercise, and sympathetic impulses constrict the arteries and veins throughout the body. This squeezes the blood and elevates the mean systemic filling pressure. Second, the tightening of the muscles themselves also squeezes the blood vessels, thus further raising the mean systemic filling pressure. Therefore, these effects also help to increase the cardiac output during exercise.

Thus, utilizing these mechanisms for dilating the muscle blood vessels and elevating mean systemic filling pressure during exercise, the venous return and cardiac output can be increased to as high as 20 to 25 liters per minute in the normal young adult and to as high as 30 to 35 liters per minute in the well-trained athlete.

Summary of Cardiac Output Regulation— Role of Tissue Need in Controlling Cardiac Output. Under normal conditions the heart is capable of pumping far more blood than the body needs. Therefore, the cardiac output is normally controlled to almost no extent by the heart. Instead, it is determined almost entirely by the rate at which blood flows into the heart from the peripheral circulation. That is, cardiac output is determined by the venous return to the heart.

In turn, the venous return is determined by the sum of the blood flows through all the different tissues of the body. And each tissue is, in general, capable of controlling its own blood flow in relation to its need for flow. Most tissues control this flow in proportion to their need for oxygen, but this is not true of all tissues. For instance, the brain controls its blood flow in proportion to the amount of carbon dioxide that it needs to remove from the brain tissue; the kidneys control their blood flow in proportion to the excretory products in the blood that need to be excreted; and the skin blood flow is controlled in proportion to the need of the body to eliminate heat.

Therefore, in effect, the cardiac output is normally controlled in proportion to the needs for blood flow in all the respective tissues of the body. When the needs are great, the local tissue control mechanisms increase the flow of blood from the arteries to the veins and this in turn increases the venous return and cardiac output simultaneously.

Conditions Under Which the Heart Plays a Major Role in Controlling Cardiac Output. At times the heart is unable to pump all of the blood that attempts to return to it. This occurs when too much blood attempts to flow from the arteries to the veins through a very large arteriovenous fistula or occasionally when the metabolism of the tissues is so great that venous return is greater than the amount the heart can pump. Under these conditions the heart becomes overloaded. Obviously, it is the heart then, instead of the venous return, that determines the cardiac output. This, however, is a rare state of events so long as the heart is normal.

A much more common condition in which the heart becomes the controller of cardiac output is cardiac failure. Cardiac failure simply means reduced effectiveness of the heart as a pump, usually occurring because of damaged heart muscle or heart valves. In cardiac failure, even the normal venous return might be too great for the heart to pump. Therefore, the limiting factor once again becomes the heart and not the venous return. The subject of heart failure is so important that it will be discussed in much more detail later in this chapter.

MEASUREMENT OF CARDIAC OUTPUT

A *flowmeter* can be applied to the major arteries or veins of an experimental animal in such a manner that the rate of blood flow through the heart will be recorded. Such procedures, however, are not very useful even in animal experimentation because of the extreme disruption of normal function caused by the necessary surgery. And, obviously, in the human being they can hardly be used at all. Therefore, for measuring cardiac output in human beings and in intact animals the following indirect procedures have been developed.

The Fick Method. One indirect method for measuring cardiac output is based on the so-called Fick principle, which is explained by the example in Figure 15–4. The volume of oxygen in a sample of venous blood is analyzed chemically to be 160 ml. of oxygen per liter of blood, and in the arterial blood, 200 ml. Thus, as each liter of blood passes through the lungs it picks up 40 ml. of oxygen. At the same time, using appropriate respiratory apparatus, the amount of oxygen absorbed into the blood through the lungs is measured to be 200 ml. per minute. From these values, one can calculate the cardiac output in the following manner: If the amount of oxygen absorbed per minute is 200 ml. and the amount of oxygen picked up by each liter of blood flowing through the lungs is 40 ml., by dividing 200 by 40 one can conclude that five 1-liter portions of blood must have passed through the lungs each minute. Thus, the total cardiac output is calculated to be 5×1 liter or 5 liters per minute.

Expressing the Fick principle mathematically the following formula applies:

$$\text{Cardiac output in liters per minute} = \frac{\text{Total oxygen absorbed per minute}}{\text{Oxygen absorbed by each liter of blood}}$$

The Dye Method. Another simple procedure for measuring cardiac output is to inject rapidly into a vein a small quantity of foreign substance such as a brightly colored dye. As the substance passes through the heart and lungs into the arterial system, its concentration in the arterial blood is recorded by a photoelectric "densitometer" or by taking pe-

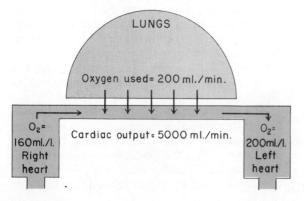

Figure 15–4. The Fick principle for indirectly measuring cardiac output.

riodic blood samples for the next 20 to 30 seconds and analyzing these for their content of dye. The more rapidly the blood flows through the heart, the more rapidly the dye appears in the arteries and the sooner it will disappear. From the area under the recorded time-concentration curve of the dye as it passes through the arteries one can calculate the cardiac output quite accurately. That is, the less the area under this curve, the greater is the cardiac output. If the reader will think about this for a moment, he will be able to understand intuitively the principle of this measurement, though the mathematics of the calculation is more than can be discussed here.

VENOUS PRESSURE

RIGHT ATRIAL PRESSURE

The pressure in the venous system is determined mainly by the pressure in the right atrium. Normally, the right atrial pressure is approximately zero — that is, it is almost exactly equal to the pressure of the air surrounding the body. This does not mean, though, that the force distending the walls of the right atrium is zero, because the pressure in the chest cavity surrounding the heart is about 4 mm. Hg less than atmospheric pressure. This partial vacuum actually pulls the walls of the atrium outward so that blood is normally sucked from the veins into the atrium.

If the heart becomes weakened, blood begins to dam up in the right atrium, thereby causing the right atrial pressure to rise. Or, if excessive quantities of blood attempt to flow into the heart from the veins, again the right atrial pressure rises. Therefore, the right atrial pressure, though normally almost exactly zero, often rises above zero in abnormal conditions.

PERIPHERAL VENOUS PRESSURE

The pressure in a peripheral vein is determined by five major factors: (1) the right atrial pressure, (2) the resistance to blood flow from the vein to the right atrium, (3) the rate of blood flow along the vein, (4) pressure caused by weight of the blood itself, called "hydrostatic pressure," and (5) the venous pump. The effects of most of these factors are self-evident, but some of them need special explanation.

Effect of Right Atrial Pressure on Peripheral Venous Pressure. Since blood flows in the veins always toward the heart, the pressure in any peripheral vein of a person in the lying position must be as great as or greater than the pressure in the right atrium. If considerable resistance exists between the peripheral vein and the heart, and the flow of blood is of reasonable quantity, then the peripheral venous pressure will be considerably higher than right atrial pressure. Many of the veins are compressed where they pass abruptly over ribs, between muscles, between organs of the abdomen, or so forth. At these points the venous resistance is usually great, impeding blood flow and increasing the pressure in the distal veins. On the average, the pressure in a vein of the arm or leg of a person lying down is approximately 6 to 8 mm. Hg in contrast to zero pressure in the right atrium.

Effect of Hydrostatic Pressure on Peripheral Venous Pressure. Hydrostatic pressure is the pressure that results from the weight of the blood itself. Figure 15–5 illustrates the human being in a standing position, showing that for blood to flow from the lower veins up to the heart, considerable extra pressure must develop in these veins to drive the blood uphill. The weight of the blood from the level of the heart down to the bottom of the foot is great enough that when all other factors affecting venous pressure are nonoperative, the venous pressure in the foot will be 90 mm. Hg.

The Venous Pump. To prevent the extremely high venous pressures that hydrostatic pressure can cause, the venous system is provided with a special mechanism for propelling blood toward the heart. This is called the *venous pump* or sometimes the *muscle pump.* All peripheral veins contain valves which allow blood to flow only toward the heart. Every time a muscle contracts or every time a limb moves, the moving tissues compress at least some of the veins. Since the valves prevent backward flow, this always pushes the blood toward the heart, in this way emptying the veins and decreasing the peripheral venous pressure.

The hydrostatic pressures shown in Figure 15–5 occur only when a person is standing completely still or when some disease condition has destroyed the valves of his veins. Ordinarily, the venous pump is so effective even when a person walks very slowly that

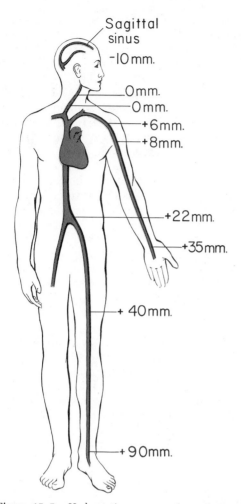

Sagittal
sinus
-10 mm.
0 mm.
0 mm.
+6 mm.
+8 mm.
+22 mm.
+35 mm.
+40 mm.
+90 mm.

Figure 15–5. Hydrostatic pressures in various parts of the venous system of a person standing quietly so that the venous pump is inactive.

pressures in the leg veins are only 15 to 30 mm. Hg. But when the valves are destroyed, such high pressures (80 to 90 mm. Hg) develop in the leg veins that they become progressively distended to diameters four and five times normal, causing the condition known as *varicose veins*.

MEASUREMENT OF VENOUS PRESSURES

The most important of all venous pressures is the right atrial pressure, for one can usually tell from this how well the heart is pumping. Right atrial pressure is usually measured through a cardiac catheter in the manner illustrated in Figure 15–6A. The catheter is threaded into a vein of the arm, then upward

through the veins of the shoulder, into the thorax, and finally into the right atrium. The pressure transmitted through the catheter is recorded by a water manometer located at the patient's side. This same technique can be used for recording pressures in other parts of the heart or pulmonary circulation simply by sliding the catheter through the tricuspid valve into the right ventricle, or on through the pulmonary valve into the pulmonary artery.

For measuring peripheral venous pressures, either a catheter or a needle can be inserted directly into a peripheral vein and the pressure measured as just described. However, one can usually estimate peripheral venous pressures quite satisfactorily using the simple technique illustrated in Figure 15–6B of raising and lowering the arm above or below the level of the heart. When the veins are lower than the heart they become full of blood and stand out beneath the skin. Then, as the arm is raised, the veins normally collapse at a level approximately 9 cm. above the level of the heart. This means that the pressure in the arm veins, when the arm is at heart level, is about 9 cm. of water (or about 7 mm. of mercury, because 1 cm. of water equals .74 mm. Hg). If the veins do not collapse until they rise to a level 20 cm. above the heart, then the peripheral venous pressure is approximately +20 cm. of water (or 15 mm. Hg).

CARDIAC FAILURE

The term *cardiac failure* means *depressed pumping effectiveness of the heart.* The most frequent cause of cardiac failure is actual damage to the heart itself by some disease process. For instance, the coronary arteries might become blocked because of *atherosclerosis,* or sometimes the valves of the heart are destroyed by rheumatic heart disease. In either case, the effectiveness of the heart as a pump is decreased so that even the normal amount of blood is not pumped as well as usually.

Low Cardiac Output Failure. A failing heart often, but not always, fails to pump adequate quantities of blood to the tissues. This is called *low cardiac output failure.* It can be caused by weakness of any part of the heart or of the whole heart, because failure of any one part can often hinder satisfactory pumping.

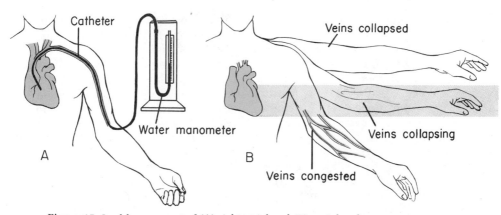

Figure 15-6. Measurement of (A) right atrial and (B) peripheral venous pressures.

Pulmonary Congestion Resulting from Cardiac Failure. When it is primarily the left side of the heart that is failing, the right heart continues to pump blood into the lungs with normal vigor while the left heart is unable to move the blood on into the systemic circulation. The resulting accumulation of blood in the lungs increases the pressures in all the pulmonary vessels and engorges them with blood. If the pulmonary capillary pressure rises especially high, fluid will then leak out of the capillaries into the lung tissues and even into the air spaces of the lungs (the alveoli), resulting in *pulmonary edema*. Indeed, many patients who die from failure of the left heart do not die because of decreased cardiac output, but because of failure of the water-soaked lungs to aerate the blood.

Peripheral Congestion and Peripheral Edema Resulting from Cardiac Failure. If the right side of the heart fails or if the entire heart fails (which is the usual case), the right atrial pressure rises, causing much of the blood attempting to return to the heart to be dammed in the peripheral veins. As a result, the pressures throughout the entire venous system rise. The veins of the neck become greatly distended, as illustrated in Figure 15-7, and the venous reservoirs such as the liver and spleen become engorged with blood. Also, the capillary pressure throughout the entire systemic circulatory system may become so great that fluid leaks continually into the tissue spaces, resulting in extreme *generalized peripheral edema*. One of the conditions occasionally encountered in severe heart disease is swelling from head to toe as the result of excessive interstitial fluid, a condition called *dropsy*.

The edema in cardiac failure is caused not only by high pressure in the capillaries but also by two other factors: First, decreased cardiac output decreases the amount of urine formed by the kidneys, causing excessive quantities of water and electrolytes to remain in the body, thus enhancing the volume of extracellular fluid. Second, diminished cardiac output also causes the adrenal cortex to secrete large quantities of aldosterone, which is still another factor that makes the kidneys retain water and salt, thus further increasing the total extracellular fluid volume.

One of the most essential features in the treatment of cardiac failure is to control the amount of salt that the person eats and the amount of water that he drinks. Also, very powerful drugs called *diuretics,* which increase the output of urine by the kidneys, are frequently administered.

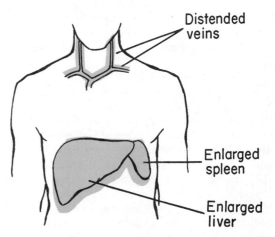

Figure 15-7. Engorgement of the peripheral venous system and blood reservoirs in cardiac failure.

CIRCULATORY SHOCK

Circulatory shock is the condition resulting when the cardiac output becomes so reduced that the tissues everywhere in the body fail to receive adequate blood supply. As a result, the tissues suffer from inadequate nutrition and inadequate removal of cellular excretory products because of the reduced blood flow.

Any condition that decreases the cardiac output to a very low level can cause circulatory shock. Therefore, shock can result from weakness of the heart itself or from diminished venous return, for which reason it can be classified into two main types, (1) *cardiac shock* and (2) *low venous return shock.*

CARDIAC SHOCK

Cardiac shock is very severe low cardiac output failure caused by greatly decreased effectiveness of the heart as a pump. This type of shock occurs most frequently immediately after a severe heart attack because this often causes the heart's ability to pump blood to decrease many-fold in only a few minutes. The person frequently dies because of diminished blood supply to his tissues before the heart can begin to recover.

SHOCK CAUSED BY LOW VENOUS RETURN

Any of the factors that decrease the tendency for blood to return to the heart can cause shock. These are (1) decreased blood volume, which causes *hypovolemic shock,* (2) increased size of the vascular bed so that even normal amounts of blood will not fill the vessels adequately — this is called *venous pooling shock,* or (3) obstruction of blood vessels, particularly veins.

Hypovolemic Shock. Hypovolemic shock most frequently results from blood loss, such as loss from a bleeding wound or a bleeding stomach ulcer. However, the blood volume can be decreased as a result of plasma loss through exuding wounds or burns; loss of plasma into severely crushed tissues; or dehydration caused by extreme sweating, lack of water to drink, or excessive loss of fluids through the gut or kidneys. In any of these conditions, the diminished blood volume decreases the mean systemic filling pressure so greatly that inadequate quantities of blood

return to the heart, and the diminished venous return causes shock.

Venous Pooling Shock. If the blood vessels lose their vasomotor tone, their diameters may increase so greatly that the blood collects, or "pools," in the highly distensible veins. As a result, the pressures in the systemic circulation fall so low that venous return becomes slight and circulatory shock ensues.

NEUROGENIC SHOCK. A special type of venous pooling shock is *neurogenic circulatory shock* caused by sudden cessation of sympathetic impulses from the central nervous system to the peripheral vascular system. The result is loss of normal vasomotor tone, diminished pressures everywhere in the systemic circulation, and consequently diminished venous return. Emotional fainting is an example of this type of shock.

ALLERGIC SHOCK. Extreme allergic reactions can also cause venous pooling shock. This type of shock is known as *anaphylactic shock,* and its probable mechanism is that illustrated in Figure 15–8, which may be explained as follows: When a person becomes immune to a foreign substance, such as the protein of a type of bacterium or a protein toxin, his body manufactures *antibodies;* these are special proteins that will destroy the bacteria or toxins, as explained in Chapter 6. However, when the person is then exposed to another large dose of the same protein to which he has become immune, the reaction between the antibodies and the protein in the circulating blood causes the release of several substances that then produce shock. One of these substances is *histamine.* The histamine circulates to all the peripheral blood vessels and promotes vasodilatation,

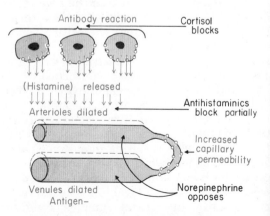

Figure 15–8. Mechanism of anaphylactic shock.

especially dilatation of the veins, thus causing venous pooling and diminished venous return, culminating immediately in a state of anaphylactic shock. Indeed, venous pooling can sometimes result so rapidly and to such an extreme extent during anaphylaxis that the person may die before therapy can be started.

STAGES OF SHOCK

The severity of shock varies tremendously, depending upon the degree of the abnormality that causes it. If it is very mild, the different control systems of the body for maintenance of normal arterial pressure and for maintenance of blood flow through the tissues can overcome the effects of the shock, and the person will recover very quickly. This is called *compensated shock*. However, if the shock is more severe it will continue to progress, giving rise to a stage of shock called *progressive shock*. And, if it is extremely severe, the shock will progress to a point that, even though the person is still alive, all types of known therapy still cannot save his life. This is called *irreversible shock*.

Compensated Stage. When circulatory shock is caused by hemorrhage, the blood loss is compensated by nervous constriction of the veins and venous reservoirs. Consequently, even though blood is lost, venous return of blood to the heart continues almost as usual, and the cardiac output remains essentially normal until 15 per cent of the total blood volume (about 750 ml.) is removed. However, beyond this point, the blood loss can no longer be compensated, and cardiac output begins to fall. This early stage of shock is called the *compensated stage*. During this stage the person is not in imminent danger of dying, and unless the cause of the shock is intensified, the condition will be corrected automatically by the usual circulatory control mechanisms, and the person will return to normal within a few hours or certainly within a day or two.

Progressive Stage of Shock. If the degree of shock becomes very severe, regardless of the initial cause of the shock, *the shock itself promotes more shock*. In other words, certain vicious cycles develop, and the shock becomes progressively more severe even though the original cause of the shock does not worsen. Some of the reasons for the increased severity are shown in Figure 15–9 and may be explained as follows:

First, if the degree of shock is so great that the heart fails to pump enough blood to supply its own coronary vessels, the heart itself becomes progressively weakened. This further diminishes the cardiac output, which weakens the heart even more and diminishes the cardiac output again, leading to a vicious cycle that eventually kills the person.

Second, very poor blood flow to the brain causes damage to the vasomotor and respiratory centers. Failure of the vasomotor center allows venous pooling, resulting in even more extensive shock. Also, respiratory failure causes diminished oxygenation of the blood and, therefore, malnutrition of the tissues, which also makes the shock worse.

Third, peripheral vascular failure causes a vicious cycle. Ischemia of the blood vessels can probably make the vascular musculature so weak that the vessels dilate, resulting in still more vascular pooling of blood, diminished venous return, increased shock, increased ischemia, and so forth, the vicious cycle repeating itself again and again until the vessels are fully dilated or the person is dead.

Fourth, recent experiments have shown that greatly diminished blood flow causes

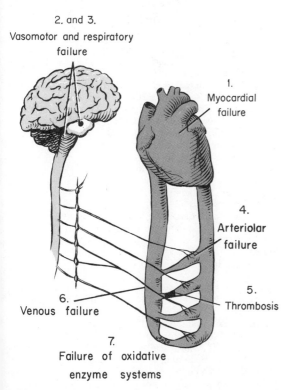

2. and 3.
Vasomotor and respiratory failure

1. Myocardial failure

4. Arteriolar failure

5. Thrombosis

6. Venous failure

7. Failure of oxidative enzyme systems

Figure 15–9. Factors that produce progression of shock and finally cause it to become irreversible.

minute blood clots to develop in the small vessels. As a consequence of the plugged vessels, venous return decreases, and the progressively more sluggish flow of blood through the peripheral vessels causes more and more clotting, creating still another vicious cycle.

Fifth, very severe shock also damages the oxidative and energy systems of the tissues, particularly the adenosine triphosphate system discussed in Chapter 3, which causes all the circulatory functions to diminish further.

Thus, because of several vicious cycles that develop once shock has reached a certain degree of severity, progressive deterioration of the circulatory system causes the shock to become more and more severe until death ensues, unless appropriate therapy is instituted.

IRREVERSIBLE SHOCK. In the early stages of progressive shock, the person's life can usually be saved by appropriate and rapid treatment. For instance, if the shock is initiated by hemorrhage, rapid transfusion of blood can restore normal circulatory dynamics unless the damage to the circulation has already become too great. However, if the shock has progressed for a long period of time, such extreme damage may have occurred that no amount of treatment can be successful in restoring the cardiac output enough to sustain life. When this state has been reached, the patient is said to be in *irreversible shock*. The heart, for instance, may have become damaged beyond repair by this time so that it is incapable of continued pumping at an adequate rate of cardiac output regardless of what therapy is instituted. Or so many peripheral vessels may have become plugged with small clots that blood can no longer flow rapidly enough to repair the damage. In these instances, despite any amount of therapy, the vicious cycles of the progressive stage of shock continue on and on until death of the person.

TREATMENT OF SHOCK

The treatment of shock depends on the cause. In the case of cardiac shock, the best therapy is to increase the pumping effectiveness of the heart. Usually this is not easily accomplished, and almost 85 per cent of persons with this type of shock die. One can sometimes help a person with cardiac shock by judiciously transfusing blood or by administering norepinephrine to constrict the peripheral arterioles and increase the arterial pressure. The increased pressure then promotes greater flow of blood in the coronary arteries to the heart muscle.

If the shock results from hypovolemia, the blood volume can be increased by administering blood, plasma, or sometimes salt solution. In fact, even drinking water is often lifesaving; and fortunately the person with shock, if he is still conscious, experiences intense thirst to make him drink. Return of the blood volume to normal increases the pressures throughout the systemic vessels, which brings the venous return back to normal and thereby alleviates the shock.

In both neurogenic and anaphylactic shock, the main difficulty is venous pooling. Administration of drugs that act like the sympathetic nervous system in constricting the blood vessels — such as norepinephrine — compresses the blood in the vessels and forces it along the veins toward the heart, usually bringing the person out of shock.

REFERENCES

Baltaxe, H. A., Amplatz, K., and Levin, D. C.: Coronary Angiography. Springfield, Ill., Charles C Thomas, Publisher, 1974.

Berne, R. M., and Rubio, R.: Regulation of coronary blood flow. *Adv. Cardiol., 12*:303, 1974.

Gregg, D. E.: The natural history of coronary collateral development. *Circ. Res., 35*:335, 1974.

Guyton, A. C., Coleman, T. G., and Granger, H. J.: Circulation: overall regulation. *Ann. Rev. Physiol., 34*:13, 1972.

Guyton, A. C., Jones, C. E., and Coleman, T. G.: Circulatory Physiology: Cardiac Output and Its Regulation. 2nd ed. Philadelphia, W. B. Saunders Company, 1973.

Kones, R. J.: Cardiogenic Shock: Mechanisms and Management. New York, Futura Publishing Company, Inc., 1975.

Mair, D. D., and Ritter, D. G.: The physiology of cyanotic congenital heart disease. *Intern. Rev. Physiol., 9*:275, 1976.

Pomerance, A., and Davies, M. J. (eds.): Pathology of the Heart. Philadelphia, J. B. Lippincott Company, 1975.

Schumer, W., and Nyhus, L. M. (eds.): Treatment of Shock. Philadelphia, Lea & Febiger, 1974.

Swan, H. J. C., and Parmley, W. W.: Congestive heart failure. *In* Sodeman, W. A., Jr., and Sodeman, W . A. (eds.): Pathologic Physiology: Mechanisms of Disease. 5th ed. Philadelphia, W. B. Saunders Company, 1974, p. 273.

Wilkinson, A. W.: Body Fluids in Surgery. 4th ed. New York. Churchill Livingston, Div. of Longman, Inc., 1973.

Zweifach, B. W.: Mechanisms of blood flow and fluid exchange in microvessels: hemorrhagic hypotension model. *Anesthesiology, 41*:157, 1974.

QUESTIONS

1. What is the normal resting cardiac output, and how is this affected by exercise?
2. Explain the mechanisms by which the cardiac output can be increased?
3. Explain the role of "venous return" in controlling cardiac output.
4. What is meant by "mean systemic filling pressure," and how does it affect venous return and cardiac output?
5. Discuss the role of tissue need for blood flow as the basic stimulus for control of cardiac output.
6. Explain the Fick method for measuring cardiac output.
7. What is the effect of hydrostatic pressure on peripheral venous pressure?
8. Why does cardiac failure often cause low cardiac output, pulmonary congestion, or peripheral congestion, or any combination thereof?
9. What is meant by circulatory shock, and what factors can cause low venous return shock?
10. What is meant by the following stages of shock: compensated, progressive, and irreversible?

THE BODY FLUIDS AND THE URINARY SYSTEM

V

BODY FLUIDS, CAPILLARY MEMBRANE DYNAMICS, AND THE LYMPHATIC SYSTEM   16

In Chapter 8, it was pointed out that there are two major types of fluid in the body, *extracellular fluid* and *intracellular fluid,* each of which has a chemical composition different from that of the other. In the present chapter, we will discuss the relationships of these fluids to overall body function, especially movement of fluids back and forth between the blood and interstitial spaces, and the function of some of the special fluids such as cerebrospinal fluid, ocular fluid, and lymph.

BODY WATER, INTRACELLULAR FLUID, AND EXTRACELLULAR FLUID

The total water in the average 70 kilogram man is about 40 liters, or 57 per cent of his total body weight. Of this, about 25 liters is intracellular fluid in the *intracellular compartment* and about 15 liters is extracellular fluid in the *extracellular compartment.* These relationships are illustrated in Figure 16–1.

BLOOD

Blood is composed of two portions, the *cells* and the fluid between the cells which is called *plasma.* The plasma portion of the blood is typical extracellular fluid except that it has a higher concentration of protein than do the extracellular fluids elsewhere in the body; this protein is very important for keeping the

plasma inside the circulatory system, as is discussed later in this chapter. The *cells* are of two types, *red blood cells* and *white blood cells.* The number of white blood cells is only $1/500$ the number of red blood cells so that, for many purposes, one can consider blood to be mainly a mixture of plasma and red blood cells. The normal total volume of plasma is 3 liters and of red cells 2 liters.

INTERSTITIAL FLUID

The interstitial fluid is the extracellular fluid that lies outside the capillaries and between the tissue cells. The volume of the interstitial fluid is equal to the total extracellular fluid volume minus the plasma volume, or $15 - 3 = 12$ liters of interstitial fluid volume in the normal, average adult man. Interstitial fluid is usually considered to include such special fluids as those in the cerebrospinal fluid system, in the fluid in the eyes, in the intrapleural space, in the peritoneal cavity, in the pericardial cavity, in the joint spaces, and in the lymph.

MEASUREMENT OF FLUID VOLUMES OF THE BODY

The Dilution Principle. Frequently it is important for the physiologist to measure the fluid volumes in different compartments of

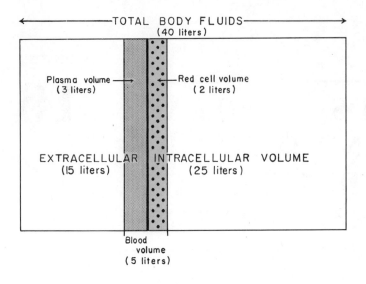

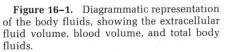

Figure 16–1. Diagrammatic representation of the body fluids, showing the extracellular fluid volume, blood volume, and total body fluids.

the body. To do this, he utilizes a basic principle called the *dilution principle* as follows:

Figure 16–2 illustrates a fluid chamber in which a small quantity of dye or other foreign substance is injected. After the substance mixes with the fluid in the chamber, its concentration becomes very dilute and equal in all areas of the chamber (Fig. 16–2B). Then a sample of the fluid is removed and the concentration of the substance analyzed by chemical, photoelectric, or any other means. The volume of the chamber can then be calculated by the following formula:

Volume in ml.

$$= \frac{\text{Quantity of test substance injected}}{\text{Concentration per ml. of sample fluid}}$$

It should be noted that all one needs to know is (1) the *total quantity of the test substance* injected into the chamber and (2) the *concentration of the substance in the fluid after dispersion.*

Measurement of Total Body Water. To measure total body water one can inject either *heavy water* or *radioactive water* into a person and then 30 minutes to an hour later remove a sample of blood and measure the concentration of heavy water or radioactive water in the blood or plasma. Since either one of these types of water disperses throughout the body, into the cells and everywhere else, in exactly the same way as normal water, one can use this measurement to calculate the total body water. Let us assume that 50 ml. of heavy water has been injected into the person

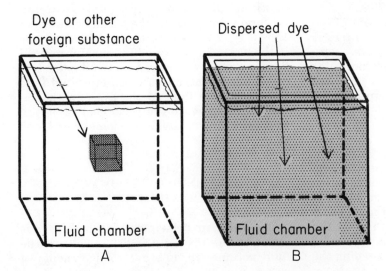

Figure 16–2. The dilution principle for measuring the volume of a fluid compartment.

and that its concentration in each milliliter of plasma an hour later, after dispersing throughout the body, is found to be 0.001 ml. Dividing this concentration into the total amount injected (using the above formula), we find that the total body water of this person is 50 liters, which is a value slightly higher than the value of 40 liters in the normal adult man.

Measurement of Extracellular Fluid Volume. To measure the extracellular fluid volume one simply uses a different test substance that will diffuse everywhere in the extracellular fluid compartment but that will not go through the cell membranes into the intracellular fluid compartment. Substances of this type include *radioactive sodium, thiocyanate ions,* and *insulin,* all of which can be used for determining the extracellular fluid volume. The procedure is almost identical with that used for measuring total body water. The test substance is injected intravenously, and a sample of plasma is removed about 30 minutes later, after appropriate mixing throughout the entire extracellular compartment has occurred. The same formula as above is used to calculate the extracellular fluid volume.

CALCULATION OF INTRACELLULAR FLUID VOLUME. Once the total body water has been determined using heavy water or radioactive water and the extracellular fluid volume has been determined using radioactive sodium or one of the other test substances, one can calculate intracellular fluid volume by subtracting extracellular fluid volume from total body water.

Measurement of Plasma Volume. The plasma volume is measured by injecting some substance that will stay in the plasma compartment of the circulating blood. Such a substance frequently used is a dye called *T-1824* which, upon intravenous injection, combines almost immediately with the plasma proteins. Since plasma proteins do not leak readily out of the plasma compartment of the blood, neither will the dye. After appropriate mixing of the dye in the blood has occurred, within about 10 minutes, a sample of blood is removed, the plasma is separated from the red cells, and the concentration of dye in the plasma is measured. Then, again using the dilution formula, one determines the plasma volume.

CALCULATION OF INTERSTITIAL FLUID VOLUME. Once the plasma and extracellular fluid volumes have been measured, one can

calculate the interstitial fluid volume by subtracting plasma volume from extracellular fluid volume.

CALCULATION OF BLOOD VOLUME. If one knows the percentage of the blood that is blood cells, he can calculate the total blood volume from the plasma volume. To determine the percentage of the blood that is plasma, blood is centrifuged at a high speed for 15 to 30 minutes, which separates the plasma from the cells so that the percentages of these two can be measured directly. The percentage of blood cells is called the *hematocrit* of the blood. From the hematocrit and the measured plasma volume one calculates total blood volume using the following formula:

Blood volume

$$= \frac{100}{100 - \text{Hematocrit}} \times \text{Plasma volume}$$

CAPILLARY MEMBRANE DYNAMICS — EXCHANGE OF FLUID BETWEEN THE PLASMA AND THE INTERSTITIAL FLUID

The whole purpose of the blood circulation is to transport substances to and from the tissues. Therefore, it is important for both water and dissolved substances to pass interchangeably between the plasma and the interstitial fluids and to bathe the tissue cells. This is achieved by diffusion of water and dissolved molecules in both directions through the capillary membrane. Usually this diffusion process is so effective that any nutrient that enters the blood will be distributed evenly among all the interstitial fluids within 10 to 30 minutes.

Fortunately, the rates of fluid diffusion in the two directions through the capillary membranes are almost exactly equal, so that the volumes of plasma and interstitial fluid remain essentially constant. Yet, under special circumstances the rate of fluid diffusion in one direction may become greater than in the other direction, in which case the plasma and interstitial volumes change accordingly, often becoming very abnormal. Therefore, it is important that we consider the dynamics of this net exchange of fluid between the plasma and the interstitial fluid and the mechanisms by which normal volumes are ordinarily maintained.

Anatomy of the Capillary System. Figure 16–3 illustrates a typical capillary bed, showing blood entering the capillaries from the arteriole, passing through the metarteriole, then into the capillaries, and finally to the venule. The *arteriole* has a muscular coat that allows it to contract or relax in response to stimuli which come mainly from the sympathetic nerves. The *metarteriole* has sparse muscle fibers, and the openings to the *true capillaries* are usually guarded by small *muscular precapillary sphincters* which can open and close the entryway into the capillaries. The muscles of the metarterioles and the precapillary sphincters are controlled mainly in response to local conditions in the tissues. For instance, lack of oxygen allows these muscles to relax, which increases the flow of blood through the capillary bed and in turn increases the amount of oxygen available to the tissues.

Pressures in the Capillary System. Though the pressure at the arterial end of a capillary varies tremendously depending upon the state of contraction or relaxation of the arteriole, metarteriole, and precapillary sphincters, the normal mean pressure at the arterial end of the capillary is usually between 25 and 35 mm. Hg. In the present discussion we will use an average value of about 30 mm. Hg. Likewise, the average pressure at the venous end of the capillary is about 10 mm. Hg, and the average mean pressure in the capillary is about 18 mm. Hg. Pressures have been measured at different points in capillaries using minute pipets protruding into the capillary lumens connected to special micro pressure manometers.

INTERSTITIAL FLUID PRESSURE

Interstitial fluid pressure is the pressure of fluid in the spaces between the cells. The quantitative level of interstitial pressure has been difficult to measure because the interstitial spaces generally have a width of less than 1 micron (micrometer), and introduction of any needle or pipet to measure the pressure can cause an abnormal reading. However, during the past few years we have studied this problem in our own laboratory in a completely different way, by implanting into tissues perforated capsules such as the one illustrated in Figure 16–4. The tissue grows inward through the holes, and new tissue then lines the inner wall of the capsule. A blood vascular system also develops in the new tissue. Yet, a large fluid space remains in the middle of the capsule, and special experiments have shown that the fluid in this space communicates directly through the capsule perforations with the surrounding interstitial fluid. A needle connected to a manometer system can be inserted through one of the perforations into the cavity and the pressure measured. Interstitial fluid pressures measured in this manner have been about −6 mm. Hg — that is, 6 mm. Hg less than atmospheric pressure — which demonstrates that the tissue spaces actually have a partial vacuum in them.

Pressure Difference Across the Capillary Membrane. If the average mean pressure in the capillary is 18 mm. Hg and the average pressure outside the capillary, the interstitial fluid pressure, is −6 mm. Hg, the total pressure difference between the two sides of the membrane would be 18 − (−6) or a total of 24 mm. Hg. That is, the pressure inside the capillary is 24 mm. Hg greater than the pressure outside the capillary. This pressure difference between the two sides of the membrane makes fluid tend to move out of the capillary into the tissue spaces. Fortunately, however, *colloid osmotic pressure*, another mechanism operating at the capillary membrane, opposes this tendency for fluid to move out of the capillaries, as is discussed in the following section.

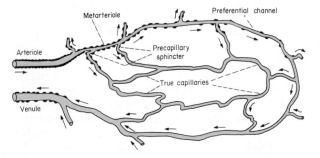

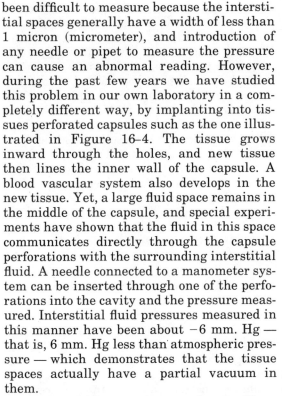

Figure 16–3. A typical capillary system. (Drawn from a figure by Zweifach: *Factors Regulating Blood Pressure.* Josiah Macy, Jr., Foundation, 1950.)

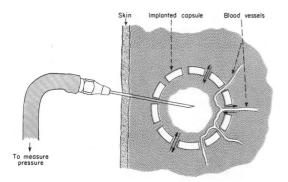

Skin   Implanted capsule   Blood vessels

To measure pressure

Figure 16–4. The perforated capsule method for measuring tissue fluid pressure.

COLLOID OSMOTIC PRESSURE AT THE CAPILLARY MEMBRANE

The Principle of Osmotic Pressure. The principles of osmosis and osmotic pressure were discussed in Chapter 8. To review briefly, if two solutions are placed on either side of a semipermeable membrane so that water molecules can go through the membrane pores but solute molecules cannot, water will move by the process of osmosis from the more dilute solution into the more concentrated solution. However, if pressure is applied to the more concentrated solution, this movement of water by osmosis can be slowed or halted. The amount of pressure that must be applied to stop completely the process of osmosis is called the *osmotic pressure*. For a much deeper understanding of why these effects take place, the student is referred back to Chapter 8.

Colloid Osmotic Pressure Caused by Plasma Proteins. The amount of osmotic pressure that develops at a membrane is determined to a great extent by the size of the pores in the membrane, because osmosis occurs only when dissolved particles cannot go through the pores. In the case of the cell membrane, such particles include sodium ions, chloride ions, glucose molecules, urea molecules, and others, all of which cause osmotic pressure. However, at the capillary membrane the pore diameter is roughly ten times that of the cell membrane — 80 Ångstroms (8 nanometers) in contrast to 8 Ångstroms) (0.8 nanometers) — so that sodium ions, glucose molecules, and essentially all other constituents of extracellular fluid pass directly through the capillary pores without causing osmotic pressure. However, the proteins in the plasma and in the interstitial

fluids are an exception, because their molecular sizes are large enough that most of them fail to penetrate even the large pores of the capillary membrane. Therefore, the proteins are the only solutes in the plasma and interstitial fluid that cause osmotic effects at the capillary membrane. Because solutions of plasma protein appeared to early chemists to be colloidal suspensions rather than true solutions, this protein osmotic pressure at the capillary membrane is called *colloid osmotic pressure*. However, some physiologists also call this pressure *oncotic pressure*.

Plasma has a normal protein concentration of 7.3 gm. per cent, while interstitial fluid has a concentration of about 2.0 gm. per cent, making a difference between the two sides of the capillary membrane of about 5.3 gm. per cent. Because the larger concentration is inside the capillary, osmosis of fluid tends to occur always from the interstitial fluid into the capillaries.

Colloid Osmotic Pressure of Plasma and Interstitial Fluids. If pure plasma were placed on one side of a capillary membrane and pure water on the other side, the colloid osmotic pressure that would develop would be about 28 mm. Hg. If pure interstitial fluid were placed on one side of the membrane and pure water on the opposite side, the colloid osmotic pressure that would develop would be about 4 mm. Hg. Therefore, we say that the colloid osmotic pressure of plasma is 28 mm. Hg and of interstitial fluid 4 mm. Hg.

The *colloid osmotic pressure difference* between the two sides of the membrane is equal to the difference between the colloid osmotic pressures of the two fluids on the two sides of the membrane. Therefore, the colloid osmotic pressure difference at the capillary membrane is 28 − 4, or 24 mm. Hg.

Equilibration of Pressures at the Capillary Membrane—the Law of the Capillaries. One will note from the foregoing discussion that under normal conditions the fluid pressure difference across the capillary membrane (24 mm. Hg) is equal to the colloid osmotic pressure difference across the membrane (24 mm. Hg). However, the fluid pressure tends to move fluid out of the capillary, while the colloid osmotic pressure tends to move fluid into the capillary. This balance between the two forces explains how it is possible for the circulation to keep its blood volume constant even though the capillary pressure is considerably higher than the interstitial fluid pressure. Were it not for the

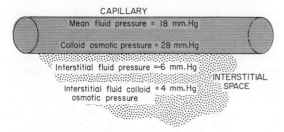

Figure 16–5.   Mean capillary dynamics.

colloid osmotic pressures, fluid would be lost continually from the circulation until eventually the blood volume would be insufficient to maintain cardiac output.

The normal state of equilibrium between the pressures tending to make fluid leave the capillaries and those tending to return fluid to the capillaries is called the *law of the capillaries,* and it is illustrated mathematically in Figure 16–5. The mean fluid presure in the capillary is shown to be 18 mm. Hg and the interstitial fluid pressure −6 mm. Hg, making a total fluid pressure difference between the inside and the outside of 24 mm. Hg tending to move fluid out of the capillary. On the other hand, the colloid osmotic pressure in the capillary is 28 mm. Hg and in the interstitial fluid 4 mm. Hg, making a net difference also of 24 mm. Hg, but this time tending to move fluid into the capillary. Thus, the two pressures are mathematically in balance so that the fluid volume of neither the plasma nor interstitial spaces will be changing.

Effect of Nonequilibrium at the Capillary Membrane.   Occasionally the pressures at the capillary membrane lose their state of equilibrium because one or more of them changes to a new value. When this occurs, fluid transudes through the membrane very rapidly until a new state of equilibrium develops.

Let us consider, for example, an increase in capillary pressure from the normal value of 18 up to 25 mm. Hg. This would increase the net pressure across the capillary membrane from zero up to a value of 7 mm. Hg in the outward direction, which would cause rapid transudation of fluid into the tissue spaces. The loss of fluid (but not of plasma proteins) out of the plasma would cause the *capillary pressure to fall* because of decreasing blood volume and would cause the *plasma colloid osmotic pressure to rise* because of increasing concentration of the plasma proteins. In addi-

tion, the *interstitial fluid pressure would rise* because of the increasing volume of interstitial fluid, and the interstitial fluid *colloid osmotic pressure would fall* because of dilution of the interstitial fluid proteins. Thus, after a few minutes the capillary pressure would be 22 mm. Hg, the plasma colloid osmotic pressure 29 mm. Hg, the interstitial fluid colloid osmotic pressure 3 mm. Hg, and the interstitial fluid pressure −4 mm. Hg. If the student will add these pressures, he will see that a new state of equilibrium has been established.

In a similar manner, any other change in any one of the pressure values on the two sides of the membrane will cause rapid movement of fluid through the capillary membrane until equilibrium is reestablished, usually within a few minutes to an hour or more. However, this fluid movement can sometimes be so great that the blood volume or the interstitial fluid volume becomes either abnormally large or abnormally small.

FLOW OF FLUID THROUGH THE TISSUE SPACES

In addition to very rapid diffusion of water and dissolved substances through the capillary membrane, there is a small amount of actual *flow* of fluid through the capillary membrane and the tissue spaces. The distinction between diffusion and flow is the following: Diffusion means movement of each molecule along its own pathway as a result of its kinetic motion irrespective of all other molecules, whereas flow means that large quantities of molecules all move together in the same direction.

Flow of fluid through the tissue spaces is caused by fluid movement through the arterial and venous ends of the capillaries as follows: If we observe Figure 16–6, we will see that the fluid pressure in the capillary at the arterial end is 30 mm. Hg, while at the venous end it is 10 mm. Hg. Now, let us study the pressures that cause fluid movement through the capillary membrane at both ends of the capillary. A sum of all the pressures acting at the arterial end of the capillary gives a net pressure of 12 mm. Hg in favor of movement of fluid *out of* the capillary, which is called the *filtration pressure* at the arterial ends of the capillaries.

At the venous ends of the capillaries, the greatly decreased fluid pressure in the capil-

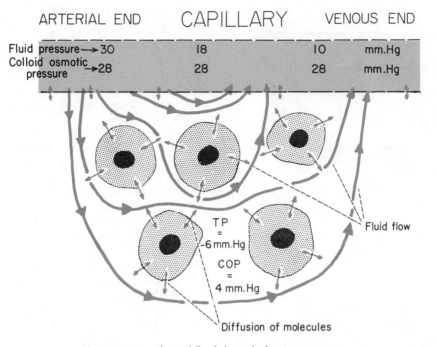

Figure 16–6.   Flow of fluid through the tissue spaces.

lary causes a net pressure in the opposite direction. This time the sum of the pressures is 8 mm. Hg tending to move fluid *inward* rather than outward. This net pressure difference is called the *absorption pressure*. Thus, it is clear that a small amount of fluid flows continually from the arterial ends of the capillaries through the tissue spaces toward the venous ends of the capillaries.

Note that the filtration and absorption pressures are not equal, but the difference is mainly counterbalanced by the fact that the venous ends of the capillaries are larger and therefore have about 50 per cent more surface area. As a result, almost equal amounts of fluid pass out of the arterial end of the capillary and return by the venous end. A small portion of the fluid, however, between $1/10$ and $1/100$ of the total, returns to the circulation by way of the lymphatics rather than by absorption into the venous ends of the capillaries.

THE LYMPHATIC SYSTEM

In addition to the blood vessels, the body is supplied with an entirely separate set of very small and thin-walled vessels called *lymphat-* ics. These originate in nearly all the tissue spaces as very minute *lymphatic capillaries.* The relationship of the lymphatic capillary to the cells and to the blood capillary is shown in Figure 16–7. The lymphatic capillaries then coalesce into progressively larger and larger lymphatic vessels that eventually lead to the neck of the person, as shown in the diagram of the lymphatic system in Figure 16–8. The lymph vessels then empty into the blood circulation at the junctures of the internal jugular and subclavian veins.

The lymphatics are an accessory system for flow of fluid from the tissue spaces into the circulation. The lymphatic capillaries are so permeable that even very large particles and protein molecules can pass directly into them along with tissue fluids. Therefore, in effect, the fluid that flows up the lymphatics is actually overflow fluid from the tissue spaces; it is known as *lymph,* and it has the same constituents as normal interstitial fluid.

At many points along the lymphatics, particularly where several smaller lymphatics combine to form larger ones, the vessels pass through *lymph nodes,* which are small organs that *filter* the lymph, taking all particulate matter out of the fluid before it empties into the veins.

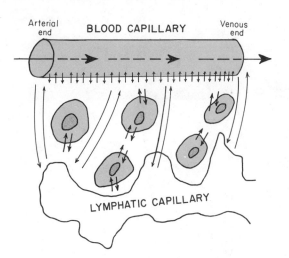

Figure 16–7. Relation of the lymphatic capillary to the tissue cells and to a blood capillary.

FUNCTION OF THE LYMPHATICS

Return of Proteins to the Circulation. The single most important function of the lymphatics is to return proteins to the circulation when they leak out of the blood capillaries. Some of the pores in the capillaries are so large that a small amount of protein leaks out of the plasma all of the time, amounting each hour to approximately $1/25$ of the total protein in the circulation. If these proteins were not returned to the circulation, the person's plasma colloid osmotic pressure would fall so low and he would lose so much blood volume into the interstitial spaces that he would die within 12 to 24 hours. Furthermore, no other means is available by which proteins can return to the circulation except by way of the lymphatics. Later in the chapter, when we discuss edema, we will see how serious it is for proteins not to be returned to the circulation even from a single part of the body.

Flow of Lymph Along the Lymphatics. Two principal factors determine the rate of lymph flow along the lymphatics: (1) the interstitial fluid pressure and (2) the degree of "pumping" by the lymphatic vessels, a mechanism called the *lymphatic pump*.

Whenever the interstitial fluid pressure rises above normal, which occurs when the interstitial fluid volume becomes too great, fluid flows easily in the lymphatic capillaries which have wide open communications with these spaces. Therefore, the greater the tissue pressure, the greater also is the quantity of lymph formed each minute.

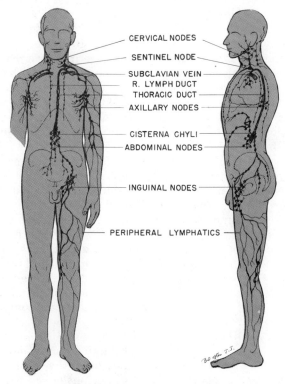

Figure 16–8. The lymphatic system.

THE LYMPHATIC PUMP. The *lymphatic pump,* the second factor that affects lymph flow, is a mechanism of the lymph vessels for pumping lymph along the vessels. To explain this mechanism, it is first necessary to point out that all lymph vessels contain *lymphatic valves* of the type illustrated in Figure 16–9, and these values are all oriented centrally so that lymph will flow only toward the point where the lymph vessels empty into the circulation and never backward toward the tissues.

A major cause of lymphatic pumping is periodic contraction of the lymph vessels themselves — about once every 6 to 10 seconds. When a lymph vessel becomes stretched with excess lymph, it automatically contracts. The contraction pushes the lymph past the next lymphatic valve. The lymph then distends the new segment of the lymph vessel, causing it also to contract. This effect reoccurs at each successive segment of the vessel until the lymph is pumped all the way back into the circulation.

Contraction of the lymphatic capillary also plays a major role in lymph flow. The lymphatic capillary has a special structure, as illustrated in Figurc 16–10, that is especially important to the pumping mechanism. Note that the endothelial cells are anchored to the surrounding tissues by *anchoring ligaments.* When a tissue fills with excess fluid, the entire tissue swells, and the swelling tissue pulls on the anchoring ligaments to open the lymphatic vessel; thus, fluid is also pulled into the vessel. Note also in the figure that the endothelial cells lining the lymphatic capillary overlap each other. Because of this arrangement, fluid flows into the lymphatic very easily because the edge of each endothelial cell simply flaps inward allowing a direct opening into the vessel. Then when the lymphatic capillary contracts, the overlapping edge of the endothelial cell flaps backward to close the opening. Thus, the overlapping edges of the endothelial cells are actually

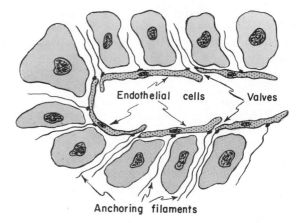

Figure 16–10. Special structure of the initial lymphatic capillaries that permits passage of substances of high molecular weight into the lymph.

valves that allow fluid to flow into the lymphatic capillary but not to flow backward. Consequently, contraction of the lymphatic capillary squeezes the lymph forward along the lymph vessel.

In addition to intrinsic contraction of the lymph vessels themselves, lymph pumping can also be caused by motion of the tissues surrounding the lymph vessels. For instance, skeletal muscle contraction adjacent to a lymph vessel can squeeze the vessel and move lymph forward. Likewise, arterial pulsations or movement of tissues because of active or passive movement of the limbs can squeeze lymphatics and cause fluid to move forward along the lymph channels. Under some conditions, these mechanisms are possibly as important as the intrinsic contraction of the lymph vessels themselves, especially during exercise.

The combination of all these lymph pumping mechanisms gives the lymphatic system its primary characteristic, namely, a tendency to return at all times any excess fluid that collects in the tissues back to the circulation. Indeed, under normal conditions, the pump actually creates a partial vacuum in the tissues, which is the basic cause of the negative pressure in the interstitial fluids.

RATE OF LYMPH FLOW, AND RETURN OF PROTEIN TO THE CIRCULATION. Lymph flow varies tremendously from time to time, but in the average person, the total lymph flow in all vessels is approximately 100 ml. per hour, or 1 to 2 ml. per minute. One can readily see that this is a very low rate of lymph flow, but it nevertheless is still sufficient to remove the excess fluid and especially the excess protein

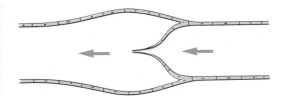

Figure 16–9. A lymphatic vessel, showing a lymphatic valve.

that tends to accumulate in the tissue spaces. The average lymph contains 3 to 4 per cent protein (2 per cent from peripheral tissues and as high as 5 to 6 per cent from the liver, where a major share of the lymph is formed).

Filtration of the Lymph by the Lymph Nodes. Another function of the lymphatic system is to filter the lymph as it returns from the tissues before emptying back into the circulation. Since large molecules or particulate matter cannot be absorbed directly into the blood capillaries, it is only through the lymphatic vessels that these substances can leave the tissues. Yet, many of these — for instance, bacteria or high molecular weight protein toxins such as tetanus toxin — can also be very dangerous to the person if they reach the blood. Fortunately, as the lymph passes through the lymph nodes, special cells, so-called *reticuloendothelial cells,* phagocytize both foreign protein molecules and particles, then digest both of these and release the products into the lymph in the form of amino acids or other small molecular breakdown products. Other cells in the lymph nodes form antibodies which, as we have seen in Chapter 6, play a major role in protecting the body against foreign toxins and bacteria.

THE NORMAL "DRY" STATE OF THE TISSUES — DEVELOPMENT OF EDEMA WHEN THIS FAILS

Importance of Negative Pressure in the Interstitial Spaces. Up to the present point very little has been said about the importance of the negative pressure that has been found to be present in the interstitial fluid spaces of normal tissues. Its main importance is that it serves to compact the tissues and, consequently, to keep the amount of fluid in the tissues to a minimum. That is, the lymphatic system keeps pulling fluid out of the tissue spaces until they become almost as small as possible, and it is in this state that the tissues normally operate. Therefore, we can say that,

from a relative point of view, the interstitial spaces are normally "dry," even though they still have 12 liters of fluid in them. This statement will become more meaningful as we discuss edema in the succeeding paragraphs. This stage of minimal fluid in the interstitial spaces is particularly beneficial to the diffusion of nutrients from the capillaries to the cells, for one of the basic principles of diffusion is that the rate of transport of substances by this mechanism becomes progressively decreased the greater the distance that the substances must diffuse.

When the body fails to maintain the "dry" state in the interstitial spaces, but instead collects large quantities of *extra* fluid in these spaces, the condition is called *edema*. The transport of nutrients to the tissues then becomes impaired, sometimes seriously enough to cause gangrene of the tissues. For instance, prolonged swelling of the feet frequently causes gangrenous ulcers of the skin, this resulting at least partially from too great a distance for the nutrients to diffuse from the capillaries to the tissue cells.

THE BASIC CAUSES OF EDEMA

Positive Interstitial Fluid Pressure as the Cause of Edema. Figure 16–11 illustrates the compactness of the cells in normal tissues and spreading apart of the cells when the interstitial fluid pressure rises and causes edema. Recent experiments have shown that as long as the pressure remains negative, the cells remain sucked together in a compact state, but just *as soon as the interstitial fluid pressure rises above the zero level and becomes positive, the cells spread apart*. At pressures of +3 mm. Hg. the interstitial fluid volume is often increased to as much as 3 to 4 times the normal amount, and at +8 mm. Hg the interstitial fluid volume, at least in some tissues, often rises to as much as 20 times normal, causing very great increases in sizes of the interstitial spaces, with resultant increase in

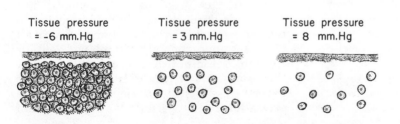

Tissue pressure = -6 mm.Hg   Tissue pressure = 3 mm.Hg   Tissue pressure = 8 mm.Hg

Figure 16–11. Effect of various tissue pressures on the amount of edema developing in the interstitial spaces.

distances for nutrients to diffuse from the capillaries to the cells.

Since it is positive interstitial fluid pressure that causes edema, one can readily understand that any change in the dynamics of fluid transfer at the capillary membrane that will increase the interstitial fluid pressure from its normal negative value of −6 mm. Hg to a positive value can cause edema. Four basic abnormalities of capillary membrane dynamics often result in edema. These are: (1) elevated fluid pressure in the capillaries, (2) decreased colloid osmotic pressure in the capillaries, (3) increased interstitial fluid colloid osmotic pressure, and (4) increased permeability of the capillaries.

EDEMA CAUSED BY ELEVATED FLUID PRESSURE IN THE CAPILLARIES. Though the normal mean capillary pressure is about 18 mm. Hg, this can rise to values as high as 40 to 50 mm. Hg in some abnormal states of the circulation. For instance, when the *heart fails,* the blood volume increases because of poor blood flow to the kidneys and the blood also dams up in the venous system, causing back pressure in the capillaries and thereby raising the capillary pressure. Also, *blockage of the veins* can do the same thing, or *overdilation of the arterioles* can cause such rapid blood flow into the capillaries that their pressure rises. In any of these conditions, a rise in the mean pressure in the capillaries will cause increased interstitial fluid pressure and resultant edema.

EDEMA CAUSED BY LOW PLASMA COLLOID OSMOTIC PRESSURE. Another cause of edema is a decrease in plasma colloid osmotic pressure, which results from diminished plasma protein concentration. This occurs very frequently in persons with some types of *kidney disease* who lose tremendous quantities of plasma proteins into the urine or in persons with *severe burns* who exude large amounts of proteinaceous fluid through their denuded skin. Also, lack of adequate protein in the diet, such as occurs in *famine areas* of the world, can cause failure of formation of adequate plasma proteins. When the plasma protein concentration falls below about 2.5 grams per cent, the normal negative pressure in the interstitial spaces becomes lost and, instead, the pressure rises to a positive value. Here again, very severe edema can result.

EDEMA CAUSED BY INCREASED INTERSTITIAL FLUID COLLOID OSMOTIC PRESSURE. Increased interstitial colloid osmotic pressure can also cause high interstitial fluid pressure and edema. The most severe cause of this is *blockage of the lymphatics,* which prevents return of proteins to the circulation. The proteins that leak through the capillary walls gradually accumulate in the tissue spaces until the interstitial colloid osmotic pressure approaches the plasma colloid osmotic pressure. As a result, the capillaries lose their normal osmotic advantage of holding fluid in the circulation so that fluid now accumulates abundantly in the tissues. This condition can cause edema of the greatest proportions.

Lymphatic blockage commonly occurs in the South Sea Island disease called *filariasis,* in which *filariae* become entrapped in the lymph nodes and cause so much growth of fibrous tissue in the nodes that lymph flow through the nodes becomes totally or almost totally blocked. As a result, certain areas of the body, such as a leg or an arm, swell so greatly that the swelling is called "elephantiasis." A single leg with this condition can weigh as much as the entire remainder of the body, all because of the extra fluid in the tissue spaces.

EDEMA CAUSED BY INCREASED CAPILLARY PERMEABILITY. Some abnormal conditions cause tremendous increase in permeability of the capillaries; that is, the capillary pores become greatly enlarged. When this occurs, not only does fluid leak rapidly from the capillaries into the tissue spaces, but also proteins are lost from the plasma while at the same time accumulating in great excess in the interstitial spaces. Therefore, increased capillary permeability can cause severe edema for at least three different reasons: (1) excess leakage of fluid through the enlarged pores into the interstitial spaces, (2) low colloid osmotic pressure in the plasma owing to loss of protein, and (3) increased colloid osmotic pressure in the interstitial fluid because of protein accumulation.

Increased capillary permeability frequently occurs in toxic states. For instance, in anaphylaxis, the condition described in the previous chapter that results from reaction of antibodies with toxic proteins, capillary permeability sometimes increases so greatly that localized edema occurs at many places in the body within minutes. Another example is the effect of some poisonous war gases which, when breathed, increase the permeability of the pulmonary capillaries so greatly that severe pulmonary edema occurs and causes the person to die because his lungs fill with fluid.

SPECIAL FLUID SYSTEMS OF THE BODY

THE POTENTIAL FLUID SPACES

There are a few spaces in the body that normally contain only a few milliliters of fluid under abnormal conditions. These spaces, called *potential spaces,* include especially the *pleural space,* illustrated in Figure 16–12, which is the space between the lungs and the chest wall; the *pericardial space,* which is the space between the heart and the pericardial sac in which it resides; the *peritoneal space,* which is the space between the gut and the abdominal wall; and the *joint spaces.* Ordinarily, all of these spaces are totally collapsed except for the presence of a highly viscid lubricating fluid that enables the tissue surfaces that surround the spaces to slip over each other, allowing free movement of the lungs in the chest cavity, free movement of the heart in the pericardial cavity, free movement of the joints, and so forth.

The Viscid Fluid in the Potential Spaces. If one will think for a few moments about the function of the potential spaces, these all exist where two tissue surfaces slide back and forth over each other — for instance, sliding of the lungs in the chest cavity during the process of breathing, sliding of the surfaces of the heart in the pericardial cavity during the heart beat, sliding of the joint surfaces over each other, and so forth. To increase the ease with which this sliding effect occurs, the potential spaces contain minute quantities of a very slick, highly viscid lubricating fluid.The essential ingredient of the fluid is *hyaluronic acid,* a mucopolysaccharide that gives the fluid a consistency almost the same as that of the slick mucus found in saliva.

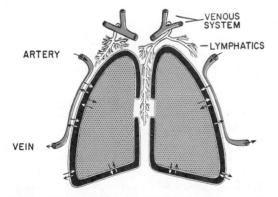

Figure 16–12. Dynamics of fluid exchange in the intrapleural spaces.

The concentration of hyaluronic acid in the fluids of the potential spaces is usually 1 to 2 per cent. It is interesting that this substance is also present in essentially all interstitial spaces of the body, not merely in the potential spaces. One of its roles in the interstitial spaces is perhaps the same as that in the potential spaces, namely, to allow the different segments of the tissues to slide over each other with ease, such as sliding of the skin over the loose subcutaneous tissues.

Dynamics of Fluid Exchange Through the Surfaces of the Potential Spaces. The dynamics of fluid exchange between the potential spaces and the capillaries in the tissues that surround the spaces are almost identical to the dynamics of fluid exchange at the usual capillary membrane. The reason for this is that the linings of these spaces are mainly permeable to both proteins and fluids so that fluid flows with ease between the potential spaces and the spaces of the adjacent tissues where the capillaries lie.

Figure 16–12 illustrates the intrapleural space, showing a capillary in the tissue adjacent to the cavity. This figure shows diffusion of fluid back and forth between the cavity and the capillary and also diffusion of fluid through the visceral pleura into the tissue of the lungs. Obviously, with such free diffusion, the pleural space is actually nothing more than an enlarged tissue space. There is negative pressure in the intrapleural space just as there is negative pressure elsewhere in the tissues. Therefore, normally, the intrapleural space is kept "dry," except for a few milliliters of viscid fluid in the whole chest cavity, in exactly the same way that the other tissue spaces of the body are kept "dry."

"Edema" of the Potential Spaces. Under abnormal conditions, a person can develop "edema" of the potential spaces in exactly the same way that he develops edema in his tissue spaces. Thus, if the capillary pressure rises too high, if the plasma colloid osmotic pressure falls too low, if the colloid osmotic pressure in the pleural cavity rises too high, or if the capillaries become too permeable, fluids will transude into the space in exactly the same way that they transude into the tissue spaces in edema. "Edema" of a potential space is called an *effusion*, though physiologically it is exactly the same process as edema.

Lymphatic Drainage of the Potential Spaces. All potential spaces have lymphatic drainage systems similar to that of the tissue spaces. Figure 16–12 illustrates the lymphatic drainage from the pleural cavity. This drainage normally removes the proteins that leak into the intrapleural space and, therefore, keeps the colloid osmotic pressure in the space at a low level.

One of the most serious causes of effusion in a potential space is *infection,* for this usually causes the lymphatics to become plugged with large masses of white blood cells as well as tissue debris caused by the infection. Also, infection increases the porosity of the surrounding capillaries, making them leak large amounts of protein. Thus, the proteins cannot leave the space even though excessive quantities of proteins are leaking into the space. As a result, the colloid osmotic pressure increases tremendously, and, as one would expect, fluid transudes into the space in massive amounts. Therefore, an infected joint swells tremendously, an infected pleural cavity develops large amounts of pleural effusion, and an infected abdominal cavity develops large amounts of abdominal effusion which in this case is called *ascites.*

THE FLUID SYSTEM OF THE EYE

The fluid system of the eye maintains a constant pressure in the eyeball of almost exactly 15 mm. Hg. This pressure, along with the very strong wall of the eyeball (the sclera), maintains the eyeball in a relatively rigid shape, keeping the distances between *cornea, lens,* and *retina* always constant within fractions of a millimeter. This is essential to the functioning of the eye's optical system for focusing images on the retina, as will be discussed in Chapter 28.

Aqueous Humor and Vitreous Humor. The fluids of the eye are divided, as shown in Figure 16–13, into two separate compartments: the *aqueous humor,* which is present in the *anterior chamber* of the eye between the front of the lens and the cornea, and the *vitreous humor,* which is in the *posterior chamber* between the lens and the retina. The vitreous humor is a gelatinous mass, and fluid in it can *diffuse* through the mass but cannot *flow* through it with ease. On the other hand, the aqueous humor flows freely.

Formation of Aqueous Humor by the Ciliary Body. Fluid is continually formed by the *ciliary body,* a small glandlike protrusion of epithelial cells that lies to the sides and slightly behind the lens, as illustrated in Figure 16–13. From here the fluid passes between the ligaments of the lens and into the anterior chamber of the eye. Then it flows, as illustrated by the arrows in the figure, into the angle between the cornea and the iris. There it leaves the eye through minute spaces called the *spaces of Fontana* and enters the

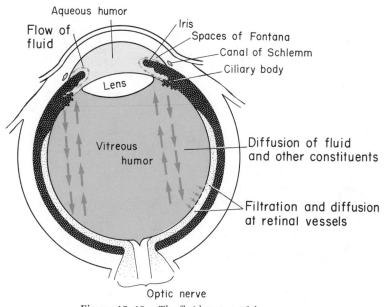

Figure 16–13.   The fluid system of the eye.

canal of Schlemm, a circular vein that passes all the way around the eye at the juncture of the cornea and the sclera.

The ciliary body is made up of a large number of small folds of epithelium which have a total surface area in each eye of about 6 sq. cm. This epithelium secretes sodium ions continually into the aqueous humor, which in turn produces the following effects: (1) The sodium ions cause a positive electrical charge to develop in the aqueous humor. This positive charge in turn pulls negative ions, especially chloride ions, through the epithelium, thus increasing the number of these ions in the aqueous humor as well. (2) The increased quantity of sodium and chloride ions in the aqueous humor causes osmosis of water through the epithelium. Thus, indirectly, the secretion of sodium causes a continuous but very slow flow of fluid, about 3 *cu.mm. per minute,* into the aqueous humor from the ciliary body.

Control of Pressure in the Eyeball. The pressure in the eye is controlled by the minute structures lining the spaces of Fontana at the angle between the cornea and the iris, which is where the fluid leaves the eyeball and enters the canal of Schlemm. If the pressure rises too high, increased amounts of fluid leave the eye. On the other hand, if the pressure falls too low, the outflow of fluid from the eyeball automatically decreases. That is, the spaces of Fontana act like a release valve on a steam boiler. When the pressure is above a critical value, the valve structures open up, and when the pressure is below the critical value, the spaces close. It is in this way that the pressure in the eyeball is regulated to almost exactly 15 mm. Hg, rarely rising above or falling below this value more than a few millimeters Hg.

GLAUCOMA. Glaucoma is a condition in which the pressure in the eye becomes very high, often high enough to cause blindness. The usual cause of the condition is failure of fluid to flow normally through the spaces of Fontana into the canal of Schlemm. When this happens, the continued formation of fluid by the ciliary body, without proper removal of the fluid into the canal of Schlemm, causes the eye pressure to rise sometimes to tremendous levels. For instance, *infection* or *inflammation* can cause debris or white blood cells to plug the openings of the spaces of Fontana. Also, many people develop narrowing of these spaces for reasons that are unknown.

In rare instances the pressure in the eye rises to as high as 60 to 80 mm. Hg. These extremely high pressures can cause blindness within a few days by damaging the optic nerve where it enters the eyeball and by compressing the blood vessels against the retina so severely that the nutritive blood flow is cut off. When the eyeball pressure is elevated only moderately, on the other hand, blindness can develop gradually over a period of years.

THE CEREBROSPINAL FLUID SYSTEM

The cerebrospinal fluid system, illustrated by the dark areas of Figure 16–14, consists of spaces inside the brain as well as the spaces surrounding the outside of the brain and spinal cord. This system is similar to that of the eye in that most of the fluid is formed in one area and is reabsorbed in an entirely different area. The fluid of the cerebrospinal fluid cavity plays a special role as a cushion for the brain; the density of the brain is almost the same as that of the cerebrospinal fluid, so that

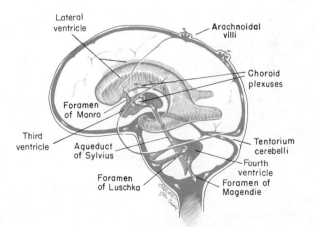

Figure 16–14.   The cerebrospinal fluid system.

the brain literally *floats* in this fluid. Furthermore, the cranial vault in which the brain and cerebrospinal fluid lie is a very solid structure so that when it is hit on one side the whole vault moves as a unit, both the fluid and the brain being propelled in the same direction at the same time. And, because of the cushioning effect of the cerebrospinal fluid, no damage to the brain usually results despite the fact that the brain is among the softest of tissues in the entire body.

Formation, Flow, and Absorption of Cerebrospinal Fluid.   Essentially all of the cerebrospinal fluid is formed by the *choroid plexuses,* which are cauliflowerlike growths that protrude into all the large spaces inside the brain, spaces called *ventricles.* The locations of the choroid plexuses are shown in Figure 16–14 — in the two *lateral ventricles* in the two cerebral hemispheres, the *third ventricle* in the diencephalic region, and the *fourth ventricle* between the hindbrain and the cerebellum. Fluid formed in all the ventricles flows finally into the fourth ventricle. From here the fluid flows out of the ventricles, through several small openings into the *subarachnoid space* which is the fluid space between the brain and the cranial vault. The fluid then flows around the brain and upward over its upper surfaces where it is absorbed into venous sinuses through *arachnoidal villi* that are valvelike openings into the veins.

The choroid plexuses, like the ciliary processes of the eye, continually secrete fluid by essentially the same mechanism as that described for the eye. Likewise, the arachnoidal villi function in a manner similar to that of the spaces of Fontana in the angle between the iris and cornea of the eye, for these villi also act as an overflow valve system. When the pressure in the cerebrospinal fluid is about 7 mm. Hg above that in the veins, fluid flows through the villi, but when the pressure in the cerebrospinal fluid system is less than this level, fluid does not leave the system.

The only major difference between the fluid system of the eye and the cerebrospinal fluid system is the level of the pressure at which fluid leaves the system, for the pressure required to open the overflow valves in the eye is about 15 mm. Hg, while the pressure required to release fluid from the cerebrospinal fluid system into the veins is about 7 mm. Hg.

Blockage of Flow in the Cerebrospinal Fluid System, and Development of Hydrocephalus.   Referring again to Figure 16–14,

one can see by the small white arrows that flow of fluid in the cerebrospinal fluid system can be blocked at many places, especially at the small openings between the ventricles, and also at the arachnoidal villi. For instance, many babies are born with congenital blockage between the ventricles so that fluid formed in the lateral and third ventricles cannot escape to the fourth ventricle and thence to the surface of the brain to be absorbed. As a result, the lateral and third ventricles swell larger and larger, compressing the brain against the cranial vault and destroying much of the neuronal tissue. This condition is called *hydrocephalus,* meaning simply "water head."

Another common cause of blockage is infection in the cerebrospinal fluid cavity, which causes so much debris and white blood cells in the fluid that the arachnoidal villi become plugged. In this case, fluid accumulates in the entire cerebrospinal fluid system, and the pressure rises very high. In rare instances the pressure in the cerebrospinal fluid system can rise so high that it actually impedes blood flow in the brain so that the brain cannot receive adequate nutrition.

REFERENCES

Andersson, B.: Regulation of body fluids. *Ann. Rev. Physiol., 39*:185, 1977.

Baez, S.: Microcirculation. *Ann. Rev. Physiol., 39*:391, 1977.

Bill, A.: Blood circulation and fluid dynamics in the eye. *Physiol. Rev., 55*:383, 1975.

Charm, S. E., and Kurland, G. S.: Blood Flow and Microcirculation. New York, John Wiley & Sons, Inc., 1974.

Comper, W. D., and Laurent, T. C.: Physiological function of connective tissue polysaccharides. *Physiol. Rev., 58*:255, 1978.

Cserr, H. F.: Physiology of the choroid plexus. *Physiol. Rev., 51*:273, 1974.

Davson, H.: The Physiology of the Cerebrospinal Fluid. Boston, Little, Brown and Company, 1967.

Guyton, A. C.: Concept of negative interstitial pressure based on pressures in implanted perforated capsules. *Circ. Res., 12*:399, 1963.

Guyton, A. C., and Lindsey, A. W.: Effect of elevated left atrial pressure and decreased plasma protein concentration on the development of pulmonary edema. *Circ. Res., 7*:649, 1959.

Guyton, A. C., Taylor, A. E., and Granger, H. J.: Circulatory Physiology. II. Dynamics and Control of the Body Fluids. Philadelphia, W. B. Saunders Company, 1975.

Guyton, A. C., Taylor, A. E., Granger, H. J., and Coleman, T. G.: Interstitial fluid pressure. *Physiol. Rev., 51*:527, 1971.

Haddy, F. J., Scott, J. B., and Grega, G. J.: Peripheral circulation: fluid transfer across the microvascular membrane. *Intern. Rev. Physiol., 9*:63, 1976.

Nicoll, P. A., and Taylor, A. E.: Lymph formation and flow. *Ann. Rev. Physiol., 39*:73, 1977.

Podos, S. M.: Glaucoma. *Invest. Ophthalmol., 12*:3, 1973.

Shoemaker, V. H., and Nagy, K. A.: Osmoregulation in amphibians and reptiles. *Ann. Rev. Physiol., 39*:449, 1977.

QUESTIONS

1. Explain the dilution principle for measuring body fluid volumes.
2. Give the average pressures in the capillary and in the interstitial fluid surrounding the capillary. Give also the colloid osmotic pressures inside and outside the capillary.
3. If the capillary pressure is 25 mm. Hg, the interstitial fluid pressure -2 mm. Hg, the plasma colloid osmotic pressure 35 mm. Hg, and the interstitial fluid colloid osmotic pressure 2 mm. Hg, which way will fluid be moving as a result of the pressure balance at the capillary membrane?
4. Explain why there is net movement of fluid out of the arterial ends of the capillaries and net absorption of fluid at the venous ends.
5. Why is the most important function of the lymphatic system to return protein to the circulation?
6. Explain the mechanism of lymphatic pumping.
7. Why does edema not occur in normal tissues?
8. What causes edema to develop, and what are some of its common causes?
9. Explain what is meant by a "potential fluid space" and discuss the dynamics of the fluids in these spaces.
10. Explain the mechanism for secretion of fluid into the eye.
11. How is intraocular pressure controlled?
12. Describe formation, flow, and absorption of cerebrospinal fluid.

FORMATION OF URINE BY THE KIDNEY, AND MICTURITION

The kidney forms urine and while doing so also regulates the concentrations of most of the substances in the extracellular fluid. It accomplishes this by removing those materials from the blood plasma that are present in excess, while conserving those substances that are present in normal or subnormal quantities.

PHYSIOLOGIC ANATOMY OF THE KIDNEY

Figures 17–1 and 17—2 illustrate respectively the gross and microscopic structures of the kidney which are responsible for this fluid purifying function. Figure 17–1 shows the renal artery entering the substance of the kidney and the renal vein returning from it. Urine is formed from the blood by the *nephrons,* one of which is shown in detail in

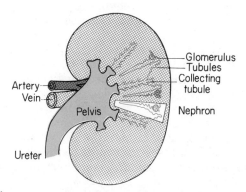

Figure 17–1. Principal anatomic structures of the kidney.

Figure 17–2. From these, urine flows into the *renal pelvis* and then out through the *ureter* into the *urinary bladder.* The two kidneys contain approximately two million nephrons, and because each nephron operates almost exactly the same as all others, we can characterize most of the functions of the kidney as a whole by explaining the function of a single nephron.

The Nephron. The nephron is composed of two major parts, the *glomerulus* and the *tubules.* The glomerulus is a tuft of capillaries surrounded by a capsule called *Bowman's capsule.* Fluid filters out of the capillaries into this capsule and then flows from here, first, into the *proximal tubule;* second, into a long loop called the *loop of Henle;* third, into the *distal tubule;* fourth, into a *collecting tubule;* and, finally, into the *renal pelvis.* As the filtrate passes through the tubules, most of the water and electrolytes are reabsorbed into the blood, but almost all of the end-products of metabolism pass on into the urine. In this way, water and electrolytes are not depleted from the body, though the waste products of metabolism are removed constantly.

FUNCTION OF THE NEPHRON

In discussing the function of the nephron it is desirable to use a simplified diagram such as that shown in Figure 17–3. This shows the "functional nephron," with an *afferent arteriole* supplying blood to the glomerulus through an *efferent arteriole* to flow into the *peritubular capillaries* and finally into the

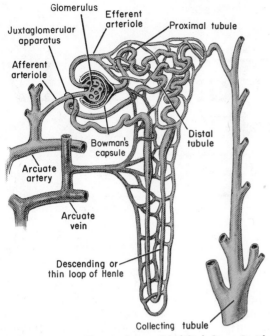

Figure 17–2. The nephron. (Modified from Smith: *The Kidney: Structure and Function in Health and Disease.* Oxford University Press.)

vein. Also shown are the *glomerular membrane, Bowman's capsule,* the *tubules,* and the *kidney pelvis.*

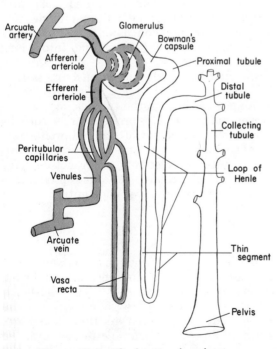

Figure 17–3. The functional nephron.

BASIC THEORY OF NEPHRON FUNCTION

The basic function of the nephron is to clean, or "clear," the blood plasma of unwanted substances as it passes through the kidney while retaining in the blood those substances that are still needed by the body. For instance, the end-products of metabolism such as urea and creatinine are especially cleared from the blood. And sodium ions, chloride ions, and any other ions are also cleared when they accumulate in excess in the plasma.

The nephron clears the plasma of unwanted substances in two different ways: (1) It filters a large amount of plasma, usually about 125 ml. per minute, through the glomerular membranes of the nephrons. Then as this filtered fluid flows through the tubules, the unwanted substances fail to be reabsorbed while the wanted substances are selectively returned to the plasma. (2) Some substances are cleared by the process of secretion. That is, the tubular walls actively remove substances from the blood and secrete them into the tubules.

Thus, the urine that is eventually formed is composed of *filtered* substances and *secreted* substances.

GLOMERULAR FILTRATION

The membranes of the capillaries in the glomerulus are collectively called the *glomerular membrane.* This membrane is several hundred times as permeable to water and small molecular solutes as the usual capillary membrane elsewhere in the body, but otherwise the same principles of fluid dynamics apply as to other capillary membranes. Like other capillary membranes, the glomerular membrane is almost completely impermeable to plasma protein, and it is also impermeable to the blood cells.

However, the pressure in the glomerulus is very high, believed to be about 60 mm. Hg, in contrast to the low pressures, between 15 and 20 mm. Hg, in capillaries elsewhere in the body. Because of this high pressure, fluid leaks continually out of all portions of the glomerular membrane into Bowman's capsule. We shall see that most of the fluid that leaks out of the glomerular membrane is later reabsorbed from the renal tubules into the peritubular capillaries. That which is not reabsorbed becomes urine.

Fluid Dynamics at the Glomerular Membrane, and the Filtration Pressure. The fluid pressures in the normal nephron are illustrated in Figure 17–4. This figure shows that the *glomerular pressure* is normally 60 mm. Hg, while the colloid pressure in the glomerulus is normally 32 mm. Hg. The pressure in Bowman's capsule is about 18 mm. Hg, and the colloid osmotic pressure in this capsule is essentially zero. Therefore, the pressure tending to force fluid out the the glomerulus is 60 mm. Hg, while the total pressure tending to move fluid into the glomerulus from Bowman's capsule is 32 + 18, or 50 mm. Hg. The difference between the outward pressure of 60 and the inward pressure of 50, or 10 mm. Hg, is the net pressure pushing fluid into Bowman's capsule; this is called the *filtration pressure.*

The Glomerular Filtration Rate. The rate at which fluid flows from the blood into Bowman's capsule, called the *glomerular filtration rate,* is directly proportional to the filtration pressure. Therefore, any factor that changes any one of the pressures on the two sides of the glomerular membrane will also change the glomerular filtration rate. Thus, an increase in glomerular pressure increases the rate of glomerular filtrate formation, while an increase in either the glomerular colloid osmotic pressure or the pressure in Bowman's capsule decreases the rate of filtrate formation.

The normal rate of formation of glomerular filtrate is 125 ml. per minute. This is approximately 180 liters each day, or 4.5 times the amount of fluid in the entire body, which illustrates the magnitude of the renal mechanism for purifying the body fluids. Fortunately, almost 179 liters of this fluid is reabsorbed by the tubules so that only slightly more than 1 liter of this total filtrate is lost into the urine.

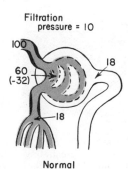

Filtration
pressure = 10

Normal

Figure 17–4. Normal pressures at different points in the nephron, and the normal filtration pressure.

EFFECT OF AFFERENT ARTERIOLAR CONSTRICTION ON FILTRATION. The major effect of constricting the *afferent arteriole* is a drastic decrease in the pressure in the glomerulus. And this in turn causes an even more drastic decrease in filtration pressure and glomerular filtration rate. For instance, a 10 mm. decrease in glomerular pressure can almost stop glomerular filtration.

The afferent arterioles are controlled partly by sympathetic nerves, and partly by an automatic control mechanism intrinsic in the nephron itself, called *autoregulation,* which will be discussed in detail later in the chapter. Sympathetic stimulation constricts the arterioles and lowers the glomerular pressure, thereby decreasing glomerular filtration. On the other hand, diminished sympathetic stimulation allows afferent arteriolar dilatation and, consequently, increased glomerular filtration.

Characteristics of Glomerular Filtrate. The filtrate entering Bowman's capsule, the *glomerular filtrate,* is an ultrafiltrate of plasma. The glomerular membrane is porous enough so that water and essentially all of the dissolved constituents of plasma except proteins can filter through. Therefore, glomerular filtrate is almost identical with plasma except that only a very minute quantity of protein (0.03 per cent) is present in the filtrate, while the protein concentration in plasma is approximately 7 per cent, or more than 200 times as great.

TUBULAR REABSORPTION

After the glomerular filtrate enters Bowman's capsule, it passes into the tubular system where each day all but slightly greater than 1 liter of the 180 liters of glomerular filtrate is reabsorbed into the blood, the remaining 1 liter passing into the renal pelvis as urine.

Figure 17–5 illustrates a microscopic cross-section of a tubular region of the kidney, showing the close proximity of the tubules to the peritubular capillaries. The tubular fluid is reabsorbed first into the interstitial spaces *ECF* and then from these spaces into the capillaries. Some of the substances are reabsorbed through the tubular epithelium by the process of *active reabsorption,* while other substances are reabsorbed by the process of *diffusion and osmosis.*

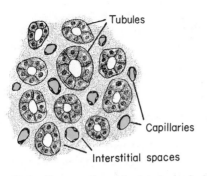

Figure 17–5. Cross-section of the tubules and adjacent capillaries in the kidney.

Active Reabsorption. The term *active reabsorption,* which was discussed in Chapter 8, means transport of substances through the tubular epithelial cells by means of special chemical transport mechanisms and also against a concentration difference between the tubular and interstitial fluids. From the interstitial spaces, the substances then diffuse into the peritubular capillaries.

Some of the substances reabsorbed by active transport are glucose, amino acids, proteins, uric acid, and most of the electrolytes — sodium, potassium, magnesium, calcium, chloride, and bicarbonate.

ACTIVE REABSORPTION OF NUTRIENTS FROM THE TUBULAR FLUID. Obviously, it is important to preserve the nutrients in the body fluids and not allow these to be wasted in the urine. To achieve this, glucose, amino acids, and proteins are all almost entirely reabsorbed even before the tubular fluid has passed all the way through the proximal tubules, the first portion of the tubular system. The active reabsorption processes for glucose, amino acids, and proteins are so powerful that ordinarily almost none of these substances are lost in the urine.

ACTIVE REABSORPTION OF IONS — ESPECIALLY SODIUM CHLORIDE (SALT). Reabsorption of the ions from the tubular fluid is somewhat different from reabsorption of the nutrients. The body needs to conserve a certain proportion of the ions but also to eliminate the excesses. Fortunately, special control mechanisms, several of which will be discussed in the following chapter, determine the amount of each ion that is to be reabsorbed. When the quantity already in the blood is too great, the ion is mainly excreted; but when the quantity in the blood is too low, much larger proportions of the ion will be reabsorbed.

The substance that is actively reabsorbed from the tubules to the greatest extent of all is sodium chloride, the total quantity reabsorbed each day being approximately 1200 grams, which is about three-fourths of all the substances actively reabsorbed in the entire tubular system. The reabsorption of sodium chloride is regulated partially by the hormone *aldosterone,* which is secreted by the adrenal cortex. This regulatory mechanism is discussed in the following chapter.

Figure 17–6 illustrates the mechanism of active sodium reabsorption. The serrated border of the tubular epithelial cell, called the *brush border,* is extremely permeable to sodium and allows sodium to *diffuse* rapidly from the lumen of the tubule to the inside of the cell. At the base and sides of the cell, on the other hand, the membrane has entirely different properties. Here the membrane is almost completely impermeable to diffusion of sodium, but it does *actively transport* sodium in the outward direction from the cell into the peritubular fluid. This active transport probably occurs in the manner explained in Chapter 8 as follows: It is believed that the sodium ion combines with a *carrier* which is dissolved in the cell membrane. The combined sodium-plus-carrier then diffuses to the opposite side of the membrane where the sodium is released into the peritubular fluid. Enzymes in or adjacent to the inner surface of the cell membrane cause the necessary reactions to take place, and the energy system in the cytoplasm of the cell supplies the energy required to make the reactions occur.

Active transport of other ions occurs in the same manner. However, each ion has its own specific carrier and its own set of enzymes for catalyzing the reactions.

ABSORPTION AGAINST A CONCENTRATION DIFFERENCE. One of the most important fea-

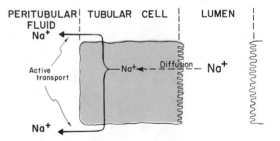

Figure 17–6. Mechanism of active reabsorption of sodium from the tubule, illustrating diffusion into the tubular epithelial cell from the tubular lumen and then active transport through the sides of the cell into the peritubular fluid.

tures of active reabsorption is that it can cause absorption of a substance even when its concentration is less in the tubule than in the peritubular fluid. To do this, however, the tubular epithelial cells must expend much energy. Therefore, these cells require tremendous amounts of nutrition, and their metabolic systems are so geared that they can transform the potential energy of their nutrients into the energy required to transport substances against the concentration difference. Indeed, these cells sometimes expend as much as 50 to 75 per cent of their energy for active transport alone.

Reabsorption by Diffusion. At this point we need to recall the basic principles of diffusion because this plays a major role in the reabsorption of water and a few other substances from the tubules. Diffusion means the random movement of molecules in a fluid, and it is caused by kinetic movement of all fluid molecules. In other words, each water molecule or each dissolved molecule in the water is constantly bouncing among all the others, wending its way from place to place, going first in one direction and then another. If a large enough pore is present in a membrane, a molecule can pass through the membrane — that is, the membrane is permeable to the molecule. The tubular epithelium is permeable to certain types of molecules, including water molecules. Therefore, water can diffuse from the tubules into the interstitial spaces of the kidney.

REABSORPTION OF WATER FROM THE TUBULES. The principal method for reabsorption of water from the tubules is by *osmotic diffusion,* which may be explained as follows:

First, let us recall what is meant by *osmosis.* Osmosis means net diffusion of water through a membrane caused by a greater concentration of nondiffusible substances on one side of the membrane than on the other side. The basic principles of osmosis were discussed in Chapter 8. However, we need to see now how the principles of osmosis apply to the transport of water from the tubular lumen.

When ions, glucose, and other substances are actively transported from the tubules into the interstitial spaces of the kidneys, the concentrations of these substances become decreased in the tubular fluid and increased in the peritubular capillary blood. Consequently, a very large total concentration difference of these substances develops across the epi-thelial membrane. The low concentration of solutes in the tubular fluid means that the water concentration in the tubule is relatively high, while the higher concentration of the solutes in the peritubular capillaries means that the water concentration is somewhat less. Obviously, therefore, water will now diffuse from its high concentration area in the tubule toward its lower concentration area in the peritubular capillaries.

Thus, when the dissolved substances are actively transported across the epithelial membrane, this automatically causes osmotic transport of water across the membrane as well. That is, the water "follows" the solutes.

FAILURE OF REABSORPTION OF UNWANTED SUBSTANCES — UREA, CREATININE, URIC ACID, PHOSPHATES, SULFATES, AND NITRATES

Some of the substances in the glomerular filtrate are undesirable in the body fluids; these substances, in general, are reabsorbed either not at all or very poorly by the tubules. For instance, *urea,* an end-product of protein metabolism, has no functional value to the body and must be removed constantly if protein metabolism is to continue. This substance is not actively reabsorbed, and the pores of the tubule are so small that urea diffuses through the tubular membrane several hundred times less easily than water. Therefore, while water is being osmotically reabsorbed, only about 50 per cent of the urea is reabsorbed. The remaining half of the urea remains behind and passes on into the urine.

Thus, the primary function of the kidney is this separation process in the tubules, the tubules reabsorbing those substances such as amino acids, electrolytes, and water which are needed by the body while, at the same time, allowing urea, a substance that is not needed by the body, to pass on into the urine.

Other substances that have a fate similar to that of urea include *creatinine, phosphates, sulfates, nitrates, uric acid,* and *phenols,* all substances that are end-products of metabolism and would damage the body if they remained in the body fluids in excessive amounts.

ACTIVE TUBULAR SECRETION

A few substances are actively secreted from the blood into the tubules by the tubular epithelium. Active secretion occurs by the same mechanism as active reabsorption but in the reverse direction.

Substances that are actively secreted include *potassium ions, hydrogen ions, ammonia,* and many toxic substances that often enter the body. Also, a number of drugs such as penicillin, Diodrast, Hippuran, and phenolsulfonphthalein are removed from the blood primarily by active secretion rather than by glomerular filtration. However, in normal function of the kidneys, tubular secretion is important only to help in regulation of potassium and hydrogen ion concentrations in the body fluids, which will be discussed in the following chapter.

RECAPITULATION OF NEPHRON FUNCTION – CONCENTRATION OF SUBSTANCES IN THE URINE

Now that both glomerular filtration and tubular reabsorption have been discussed it will be valuable to review in still a different way the total function of the nephron as follows:

The total blood flow into all the nephrons of both kidneys is approximately 1200 ml. per minute. Approximately 650 ml. of this is plasma, and about one-fifth of the plasma filters through the glomerular membranes of all the nephrons into Bowman's capsules, forming an average of 125 ml. of glomerular filtrate per minute. The glomerular filtrate is actually plasma minus the proteins. As the glomerular filtrate passes downward through the tubules, approximately 70 per cent of the water and ions are reabsorbed in the proximal tubules, while essentially all of the glucose, proteins, and amino acids are reabsorbed. As the remaining 30 per cent of the glomerular filtrate passes through the loop of Henle, distal tubules, and collecting tubules, variable amounts of the remaining water and ions are absorbed, depending on the need of the body for these substances, as discussed in the following chapter. The pH of the tubular fluid may rise or fall depending on the relative amounts of acidic and basic ions reabsorbed by the tubular walls. Also, the osmotic pressure of the tubular fluid may rise or fall depending on whether large quantities of ions or great amounts of water are reabsorbed. Thus,

the pH of the finally formed urine may vary anywhere from 4.5 to 8.2, while the total osmotic pressure may be as little as one-fourth that of plasma or as great as four times that of plasma.

Rate of Urine Flow. The final quantity of urine formed is normally about 1 ml. per minute or $1/125$ of the amount of glomerular filtrate filtered each minute. This 1 ml. of urine contains about one-half of the urea that is in the original glomerular filtrate, all of the creatinine, and large proportions of the uric acid, phosphate, potassium, sulfates, nitrates, and phenols. Thus, even though almost all the water and salt in the tubular fluid is reabsorbed, a very large proportion of the waste products in the original glomerular filtrate is never reabsorbed, but instead passes into the urine in a highly concentrated form.

REGULATION OF RATE OF FLUID PROCESSING BY THE TUBULES — THE PHENOMENON OF GLOMERULAR FILTRATION AUTOREGULATION

The reabsorption of water, salts, and other substances from the tubules depends greatly upon the rate at which glomerular filtrate flows into the tubular system. If the rate is very fast, none of the constituents is reabsorbed adequately before the fluid empties into the urine. On the other hand, when very little glomerular filtrate is formed each minute, essentially everything is reabsorbed, including urea and the other end-products of metabolism. Therefore, for optimum effectiveness in reabsorbing water and salts while not reabsorbing too much urea and other endproducts of metabolism, the glomerular filtration rate in each nephron must be very exactly controlled. This is called *glomerular filtration autoregulation.* And the need for very precise autoregulation of glomerular filtration is so important that there are two separate mechanisms for controlling the rate of glomerular filtration. These are (1) an afferent arteriolar vasoconstrictor feedback mechanism to decrease filtration and (2) an efferent arteriolar vasoconstrictor feedback mechanism to increase filtration. However, before we can explain these mechanisms, it is first necessary to discuss more fully the anatomy of the nephron. Referring back to Figure 17–2, note that the distal tubule, where it comes up from the loop of Henle, passes adjacent to the glomerulus and lies

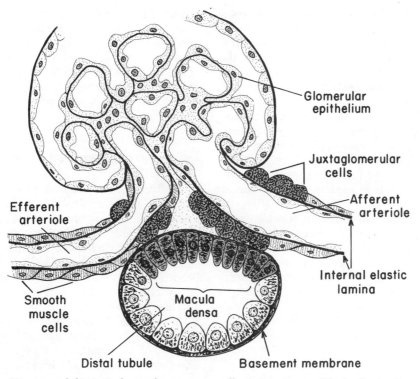

Figure 17-7. Structure of the juxtaglomerular apparatus, illustrating its possible feedback role in the control of nephron function. (Modified from Ham: *Histology*. J. B. Lippincott Co.)

between the afferent and efferent arterioles. This area of the nephron, called the *juxtaglomerular apparatus*, is illustrated in much greater detail in Figure 17-7, which shows the distal tubule lying in intimate contact with both the afferent and efferent arterioles. Where this contact is made, the epithelial cells of the distal tubule are dense and increased in number. Therefore, this portion of the distal tubule is called the *macula densa*. Some of the smooth muscle cells of the adjacent afferent arteriole, and to a lesser extent of the efferent arteriole as well, contain granules; these granulated cells are called *juxtaglomerular cells*. The granules contain a precursor form of the hormone *renin* which plays an important role in efferent arteriolar constriction during the glomerular filtration autoregulation process, as we shall see below. This renin also plays a role in the control of arterial pressure, as was discussed in Chapter 14.

The Afferent Arteriolar Vasoconstrictor Mechanism. When the rate of formation of glomerular filtrate is too great, the filtered fluid flows through the tubular system too rapidly for it to be processed adequately in the proximal tubule and the loop of Henle. As a result, the concentration of chloride ions in the fluid flowing from the loop of Henle into the distal tubule becomes too high. And this high concentration of chloride ions then, in some way that is not yet understood, causes the afferent arteriole to constrict. This in turn reduces the glomerular filtration rate back toward normal and also decreases the chloride ion concentration at the macula densa back toward normal. Thus, this afferent vasoconstrictor mechanism automatically helps to limit the rate of formation of glomerular filtrate to that amount of fluid that can be properly processed by the tubular system.

The Efferent Arteriolar Vasoconstrictor Mechanism. Constriction of the efferent arterioles causes exactly the opposite effects on glomerular filtration rate as afferent arteriolar constriction causes. That is, efferent constriction *increases* the glomerular pressure and therefore *increases* glomerular filtration rate. When glomerular filtration into the tubular system is too low, an automatic mechanism causes constriction of the efferent arterioles, thereby raising the glomerular filtration rate back toward normal. This mechanism is the following: Slow flow of fluid from the loop of Henle up to the macula densa of the distal tubule, for reasons that are not yet understood, causes the juxtaglomerular cells in the afferent and efferent arterioles to secrete renin. This then acts on renin substrate (a globulin type of protein) in the surrounding interstitial fluids of the kidney to form *angiotensin,* and the angiotensin in turn causes efferent arteriolar constriction. This increases the glomerular filtration rate, thereby returning the flow of fluid through the tubular system back toward the normal mean value.

Thus, there are two different mechanisms that work together to maintain a rate of glomerular filtration that is almost exactly right — not too much, not too little — for the filtrate to be processed properly by the tubular system.

EFFECT OF ARTERIAL PRESSURE ON RATE OF URINE FORMATION

An increase in arterial pressure steps up the rate of urine formation in two different ways. First, despite the autoregulation mechanism, increased arterial pressure still causes a slight rise in glomerular pressure. This, in turn, *increases* the glomerular filtration rate. Second, an elevation in arterial pressure also increases very slightly the peritubular capillary pressure, which *decreases* the rate of reabsorption of fluid from the tubules. Therefore, because of increased flow of glomerular filtrate into the tubules and yet decreased reabsorption of this fluid, the effect on urine output is multiplied, so that an increase in arterial pressure has a marked effect on urinary output.

An increase in arterial pressure from the normal value of 100 mm. Hg up to 120 mm. Hg approximately doubles urinary output, and an increase to 200 mm. Hg boosts urinary output six- to eightfold. Conversely, a decrease in arterial pressure to 60 mm. Hg almost completely stops urinary output.

This effect of arterial pressure on urinary output is exceedingly important for the control of arterial pressure itself, as was pointed out in Chapter 14. That is, when the arterial pressure rises too high, this automatically increases the rate of urine flow, a process that continues until the person is dehydrated enough so that his arterial pressure returns to its original value. Conversely, a decrease in

arterial pressure below normal causes retention in the body of ingested fluids and electrolytes until the arterial pressure increases enough to make the kidneys once again excrete an amount of fluid equal to the daily intake. Recent experiments demonstrate that this is the most important of all the pressure control mechanisms of the body.

THE CONCEPT OF CLEARANCE

The function of the kidney is actually to *clean* or to *clear* the extracellular fluids of substances. Every time a small portion of plasma filters through the glomerular membrane, passes down the tubules, and then is reabsorbed into the blood, the plasma is "cleared" of the unreabsorbed substances. For instance, out of the 125 ml. of glomerular filtrate formed each minute, approximately 60 ml. of that reabsorbed leaves its urea behind. In other words, 60 ml. of plasma is cleared of urea each minute by the kidneys. In the same way, 125 ml. of plasma is cleared of creatinine each minute; 12 ml., of uric acid; 12 ml., of potassium; 25 ml., of sulfate; 25 ml., of phosphate; and so forth.

Calculation of Renal Clearance. The method by which one determines how much plasma is cleared of a particular substance each minute is to take simultaneous samples of blood and urine while also measuring the volume of urine excreted each minute. From the samples the quantity of the substance in each milliliter of blood is analyzed chemically, and the quantity of the substance appearing in the urine each minute is also determined. By dividing the quantity of the substance in each milliliter of plasma into the quantity passing into the urine each minute, one can calculate the milliliters of plasma cleared per minute. That is,

Plasma clearance =

$$\frac{\text{Milligrams secreted in urine per minute}}{\text{Milligrams in each milliliter of plasma}}$$

As an example, if the concentration of urea in the plasma is 0.2 mg. in each milliliter and the quantity of urea entering the urine per minute is 12 mg., then the amount of plasma that loses its urea during that minute is 60 ml. To express this another way, the *plasma clearance* of urea is 60 ml. per minute.

Renal Clearance As a Test of Kidney Function. Since the primary function of the kidneys is to clear the plasma of undesired substances, one of the best means for testing overall kidney function is to measure the clearance of these substances. The clearance of urea has been one of the most widely used tests of renal function. As calculated above, normal plasma clearance of urea is approximately 60 ml. per minute; the clearance is less than this if the kidneys are damaged; the amount that it is depressed below normal gives one a measurement of the degree of kidney damage.

Measurement of Glomerular Filtration Rate by Measuring Renal Clearance of Inulin. The rate of clearance of the substance *inulin* by the kidneys is exactly equal to the glomerular filtration rate for the following reasons: Inulin filters through the glomerular membrane as easily as water, so that the concentration of inulin in the glomerular filtrate is exactly equal to that in the plasma. However, inulin is not reabsorbed or secreted even in the minutest degree by the tubules. Therefore, all of the inulin of the glomerular filtrate appears in the urine. In other words, all of the originally formed glomerular filtrate is cleared of inulin, which means that the rate of inulin clearance is equal to the rate of glomerular filtrate formation. As an example, a small quantity of inulin is injected into the blood stream of a person, and after mixing with the plasma its concentration is found to be 0.001 mg. in every milliliter of plasma. The amount of inulin appearing in the urine is 0.125 mg. per minute. On dividing 0.125 by 0.001 we find that the plasma clearance of inulin is 125 ml. per minute. This, therefore, is also the amount of glomerular filtrate formed each minute.

Estimation of Renal Blood Flow by Clearance Methods. The amount of blood flow through the two kidneys can be estimated from the clearance of either Diodrast or para-aminohippuric acid. These two substances, when injected into the blood in small quantities, are about 90 per cent cleared by active tubular secretion. Therefore, if the plasma clearance of Diodrast is found to be 600 ml. per minute, one can estimate that 600 ml. is equal to 90 per cent of the plasma that flowed through the kidneys during that minute, or a plasma flow of $600 \times {}^{100}/_{90} = 667$ ml. per minute. And if the blood hematocrit is 40, then the renal blood flow is $667 \times \dfrac{100}{100 - 40} = 1111$ ml. per minute.

ABNORMAL KIDNEY FUNCTION

Almost any type of kidney damage decreases the ability of the kidney to cleanse the blood. Therefore, kidney abnormalities usually cause an excess of unwanted metabolic waste products in the body fluids as well as poor regulation of the electrolyte and water composition of the fluids.

Kidney Shutdown. Several types of kidney damage can cause the kidneys to stop functioning suddenly and completely. Three of the most common are, first, poisoning of the nephrons by mercury, uranium, gold, or other heavy metals; second, plugging of the kidney tubules with hemoglobin following a transfusion reaction; and, third, destruction of the kidney tubules because of prolonged circulatory shock. In addition to these specific causes of shutdown, almost any other disease of the kidneys can cause either gradual or rapid shutdown.

Following kidney shutdown the concentrations of urea, uric acid, and creatinine, all of which are metabolic waste products, may reach ten times normal levels. Also, the body fluids may become extremely acidotic because of failure of the kidneys to excrete sufficient quantities of acid; or a person may develop hyperkalemia (excess potassium in the body fluids) because of failure to excrete potassium, and if the person continues to drink water and eat salt, he will become very edematous because of failure to rid himself of the ingested fluid and electrolytes. The person passes into coma within a few days, mainly because of acidosis. If the shutdown is complete, he will die in 8 to 14 days.

Kidney Abnormalities That Cause Loss of Nephrons. Many kidney diseases destroy large numbers of whole nephrons at a time. For instance, infection of the kidney can destroy large areas of the kidney; trauma can destroy part of or an entire kidney; occasionally a person is born with congenitally abnormal kidneys in which many of the nephrons are already destroyed; or nephron destruction can result from poisons, toxic diseases, or arteriosclerotic blockage of renal blood vessels.

As many as two-thirds of the nephrons in the two kidneys can usually be destroyed before the composition of the person's blood becomes greatly abnormal. The reason for this large margin of safety is that the undamaged nephrons can then function much more rapidly than usual. The amount of blood flowing into each nephron becomes greatly increased, and the glomerular filtration rate per nephron can rise to two times normal. This increased activity compensates to a great extent for the lost nephrons, allowing the metabolic waste products to be removed in sufficient quantity to maintain normal body fluid composition. However, these persons usually are treading a thin line of safety because bouts of excess metabolism caused by exercise, fever, or even ingestion of too much food may present the kidneys with far more waste products than they can handle.

Obviously, as the degree of kidney destruction progresses, the derangements of the extracellular fluid become progressively more severe until finally the condition approaches that of kidney shutdown. The person then develops extreme edema, acidosis, and eventually coma and death.

Acute and Chronic Glomerulonephritis. A very common kidney disease is acute glomerulonephritis. This is an allergic disease caused by toxins of certain types of streptococcic bacteria. Almost all episodes of acute glomerulonephritis occur approximately 2 weeks after a severe streptococcic sore throat or some other streptococcic infection. The glomeruli become acutely inflamed, swollen, and engorged with blood. Blood flow through the glomeruli almost ceases, and the glomerular membranes become extremely porous, allowing both red blood cells and protein to flow freely into the tubules. A person with acute glomerulonephritis has decreased kidney function, sometimes to the extent of complete kidney shutdown. If any urine is still formed, it usually contains large quantities of red blood cells and proteins.

In many instances of acute glomerulonephritis, the inflammation of the glomeruli regresses within 2 to 3 weeks, but even so the disease usually permanently destroys or damages a large number of nephrons. Repeated small bouts of acute glomerulonephritis may damage more and more nephrons, causing *chronic glomerulonephritis*. This can run a course of a few to many years, leading eventually to edema, coma, and death.

USE OF THE ARTIFICIAL KIDNEY

When the kidneys are damaged so severely that they can no longer maintain normal composition of the extracellular fluid, it is sometimes desirable to remove waste products from the extracellular fluid by use of an artificial kidney, such as the one shown in Figure 17–8. The artificial kidney is nothing more than a semiporous cellophane membrane arranged so that blood flows on one side of the membrane surface and a *dialyzing solution* flows on the other side. The membrane is porous to all substances in the blood except the plasma proteins and red blood cells. Therefore, almost all the blood substances can diffuse into the dialyzing solution, and the substances in the solution can also diffuse into the blood.

The dialyzing fluid contains none of the waste products of metabolism. Consequently, the waste products of metabolism diffuse from the blood into the fluid. On the other hand, the dialyzing fluid contains approximately the same concentrations of ions such as sodium, chloride, and so forth as those found in normal plasma. Therefore, these ions diffuse in both directions approximately equally, which prevents the blood from losing its normal ions.

Unfortunately, the artificial kidney cannot be used with impunity because the person's blood must be rendered incoagulable during its use and because a very large amount of blood must flow through the artificial kidney to make it work. For these reasons it can be applied only for a few hours out of every several days.

MICTURITION

The term *micturition* means emptying the urinary bladder of urine. However, before this can happen, the urine must be transported from the kidneys to the bladder itself.

Transport of Urine to the Bladder. Urine, formed by each kidney, first passes into the *pelvis* of the kidney, shown in Figure 17–1, and then through the *ureter* to the *urinary bladder,* shown in Figure 17–9. Urine is forced along the ureter by *peristalsis,* an intermittent wavelike constriction beginning at the pelvis and spreading downward along the ureter toward the bladder. The constriction squeezes the urine ahead of it. Ordinarily, the urine is transported the entire distance from the pelvis to the bladder in less than one-half minute.

Occasionally, severe infection or congenital abnormalities destroy the ability of the ureteral wall to contract. As a result, urine begins to collect in the kidney pelvis, causing it to swell and promoting infection that may extend into the kidney. Also, the stagnation

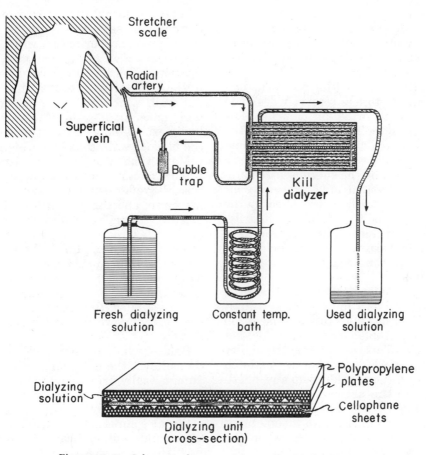

Figure 17–8.   Schematic diagram of the artificial kidney.

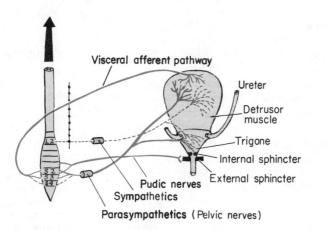

Figure 17–9.   The urinary bladder, and nervous pathways for control of micturition.

of urine may lead to precipitation of crystalline substances, the most prominent of which are various calcium compounds, the crystalline precipitates of which can grow eventually into large *calculi,* or *renal stones,* that partially or totally fill the pelvis. These stones in turn often cause extreme pain and further obstruction to urine flow.

Storage of Urine in the Bladder. The urinary bladder is a storage reservoir designed to prevent constant dribbling of urine. The exit of urine from the bladder is through the *urethra.* Two muscular sphincters surround the urethra, the *internal urethral sphincter* which is controlled by the subconscious autonomic nervous system and the *external urethral sphincter* which is controlled by the conscious portion of the brain. Ordinarily, both of these sphincters remain contracted so that urine cannot flow out of the bladder except when the person needs to urinate, as is explained below.

The urinary bladder itself can expand from a volume as small as 1 ml. to almost a liter. Until the bladder has filled to a volume of 200 to 400 ml., the intrabladder pressure does not increase greatly. This results from the ability of the smooth-muscle bladder wall to stretch tremendously without building up any significant tension in the muscle. However, once the bladder fills beyond 200 to 400 ml., the pressure does begin to rise, and it sometimes reaches as much as 40 mm. Hg when the bladder fills to 600 to 700 ml.

Emptying of the Bladder — the Micturition Reflex. In the human being micturition is caused by a combination of involuntary and voluntary nervous activity which may be explained as follows: When the volume in the bladder is greater than 200 to 400 ml., special nerve endings in the bladder wall called "stretch receptors" become excited. These transmit nerve impulses through the *visceral afferent nerve pathway* into the spinal cord (see Fig. 17–9), initiating a subconscious reflex called the *micturition reflex.* The nervous centers for the subconscious reflex are in the lower tip of the cord. The reflex signal is then transmitted from here by the parasympathetic nerves to both the bladder wall and the internal urethral sphincter. The bladder wall contracts to build up pressure in the bladder, and this in turn creates a conscious desire to urinate. At the same time, the micturition reflex also relaxes the internal sphincter. Then, the only impediment to urination is the still contracted external urethral sphincter. If the time and place are propitious for urination, the conscious portion of the brain will relax this external sphincter by inhibiting the normal impulses to this sphincter through the pudic nerve, and urination will take place.

If a person wishes to urinate before a micturition reflex has occurred, he can usually initiate the reflex by contracting the abdominal wall, which pushes the abdominal contents down against the bladder and momentarily excites some of the stretch receptors in the bladder wall, thus initiating the reflex.

Many times it is not convenient to urinate when a micturition reflex takes place. In this case, the reflex usually subsides within a minute or so, and the person loses his desire to urinate. The reflex then remains inhibited for another few minutes to as much as an hour before it returns again. If the reflex is again prevented from causing urination, it will once again become dormant for another period of time. However, as the bladder becomes more and more filled, the micturition reflex finally becomes so powerful that it is essential to urinate.

Babies, who have not developed voluntary control over the external urethral sphincter, automatically urinate every time the bladder fills. Also, adults whose spinal cords have been severed from the brain cannot keep the external sphincter contracted, so that they, too, urinate automatically when the bladder fills. However, these persons can often initiate this reflex by scratching the genital region, in this way controlling the time that the bladder will empty rather than waiting for the automatic and unannounced emptying.

REFERENCES

Aukland, K.: Renal blood flow. *Intern. Rev. Physiol., 11*:23. 1976. Brenner, B. M., Baylis, C., and Deen, W. M.: Transport of molecules across renal glomerular capillaries. *Physiol. Rev., 56*:502, 1976.

Brenner, B. M., Deen, W. M., and Robertson, C. R.: Determinations of glomerular filtration rate. *Ann. Rev. Physiol., 38*:9, 1976.

Burg, M., and Stoner, L.: Renal tubular chloride transport and the mode of action of some diuretics. *Ann. Rev. Physiol., 38*:37, 1976.

Gilmore, J. P.: Renal Physiology. Baltimore, The Williams & Wilkins Company, 1972.

Glynn, L. M., and Karlish, S. J. D.: The sodium pump. *Ann. Rev. Physiol., 37*:13, 1975.

Grantham, J. J., Irish, J. M. III, and Hall, D. A.: Studies of isolated renal tubules in vitro. *Ann. Rev. Physiol., 40*:249, 1978.

Greger, R., Lang, F., and Deetjen, P. Renal excretion of purine metabolites, urate and allantoin, by the mammalian kidney. *Intern. Rev. Physiol., 11*:257, 1976.

Kinne, R.: Membrane-molecular aspects of tubular transport. *Intern. Rev. Physiol., 11*:169, 1976.

Lassiter, W. E.: Kidney. *Ann. Rev. Physiol., 37*:371, 1975.

Maunsbach, A. B.: Cellular mechanisms of tubular protein transport. *Intern. Rev. Physiol., 11*:145, 1976.

Pitts, R.: Physiology of the Kidney and Body Fluids. 3rd ed. Chicago, Year Book Medical Publishers, Inc., 1974.

Renkin, E. M., and Robinson, R. R.: Glomerular filtration. *N. Engl. J. Med., 290*:785, 1974.

Stoff, J. S., Epstein, F. H., Narins, R., and Relman, A. S.: Recent advances in renal tubular biochemistry. *Ann. Rev. Physiol., 38*:46, 1976.

Sullivan, L. P.: Physiology of the Kidney. Philadelphia, Lea & Febiger, 1974.

Wright, F. S.: Intrarenal regulation of glomerular filtration rate. *N. Engl. J. Med., 291*:135, 1974.

QUESTIONS

1. Give the structure of a nephron.
2. Explain the mechanism for formation of glomerular filtrate and control of the glomerular filtration rate.
3. Explain the role of active reabsorption from the tubules.
4. Explain the role of absorption by diffusion from the tubules.
5. What substances are reabsorbed from the tubules by active reabsorption, and what substances by diffusion?
6. What substances are actively secreted by the tubules, and in what parts of the tubules are these substances secreted?
7. Explain the afferent arteriolar and the efferent arteriolar mechanisms for autoregulation of glomerular filtration rate.
8. If urea is present in the plasma in a concentration of 0.5 mg. per milliliter and the quantity of urea excreted into the urine per minute is 25 mg., how much plasma is cleared of urea each minute?
9. Describe the structure and function of the artificial kidney.
10. How is urine transported from the kidneys to the bladder, and what is the mechanism of bladder emptying?

18 REGULATION OF BODY FLUID CONSTITUENTS AND VOLUMES

Now that the function of the kidneys has been explained, it is possible to discuss the mechanisms by which most of the body fluid constituents are regulated. The kidneys play a special role in the regulation of (a) ion concentrations of the extracellular fluid, (b) osmotic pressure of all body fluid, (c) acidity of the fluid and (d) volumes of both extracellular fluid and blood. In the regulation of acidity, the respiratory system also plays a major role. All of these interrelationships are discussed in this chapter.

REGULATION OF THE ION CONCENTRATIONS AND BODY FLUID OSMOLALITY

REGULATION OF SODIUM ION CONCENTRATION AND BODY FLUID OSMOLALITY IN THE EXTRACELLULAR FLUID

Over 90 per cent of the positively charged ions (which are also called "cations") in the extracellular fluid are sodium. Furthermore, the total concentration of the positively charged ions automatically controls that of the negatively charged ions (the "anions") as well, because for each positive ion in the body fluids there must also be a negative ion. Therefore, for all practical purposes, when the sodium ion concentration is regulated, the total concentration of more than 90 per cent of all the ions in the body fluids is determined almost entirely by the total sodium ion concentration. It thus follows that sodium ion

concentration and osmolality of the fluids usually go hand in hand. That is, whenever the concentration of sodium ions increases, there is an almost exact corresponding increase in extracellular fluid osmolality. Conversely, whenever the sodium ion concentration becomes greatly decreased, there is an almost exact corresponding decrease in extracellular fluid osmolality.

Now, finally, to place the whole picture in perspective, it should be recalled from the discussion in Chapters 8 and 16 that a change in osmolality of the extracellular fluids causes a simultaneous and equal change in osmolality of the intracellular fluids.

For all of these reasons, one can see that the mechanisms that control the sodium ion concentration of the extracellular fluids are the same that control the osmotic pressure of both the extracellular and the intracellular fluids.

Water Dilution of the Body Fluids — the Mechanism for Controlling Sodium Ion Concentration and Body Fluid Osmolality. The *concentration* of sodium ions in the body fluids as well as the fluid osmolality is controlled mainly by increasing or decreasing the quantity of water in the body fluids. That is, whenever the fluids become too concentrated, water is automatically accumulated, thus diluting the body fluids and thereby decreasing both the sodium ion concentration and the body fluid osmolality. Conversely, when the concentration of the ions becomes too little, the water accumulated in the body is decreased, and this automatically increases both the sodium ion concentration and the body fluid osmolality back toward normal.

There are two separate mechanisms for regulating the degree of water dilution of the body fluids. These are (1) the antidiuretic hormone mechanism for controlling the rate of excretion of water through the kidneys and (2) the thirst mechanism for controlling the rate of intake of water. These two mechanisms work hand in hand, and they are regulated by an integrated neurologic mechanism located in the hypothalamus of the brain. Let us first discuss the antidiuretic mechanism for controlling the rate of excretion of water by the kidneys.

Control of Water Excretion by the Kidneys

Function of Antidiuretic Hormone to Control Water Reabsorption by the Distal Tubules and the Collecting Tubules. Before we can discuss the overall mechanism for controlling water excretion, it is first necessary to understand the function of antidiuretic hormone in controlling water reabsorption by the tubules. In the previous chapter it was pointed out that when ions and other solutes are absorbed by the tubular epithelium, osmotic forces that tend to cause water absorption at the same time are created across the tubular wall. However, for water reabsorption to occur, it is also essential for the tubular membrane to be permeable to water. In the proximal tubule and in the early portions of the loop of Henle, the tubular membrane is very permeable to water. Therefore, in these areas of the tubular system, water is reabsorbed in direct proportion to the reabsorption of solutes from the tubules. However, the distal tubule and collecting tubule are sometimes permeable and sometimes not; this depends upon the concentration of antidiuretic hormone in the body fluids.

Antidiuretic hormone, which is secreted by the posterior pituitary gland, controls the permeability of the distal and collecting tubules to water. In the absence of antidiuretic hormone, these tubules are almost entirely impermeable to water. Therefore, water fails to be reabsorbed and instead passes on through these tubules and is expelled into the urine. In the presence of antidiuretic hormone, however, the distal and collecting tubules become highly permeable to water so that water is then easily reabsorbed back into the blood

from the tubules; consequently, the amount of water that passes into the urine becomes greatly diminished. The rate of loss of water from the body fluids is thus determined by the presence or absence of antidiuretic hormone.

Mechanism for Excreting a Dilute Urine. Now that we have explained the effect of antidiuretic hormone on the distal and collecting tubules, it becomes very easy also to explain how the kidney can excrete a dilute urine and thereby remove excess water from the body. This is achieved in the following way: First, let us refer to Figure 18–1, which shows to the right the glomerulus and the tubular system of the nephron. Note that the tubular fluid passes downward into a *descending limb* of the loop of Henle and then back upward through an *ascending limb*. This ascending limb of the loop of Henle is very impermeable to water. On the other hand, it has very powerful active transport mechanisms for absorbing ions from the tubular fluid, especially chloride and sodium ions, which are the most abundant ions. The greater proportion of the ions is thus absorbed back into the blood while the water remains in the tubular fluid. When the posterior pituitary gland is not secreting antidiuretic hormone, this dilute fluid, with its great excess of water, passes on through the distal and collecting tubules directly into the urine. Under these conditions, then, the kidney excretes a very dilute urine, thereby removing large quantities of excess water from the body fluids. Note again, however, that this occurs only in the absence of antidiuretic hormone, which explains why the absence or presence of antidiuretic hormone is so important in determining whether or not excess water will be excreted.

Renal Mechanism for Excreting a Concentrated Urine — the "Countercurrent" Mechanism for Conserving Water in the Body. The kidney can also, at appropriate times, excrete a concentrated urine, thereby retaining water in the body while excreting large quantities of ions and other solutes into the urine. However, this mechanism is somewhat more difficult to understand than is the mechanism by which the kidney excretes a dilute urine. Let us return again to Figure 18–1 to help explain this mechanism.

Note in the figure that the upper part of the diagram represents the cortex of the kidney where the glomeruli, the proximal tubules, and the distal tubules are located. The lower part of the diagram represents the medulla

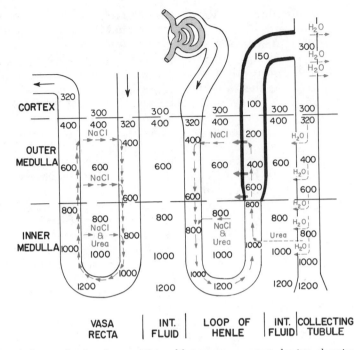

Figure 18-1. The tubular mechanism for excreting a dilute or a concentrated urine, showing especially the "counter-current" mechanism. (Numerical values are in milliosmols per liter.)

into which many of the loops of Henle protrude and through which the collecting tubule must also pass on its way to the pelvis of the kidney. The numbers on the diagram represent the total osmolar concentration of solutes at different points in the kidney in milliosmols per liter. These concentrations are very high in the medulla, and they become progressively greater toward the lower medulla where the lower tips of the loops of Henle lie and where the collecting tubules empty into the renal pelvis. The cause of this very high concentration of solutes in the medulla is twofold. First, large quantities of solutes are absorbed into the medulla from the ascending limbs of the loops of Henle and from the collecting tubules. Second, a so-called "countercurrent" mechanism operating in the medulla makes it difficult for the blood flowing through the medulla to remove these solutes, thus causing them to accumulate in the medulla until their concentrations rise very high. The countercurrent mechanism may be explained as follows:

Note to the left in Figure 18-1 the loop called *vasa recta*. This represents capillary blood vessels that pass down from the cortex deep into the medulla and then return again to the cortex. When the blood first enters the vasa recta, it has a solute concentration of about 300 milliosmols. However, because the vasa recta are very permeable, the osmolar concentration of blood in the vasa recta quickly approaches that in the medulla, becoming higher and higher as the blood passes farther down the vasa recta into the medulla and reaching a concentration of 1200 milliosmols at the tip of the medulla. Then as the blood passes back up the ascending limb of the vasa recta, rapid diffusion of solutes out of and water into the vasa recta causes the concentration to fall once again toward 300 milliosmols. Thus, the rapid diffusion of solutes back and forth between the medullary interstitium and the blood makes it difficult for the blood to carry many of the solutes away from the medulla. This is called the countercurrent mechanism for maintaining a high concentration of solutes in the medulla. It is represented by the arrows in the vasa recta.

A similar countercurrent mechanism occurs in the loop of Henle, which is also illustrated by arrows to the right in the figure.

Now let us understand the importance of these concentrated fluids in the medulla for excreting a concentrated urine.

Role of Antidiuretic Hormone in Excreting a Concentrated Urine. It was noted above that in the absence of antidiuretic hormone the kidney will excrete a very dilute urine.

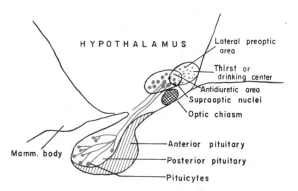

Figure 18-2. The supraoptico-pituitary antidiuretic system and its relationship to the thirst center in the hypothalamus.

We shall now see that in the presence of large quantities of antidiuretic hormone the kidney will excrete a very concentrated urine. This results in the following way:

It will be recalled that antidiuretic hormone makes the collecting tubule very permeable to water. Therefore, in the presence of antidiuretic hormone, the high concentration of solutes in both the medullary interstitium and the vasa recta will cause rapid osmosis of water out of the collecting tubules into the interstitium and the blood. Consequently, the solutes in the tubule become progressively more concentrated until they approach the concentration in the medullary interstitium and vasa recta. A concentrated urine is thus excreted, with loss of large quantities of solutes but very little loss of water. This obviously increases the ratio of water to solutes in the body fluids and thereby dilutes the body fluids, at the same time decreasing the sodium ion concentration as well.

The Osmo-Sodium Receptors of the Hypothalamus and the Antidiuretic Hormone Feedback Control System. Located in the anterior part of the hypothalamus, as illustrated in Figure 18-2, are the *supraoptic nuclei,* which contain nerve cells sensitive to the concentration especially of sodium ions in the extracellular fluid, but to a slight extent to that of other ions as well. These cells transmit large numbers of impulses down the pituitary stalk to the posterior pituitary gland when the concentration of sodium or other osmotically active substances becomes too great. On reaching the posterior pituitary gland, the nerve impulses cause the release of antidiuretic hormone (ADH) into the circulating blood. The blood in turn carries the antidiuretic hormone to the kidneys. (The hormonal mechanism of the posterior pituitary gland will be presented in more depth in Chapter 34.)

Figure 18-3 illustrates the effect of antidiuretic hormone once it reaches the kidneys. In the presence of antidiuretic hormone, as already explained, the kidney excretes a very concentrated urine, thus conserving water in the body. In the absence of antidiuretic hormone, the kidney excretes a very dilute urine, thus losing large quantities of water from the body fluids.

The antidiuretic hormone system, there-

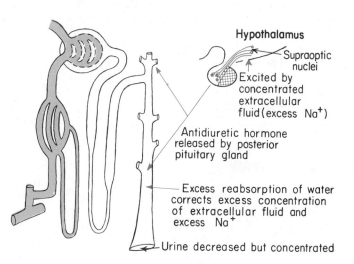

Figure 18-3. Control of extracellular fluid osmolality and sodium ion concentration by the osmo-sodium receptor–antidiuretic hormone feedback control system.

fore, represents a very important feedback control system for controlling both the sodium ion concentration and the osmotic concentration of the body fluids. When the sodium and osmotic concentrations become too great, increased secretion of antidiuretic hormone causes water retention in the body and thereby dilutes these concentrations. When the concentration of sodium or other osmotic elements becomes too low, decreased secretion of antidiuretic hormone causes water to be lost from the fluid, and their concentrations rise back toward normal level.

The Thirst Mechanism for Control of Sodium and Osmotic Concentrations

The thirst mechanism controls the intake of water at the same time that the antidiuretic mechanism controls the output of water. Returning again to Figure 18–2 one can see located lateral to, slightly above, and forward of the antidiuretic area in the supraoptic nuclei a center called the *thirst or drinking center*. At the same time that the osmo-sodium receptors of the supraoptic nuclei are stimulated, so also are neurons stimulated in the thirst or drinking center. Therefore, when antidiuretic hormone causes the kidneys to conserve water in the body, the drinking center simultaneously causes the person to drink large amounts of water. For both of these reasons the quantity of body water increases.

Conversely, when the osmo-sodium receptors of the supraoptic nuclei are not stimulated, so also are the neurons of the drinking center unstimulated, and the person has no desire to drink. Therefore, there is essentially no intake of water even though at the same time the kidneys excrete large amounts of water.

The thirst and drinking mechanisms thus operate synergistically with the kidney mechanism for control of body water, and thereby also for control of extracellular sodium ion and osmotic concentrations.

REGULATION OF POTASSIUM ION CONCENTRATION

From earlier discussions in this text, it will be recalled that the function of both nerve and muscle is highly dependent upon a well-regulated potassium ion concentration in the extracellular fluid. For instance, when the potassium concentration becomes too high, the membrane potential of both nerve and muscle decreases, thereby diminishing the voltage of the action potential and reducing the effectiveness of nerve impulse transmission as well as the strength of muscle contraction. Indeed, a high potassium ion concentration can sometimes make the heart so weak that the person dies of heart failure.

Potassium ion concentration in the extracellular fluids is regulated in a manner similar to that of sodium ion concentration, except that the hormone aldosterone, rather than antidiuretic hormone, plays the major role.

Effect of Aldosterone on the Transport of Sodium and Potassium in the Distal and Collecting Tubules. Aldosterone directly affects the epithelium of both the distal tubule and the collecting tubule to promote, simultaneously, *active absorption of sodium* from the tubules and *active secretion of potassium* into the tubules. These effects result from the fact that aldosterone increases the tubular epithelial cell enzymes and possibly other components of the carrier mechanisms for sodium absorption and potassium secretion.

Ordinarily one would think that since aldosterone causes increased tubular absorption of sodium, this hormone would be important for controlling the sodium ion concentration in the extracellular fluid. However, this turns out not to be true, the reason being that the antidiuretic hormone–thirst mechanism is about ten times as potent for control of sodium ion concentration as is the aldosterone mechanism and therefore normally overrides this mechanism. The quantity of aldosterone in the body thus can increase and decrease rather markedly with only small changes in the sodium ion concentration of the extracellular fluid.

On the other hand, the effect of aldosterone on potassium secretion does play an extremely important role in the control of potassium ion concentration.

The Aldosterone Feedback Mechanism for Control of Extracellular Fluid Potassium Ion Concentration. Figure 18–4 illustrates the aldosterone mechanism for control of potassium ion concentration. An increase in potassium ion concentration causes the adrenal cortex to boost its rate of secretion of aldosterone. The aldosterone then circulates in the blood until it reaches the kidneys. Here the hormone directly affects the distal and collecting tubules to increase potassium secretion. Con-

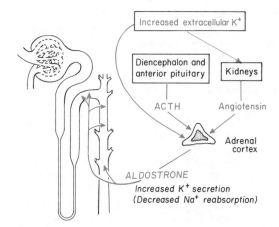

Figure 18–4. Postulated mechanisms for control of potassium concentration in the extracellular fluid.

sequently, large quantities of potassium are lost into the urine, and the potassium ion concentration in the extracellular fluid decreases back toward normal. Then, when the potassium concentration becomes too low, aldosterone secretion by the adrenal cortex becomes greatly decreased. Now the rate of potassium secretion into the urine is diminished almost to zero. And since the daily food intake contains a large amount of potassium, potassium accumulates in the extracellular fluid, thereby returning the potassium ion concentration once again toward normal.

Thus, the aldosterone mechanism serves as a typical feedback mechanism for control of the potassium ion concentration.

In addition to this feedback control of potassium ion concentration, an increase in potassium ions in extracellular fluids also has a direct effect on the tubules, independent of the aldosterone effect, of causing increased potassium ion secretion. This, too, helps in the feedback control of potassium ion concentration in the extracellular fluid.

REGULATION OF OTHER IONS IN THE EXTRACELLULAR FLUIDS

Regulation of Chloride and Bicarbonate Ions. The regulation of chloride and bicarbonate ion concentrations is mainly secondary to that of sodium concentration. The reason for this is the following: When sodium is reabsorbed from the kidney tubules, the positively charged sodium ions are transferred out of the tubular fluid into the interstitial fluids of the kidneys, creating a state of elec-

tronegativity in the tubules and electropositivity in the interstitial fluids. The negativity in the tubules repels negative ions from the tubules, while the positivity of the interstitial fluids attracts the negative ions. Therefore, when sodium ions are reabsorbed, chloride and bicarbonate ions are usually also reabsorbed. Sometimes more chloride than bicarbonate ions are reabsorbed, while at other times more bicarbonate ions are reabsorbed; which one it will be is determined by the acid-base balance of the extracellular fluid, which is discussed later in the chapter.

Regulation of Calcium, Magnesium, and Phosphate Concentrations. The details of the mechanisms by which calcium, magnesium, and phosphate concentrations are regulated by the kidneys are not very clear. Yet, it is known that too high a concentration of any one of these substances in the extracellular fluid causes the tubules to reject it and to pass it on into the urine. On the other hand, a low concentration causes the opposite effect, that is, rapid reabsorption of the substance until its concentration in the extracellular fluids returns to normal.

REGULATION OF ACID-BASE BALANCE

Regulation of acid-base balance actually means regulation of hydrogen ion (H^+) concentration in the body fluids. When the hydrogen ion concentration is great, the fluids are *acidic;* when hydrogen concentration is low, the fluids are *basic* (or *alkaline*). However, before we discuss acid-base balance, it is necessary to present some of the terminology that is often used.

First, an *acid* is a substance that has large numbers of free hydrogen ions (H^+) when dissolved in water. On the other hand, a *base* is a substance that has large numbers of hydroxyl ions (OH^-).

Second, a base is frequently called an *alkali*. Thus, when a person has excess base in his blood, he is said to have *alkalosis*. Conversely, when he has excess acid in his blood, he is said to have *acidosis*.

Third, acids and bases tend to neutralize each other. The reason for this is that the hydrogen ion combines with the hydroxyl ion to form water, as expressed by the following equation:

$$H^+ + OH^- \rightarrow H_2O$$

When there is a great excess of hydrogen ions in a solution, they will neutralize essentially all of the hydroxyl ions so that almost none of these will be present. Conversely, when there is an excess of hydroxyl ions, they will neutralize essentially all of the hydrogen ions so that the hydrogen ion concentration becomes extremely low. To state this another way, hydrogen ions and hydroxyl ions are mutually exclusive in the same fluid. When there are approximately equal amounts of acidic and basic substances in the body fluids, the acids and bases almost completely neutralize each other, and the fluids are said to be *neutral*. The normal body fluids are very near neutral, but are actually slightly on the basic side.

Fourth, the normal concentration of hydrogen ions in the body fluids is only one part in approximately 9 billion, or to express this in chemical terms, it is 4×10^{-8} equivalents per liter. However, this is such a difficult expression that the hydrogen ion concentration is usually expressed in terms of pH, which is the logarithm of the reciprocal of the hydrogen ion concentration according to the following formula:

$$pH = \log_{10}\left(\frac{1}{H^+ \text{ Conc.}}\right)$$

The normal pH of extracellular fluid and blood, therefore,

$$\left(\text{the logarithm of } \frac{1}{4 \times 10^{-8}}\right)$$

is 7.4. A pH of less than 7.4 is on the acidic side, and a pH greater than 7.4 is on the basic side. The acid-base regulatory systems of the body are all geared toward maintaining a normal pH of approximately 7.4 in the extracellular fluid. Even in disease conditions it almost never becomes more acidic than 7.0 or more basic than 7.8. ·

EFFECTS OF ACIDOSIS AND ALKALOSIS ON BODILY FUNCTIONS

The hydrogen ion concentration, despite its very low level, is one of the most important controlling factors in most metabolic reactions of the cells. Therefore, an increase or a decrease in H^+ concentration above or below normal causes serious derangements in the overall function of the body.

Acidosis generally depresses mental activity, and in severe states can lead to coma and death. This occurs in many instances of acidosis resulting from severe diarrhea or from diabetes mellitus, as will be discussed in more detail later in the chapter. Usually the afflicted person will pass into coma when the pH of the extracellular fluid falls below approximately 6.9, which represents a hydrogen ion concentration almost four times the normal level.

On the other hand, alkalosis in which the H^+ concentration has been decreased to less than one-half normal frequently leads to very severe overexcitability of the nervous system, often resulting in excessive initiation of signals in many areas of the brain and peripheral nerves. These signals can produce tetanic contraction of the muscles, or they can actually kill a person by causing convulsions or other derangements of nervous activity.

REGULATION OF ACID-BASE BALANCE BY CHEMICAL BUFFERS

All of the body fluids contain *acid-base buffers*. These are chemicals that can combine readily with any acid or base in such a way that they keep the acid or base from changing the pH of the fluids greatly. One of the chemical systems that perform this function is the *bicarbonate buffer,* which is present in all body fluids. The bicarbonate buffer is a mixture of carbonic acid (H_2CO_3) and bicarbonate ion (HCO_3^-). When a strong acid is added to this mixture its hydrogen ion (H^+) combines immediately with the bicarbonate ion to form carbonic acid. Carbonic acid is an extremely weak acid in which the hydrogen atom is tightly bound in the molecule so that relatively few hydrogen ions remain free in the solution. That is, "ionization" of carbonic acid as represented by the following equation

$$H_2CO_3 \rightleftharpoons H^+ + HCO_3^-$$

is very slight; the reaction proceeds mainly in the left-hand direction. Therefore, the buffer system changes the strong acid (many hydrogen ions) into a weak one (few hydrogen ions) and keeps the fluids from becoming strongly acid. On the other hand, when a strong base—a substance that contains large numbers of hydroxyl ions (OH^-) but very few hydrogen ions — is added to this mixture, the OH^- ions immediately combine with the carbonic acid to form water and bicarbonate ions.

Loss of the weak acid (carbonic acid) hardly affects the hydrogen ion concentration in the body fluids. Thus, it can be seen that this mixture of carbonic acid and bicarbonate ion protects the body fluids from becoming either too acidic or too basic.

Other important buffers are *phosphate* and *protein buffers*. These are especially important for maintaining normal hydrogen ion concentration in the intracellular fluids, because their concentrations inside the cells are many times as great as the concentration of the bicarbonate buffer.

In essence, the buffers of the body fluids are the first line of defense against changes in hydrogen ion concentration, for any acid or base added to the fluids immediately reacts with these buffers to prevent marked changes in the acid-base balance.

RESPIRATORY REGULATION OF HYDROGEN ION CONCENTRATION

Carbon dioxide is continually formed by all cells of the body as one of the end-products of metabolism. This carbon dioxide combines with water to form carbonic acid in accordance with the following reaction:

$$CO_2 + H_2O \rightleftarrows H_2CO_3$$

In other words, even the normal metabolic processes are always pouring carbonic acid and its few hydrogen ions into the body fluids.

On the other hand, one of the functions of respiration is to expel carbon dioxide through the lungs into the atmosphere. Normally, respiration removes carbon dioxide at the same rate that it is being formed. If, however, pulmonary ventilation decreases below normal, carbon dioxide will not be expelled normally but instead will pile up in the body fluids, causing the concentration of carbonic acid to increase. As a result, the H^+ concentration rises. If the rate of pulmonary ventilation rises above normal, however, the opposite effect occurs; carbon dioxide is blown off through the lungs at a more rapid rate than it is being formed, thereby decreasing the carbon dioxide and carbonic acid concentrations. Complete lack of breathing for a minute will increase the H^+ about twofold; this corresponds to a decrease in the pH of the extracellular fluid from the normal level of 7.4 down to about 7.1. Conversely, very active overbreathing can decrease the H^+ to about one-half normal (an increase of the pH to 7.7) in about one minute. Thus the hydrogen ion concentration of the body can be changed greatly by over- or underventilation of the lungs.

Control of Respiration by the Hydrogen Ion Concentration of the Body Fluid. In the preceding paragraph the effect of respiration on the H^+ was discussed. In this paragraph the opposite effect, that of H^+ on respiration, is discussed. A high H^+ stimulates the respiratory center in the medulla of the brain, greatly enhancing the rate of ventilation. Conversely, a low H^+ depresses the rate of ventilation. These effects afford an automatic mechanism for maintaining a fairly constant H^+ in the body fluids. That is, an increase in H^+ boosts the rate of ventilation. This in turn removes carbonic acid from the fluids, which decreases the H^+ back toward normal.

This respiratory mechanism for regulating hydrogen ion concentration reacts within a few seconds to a few minutes when the extracellular fluids become either too acidic or too basic. Figure 18–5 illustrates the effect on the respiration of adding first a small amount of acid to the blood, and then later a small amount of alkali (base). The height of the curves represents the depth of respiration, and the rate of repetition of the curves represents the frequency of respiration. Note that acidosis greatly increases both the depth and rate of respiration, while alkalosis depresses respiratory function greatly so that the depth

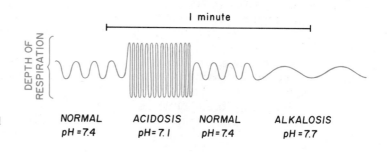

Figure 18–5.   Effect of acidosis and alkalosis on respiration.

DEPTH OF RESPIRATION

I minute

NORMAL
pH = 7.4

ACIDOSIS
pH = 7.1

NORMAL
pH = 7.4

ALKALOSIS
pH = 7.7

of respiration becomes very slight and the rate very slow. This respiratory mechanism is effective for regulating acid-base balance to the extent that it usually can return the H^+ of the body fluids about three-fourths of the way to normal within a minute after an acid or alkali has been administered.

RENAL REGULATION OF ACID-BASE BALANCE

A number of other acids besides carbonic acid are also continually being formed by the metabolic processes of the cells, and these can be eliminated from the body only by the kidneys. They include sulfuric, uric, and keto acids, all of which, on entering the extracellular fluids, can cause acidosis. The kidneys normally rid the body of these excess acids as rapidly as they are formed, preventing an excessive buildup of hydrogen ions. These acids are generally called *metabolic acids,* in contradistinction to carbonic acid, which is called a *respiratory acid.*

In rare instances too many basic compounds enter the body fluids rather than too many acidic compounds. In general, this occurs when basic compounds are injected intravenously or when the person ingests a large quantity of alkaline foods or drugs.

Mechanisms by Which the Kidneys Regulate Acid-Base Balance. The kidneys regulate acid-base balance by (1) excreting hydrogen ions into the urine when the extracellular fluids become too acidic and (2) excreting basic substances, particularly sodium bicarbonate, into the urine when the extracellular fluids become too alkaline.

SECRETION OF HYDROGEN IONS AND REABSORPTION OF SODIUM.   The distal tubular epithelium continually secretes hydrogen ions into the tubular fluid. These ions react with sodium salts in the tubular fluid to form weak acids and to free the sodium ion that is bound in the salt. The sodium in turn is absorbed through the tubular wall back into the extracellular fluid. Thus, there is a net exchange of hydrogen for sodium ions, the hydrogen ions passing into the urine and the sodium ions being reabsorbed. Obviously, loss of the hydrogen ions from the extracellular fluid and their replacement by sodium ions decreases the hydrogen ion concentration. This tends to make the fluid more basic, thus helping to overcome the continual formation of acid by the metabolic processes of the body.

CONTROL OF HYDROGEN ION SECRETION.   The rate of hydrogen ion secretion into the tubules is approximately proportional to the concentration of carbon dioxide in the extracellular fluid, and excess carbon dioxide in the extracellular fluid is normally associated with increased carbonic acid and increased hydrogen ion concentration, which is the cardinal measure of acidosis. Therefore, in effect, the rate of hydrogen ion secretion is roughly proportional to the degree of acidosis. To recapitulate, the more acidic the fluids, the greater is the rate of secretion of hydrogen ions and, therefore, the more rapidly the kidney attempts to eliminate the excess acid.

AMMONIA SECRETION AND ITS COMBINATION WITH HYDROGEN IONS IN THE TUBULES.   Sometimes so much metabolic acid is formed in the body and so many hydrogen ions are secreted into the tubules that the tubular fluid becomes very acidic (pH as low as 4.5) despite the presence of chemical buffers in the fluid. When this occurs, further secretion of hydrogen ion ceases. However, to prevent this from occurring and to allow continued secretion of the excess quantities of hydrogen ions, the tubular acidosis causes the tubular epithelial cells to secrete large quantities of ammonia. The ammonia in turn combines with the hydrogen ions to form ammonium ions, which prevents the tubular fluid from becoming highly acidic because the hydrogen ions are removed from the fluid and the ammonium ions form neutral salts in their place.

EXCRETION OF LARGE AMOUNTS OF BICARBONATE ION IN ALKALOSIS.   In alkalosis the alkaline substances in the body fluids combine with carbonic acid to form bicarbonate salts. Therefore, the extracellular bicarbonate ion concentration becomes greatly increased. These extra bicarbonate ions pass into the tubules in large quantities as part of the glomerular filtrate, and a large share of these pass on into the urine in the form of sodium bicarbonate. Since sodium bicarbonate is the alkaline half of the bicarbonate buffer system, its excretion actually represents loss of alkaline substances from the body.

To recapitulate, when the body fluids become too alkaline, the alkaline substances react with carbonic acid to form mainly the neutral salt sodium bicarbonate. The sodium bicarbonate in turn is lost into the urine, which represents loss of alkaline substances from the body. Thus, this mechanism helps to correct alkalosis.

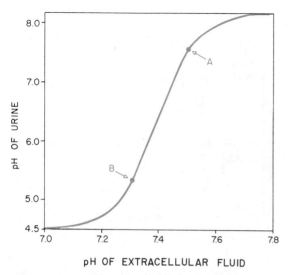

Figure 18–6.   Effect of extracellular fluid pH on urine pH.

SUMMARY OF RENAL REGULATION OF ACID-BASE BALANCE.   In summary, the renal mechanisms for regulating acid-base balance remove hydrogen ions from the extracellular fluids when the hydrogen ion concentration becomes too great, and they remove sodium and bicarbonate ions when the hydrogen ion concentration becomes too low. This principle is illustrated in Figure 18–6, which shows at point A a pH of about 7.5 in the extracellular

fluid. Since this is on the alkaline side, the pH of the urine becomes alkaline (pH 7.7) because of loss of alkaline substances from the body fluids. On the other hand, at point B the extracellular pH has fallen to 7.3, and the pH of the urine has become very acidic (pH 5.6) because of loss of large quantities of acidic substances from the body fluids. In both of these instances the loss of alkaline or acidic substances returns the pH toward normal.

ABNORMALITIES OF ACID-BASE BALANCE

Many disorders of the respiratory system, of the kidneys, or of the metabolic systems for forming acids and bases can cause serious derangement of the acid-base balance. Some of the effects of these on extracellular pH are shown in Figure 18–7. For instance, the normal pH of the blood is 7.4, but intense over-ventilation (overbreathing) can cause the pH to rise sometimes to as high as 7.8 because of excess expiration of carbon dioxide through the lungs. On the other hand, asphyxia, which means extreme decrease in ventilation, causes a buildup of carbon dioxide and carbonic acid in the body fluids, thereby promoting acidosis in which the pH of the blood sometimes falls to as low as 7.0.

A common cause of alkalosis is ingestion of

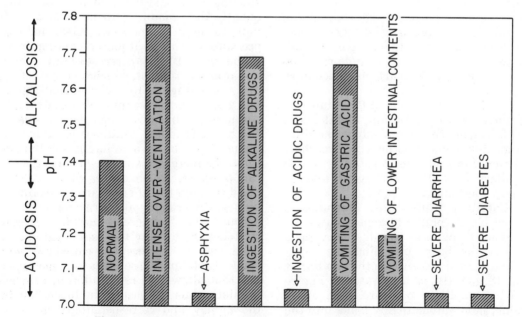

Figure 18–7.   pH of the extracellular fluid in various acid-base disorders.

alkaline drugs used for treatment of gastritis or stomach ulcers. The drugs sometimes are absorbed into the body fluids in quantities greater than the kidneys can remove, resulting in alkalosis.

Loss of large quantities of fluids from the intestinal tract at times causes acidosis but at other times alkalosis. For instance, vomiting large quantities of hydrochloric acid from the stomach gradually depletes the acid reserves of the extracellular fluid and causes alkalosis. On the other hand, vomiting of fluids from the lower intestine or loss of fluids as a result of diarrhea usually causes acidosis. This is because secretions from the lower intestine contain large quantities of sodium bicarbonate. When this basic salt is lost from the body it is mainly replaced by carbonic acid formed by the reaction of carbon dioxide with water. Therefore, the net effect is loss of sodium ions and gain of hydrogen ions, causing acidosis.

Finally, Figure 18–7 shows extreme acidosis that sometimes results from severe diabetes mellitus. In diabetes many of the fats normally used by the body for energy are not completely metabolized but instead are broken into substances called *keto acids* that build up in the body fluids, producing extreme acidosis.

REGULATION OF BLOOD VOLUME

In earlier chapters in which the circulatory system was discussed, the importance of blood volume as one of the major determinants of both cardiac output and arterial pressure regulation was emphasized. Indeed, only a small long-term increase in blood volume can lead to a very large increase in arterial blood pressure.

The normal blood volume of the adult person is almost exactly 5000 ml., and it rarely rises or falls more than a few hundred milliliters from this value. Two principal mechanisms for maintaining this constancy are (1) the capillary fluid shift mechanism and (2) the kidney mechanism.

The Capillary Fluid Shift Mechanism. When the blood volume becomes too great, the pressures increase in all the vessels throughout the body, including the capillaries. The normal pressure in the capillaries is about 18 mm. Hg, as illustrated in Figure 18–8. When capillary pressure rises above this value, fluid automatically leaks into the tissue spaces, as we explained in Chapter 16.

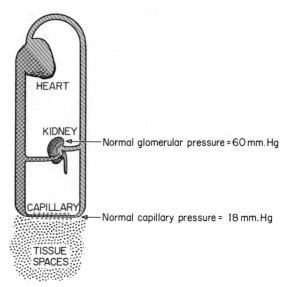

Figure 18–8. The capillary and kidney systems for regulating blood volume.

Therefore, the blood volume decreases back toward normal. When the capillary pressure has returned to normal, further loss of fluid into the tissue spaces ceases. Conversely, when the blood volume falls too low, the capillary pressure falls, and fluid is then absorbed from the interstitial spaces, increasing the blood volume again back toward normal.

The Kidney Mechanism. As was explained in the previous chapter, when the glomerular pressure in the kidneys rises, the amount of glomerular filtrate and consequently the amount of urine formed by the kidneys greatly increase. Also, increased pressure in the peritubular capillaries surrounding the tubules *decreases* fluid reabsorption from the tubules, which further *increases* urine flow. Figure 18–8 shows that the normal glomerular pressure is about 60 mm. Hg. An increased blood volume raises this as well as the peritubular capillary pressure to a higher level by two different mechanisms: First, the increased volume elevates the arterial pressure, which causes greater flow of blood through the afferent arterioles into the kidney, thus raising all intrarenal pressures. Second, the increased vascular pressure stretches the atria of the heart that are supplied with stretch receptors called *volume reflexes*. Stretching these initiates a nervous reflex that allows the renal afferent arterioles to dilate, thus increasing the blood flow into the kidney and correspondingly increasing the amount of urine formed. The volume receptors also inhibit antidiuretic hormone secretion by the posterior pituitary gland, and this increases the urine output as well.

One will immediately recognize all of these effects to be feedback mechanisms by which the blood volume can be regulated; that is, an increase in blood volume initiates an increase in urinary output, which automatically decreases the blood volume back toward normal.

REFERENCES

Actions of hormones on the kidney. *Ann. Rev. Physiol.*, *39*:97, 1977.

Arruda, J. A. L., and Kurtzman, N. A.: Relationship of renal sodium and water transport to hydrogen ion secretion. *Ann. Rev. Physiol.*, *40*:43, 1978.

Blair-West, J. R.: Renin-angiotensin system and sodium metabolism. *Intern. Rev. Physiol.*, *11*:95, 1976.

Cross, B. A., and Wakerley, J. B.: The neurohypophysis. *Intern. Rev. Physiol.*, *16*:1, 1977.

Davis, J. O., and Freeman, R. H.: Mechanisms regulating renin release. *Physiol. Rev.*, *56*:1, 1976.

Gauer, O. H., and Henry, J. P. Neurohormonal control of plasma volume.: *Intern. Rev. Physiol.* *9*:145, 1976.

Goldberg, M., Agus, Z. S., and Goldfarb, S. Renal handling of calcium and phosphate. *Intern. Rev. Physiol.*, *11*:211, 1976.

Grantham, J. J.: Fluid secretion in the nephron: relation to renal failure. *Physiol. Rev.*, *56*:248, 1976.

Guyton, A. C., Langston, J. B., and Navar, G.: Theory for renal autoregulation by feedback at the juxtaglomerular apparatus. *Circ. Res.*, *14*:187, 1964.

Guyton, A. C., Taylor, A. E., and Granger, H. J.: Circulatory Physiology. H. Dynamics and Control of Body Fluids. Philadelphia, W. B. Saunders Company, 1975.

Hayward, J. N.: Neural control of the posterior pituitary. *Ann. Rev. Physiol.*, *37*:191, 1975.

Knox, F. G., and Diaz-Buxo, J. A.: The hormonal control of sodium excretion. *Intern. Rev. Physiol.*, *16*:173, 1077.

Malnic, G., and Giebisch, G.: Symposium on acid-base homeostasis. Mechanism of renal hydrogen ion secretion. *Kidney Int.*, *1*:280, 1972.

Mason, E. E.: Fluid, Electrolyte, and Nutrient Therapy in Surgery. Philadelphia, Lea & Febiger, 1974.

Stein, J. H., and Reineck, H. J.: Effect of alterations in extracellular fluid volume on segmental sodium transport. *Physiol. Rev.*, *55*:127, 1975.

QUESTIONS

1. How are the sodium ion concentration and osmolality of the extracellular fluid both regulated at the same time?
2. Explain the role of antidiuretic hormone in the control of water excretion by the kidneys.
3. Explain the countercurrent mechanism for excreting a concentrated urine.
4. What is the role of the hypothalamus in the control of urinary output? In the control of fluid intake?
5. Why is aldosterone especially important for control of potassium ion concentration but less important for control of sodium ion concentration?
6. The concentration of hydrogen ions in a solution is 8×10^{-8} equivalents per liter. Calculate the pH of the solution.
7. Explain the relative roles of the body fluid buffers, the respiratory system, and the kidneys in the regulation of acid-base balance.
8. How does the carbon dioxide-bicarbonate buffer system function in respiratory control of acid-base balance and in renal control of acid-base balance?
9. What is the role of ammonia in renal control of acid-base balance?
10. Explain how the capillary fluid shift mechanism and the kidney mechanism operate together to regulate blood volume.

RESPIRATION VI

MECHANICS OF RESPIRATION; AND TRANSPORT OF OXYGEN AND CARBON DIOXIDE

19

The function of the respiratory system is, first, to supply oxygen to the tissues and, second, to remove carbon dioxide. Figure 19–1 illustrates the principal structures of this system, showing the lungs, trachea, glottis, and nose. The lungs contain millions of small air sacs, called *alveoli,* connected by the bronchioles and trachea with the nose and mouth. With each intake of breath the alveoli expand, and during expiration air is forced out of the alveoli again to the exterior. Thus, there is continual renewal of air in the alveoli, a process called *pulmonary ventilation.* Later in the chapter, in Figure 19–10, we will see more details of the structure of these terminal air sacs in the lungs.

To the upper right in Figure 19–1 is illustrated the functional relationship of an alveolus to a pulmonary capillary. Each alveolus has in all its walls a network of capillaries, a surface view of which is illustrated in Figure 19–2. Also, the membrane between the air in the alveolus and the blood in these capillaries is so thin that oxygen can diffuse into the blood with extreme ease and carbon dioxide out with even greater ease. Therefore, the role of the basic structure of the lungs is simply to aerate the blood, and to allow replenishment of oxygen and removal of carbon dioxide. And the object of breathing is to move air continually into and out of the alveoli.

FUNCTIONS OF THE RESPIRATORY PASSAGEWAYS

FUNCTION OF THE NOSE

The nose is not merely a passageway for movement of air into the lungs, but it also preconditions the air in several ways, including (a) warming the air, (b) humidifying the air, and (c) cleansing the air. These functions may be explained as follows:

The inside surface of the nose is extensive. The nasal cavity is divided by a central septum, and several projections called *turbinates,* illustrated in Figure 19–1, extend into each cavity from the lateral side. Air passing through the nose, on coming in contact with all of these nasal surfaces, becomes warmed and humidified. The turbinates also cause turbulence in the flowing air, forcing it to rebound in many different directions before finally completing its passage through the nose. This causes dust or other suspended particles in the air to precipitate against the nasal surfaces by the following mechanism: When air containing a foreign particle travels toward a surface and then suddenly changes its direction of movement, the momentum of the particle causes it to continue to travel in the original direction, while the air, which has little momentum because of its low mass, takes off in the new direction. The particles

233

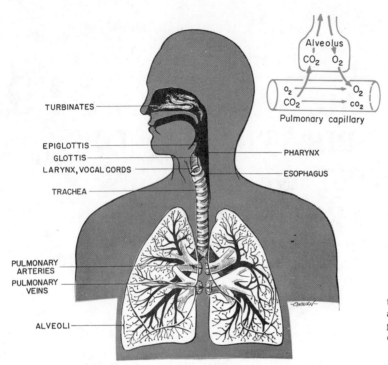

TURBINATES

Alveolus
CO_2 O_2

O_2 O_2
CO_2 CO_2
Pulmonary capillary

EPIGLOTTIS
GLOTTIS
LARYNX, VOCAL CORDS
TRACHEA

PHARYNX

ESOPHAGUS

PULMONARY
ARTERIES
PULMONARY
VEINS

ALVEOLI

Figure 19–1. The respiratory system, showing the respiratory passages and function of the alveolus to oxygenate the blood and to remove carbon dioxide.

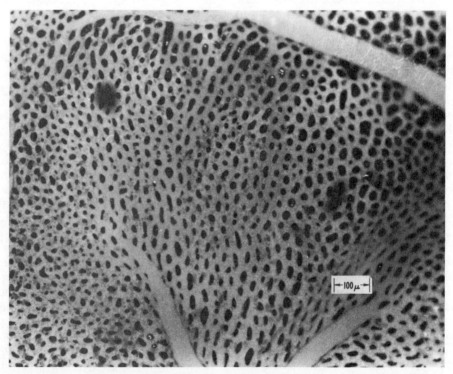

$\leftarrow$ 100μ $\rightarrow$

Figure 19–2. Surface view of capillaries in an alveolar wall. (From Maloney and Castle: *Resp. Physiol.*, 7:150, 1969. Reproduced by permission of ASP Biological and Medical Press, North-Holland Division.)

impinge on a turbinate or on another surface of the nasal passageways and are entrapped in the layer of mucus covering the surface. The surface, in turn, is lined with ciliated epithelial cells, whose cilia protrude into the mucus and beat toward the pharynx, slowly moving the mucus and entrapped particles into the throat to be swallowed. This method of removing foreign particles from the air is so efficient that very rarely do any particles greater in size than 3 to 5 microns (micrometers), about half the size of a red blood cell, pass through the nose into the lower respiratory passageways.

FUNCTIONS OF THE PHARYNX AND LARYNX

The *pharynx*, which is commonly called the throat, separates posteriorly into the trachea and esophagus. Here food is separated from air, the air passing through the *larynx* (the "voice box") into the *trachea* while the food passes into the *esophagus*. This separation of food and air is controlled by nerve reflexes. Whenever food touches the surface of the *pharynx*, the *vocal cords* close together and the *epiglottis* also closes automatically over the opening of the larynx, allowing the food to slide on into the esophagus.

Function of the Vocal Cords. The *vocal cords* are the portion of the larynx that makes sound. They are two small vanes located on either side of the air passageway, as shown in Figure 19–3. Contraction of muscles in the larynx can bring these vanes close to each other or can spread them apart. They can also be stretched or relaxed, and their edges can be flattened or thickened by muscles actually in the cords themselves. When the cords are together and air is forced between them, they vibrate to generate sound, and the different *pitches* of sound are controlled by the degree

to which the cords are stretched and by the degree of flattening or thickening of the vocal cord edges. The formation of words or other complicated sounds is a function of the mouth as well as the larynx because the *quality* of a sound depends upon the momentary position of the lips, cheeks, teeth, tongue, and palate.

For speech or other sounds to be emitted, the respiration, the vocal cords, and the mouth must all be controlled at the same time. This is effected by a special brain center, called *Broca's area,* located in the left frontal lobe of the brain. The organization and function of this center will be discussed in Chapter 25.

THE COUGH AND SNEEZE REFLEXES

A means for keeping the respiratory passages clean is to force air very rapidly outward by either coughing or sneezing. The cough reflex is initiated by any irritant touching the surface of the glottis, the trachea, or a bronchus. Sensory signals are transmitted to the medulla of the brain, and, in turn, appropriate motor signals are transmitted back to the respiratory system and larynx to cause the cough. The respiratory muscles first contract very strongly, building up high pressure in the lungs, while at the same time the vocal cords remain clamped tightly closed. Then suddenly the vocal cords open, allowing the pressurized air in the lungs to flow out with a blast. Sometimes the air flows as rapidly as 70 miles per hour. In this way unwanted foreign matter such as particles, mucus, or other substances is expelled from the respiratory passageways.

The sneeze reflex is very similar to the cough reflex except that it is initiated by irritants in the nose. Impulses pass from the nose to the medulla and then back to the respiratory system. A sudden forceful expiration

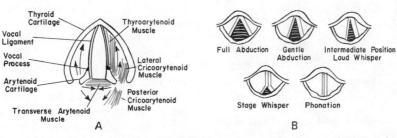

Figure 19–3.   Laryngeal function in phonation. (Modified from Greene: The Voice and Its Disorders. Pitman Medical Publishing Co.)

blows air outward while the soft palate varies its position to allow rapid flow of air successively through the nose and mouth. In this way the sneeze is capable of clearing the nasal passageways in the same manner that the cough reflex clears many of the lower passageways.

FLOW OF AIR INTO AND OUT OF THE LUNGS

The lungs are enclosed in the *thoracic cage,* which is illustrated in Figure 19–4 and is composed of the *sternum* in front, the *spinal column* in back, the *ribs* encircling the chest, and the *diaphragm* below. The act of breathing is performed by enlarging and contracting the thoracic cage. The cavity formed by the thoracic cage is called the *pleural cavity,* and the lungs normally fill this cavity entirely. The lungs are covered with a lubricated membrane called the *visceral pleura,* and the inside of the pleural cavity is lined with a similar membrane called the *parietal pleura.* The lungs slide freely inside the pleural cavity, so that any time the cavity enlarges, the lungs must also enlarge. In other words, any change in the volume of the thoracic cage is immediately reflected by a similar change in the volume of the lungs.

The Muscles of Respiration. THE INSPIRATORY MUSCLES. Figure 19–4 also illustrates several of the muscles of respiration, as well as the change in shape of the thoracic cage between expiration and inspiration. The major muscles of inspiration are the *dia-*

phragm, the *external intercostals,* and a number of small muscles in the neck which pull upward on the front of the thoracic cage. The inspiratory muscles cause the pleural cavity to enlarge in two ways. First, contraction of the diaphragm pulls the bottom of the pleural cavity downward, thus elongating it, which is shown to the right in Figure 19–4. Second, the external intercostals and neck muscles lift the front of the thoracic cage, causing the ribs to angulate more directly forward than previously, increasing the thickness of the cage, which is also shown in the right half of the figure.

THE EXPIRATORY MUSCLES. The major muscles of expiration are the *abdominals* and, to a lesser extent, the *internal intercostals.* The abdominal muscles cause expiration in two ways. First, they *pull downward on the chest cage,* thereby decreasing the thoracic thickness. Second, they *force the abdominal contents upward,* thus pushing the diaphragm upward as well and thereby decreasing the longitudinal dimension of the pleural cavity. The internal intercostals help in the process of expiration by pulling the ribs downward. When they are in this downward position, the thickness of the chest is considerably decreased, as can be seen to the left in Figure 19–4.

PULMONARY PRESSURES

Alveolar Pressure. During inspiration the thoracic cage enlarges, which also enlarges the lungs. One will recall from the basic laws of physics that when a volume of gas is suddenly enlarged, its pressure falls. Thus, during inspiration the enlargement of the thoracic cage decreases the pressure in the alveoli to about −3 mm. Hg, and it is this negative pressure that pulls air through the respiratory passageways into the alveoli.

During expiration exactly the opposite effects occur; compression of the thoracic cage around the lungs increases the alveolar pressure to approximately +3 mm. Hg, which obviously pushes air out of the alveoli to the atmosphere.

If a person breathes with maximal effort but with his nose and mouth closed so that air cannot flow into or out of his lungs, the alveolar pressure can be decreased to as low as −80 mm. Hg or increased to as high as +100 mm. Hg. Thus, the maximum strength of the muscles of respiration is far greater than that

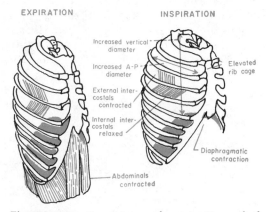

Figure 19–4. Expansion and contraction of the thoracic cage during expiration and inspiration, illustrating especially diaphragmatic contraction, elevation of the rib cage, and function of the intercostals.

needed for normal quiet respiration. This provides a tremendous reserve respiratory ability which can be called upon in times of need for maximal respiratory activity, such as during heavy exercise.

Intrapleural Pressure. The space between the lungs and the outer walls of the pleural cavity is called the *intrapleural space,* and the pressure in this space is the *intrapleural pressure.* This pressure is always a few millimeters of mercury less than that in the alveoli. Figure 19–5 illustrates that during inspiration the intra-alveolar pressure is about −3 mm. Hg while the intrapleural pressure is about −8 mm. Hg. During normal expiration, the intra-alveolar pressure rises to + 3 mm. Hg while the intrapleural pressure rises to −2 mm. Hg. It will be noted that the intrapleural pressure remains about 5 mm. Hg less than the intra-alveolar pressure. The reason for this is that the lungs are always pulling away from the chest wall because of two effects: First, the surface tension of fluid lining the inside of the alveoli makes the alveoli try to collapse in the same manner that soap bubbles will collapse if an opening will allow air to escape. Second, elastic fibers spread in all directions through the tissues of the lungs,

and these also tend to contract the lungs. Both of these effects pull the lungs away from the outer walls of the pleural cavity, creating an average negative pressure in the intrapleural space of about −5 mm. Hg.

SURFACTANT IN THE ALVEOLI. If the alveoli were lined with water instead of normal alveolar fluid, the surface tension would be so great that it would cause the alveoli to remain collapsed almost all the time. However, a substance called a *surface active agent* or simply *surfactant* is secreted into the alveoli by the alveolar membrane. This substance is a detergent that greatly decreases the surface tension of the fluid lining the alveoli.

COLLAPSE OF THE LUNG CAUSED BY PNEUMOTHORAX. When an opening is made in the chest wall the elastic forces in the lungs cause them to collapse immediately, sucking air through the opening into the chest cavity. This is called a "pneumothorax." Then, when the person tries to breathe, instead of the lungs expanding and contracting, air flows in and out of the hole in the chest. Thus, a wound of the chest can kill a person by suffocation; yet the condition can be treated very easily by sucking air out of the pleural cavity and plugging the hole.

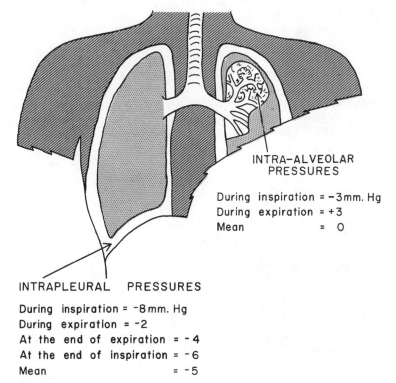

INTRA–ALVEOLAR
PRESSURES

During inspiration = −3mm. Hg
During expiration = +3
Mean   = 0

INTRAPLEURAL PRESSURES

During inspiration = −8mm. Hg
During expiration = −2
At the end of expiration = − 4
At the end of inspiration = − 6
Mean   = − 5

Figure 19–5. Alveolar and intrapleural pressures during normal breathing.

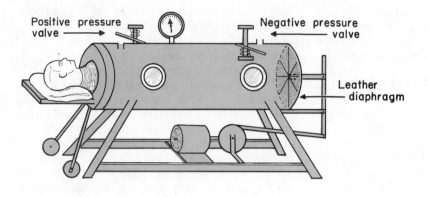

Figure 19–6. The artificial respirator.

ARTIFICIAL RESPIRATION

When the respiratory muscles fail, a person can be kept alive only by some means of artificial respiration. The simplest method is to force air into and out of his mouth or nose. Several devices have been made for this purpose, one of which is called the *resuscitator*. To use this apparatus a mask is placed over the mouth and nose, and intermittent blasts of air fill the lungs. Between blasts, air is pulled back out of the lungs into the atmosphere.

For prolonged artificial respiration the person is usually placed inside an artificial tank type of respirator, such as that shown in Figure 19–6. This is a large tank provided with some means for repetitively increasing and decreasing the pressure and for controlling the extent of the rise and fall in pressure. In the respirator of Figure 19–6 the air pressure is alternated by inward and outward movement of a large leather diaphragm, and the positive and negative pressures are controlled by relief valves shown on top of the tank. Because respiration normally is accomplished mainly by the inspiratory muscles rather than by the expiratory muscles, the pressure inside the tank is usually adjusted to give more of a vacuum cycle than a pressure cycle. A vacuum of about −10 mm. Hg causes approximately normal inspiration by pulling outward on the chest cage and abdominal wall, thereby sucking air into the lungs. Then, the respirator changes from the vacuum to the pressure cycle and applies 2 to 3 mm. Hg of positive pressure. This causes expiration by pushing against the abdomen and chest.

SPIROMETRY; AND DISTRIBUTION OF RESPIRATORY AIR IN THE LUNG

Figure 19–7 represents a *spirometer,* an apparatus that can be used to record the flow of air into and out of the lungs. It is composed of a drum inverted in a tank of water, and with a tube extending from the air space in the drum to the mouth of the person to be tested. The drum is suspended from pulleys and counterbalanced by a weight. As the person breathes

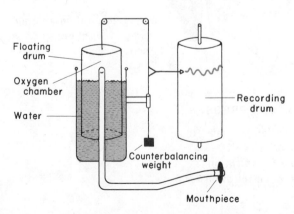

Figure 19–7. The spirometer.

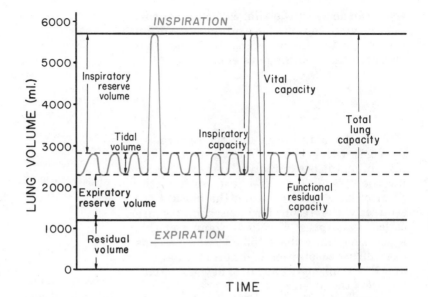

Figure 19–8. A spirogram, showing the divisions of the respiratory air.

in and out, the drum moves up and down, and the counterweight balancing the drum also rides up and down, recording on a moving paper the changing volume of the chamber. Figure 19–8 illustrates a typical spirometer recording for successive breath cycles, showing different depths of inspiration and expiration.

Tidal Volume, Respiratory Rate, and Minute Respiratory Volume. The air that passes into and out of the lungs with each respiration is called the *tidal air,* and the volume of this air in each breath is called the *tidal volume.* The normal tidal volume is about 500 ml., and the normal rate of respiration for an adult is usually about 12 times per minute. Therefore, a total of about 6 liters of air normally passes into and out of the respiratory passageways each minute. This amount is called the *minute respiratory volume.*

The Inspiratory Capacity. In Figure 19–8, after three normal breaths, the person breathes inward as deeply as possible. The amount of air that he can pull into his lungs beyond that already in his lungs at the beginning of the breath is called the *inspiratory capacity.* This quantity is approximately 3000 ml. in the normal person.

Expiratory Reserve Volume. After several more normal respirations, the subject expires as much as possible (see Fig. 19–8). The amount of air that he is capable of expiring beyond that which he normally expires is called the *expiratory reserve volume.* This is usually about 1100 ml.

Residual Volume and the Functional Residual Capacity

In addition to the expiratory reserve volume, there is air in the lungs that cannot be expired even by the most forceful exhalation. The volume of this air is about 1200 ml., and it is called the *residual volume.* The sum of the expiratory reserve volume plus the residual volume is called the *functional residual capacity;* this is the amount of air remaining in the respiratory system at the end of a normal expiration. It is this air that allows oxygen and carbon dioxide transfer into and out of the blood to continue even between breaths.

Vital Capacity. Finally, the right portion of Figure 19–8 shows the spirogram of a maximal inspiratory effort followed immediately by a maximal expiratory effort. The total change in pulmonary volume between these two extremes is called the *vital capacity.* The vital capacity of the normal person is approximately 4500 ml. A well-trained male athlete may have a vital capacity as great as 6500 ml., while a small female frequently has a vital capacity no greater than 3000 ml.

The vital capacity is a measure of the person's overall ability to inspire and expire air, and it is determined mainly by two factors: (1) the strength of the respiratory muscles and (2) the resistance of the thoracic cage and lungs to expansion and contraction. Disease processes that either weaken the muscles — such as poliomyelitis – or decrease the expansibility of the lungs — such as tuberculosis — can decrease the vital capacity. For this reason, vital capacity measurements are an invaluable tool in assessing the functional ability of the breathing system.

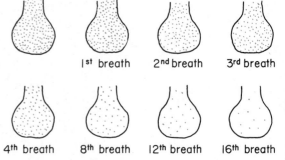

1st breath   2nd breath   3rd breath

4th breath   8th breath   12th breath   16th breath

Figure 19–9. Expiration of a foreign gas from the alveoli with successive breaths.

Much of the air pulled into the respiratory passages with each breath never reaches the alveoli, and it is expired without ever reaching the alveoli. This air is useless from the point of view of oxygenating the blood. Consequently the respiratory passageways are called *dead space*. The total volume of this space is normally about 150 ml., which means that during inspiration of a normal tidal volume of 500 ml., only 350 ml. of new air actually enters the alveoli.

Alveolar Ventilation. The most important measure of the effectiveness of a person's respiration is his *alveolar ventilation,* which is the total quantity of new air that enters his alveoli each minute. With only 350 ml. of air reaching the alveoli with each breath and with a normal respiratory rate of 12 times per minute, the alveolar ventilation is approximately 4200 ml. per minute. During maximal respiratory effort this can be increased to as high as 100 liters per minute, and at the opposite extreme a person can remain alive, at least for a few hours, with an alveolar ventilation as low as 1200 ml. per minute.

EXCHANGE OF ALVEOLAR AIR WITH ATMOSPHERIC AIR

At the end of each expiration approximately 2300 ml. of air still remains in the lungs. This is called the *functional residual capacity* as was explained previously. With each breath, about 350 ml. of new air is brought into this air, and with each expiration this same amount of air is removed. It is obvious, then, that each breath does not exchange all the old alveolar air for new air, but instead *exchanges only about one-seventh of the air.* This effect is illustrated in Figure 19–9, which shows an alveolus containing a quantity of some foreign gas at the beginning of the series of breaths. After the first breath a small portion of this gas has been removed, after the second breath a little more, after the third a little more, and so forth. Note that even after the sixteenth breath a small portion of the foreign gas is still in the alveolus.

At a normal alveolar ventilation of 4200 ml. per minute, approximately one-half the alveolar gases are replaced by new air every 23 seconds. This slow turnover of air in the alveoli keeps the alveolar gaseous concentrations from rising and falling greatly with each individual breath.

ALVEOLAR AIR

Alveolar air is a mixture of inspired air, water vapor from the respiratory passageways, and carbon dioxide excreted from the blood. Also, because oxygen is continually being absorbed into the blood, the oxygen concentration of alveolar air is considerably less than that of the atmosphere. However, before it will be possible to understand how both the oxygen and carbon dioxide are controlled in the alveolar air we will need to review briefly the principle of partial pressures.

PARTIAL PRESSURES

The partial pressure of a gas is the amount of pressure exerted by each type of gas in a mixture of gases. For instance, let us assume that we have a mixture of 50 per cent oxygen (in terms of numbers of molecules) and 50 per cent nitrogen and that the total pressure of the mixture is 100 mm. Hg. The portion of this pressure that is caused by oxygen is 50 mm. Hg, and the portion caused by nitrogen is also 50 mm. Hg. Therefore, it is said that the partial pressure of each of these two gases is 50 mm. Hg. But now let us assume that the composition is changed to 75 per cent nitrogen and 25 per cent oxygen, and at the same time the total pressure is changed to 1000 mm. Hg. Now the partial pressure exerted by the nitrogen will be 750 mm. Hg, while that of the oxygen will be 250 mm. Hg.

Thus, the partial pressure of a gas is a measure of the total force that each gas alone exerts against the walls surrounding it. Consequently, the penetrating power of a gas through a membrane, such as the membrane between the alveoli and the blood, is also directly proportional to its partial pressure. That is, the greater the partial pressure of a gas in the alveolus, the greater is its tendency to pass through the pulmonary membrane into the blood.

The partial pressures of the important respiratory gases are represented by the following symbols:

Nitrogen	P_{N_2}
Oxygen	P_{O_2}
Carbon dioxide	P_{CO_2}
Water vapor	P_{H_2O}

Composition of Alveolar Air. Table 19–1 shows the partial pressures and per cent concentrations of nitrogen, oxygen, carbon dioxide, and water vapor in atmospheric and alveolar airs at sea level. This table shows that normal atmospheric air is composed almost entirely of only two gases, about four-fifths nitrogen and one-fifth oxygen, with almost negligible quantities of carbon dioxide and water vapor. In contrast, alveolar air contains considerable quantities of both carbon dioxide and water vapor, while the oxygen content is considerably less than that of atmospheric air. These differences can be explained as follows:

Humidification of Air as it Enters the Respiratory Passages. When air is inspired it is humidified immediately by moisture from the linings of the respiratory passages. At normal body temperature the partial pressure

of water vapor in the lungs is 47 mm. Hg, and as long as the body temperature remains constant, this partial pressure also remains constant.

The mixing of water vapor with the incoming atmospheric air dilutes the air so that the pressure of the other gases becomes slightly decreased. This explains why the nitrogen partial pressure in the alveoli is slightly less than its partial pressure in atmospheric air.

Oxygen and Carbon Dioxide Partial Pressure in the Alveoli. Alveolar air continually loses oxygen to the blood, and this oxygen is replaced by carbon dioxide diffusing out of the blood into the alveoli. This explains why the oxygen pressure in alveolar air is much less than in atmospheric air, and it also explains why the carbon dioxide pressure is much greater in the alveolar air than in the atmospheric air. The normal partial pressure of oxygen in the alveoli is approximately 104 mm. Hg in comparison with 159 mm. Hg in the atmosphere. And that of carbon dioxide is 40 mm. Hg in comparison with almost zero in the atmosphere. However, these values change greatly from time to time depending on the rate of alveolar ventilation and the rate of oxygen and carbon dioxide transfer into and out of the blood. For instance, a high alveolar ventilation provides greater amounts of oxygen to the alveoli and increases the alveolar oxygen pressure. Also, a high alveolar ventilation removes carbon dioxide from the alveoli more rapidly than usual, thereby decreasing the carbon dioxide pressure.

TRANSPORT OF GASES THROUGH THE PULMONARY MEMBRANE

THE PULMONARY MEMBRANE

The pulmonary membrane is composed of all the pulmonary surfaces that are thin enough to allow gases to diffuse into the pulmonary blood. These include, as illustrated in Figure 19–10, the membranes of the respiratory bronchioles, the alveolar ducts, the atria, and the alveolar sacs.

The total area of the pulmonary membrane is approximately 70 square meters, which is about equal to the floor area of a moderate-sized classroom. And less than 100 ml. of blood is in the capillaries at any one time. If one will think for a moment of this small amount of blood spread out evenly over the entire floor of a classroom, he can quite readi-

Table 19–1. **Partial Pressures (mm. Hg) and Per Cent Concentrations of Respiratory Gases in the Atmosphere and in the Alveoli**

GAS	ATMOSPHERIC AIR	ALVEOLAR AIR
N_2	597.0 (78.62%)	569.0 (74.9%)
O_2	159.0 (20.84%)	104.0 (13.6%)
CO_2	0.15 (0.04%)	40.0 (5.3%)
H_2O	3.85 (0.5%)	47.0 (6.2%)

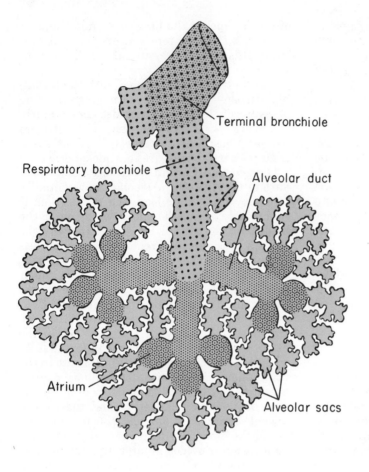

Terminal bronchiole

Respiratory bronchiole

Alveolar duct

Atrium

Alveolar sacs

Figure 19–10. The respiratory unit, illustrating the anatomy of the respiratory surfaces (From Miller: The Lung. Charles C Thomas.)

ly understand that very large quantities of gas can enter and leave this blood in only a fraction of a second.

Figure 19–11 illustrates diagrammatically an electronmicrographic cross-section of the pulmonary membrane, showing also a red blood cell in a capillary in close proximity to the membrane. This membrane is composed of several layers, including (1) a thin lining of fluid on the surface of the alveolus containing the substance called "surfactant" which was discussed earlier in the chapter, (2) a lining of epithelial cells that is the surface of the alveolus, (3) a small interstitial space containing only minute amounts of connective tissue and basement membrane, and (4) a layer of capillary endothelial cells which forms the wall of the capillary. However, all layers of this membrane are extremely thin, so much so that most parts of the membrane are less than 0.4 micron (micrometer) thick, which is only a fraction of the thickness of a red blood cell itself. The thinness of the membrane obviously allows rapid diffusion of oxygen and

carbon dioxide between the alveolar air and the blood.

SOLUTION AND DIFFUSION OF GASES IN WATER

To understand the transport of gases through the pulmonary membrane, one must first be familiar with the physical principles of gaseous solution and diffusion in water. Figure 19–12 shows a chamber containing water at the bottom and normal air at the top. Molecules of nitrogen and oxygen are continually bouncing against the surface of the water, and some of the molecules enter the water to become dissolved. The dissolved molecules bounce among the water molecules in all directions, some of them eventually reaching the surface again to bounce back into the gaseous space. After the air has remained in contact with the water for a long time, the number of molecules passing out of the solution equals exactly the number of molecules

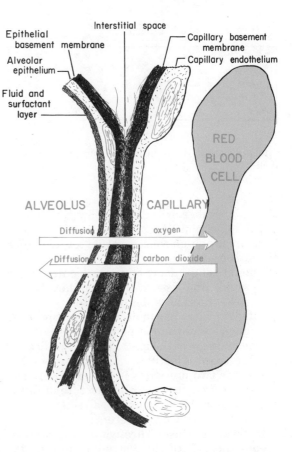

Figure 19-11. Ultrastructure of the respiratory membrane.

passing into the solution. When this state has been reached, the gases in the gaseous phase are said to be in equilibrium with the dissolved gases.

In the equilibrium state, the partial pressure tending to force nitrogen molecules into the water is 564 mm. Hg, and the pressure of nitrogen molecules pushing outward from the water is also 564 mm. Hg. If the nitrogen pressure becomes greater in the gaseous

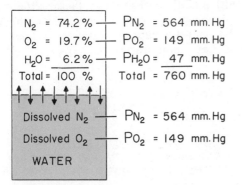

Figure 19-12. A sample of air exposed to water, showing equilibration between gases in the gaseous phase and in the dissolved state.

phase than in the dissolved phase, more nitrogen molecules will move into the solution than out. If the pressure becomes less in the gas than in the solution, more nitrogen molecules will leave the solution than will enter it. In other words, *the pressure of a gas is the force with which it attempts to move from its present surroundings.* Whenever the pressure is greater at one point than at another, whether this be in a solution or in a gaseous mixture, more molecules will move toward the low pressure area than toward the high pressure area.

Diffusion of Gases Through Tissues and Fluids. Figure 19-13 shows a chamber containing a solution. At end A of the chamber is a high concentration of dissolved gas molecules, while at end B is a low concentration. The molecules are bouncing at random in all directions because of their kinetic activity. Obviously, because of the higher concentration at A than at B, it is easier for large numbers of molecules to move from A to B than in the opposite direction, which is indicated by the lengths of the arrows, and eventually the numbers of molecules at both ends

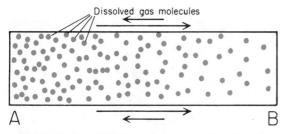

Figure 19-13. Diffusion of gas molecules through a liquid.

of the chamber will become approximately equal. Movement of molecules in this manner from areas of high concentration toward areas of low concentration is called *diffusion.*

The rate of diffusion of gas molecules in the body fluids and tissues is determined by the pressure differences between the different points. Far more gas molecules are at point A in Figure 19-13 than at point B. Consequently, the pressure of the gas at point A is also far greater than at point B. The difference between the two pressures is called simply the *pressure difference,* and the net rate of diffusion between the two points is directly proportional to the pressure difference.

Diffusion of Gases Through the Pulmonary Membrane. With the above background information, we can now discuss all the factors that determine the rate of gas diffusion through the pulmonary membrane. These are the following:

1. The greater the *pressure difference* between one side of the membrane and the other, the greater will be the rate of gaseous flow. If the pressure of a gas in the alveolus is 100 mm. Hg while that in the blood is 99 mm. Hg, the pressure difference will be only 1 mm., and the rate of gaseous flow will be very slight. If the pressure in the blood suddenly falls to 0 mm. Hg, the pressure difference rises to 100 mm., and the gaseous diffusion increases to 100 times its former rate.

2. The greater the *area of the pulmonary membrane,* the greater will be the quantity of gas that can diffuse in a given period of time. In some pulmonary diseases, such as *emphysema,* large portions of the lungs are destroyed and the total area of the pulmonary membrane is greatly decreased. Often the decrease is enough to cause continual respiratory embarrassment.

3. The *thinner the membrane,* the greater will be the rate of gaseous diffusion. The normal pulmonary membrane is sufficiently thin

so that venous blood entering a pulmonary capillary can be brought almost to complete gaseous equilibrium with the alveolar air within approximately one-fifth of a second. Occasionally, though, the thickness of the membrane and the fluid on its surface increases many-fold, especially when the lungs become edematous because of pulmonary congestion, pneumonia, or some other lung disease. The person may die simply because gases cannot diffuse through the thick membrane rapidly enough.

4. The greater the *solubility of the gas in the pulmonary membrane,* the greater also will be the rate of gas diffusion. The reason for this is that when greater quantities of gas dissolve in a given area of the membrane, larger numbers of molecules can then traverse the membrane at the same time. Considering the solubility of oxygen to be 1, the relative solubilities of the three important respiratory gases are the following:

$$Oxygen = 1$$
$$Carbon\ dioxide = 20$$
$$Nitrogen = 0.5$$

Therefore, carbon dioxide diffuses through the pulmonary membrane approximately 20 times as easily as oxygen, while oxygen diffuses approximately 2 times as easily as nitrogen.

5. A gas diffuses through the membrane approximately in *inverse proportion to the square root of its molecular weight.* The above three usual respiratory gases all have square roots of their molecular weights that are almost equivalent. However, for gases with much smaller molecular weights, such as helium, this can be a significant factor in determining the rate of diffusion.

Putting all of the above factors together, one derives the formula shown at the bottom of the page for rate of diffusion through the pulmonary membrane.

TRANSPORT OF OXYGEN TO THE TISSUES

DIFFUSION OF OXYGEN FROM THE ALVEOLI INTO THE BLOOD AND FROM THE BLOOD INTO THE TISSUES

The overall transport of oxygen from the alveolus to the tissue cell requires three different events (1) diffusion of oxygen from the

Table 19–2.   **Relative Gaseous Pressures in the Alveoli and the Blood**

Gas	Alveolar Air	Venous Blood	Arterial Blood
O_2	104	40	100
CO_2	40	45	40
N_2	569	569	569

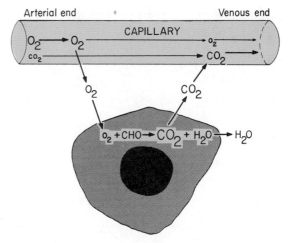

Figure 19–14. Diffusion of oxygen from the capillary to the tissue cell, formation of carbon dioxide inside the cell, and diffusion of carbon dioxide back to the tissue capillary.

alveolus into the pulmonary blood, (2) transport of the blood through the arteries to the tissue capillaries, and (3) diffusion of oxygen from the capillaries to the tissue cell.

The data of Table 19–2 explain why oxygen diffuses from the alveoli into the pulmonary blood. The oxygen pressure (Po_2) of "venous" blood entering the lungs is only 40 mm. Hg in comparison with the alveolar air Po_2 of 104 mm. Hg. The pressure difference, therefore, is 64 mm. Hg, which causes extremely rapid diffusion of oxygen into the blood. During the very short time that the blood remains in the capillary, only about 1 second, it attains a Po_2 of approximately 100 mm. Hg, almost as much as that of the alveolar air itself.

The aerated blood then flows from the lungs to the tissue capillaries, where oxygen diffuses through the capillary membrane and tissue spaces into the tissue cells. The reason for this is that the oxygen pressure in the cells is very low, which may be explained as follows: When oxygen enters a cell it reacts very readily with sugars, fat, and proteins to form carbon dioxide and water, as is illustrated in Figure 19–14. As a result, much of the oxygen is immediately removed, thus keeping the intracellular Po_2 usually between 10 and 20 mm. Hg. The high pressure of the oxygen in the arterial blood as it enters the capillaries, 100 mm. Hg, causes it to diffuse into the interstitial fluid and thence through the cellular membrane to the interior of the cells.

In summary, oxygen movement, both into the blood of the lungs and from the blood to the tissues, is caused by diffusion. The oxygen pressure difference from the lungs to the pulmonary blood is sufficient to keep oxygen always diffusing into the blood, and the pressure difference from the capillary blood to the tissue cells is also always sufficient to keep oxygen diffusing from the blood to the cells.

TRANSPORT OF OXYGEN IN THE BLOOD BY HEMOGLOBIN

When oxygen diffuses from the lungs into the blood a small proportion of it becomes dissolved in the fluids of the plasma and red cells, but approximately 60 times that much more combines immediately with the hemoglobin of the red blood cells and is carried in this combination to the tissue capillaries. Indeed, without the hemoglobin, the amount of oxygen that could be carried to the tissues would be only a fraction of that required to maintain life.

When the blood passes through the tissue capillaries, oxygen splits away from the hemoglobin and diffuses into the tissue cells. Thus, hemoglobin acts as a carrier of oxygen, increasing the amount of oxygen that can be transported from the lungs to the tissues to some 60 times more than that which could be transported only in the dissolved state.

Combination of Oxygen with Hemoglobin: the Oxygen-Hemoglobin Dissociation Curve. Figure 19–15 illustrates the so-called *oxygen-*

$$\text{Rate of Diffusion} \propto \frac{(\text{Pressure A} - \text{Pressure B}) \times \text{Surface Area} \times \text{Solubility}}{\text{Membrane Thickness} \times \sqrt{\text{Molecular Weight}}}$$

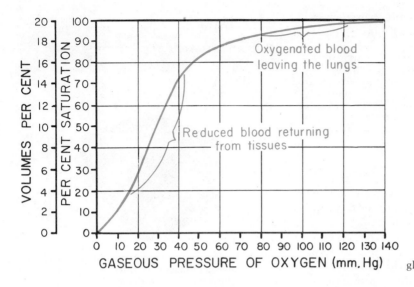

Figure 19–15. The oxygen-hemoglobin dissociation curve.

hemoglobin dissociation curve. This curve shows the per cent of the hemoglobin that is combined chemically with oxygen at each oxygen pressure level. Aerated blood leaving the lungs usually has an oxygen pressure of about 100 mm. Hg. Referring to the curve, it is seen that at this pressure *approximately 97 per cent of the hemoglobin is combined with oxygen.* As the blood passes through the tissue capillaries, the oxygen pressure falls normally to about 40 mm. Hg. At this pressure *only about 70 per cent* of the hemoglobin is combined with oxygen. Thus, about 27 per cent of the hemoglobin normally loses its oxygen to the tissue cells. Then, on returning to the lungs it combines with new oxygen and transports this once again to the tissue cells.

Utilization Coefficient and the Hemoglobin Reserve Capacity. The proportion of the hemoglobin that loses its oxygen to the tissues during each passage through the capillaries is called the *utilization coefficient*. In the normal person the utilization coefficient is 27 per cent, or expressed another way, approximately one-fourth of the hemoglobin is used to transport oxygen to the tissues under normal conditions.

When the tissues are in extreme need of oxygen, the oxygen pressure in the tissues can fall to extremely low values, allowing oxygen to diffuse from the capillary blood much more rapidly than usual. As a result, the saturation of the hemoglobin can fall to as low as 10 to 20 per cent instead of the normal level of 70 per cent, and the utilization coefficient rises to 77 to 87 per cent. Therefore, without even increasing the rate of blood flow, the amount of oxygen transported to the tissues in times of serious need can be increased more than three-fold.

If one also remembers that the cardiac output can increase as much as five- to sixfold in times of stress, then it is clear that the amount of oxygen that can be transported to the tissues can be raised to as much as 15 to 20 times normal, part of this being caused by an increase in the utilization coefficient and even more by the cardiac output.

Hemoglobin As an Oxygen Buffer. For cellular function to continue at a normal pace, the concentrations of all substances in the extracellular fluid must remain relatively constant at all times. One of the functions of hemoglobin is to keep the oxygen pressure in the tissues almost always between the limits of 20 and 45 mm. Hg. This may be explained as follows: As blood flows through the capillaries, 27 per cent of the oxygen is usually removed from the hemoglobin, making the hemoglobin saturation fall to about 70 per cent. Referring again to Figure 19–15, it can be seen that to remove this much oxygen the oxygen pressure in the tissues can never rise above 45 mm. Hg. On the other hand, whenever the oxygen pressure falls to 20 mm. Hg, more than three-fourths of the oxygen is released from the hemoglobin, and this rapid release usually keeps the tissue oxygen pressure from falling any lower except in times of extreme oxygen need. Thus, hemoglobin automatically releases oxygen to the tissues so that the tissue P_{O_2} remains almost always between the limits of 20 and 45 mm. Hg. If it were not for this buffering function of hemoglobin, a person would probably be killed by breathing pure oxygen, which would cause delivery of oxygen to the cells at lethal levels of Po_2.

Carbon Monoxide As a Hemoglobin Poison. Carbon monoxide combines with hemoglobin in a manner almost identical to that of oxygen, except that the combination of carbon monoxide with hemoglobin is approximately 210 times as tenacious as the combination of oxygen with hemoglobin. Also, because the two substances combine with hemoglobin at the same point on the molecule, they cannot both be bound to the hemoglobin at the same time. Carbon monoxide mixed with air in a concentration of 0.1 per cent, 210 times less than that of oxygen, will cause half of the hemoglobin to combine with carbon monoxide and therefore leave only half of the hemoglobin to combine with oxygen. When the concentration of carbon monoxide rises above approximately 0.2 per cent, which is still 100 times less than the normal concentration of oxygen, the quantity of hemoglobin available to transport oxygen becomes so slight that death ensues.

When death is about to occur from carbon monoxide poisoning, it can frequently be prevented by administering pure oxygen. The pure oxygen, on entering the alveoli, has a partial pressure of approximately 600 mm. Hg, six times the normal alveolar oxygen pressure. The force with which oxygen can combine with hemoglobin is therefore increased sixfold, which forces the carbon monoxide off the hemoglobin molecule six times as rapidly as would occur without treatment.

TRANSPORT OF CARBON DIOXIDE

Referring once again to Figure 19–14, it can be seen that, when oxygen is used by the cells for metabolism, carbon dioxide is formed. The pressure of carbon dioxide (Pco_2) in the cells becomes high, about 50 mm. Hg, and a pressure difference develops between the cells and the blood in the capillary. This causes carbon dioxide to diffuse out of the cell, into the interstitial fluid, and from there into the capillary blood. It is transported in the blood to the lungs. The Pco_2 of the blood entering the tissue capillaries is about 40 mm. Hg, but this

rises to 45 mm. Hg as carbon dioxide enters the blood from the cells. Therefore, the Pco_2 of the blood entering the lungs is also approximately 45 mm. Hg, while the Pco_2 in the air of the alveolus is only 40 mm. Hg. A pressure difference of 5 mm. Hg thus exists between the blood and the alveolus, causing carbon dioxide to diffuse out of the blood. Furthermore, because of the extreme solubility of carbon dioxide in the pulmonary membrane, the Pco_2 of the pulmonary blood falls to almost complete equilibrium with the alveolar Pco_2, that is, to 40 mm. Hg, before the blood leaves the capillary.

CHEMICAL COMBINATIONS OF CARBON DIOXIDE WITH THE BLOOD

Only about 5 per cent of the carbon dioxide is transported in the blood in the dissolved state. About 95 per cent of it diffuses from the plasma into the red cell where it undergoes two chemical reactions: First, carbon dioxide reacts with water to form carbonic acid. Red blood cells contain an enzyme called *carbonic anhydrase* which speeds this reaction approximately 250-fold. Therefore, very large quantities of carbon dioxide combine with water to form carbonic acid inside the red blood cells, while only small quantities are formed in the plasma. The carbonic acid in the cells immediately reacts with the acid-base buffers of the cells and becomes mainly bicarbonate ion. This reaction prevents the acidity of the cell from becoming too great.

Second, a few per cent of the total carbon dioxide entering the red cell combines directly with hemoglobin to form a compound called *carbaminohemoglobin.* This reaction does not occur at the same point on the hemoglobin molecule as the reaction between oxygen and hemoglobin. Therefore, hemoglobin can combine with both carbon dioxide and oxygen at the same time. In this way hemoglobin acts as a carrier not only of oxygen but also of carbon dioxide. However, the reaction occurs relatively slowly so that this method of carbon dioxide transport is much less important than transport in the form of bicarbonate ion.

To recapitulate, carbon dioxide is transported in the blood in three principal forms. First, only a very small quantity is transported as dissolved carbon dioxide. Second, by far the largest proportion is transported in the form of bicarbonate ion. And, third, a few per cent is transported in the form of carbamino-hemoglobin.

When the blood enters the pulmonary capillaries, all the chemical combinations of carbon dioxide with blood are reversed, and the carbon dioxide is released into the alveoli.

REFERENCES

Adamson, J. W., and Finch, C. A.: Hemoglobin function, oxygen affinity, and erythropoietin. *Ann. Rev. Physiol., 37*:351, 1975.

Bartels, H., and Baumann, R.: Respiratory function of hemoglobin. *Intern. Rev. Physiol., 14*:107, 1977.

Engel, L. A., and Macklem, P. T.: Gas mixing and distribution in the lung. *Intern. Rev. Physiol., 14*:37, 1977.

Fink, B. R.: The Human Larynx: A Functional Study, New York, Raven Press, 1975.

Forster, R. E.: Pulmonary ventilation and blood gas exchange. In Sodeman, W. A., Jr., and Sodeman, W. A. (eds.): Pathologic Physiology: Mechanisms of Disease. 5th ed. Philadelphia, W. B. Saunders Company, 1974, p. 371.

Forster, R. E., and Crandall, E. D.: Pulmonary gas exchange. *Ann. Rev. Physiol., 38*:69, 1976.

Kessler, M., Bruley, D. F., Clark, L. C., Lubbers, D. W., Silver, L. A., and Strauss, J. (eds.): Oxygen Supply. Baltimore, University Park Press, 1973.

Macklem, P. T.: Respiratory mechanics. *Ann. Rev. Physiol., 40*:157, 1978.

Murray, J. F.: The Normal Lung. Philadelphia. W. B. Saunders Company, 1975.

Piiper, J., and Scheid, P.: Comparative physiology of respiration: functional analysis of gas exchange organs in vertebrates. *Intern. Rev. Physiol., 14*:219, 1977.

Thurlbeck, W. M.: Structure of the lungs. *Intern. Rev. Physiol., 14*:1, 1977.

Wagner, P. D.: Diffusion and chemical reaction in pulmonary gas exchange. *Physiol. Rev., 57*:257, 1977.

West, J. B.: Pulmonary gas exchange. *Intern. Rev. Physiol., 14*:83, 1977.

QUESTIONS

1. Describe the functional anatomy of the respiratory system.
2. Tell how different muscles cause inspiration and expiration.
3. How does alveolar pressure change during the respiratory cycle?
4. What is the function of surfactant in the alveoli?
5. Explain what is meant by, and also give average values for, tidal volume, respiratory rate, minute respiratory volume, functional residual capacity, and vital capacity.
6. Explain why dead space occurs in the respiratory airways and tell its significance.
7. Explain what is meant by "partial pressure." What are the approximate partial pressures of the important gases in the alveoli?
8. Explain the factors that determine the rate of transport of a gas through the pulmonary membrane.
9. Discuss the overall transport of oxygen from the atmosphere all the way to the tissue cells.
10. Explain the role of hemoglobin in oxygen transport.
11. Explain the transport of carbon dioxide from the tissues to the blood and thence through the lungs to the atmosphere.

REGULATION OF RESPIRATION AND THE PHYSIOLOGY OF RESPIRATORY ABNORMALITIES

20

THE BASIC RHYTHM OF RESPIRATION

Respiration is controlled by the *respiratory center* located on both sides of the brain stem in the lateral reticular substance of the medulla and lower pons, illustrated in Figure 20–1A. Nerve signals are transmitted from this center through the spinal cord to the muscles of respiration. The most important muscle of respiration, the diaphragm, receives its respiratory signals through a special nerve, the *phrenic nerve,* that leaves the spinal cord in the upper half of the neck and passes downward through the chest to reach the diaphragm. Signals to the expiratory muscles, especially the abdominal muscles, are transmitted down the cord to the spinal nerves that innervate these muscles.

RHYTHMIC OSCILLATION IN THE RESPIRATORY CENTER

The rhythmic respiratory cycle is caused by oscillation back and forth between *inspiratory* and *expiratory* neurons intermingled in the respiratory center. Figure 20–1B illustrates the oscillating mechanism, showing to the left four expiratory neurons and to the right four inspiratory neurons. It is believed that the expiratory neurons are organized in a loop in

such a way that when one of these neurons becomes excited it sends a signal around and

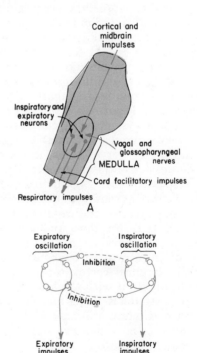

Figure 20–1. (A) The respiratory center located bilaterally in the lateral reticular substance of the medulla and lower pons. (B) Theoretical mechanism for the rhythmic control of respiration.

around the circuit, thus causing oscillation. As illustrated in the figure, fibers pass from this loop down the spinal cord to carry expiratory impulses to the expiratory muscles.

The comparable group of inspiratory neurons, shown to the right in the figure, is also capable of oscillating in a manner similar to that for the expiratory neurons. Each time oscillation occurs in this loop, impulses are sent to the inspiratory muscles.

Now let us see how the expiratory neurons interact with the inspiratory neurons. It is believed that when inspiratory oscillation occurs, this not only sends impulses down to the inspiratory muscles but simultaneously sends *inhibitory* impulses to the expiratory neurons to reduce or stop expiratory activity. Likewise, when the expiratory system oscillates, it sends collateral impulses to the inspiratory neurons to inhibit them. Thus, the two oscillating groups of neurons mutually inhibit each other. For this reason, when one is oscillating, the other stops oscillating. Furthermore, each of these sets of neurons usually oscillates for only 1 to 3 seconds, possibly because of neuronal fatigue. That is, when oscillation first begins, the neurons can pass signals from one to another with ease, but after a second or more they become fatigued so that their transmission fades. Thus, at the end of inspiratory oscillation the expiratory neurons, having rested for a few seconds, begin to oscillate. At the end of another 1 to 3 seconds, expiratory oscillation ceases and inspiratory oscillation begins again.

The arrows of Figure 20–1 illustrate several nerve pathways entering the respiratory center. Nerve impulses from the vagal and glossopharyngeal nerves cause reflex excitation or depression of the respiratory center, as will be discussed later. Signals from the spinal cord, including sensory impulses from the limbs during exercise and impulses from the skin, stimulate respiration. And signals from upper levels of the brain can affect respiration in many different ways. For instance, during speech the cerebral cortex controls respiration to coordinate it with the speech.

The Lung-Stretch Reflexes and Their Effect on the Basic Rhythm of Respiration. Located throughout the lungs are nerve endings that become stimulated when the lungs are distended and other nerve endings that are excited when the lungs are markedly deflated. Impulses from these pass through the vagus nerve to the medulla, where they inhibit inspiration upon lung expansion and expi-

ration upon lung deflation. One of the major functions of these reflexes, called the *lung-stretch reflexes* or *Hering-Breuer reflexes,* is to prevent overinflation or overdeflation, either of which can damage the lungs or seriously alter blood flow through the lungs.

The lung-stretch reflexes also aid in maintaining the basic rhythm of respiration. As the lungs distend during inspiration, the inflation reflex inhibits inspiration. Then, as the lungs deflate, the deflation reflex inhibits expiration. Therefore, this reflex helps to maintain the alternating rhythmic play between the inspiratory and expiratory centers.

It is evident, then, that at least two different feedback mechanisms help to maintain the respiratory cycle, one intrinsic in the medulla and one operating extrinsically through the lung-stretch reflexes. If either one of them becomes nonfunctional, the other is still capable of continuing the cycle. This duplication of neurogenic mechanisms is characteristic of most of the important functions of the brain, for any system that must continue to operate without a single instance of failure for up to 100 years must have a large margin of safety should some portion of the system become temporarily damaged.

Failure of the Respiratory Center. Occasionally all the oscillating mechanisms of the respiratory center fail at the same time. One of the more frequent causes of this is cerebral concussion or some other intracerebral abnormality that causes excess pressure on the brain medulla. The pressure collapses the blood vessels supplying the respiratory center, which blocks all activity of the medulla and therefore stops the respiratory rhythm. Another common cause of failure is acute poliomyelitis, which sometimes destroys neuronal cells in the reticular substance of the hindbrain, thereby depressing the respiratory center. Finally, the most frequent of all causes of respiratory failure is attempted suicide with sleeping drugs. These anesthetize the respiratory neurons and in this way stop the rhythm of respiration.

Failure of the respiratory rhythm is one of the most difficult of all abnormalities of the body to overcome. In general, very few drugs can be used to excite the respiratory center, and those that are available — caffeine, picrotoxin, and a few others — are so weak in their effects on the respiratory center that they are of little value. In general, the only effective treatment is to give prolonged artificial respiration in the hope that the cause of the respiratory depression will be rectified by the body itself. For instance, respiratory depression caused by poliomyelitis, pressure on the brain, or sleeping drugs is often reversible provided artificial respiration is maintained long enough.

REGULATION OF ALVEOLAR VENTILATION

Even though the basic rhythm of respiration is continuous, the rate of rhythm and the depth of respiration vary tremendously in response to the need of the tissues for oxygen and the need to rid the tissues of carbon dioxide. Obviously, the more oxygen needed by the body and the more carbon dioxide the body needs to expel, the greater must be the alveolar ventilation.

EFFECT OF CARBON DIOXIDE ON ALVEOLAR VENTILATION

One of the most powerful stimuli known to affect the respiratory center is *carbon dioxide.* When the carbon dioxide content of the blood increases above normal, all portions of the respiratory center become excited. The intensity of both inspiratory and expiratory signals becomes greatly enhanced, and the rapidity of the oscillation back and forth between the inspiratory and expiratory centers is also speeded up. Therefore, *both the rate and depth of respiration are increased.* Indeed, an increase of carbon dioxide in the blood can raise the rate of alveolar ventilation to as much as

10 times normal, which is illustrated by the CO_2 bar in Figure 20-2.

About one-half of the carbon dioxide effect on the respiratory center is caused by a direct action of carbon dioxide on the respiratory neurons to increase their excitability. However, the other half of the effect is caused indirectly in the following manner: When carbon dioxide increases in the blood some of it diffuses into the cerebrospinal fluid in which the brain is bathed. This carbon dioxide combines with water to form carbonic acid, thereby increasing the hydrogen ion concentration in the cerebrospinal fluid. Part of the respiratory center lies immediately beneath the anterior surface of the medulla in intimate contact with the cerebrospinal fluid, and this portion of the center is specifically excited by hydrogen ions. Thus, indirectly, the appearance of carbon dioxide in the cerebrospinal fluid excites the respiratory center as a result of increased hydrogen ions in the cerebrospinal fluid.

Control of Carbon Dioxide Concentration in the Body Fluids by the Carbon Dioxide Feedback Mechanism. It is extremely important that the respiratory system is excited by concentration of carbon dioxide in the blood because this provides a method by which the carbon dioxide concentration can in turn be controlled in all fluids of the body. This can be explained as follows:

The concentration of carbon dioxide in the blood is controlled by the rate of alveolar ventilation. That is, increased ventilation causes the lungs to blow off greater amounts of carbon dioxide from the blood. Therefore, when excess carbon dioxide excites the respiratory center, the resulting increase in ventilation

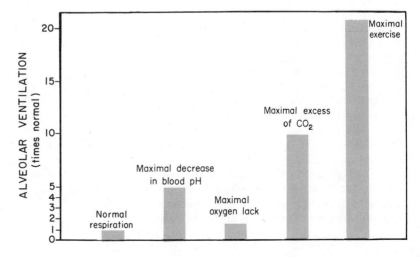

Figure 20-2. Effect on alveolar ventilation of maximal decrease in blood pH, maximal oxygen lack, maximal excess of CO_2, and maximal exercise.

causes the excess carbon dioxide concentration to return to near normal. Conversely, when carbon dioxide concentration falls too low, the resulting decrease in ventilation allows the carbon dioxide concentration to rise again back toward normal.

The body has no other significant means for controlling carbon dioxide concentration in the blood and body fluids, which makes this mechanism all-important. Should carbon dioxide concentration rise too high, it would stop essentially all chemical reactions in the body, because carbon dioxide is one of the end-products of almost all metabolic reactions. On the other hand, if the carbon dioxide concentration should fall too low, other dire consequences would occur, such as development of alkalosis caused by loss of carbonic acid from the body fluids. The alkalosis then would cause increased irritability of the nervous system, sometimes resulting in tetany or even epileptic convulsions.

REGULATION OF ALVEOLAR VENTILATION BY THE HYDROGEN ION CONCENTRATION

Another factor that has a powerful influence on alveolar ventilation, but not as powerful as that of carbon dioxide, is the hydrogen ion concentration of the blood. It will be recalled from the discussion in Chapter 18 that hydrogen ion concentration is often closely related to the blood concentration of carbon dioxide, because carbon dioxide combines with water in the body fluids to form carbonic acid and the carbonic acid in turn dissociates into hydrogen ions and bicarbonate ions. Therefore, in general, an increase in carbon dioxide concentration also increases the hydrogen ion concentration. On the other hand, when other types of acid besides carbonic acid accumulate, these, too, can increase the hydrogen ion concentration, but this time independently of the carbon dioxide concentration of the fluids.

When the hydrogen ion concentration becomes high — that is, when the pH becomes low — the respiratory neurons are excited, and alveolar ventilation increases. However, the maximum amount that ventilation can be increased by raising the hydrogen ion concentration is about fivefold. Therefore, an increase in the hydrogen ion concentration is only one-half as potent a stimulus of respiration as an increase in carbon dioxide concentration.

Importance of the Hydrogen Ion Feedback Mechanism for Regulating Respiration. Regulation of respiration by the hydrogen ion concentration helps to maintain normal acid-base balance in the body fluids, which was discussed in more detail in Chapter 18. When carbon dioxide is blown off through the lungs, loss of this carbon dioxide from the body fluids shifts the chemical equilibria of the acid-base buffers in such a way that much of the carbonic acid of the body fluids dissociates into water and carbon dioxide. This removes this carbonic acid portion of the body's acid and thereby decreases the hydrogen ion concentration. Thus, stimulation of the respiratory center by increased hydrogen ion concentration sets off a feedback mechanism that automatically decreases the hydrogen ion concentration back toward normal. Conversely, reduced hydrogen ion concentration decreases alveolar ventilation, increasing the acidity of the fluids back toward normal.

REGULATION OF ALVEOLAR VENTILATION BY OXYGEN DEFICIENCY

On first thought most students intuitively believe that respiration is regulated mainly by the need of the body for oxygen. However, the rate of ventilation normally has little effect on the amount of oxygen delivered to the tissues. The reason for this is the following: At ordinary ventilatory rates, or even at a rate as low as one-half normal, the hemoglobin in the blood becomes almost completely saturated with oxygen as the blood passes through the lungs. Increasing the ventilatory rate to an infinite level will not further saturate the hemoglobin because all of the hemoglobin that is available to combine with oxygen has already combined. Therefore, extreme increases or moderate decreases in alveolar ventilation make very little difference in the amount of oxygen carried from the lungs by the hemoglobin. There is thus no need for very sensitive regulation of respiration to maintain constant oxygen concentration in the blood.

On rare occasions, however, the oxygen concentration in the alveoli does fall too low to supply adequate quantities of oxygen to the hemoglobin. This occurs especially when one ascends to high altitudes where the oxygen

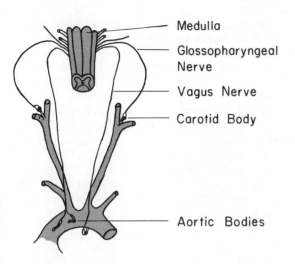

Figure 20–3. The chemoreceptor system for stimulating respiration.

concentration in the atmosphere is very low. Or it occurs when a person contracts pneumonia or some disease that reduces the oxygen in the alveoli. Under these conditions the respiratory system needs to be stimulated by *oxygen deficiency.* A mechanism for this is the *chemoreceptor system,* shown in Figure 20–3. Minute *aortic* and *carotid bodies,* each only a few millimeters in diameter, lie adjacent to the aorta and carotid arteries in the chest and neck, each of these having an abundant arterial blood supply and containing neuronlike cells called *chemoreceptors* which are sensitive to the lack of oxygen in the blood. When stimulated, these receptors send signals along the vagus and glossopharyngeal nerves to the medulla, where they stimulate the respiratory center to increase alveolar ventilation.

Figure 20–2 shows that maximum oxygen lack can increase alveolar ventilation to only 1.6 times normal in comparison with an increase of 10 times caused by excess carbon dioxide and 5 times caused by excess hydrogen ions. This illustrates that under normal conditions the oxygen lack system for control of respiration is a very weak one. However, Figure 20–2 is to some extent misleading, because in those special conditions in which oxygen lack occurs at the same time that there is even slight excess of carbon dioxide and hydrogen ions in the body fluids, the oxygen lack mechanism does then become a powerful stimulator of respiration. Furthermore, under special conditions, such as after a person has been at high altitude for several days, the carbon dioxide mechanism loses its potency

for control of respiration, and the oxygen lack mechanism becomes more potent, then frequently increasing alveolar ventilation as much as five- to sevenfold.

EFFECT OF EXERCISE ON ALVEOLAR VENTILATION

Alveolar ventilation increases almost directly in proportion to the amount of work performed by the body during exercise, reaching as high as 120 liters per minute in the most strenuous exercise. This value is about 20 times that during normal quiet respiration, as illustrated in Figure 20–2.

Physiologists still have difficulty explaining the extreme increase in pulmonary ventilation that occurs during exercise. Indeed, despite the extremely rapid production of carbon dioxide during exercise and simultaneous rapid usage of oxygen, alveolar ventilation increases so greatly that this prevents the blood concentrations of these gases from changing significantly from normal. Therefore, most physiologists believe that the increase in respiration during exercise is not caused by either or both of these two chemical factors.

If it is not chemical factors that increase

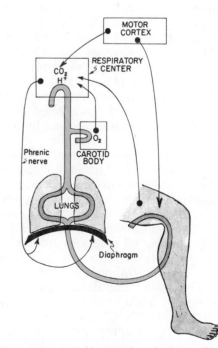

Figure 20–4. Mechanisms by which muscle activity stimulates respiration.

ventilation during exercise, it must be some stimulus entering the respiratory center by way of nerve pathways. Two such nerve pathways illustrated in Figure 20–4 have been discovered. (1) At the same time that the cerebral cortex transmits signals to the exercising muscles it also transmits parallel signals into the respiratory center to increase the rate and depth of respiration. (2) Movement of the limbs and other parts of the body during exercise sends sensory signals up the spinal cord to excite the respiratory center. Therefore, it is believed that these two signals, one from the cerebral cortex and the other from the moving parts of the body, are the factors that increase respiration during exercise. If these two factors fail to increase respiration adequately, only then do carbon dioxide and hydrgoen ions begin to collect in the body fluids and oxygen begins to diminish, and these then excite the respiratory center as a second line of defense to increase ventilation.

OTHER FACTORS THAT AFFECT ALVEOLAR VENTILATION

Effect of Arterial Pressure.   Arterial pressure is another factor that helps to regulate alveolar ventilation. This operates through the *baroreceptor system* which was described in Chapter 14. When the arterial pressure is high, signals from this system depress the respiratory center, and ventilation is reduced correspondingly. Conversely, when the pressure is low and blood flow to the tissues is poor, ventilation is increased; this partially compensates for the poor blood flow by allowing improved oxygenation and better removal of carbon dioxide in the lungs.

Effect of Psychic Stimulation.   Impulses initiated by psychic stimulation of the cerebral cortex can also affect respiration. For instance, anxiety states frequently can lead to intense hyperventilation, sometimes so much so that the person makes his body fluids alkalotic by blowing off too much carbon dioxide, thus precipitating alkalotic tetanic contraction of muscles throughout the body.

Effect of Sensory Impulses.   Sensory impulses from all parts of the body can affect respiration. The effect of entering a cold shower is well known; this causes an intense inspiratory gasp followed by a period of prolonged inspiration and then by rapid, forceful breathing. Even a pin prick can cause sudden

changes in the rate and depth of respiration. However, these factors affect alveolar ventilation only temporarily, for the chemical factors mentioned previously soon become dominant over these aberrant effects and suppress them.

Effect of Speech.   The speech centers of the brain also control respiration at times. When one talks, it is as important to control the flow of air between the vocal cords as to control the vocal cords themselves. Therefore, whenever nerve impulses are transmitted from the brain to the vocal cords, collateral impulses are sent simultaneously into the respiratory system.

THE MULTIPLE FACTOR THEORY FOR CONTROL OF RESPIRATION

It is obvious from the preceding discussions that no single factor regulates respiration, but that many different factors, including carbon dioxide concentration, hydrogen ion concentration, oxygen availability, exercise, arterial pressure, sensory phenomena, and speech, all help to regulate respiration. The most generally accepted theory for the regulation of respiration, then, is the so-called *multiple factor theory*. Some factors are known to be more important than others — carbon dioxide and exercise, for example. Others are of intermediate importance, such as hydrogen ion concentration, and others are of relatively minor importance, such as oxygen deficiency and arterial pressure.

PHYSIOLOGY OF RESPIRATORY ABNORMALITIES

HYPOXIA

One of the most important effects of most respiratory diseases is hypoxia, which means diminished availability of oxygen to the cells of the body. The different types of hypoxia that can occur are hypoxic hypoxia, stagnant hypoxia, anemic hypoxia, and histotoxic hypoxia, the principal mechanisms of which are shown in Figure 20–5.

Hypoxic Hypoxia.   Hypoxic hypoxia means failure of oxygen to reach the blood of the lungs. The obvious causes of hypoxic hypoxia are (1) *too little oxygen in the atmosphere,* (2) *obstruction of the respiratory passages,* (3) *thickening of the pulmonary membrane,* and

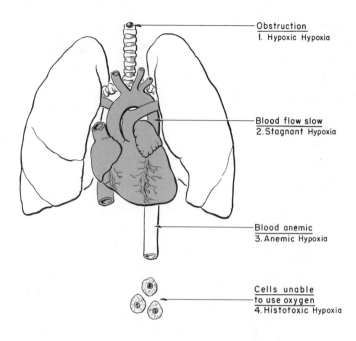

Figure 20-5. The types of hypoxia.

(4) *decreased area of the pulmonary membrane.*

Stagnant Hypoxia. Stagnant hypoxia means failure to transport adequate oxygen to the tissues because of too little blood flow. The most common cause of stagnant hypoxia is low cardiac output caused by *heart failure* or *circulatory shock*. Immediately after a heart attack the blood flow to the tissues may be so slight that the person may die of the stagnant hypoxia itself. At other times the transport of oxygen will be barely enough to keep him alive, but his tissues nevertheless will suffer severely from oxygen deficiency.

Anemic Hypoxia. Anemic hypoxia means too little hemoglobin in the blood to transport oxygen to the tissues. This may be caused by any one of three different abnormalities: First, the person may have *anemia,* which means too few red blood cells. Second, he may have a sufficient number of red blood cells but *too little hemoglobin in the cells.* Third, he may have plenty of hemoglobin, but much of it may have been *poisoned by carbon monoxide* or some other poison so that it cannot transport oxygen; for instance, breathing air with a content of only 0.2 per cent carbon monoxide can decrease the hemoglobin available to transport oxygen to as low as one-third normal, which is a lethal level.

Histotoxic Hypoxia. This means failure of the tissues to utilize oxygen even though adequate quantities are transported to them. The classic cause of serious histotoxic hypoxia is *cyanide poisoning;* this blocks the enzymes responsible for use of oxygen in the cell. Mild forms of histotoxic hypoxia frequently occur with *vitamin deficiencies,* for lack of certain of the vitamins often results in diminished quantities of oxidative enzymes in the cells.

OXYGEN THERAPY

In most types of hypoxia, breathing purified oxygen can be tremendously beneficial, the benefit occurring in three separate ways, which may be described as follows:

First, the normal partial pressure of oxygen in the alveoli is about 104 mm. Hg, but when a person breathes pure oxygen, this can rise to as high as 600 mm. Hg. Such an elevation can increase as much as sixfold the pressure that causes oxygen to diffuse through the pulmonary membrane.

Second, even when there is insufficient hemoglobin in the blood to carry enough oxygen to the tissues, oxygen therapy can still be beneficial. The reason for this is that when the partial pressure of oxygen in the alveoli becomes 600 mm. Hg, 2 ml. of oxygen dissolves in the fluid of every 100 ml. of blood and is transported in this form to the tissues. This effect is illustrated in Figure 20-6. At a normal alveolar oxygen partial pressure of about 100 mm. Hg, almost no oxygen is dissolved in the fluids. Yet as the alveolar oxygen partial pressure rises to 600 mm. Hg, one

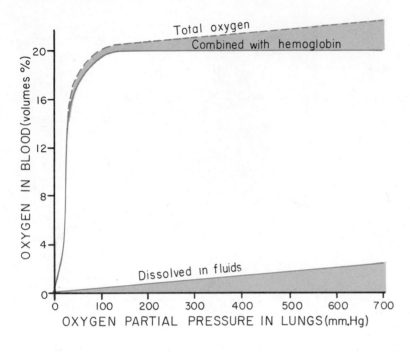

Figure 20-6. Effect of elevated alveolar partial pressure of oxygen on the total oxygen transported by each 100 ml. of blood.

sees by the shaded area of the figure that reasonable quantities of oxygen begin to dissolve in the fluids. This oxygen is the difference between life and death in some hypoxic persons.

The third means by which oxygen therapy can benefit some types of hypoxia is by decreasing the volume of gases that must flow in and out of the lungs through the trachea. If the respiratory passageways are partially obstructed, the blood can still be adequately oxygenated with only one-fifth normal alveolar ventilation if the person breathes pure oxygen rather than air because the concentration of oxygen in air is only 20 per cent. Yet the decreased alveolar ventilation will still cause a buildup of carbon dioxide in the body fluids, sometimes enough to produce serious acidosis even though the tissues are adequately oxygenated.

DYSPNEA

Dyspnea means *air hunger* or, in other words, a psychic feeling that more ventilation of the blood is needed than is being attained. Most instances of dyspnea occur when some respiratory abnormality causes the blood to become hypoxic or, even more often, when too much carbon dioxide collects in the body fluids. The respiratory center becomes overly excited, and impulses are transmitted to the conscious portions of the brain, apprising the psyche of the need for increased ventilation. This causes a feeling of air hunger or dyspnea.

However, an occasional person develops *psychic dyspnea* because of a neurosis. His blood has normal carbon dioxide and is adequately oxygenated, but the neurosis causes him to become unduly conscious of his respiration, making him feel that he is not receiving adequate quantities of air. One of the neuroses that most frequently cause this type of dyspnea is cardiac neurosis, for many who fear cardiac disease have been acquainted with cardiac patients who have had episodes of dyspnea because of pulmonary edema. Therefore, the neurotic who believes that he has heart disease often develops psychic dyspnea.

PNEUMONIA

Pneumonia is caused by infection of the lung with bacteria such as the pneumococcus bacterium or with viruses. The infection causes the walls of the alveoli to become inflamed and edematous and the spaces in the alveoli to become filled with fluid and blood cells. This is illustrated in the center of Fig-

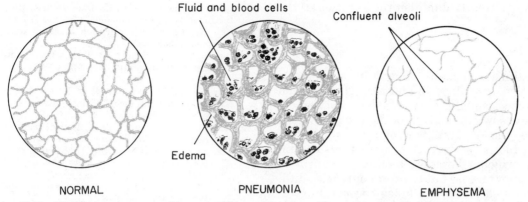

Figure 20–7.   Histologic appearance of the normal lung, of the lung with pneumonia, and of the emphysematous lung.

ure 20–7. Pneumonia causes hypoxic hypoxia for two reasons: first, because fluid and blood cells fill many alveoli so that they are not aerated at all, and, second, because the membranes of those alveoli that are still aerated are frequently so thickened with edema that oxygen cannot diffuse with ease.

PULMONARY EDEMA

Pulmonary edema means collection of fluid in the interstitial spaces of the lungs and in the alveoli; this affects respiration in the same way penumonia affects it. Generalized pulmonary edema is usually caused by failure of the left heart to pump blood from the pulmonary circulation into the systemic circulation. This can result from mitral or aortic valvular disease or from failure of the left ventricular muscle. In any event, blood is dammed up in the pulmonary circulation, and the pulmonary capillary pressure rises. When this pressure becomes greater than the colloid osmotic pressure of the blood, 28 mm. Hg, fluid transudes very rapidly from the plasma into the alveoli and interstitial spaces of the lungs, causing hypoxic hypoxia. Sometimes acute failure of the left heart causes pulmonary edema to occur so rapidly that the person dies of hypoxia in only 20 to 40 minutes.

EMPHYSEMA

Emphysema is a disease caused most frequently by chronic bronchial and alveolar infection resulting from *smoking*. In this condition, large portions of the alveolar walls are destroyed; this is illustrated to the right in Figure 20–7. Therefore, the total surface area of the pulmonary membrane becomes greatly diminished, which also diminishes the aeration of the blood. Hypoxic hypoxia results, and the quantity of carbon dioxide in the body fluids becomes increased.

The emphysematous person also usually develops pulmonary hypertension, for every time the wall of an alveolus is destroyed, the blood vessels in the wall are also destroyed. This increases the pulmonary resistance, which in turn elevates the pulmonary arterial pressure and eventually overloads the right heart.

ATELECTASIS

Atelectasis means collapse of a portion of the lung or of an entire lung. A common cause of atelectasis is a chest wound that allows air to leak into the pleural cavity. The surface tension of the fluid in the alveoli and the elastic fibers in the interstitial spaces of the lungs cause the lungs to collapse to a small size, the alveoli losing all their air. Another common cause of atelectasis is plugging of a bronchus, in which case the air in the alveoli is absorbed into the blood, thus collapsing the alveoli.

In atelectasis, not only do the alveoli collapse, but the blood vessels also collapse simultaneously. Consequently, blood flow through the collapsed lung decreases greatly, allowing the major portion of the pulmonary blood to flow through those areas of the lungs that are still aerated. Because of this shift of blood flow, atelectasis of a whole lung often does not greatly diminish the aeration of the blood.

ASTHMA

Asthma is usually caused by an allergic reaction to pollens in the air. The reaction causes spasm of the bronchioles, thus impeding air flow into and out of the lung. Furthermore, the outward flow of air through the bronchioles is obstructed more than the inflow, which allows the asthmatic person to inspire with ease but to expire only with difficulty. As a result, the lungs become progressively more distended, and with repeated asthmatic attacks year in and year out, the prolonged distention of the thoracic cage causes the chest to become barrel-shaped. Asthma rarely becomes severe enough to cause serious hypoxia, but it does cause severe dyspnea. Asthma can usually be treated by administering drugs that relax the bronchiolar musculature.

REFERENCES

Bradley, G. W.: Control of the breathing pattern. *Intern. Rev. Physiol., 14*:185, 1977.

Cohen, A. B., and Gold, W. M.: Defense mechanisms of the lungs. *Ann. Rev. Physiol., 37*:325, 1975.

Guz, A.: Regulation of respiration in man. *Ann. Rev. Physiol., 37*:303, 1975.

Hememann, H. O., and Goldring, R. M.: Bicarbonate and the regulation of ventilation. *Am. J. Med., 57*:361, 1974.

Milhorn, H. T., Jr., Benton, R., Ross, R., and Guyton, A. C.: A mathematical model of the human respiratory control system. *Biophys. J., 5*:27, 1965.

Mitchell, R. A., and Berger, A. J.: Neural regulation of respiration. *Am. Rev. Respir. Dis., 111*:206, 1975.

Paleček, F.: Control of breathing in diseases of the respiratory system. *Intern. Rev. Physiol., 14*:255, 1977.

Phillipson, E. A.: Respiratory adaptations in sleep. *Ann. Rev. Physiol., 40*:133, 1978.

Staub, N. C.: Pulmonary edema, *Physiol. Rev., 54*:678, 1974.

Tysinger, D. S., Jr.: The Clinical Physics and Physiology of Chronic Lung Disease, Inhalation Therapy, and Pulmonary Function Testing. Springfield, Ill., Charles C Thomas, Publisher, 1973.

White, F. N.: Comparative aspects of vertebrate cardiorespiratory physiology. *Ann. Rev. Physiol., 40*:471, 1978.

Widdicombe, J. G.: Reflex control of breathing *In* MTP International Review of Science: Physiology, Vol. 2. Baltimore, University Park Press, 1974, p. 273.

Wyman, R. J.: Neural generation of the breathing rhythm. *Ann. Rev. Physiol., 39*:417, 1977.

QUESTIONS

1. What causes the normal rhythmic nerve signals that promote inspiration and expiration of air?
2. Explain the role of and the mechanisms by which carbon dioxide controls alveolar ventilation.
3. Explain the relationship between hydrogen ion concentration and the regulation of alveolar ventilation.
4. Why is oxygen not the normal regulator of alveolar ventilation, and under what conditions does it play an important role?
5. What is there about exercise that causes an exceptionally great increase in alveolar ventilation?
6. What are the four major types of hypoxia?
7. Explain why oxygen therapy is of value in each of the types of hypoxia. How much so?

AVIATION, SPACE, AND DEEP SEA DIVING PHYSIOLOGY

21

One of the most important problems in aviation, space, and deep sea diving physiology is the altered barometric pressure to which the aviator, spaceman, or diver is exposed. Therefore, many of the physical principles applicable to respiration are also applicable to aviation, space flights, or dives deep beneath the sea.

Problems also arise in aviation, space, and diving physiology because of mechanical stresses on the body or because of extremes of climate. For instance, extreme acceleration during takeoff of a spaceship can cause the spaceman to weigh as much as 1500 pounds, a force that the body can withstand only when in the proper position. Also, in the upper atmosphere the air temperature is 20° to 50° C. below zero, and deep beneath the sea, pressures on the outside of the body can sometimes become so intense that the entire chest wall actually collapses.

EFFECTS OF LOW BAROMETRIC PRESSURE

OXYGEN DEFICIENCY AT HIGH ALTITUDES

At high altitudes the barometric pressure is low, as shown in Table 21–1, and the partial pressure of oxygen is reduced correspondingly. This decreases the amount of oxygen absorbed into the blood, and leads to hypoxia. Table 21–1 also shows the effect of low oxygen partial pressure on the saturation of arterial hemoglobin in the blood. For instance, at an altitude of approximately 23,000 feet only half of the arterial hemoglobin is saturated with oxygen. Obviously, this decreases the effectiveness of oxygen transport to the tissues at least 50 per cent, so that a person at this altitude will likely be debilitated because of tissue hypoxia.

Table 21–1. **Effects of Low Atmospheric Pressures on Alveolar Oxygen Concentrations and on Arterial Oxygen Saturation**

Altitude (ft.)	Barometric Pressure (mm. Hg)	P_{O_2} in Air (mm. Hg)	BREATHING AIR		BREATHING PURE OXYGEN	
			P_{O_2} in Alveoli (mm. Hg)	Arterial Oxygen Saturation (%)	P_{O_2} in Alveoli (mm. Hg)	Arterial Oxygen Saturation (%)
0	760	159	104	97	673	100
10,000	523	110	67	90	436	100
20,000	349	73	40	70	262	100
30,000	226	47	21	20	139	99
40,000	141	29	8	5	58	87
50,000	87	18	1	1	16	15

It will be noted also from Table 21–1 that the alveolar oxygen partial pressure at an altitude of 50,000 feet is only 1 mm. Hg and that the arterial oxygen saturation is only 1 per cent. Therefore, at this altitude, the small quantities of oxygen stored in the tissues actually diffuse backward into the blood and then from the blood into the lungs. Obviously, a person would have no more than a few minutes to live under these conditions.

Effects of Oxygen Deficiency.   Figure 21–1 shows graphically the time required at various altitudes for oxygen deficiency to cause either collapse or coma in a normal *unacclimatized* person. Above 30,000 feet an unacclimatized person lapses into coma in about 1 minute. At 20,000 feet the person usually does not go into coma, but after 10 or more minutes he exhibits signs of collapse such as weakness, mental haziness, and other deficiencies. He usually does not pass into coma until he ascends to altitudes of 22,000 to 24,000 feet, which is called his *ceiling*.

The first symptom of oxygen deficiency occurs in vision. Even at altitudes as low as 6000 to 10,000 feet, the sensitivity of the eyes to light begins to decrease — not enough to be noticed usually, but nevertheless enough to be measured by instruments. When the aviator rises to altitudes of 12,000 to 15,000 feet he is likely to experience alterations in psychic behavior. Some aviators tend to go to sleep while others develop euphoric exaltation. In almost all instances the acuity of reasoning becomes greatly depressed. In short, the aviator at this altitude is likely to develop a sleepy slap-happiness, and as he rises to 18,000 to 24,000 feet these symptoms often become so severe that control of the aircraft is endangered. For this reason safety regulations require that an aviator breathe oxygen when he ascends to dangerous altitudes whether he feels need for the oxygen or not. On occasion, he rises too high before applying his oxygen, and by that time his mental acuity might have become so depressed that he no longer knows his need. This often results in a vicious cycle of depressed mentality, for he rises still higher without knowing what he is doing; his mentality becomes further depressed, he rises higher, and so forth until coma develops.

Effect of Water Vapor and Carbon Dioxide on Alveolar Oxygen Partial Pressure.   Were it not for water vapor and carbon dioxide in the alveoli, the aviator could go to considerably higher heights than he actually can before developing oxygen deficiency. Regardless of the altitude, the alveolar pressure of water vapor does not change at all, and the pressure of carbon dioxide does not change greatly. Therefore, unless the barometric pressure is considerably greater than the combined pressures of these two gases, little space will be left in the alveoli for other gases. Figure 21–2 shows this effect. At sea level the combined pressure of all the gases in the alveoli is 760 mm. Hg. Carbon dioxide is continually excreted from the blood into the aveolar air, keeping the Pco_2 at a level of 40 mm. Hg, and water vapor continually evaporates from the surfaces of the alveoli, creating a water vapor pressure of 47 mm. Hg, which remains constant as long as the body temperature is normal, regardless of changes in barometric pressure. Thus, the combined pressure of water vapor and carbon dioxide at sea level is 40 plus 47 or 87 mm. Hg. Subtracting this from 760 mm. Hg, the remaining partial pressure available in the alveolus for both nitrogen and oxygen is 673 mm. Hg. This is more than adequate, so that the oxygen partial pressure is high enough to keep the arterial blood saturated with oxygen.

Now, let us see what effect the water vapor pressure and carbon dioxide partial pressure have on alveolar function at an altitude of 50,000 feet. Here, the total barometric pressure is only 87 mm. Hg. The water vapor pressure remains 47 mm. Hg because the body temperature does not change. The Pco_2

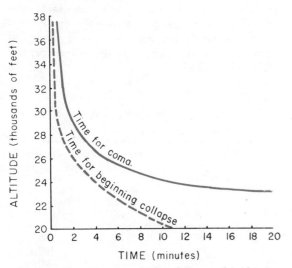

Figure 21–1.   Time required at different altitudes for oxygen deficiency to cause collapse or coma. (From Armstrong: Principles and Practice of Aviation Medicine. The Williams & Wilkins Co.)

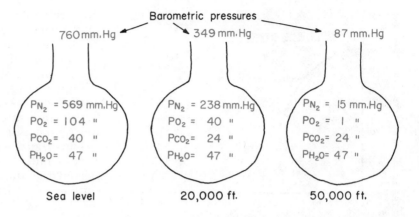

Barometric pressures

760 mm. Hg   349 mm. Hg   87 mm. Hg

P_{N_2} = 569 mm. Hg
P_{O_2} = 104 "
P_{CO_2} = 40 "
P_{H_2O} = 47 "

P_{N_2} = 238 mm. Hg
P_{O_2} = 40 "
P_{CO_2} = 24 "
P_{H_2O} = 47 "

P_{N_2} = 15 mm. Hg
P_{O_2} = 1 "
P_{CO_2} = 24 "
P_{H_2O} = 47 "

Sea level   20,000 ft.   50,000 ft.

Figure 21–2. Effects of low barometric pressures at different altitudes on alveolar gas concentrations.

has fallen to about 24 mm. Hg because of increased respiration. Therefore, the combined pressure of water vapor and carbon dioxide is now 71 mm. Hg, which when subtracted from 87 mm. Hg leaves a total partial pressure for both nitrogen and oxygen of only 16 mm. Hg. Furthermore, the person is so hypoxic that his blood absorbs oxygen out of the alveoli almost as rapidly as it enters. Therefore, the P_{O_2} falls to about 1 mm. Hg, which is so slight that the person will become unconscious in only a few seconds and will die within another minute or so. Were it not for the carbon dioxide and water vapor, it would be possible at 50,000 feet to supply far more oxygen to the blood.

Respiratory Compensation for Oxygen Deficiency — the Process of Acclimatization. The chemoreceptor mechanism described in the previous chapter increases alveolar ventilation when a person develops oxygen deficiency at high altitudes. This mechanism unfortunately is not very powerful, for it normally can increase alveolar ventilation only about 60 per cent. Yet even this amount allows the aviator to ascend several thousand feet higher than he could otherwise.

When a person is exposed to high altitudes for several days at a time, his chemoreceptor mechanism, for reasons not well understood, becomes progressively more active until his alveolar ventilation increases to as much as five to seven times normal. This slow process of adjusting to the high altitudes is called *acclimatization* to high altitudes.

Another factor that favors acclimatization is an increase in the number of red blood cells, for oxygen deficiency causes rapid production of these cells by the bone marrow, as explained in Chapter 5. Unfortunately, the formation of many new cells is a slow process,

requiring several weeks to help acclimate a person to high altitudes. Persons who live at high altitudes all the time often develop red blood cell counts as high as 7 to 8 million per cubic millimeter, which is 50 per cent or more greater than normal.

Acclimatization to high altitudes is not of special importance in aviation because the aviator rarely remains aloft long enough for acclimatization to occur. It is far more important to the mountain climber, who must become slowly acclimatized if he is to succeed in ascending to the top of the highest mountains. This explains why ascension is normally accomplished in slow stages over a period of weeks, thereby allowing the body to become progressively acclimatized. By using this procedure of acclimatization, conquerors of Mt. Everest have been able to remove their oxygen masks atop this highest mountain of the world at an altitude over 29,000 feet, though this would cause coma in a normal person in a minute or two.

OXYGEN BREATHING AT HIGH ALTITUDES

A person can ascend to far higher altitudes when he breathes pure oxygen than when he breathes air, because oxygen then occupies the space in the alveoli normally occupied by nitrogen in addition to the usual oxygen. This allows the alveolar partial pressure of oxygen to remain quite high even when the barometric pressure falls to a low value. Referring again to Figure 21–2, it will be noted that at 20,000 feet the combined alveolar pressure of oxygen and nitrogen when breathing air is 278 mm. Hg, and the alveolar P_{O_2} is only 40 mm. Hg, low enough to cause serious hypoxia.

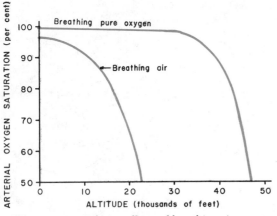

Figure 21-3. Relative effects of breathing air or pure oxygen on the saturation of arterial hemoglobin with oxygen at different altitudes.

If the nitrogen were replaced with oxygen, the oxygen pressure would be 278 mm. Hg, and the arterial hemoglobin would be 100 per cent saturated with oxygen, though when breathing air the saturation is only 67 per cent.

Figure 21-3 presents graphically the percentage saturation of hemoglobin in the arterial blood at different altitudes when breathing pure oxygen as opposed to breathing air. It is evident from this figure that essentially no desaturation occurs when breathing pure oxygen below an altitude of approximately 33,000 feet. Above that point, however, the barometric pressure becomes so low that even with nitrogen eliminated from the alveoli, the pressure of oxygen is still not sufficient to keep the blood saturated. The lowest level of arterial oxygen saturation at which a non-acclimatized person can remain alive for more than a few hours is about 50 per cent. Therefore, the ceiling for a person breathing pure oxygen is approximately 47,000 feet, in comparison with a ceiling of 23,000 feet for one breathing air.

PRESSURIZED CABINS

Obviously, all the problems of low barometric pressure at high altitudes can be overcome by pressurizing the airplane. Usually the air inside the cabin of passenger planes is pressurized to maintain approximately the same pressure as that at 5000 feet.

Explosive Decompression. One of the major problems of pressurizing high altitude equipment is the possibility that the chamber might explode. Experiments have shown, for-

tunately, that sudden decompression does not cause serious damage to the body because of the decompression itself. The danger lies in exposure to the low partial pressures of oxygen in the rarefied atmosphere. Referring to Figure 21-1, one sees that at altitudes above 30,000 feet the person has only a minute or more of consciousness, during which time he must institute appropriate lifesaving measures, such as putting on an oxygen mask or parachuting to safety. If he parachutes without the aid of a special automatic parachute opening device, it is important to open the chute as soon as possible, because he is likely to lapse into coma and fall to earth without benefit of the parachute. On the other hand, if he opens the chute he is likely to lapse into coma and die because of the altitude. Fortunately, modern high-flying jets are equipped with special devices that eject the aviator out of the plane when need be and automatically delay the opening of the chute until an appropriate altitude is reached.

ACCELERATION EFFECTS OF AVIATION

Airplanes move so rapidly and change their direction of motion so frequently that the body is often subjected to severe physical stress caused by the sudden changes in motion. When the *velocity* of motion changes, the effect is called *linear acceleration*. When the *direction* of motion is changed, the effect is called *centrifugal acceleration*. In general, the forces induced by linear acceleration during normal flight of an airplane are not sufficient to cause major physiologic effects. But when an airplane turns, dives, or loops, the centrifugal forces caused by centrifugal acceleration are frequently sufficient to promote serious derangements of bodily function.

CENTRIFUGAL ACCELERATION

Positive G. Figure 21-4 illustrates an airplane going into a dive and then pulling out. While the airplane is flying level, the downward force of the aviator against his seat is exactly equal to his weight. However, as he begins to come out of the dive, he is pushed against the seat with much greater force than his weight because of centrifugal force. At the lowest point of the dive the force is six times that which would be caused by normal gravi-

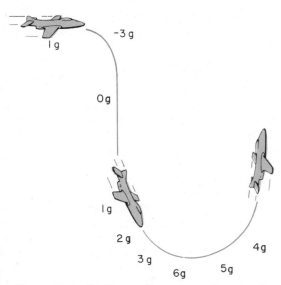

Figure 21-4. Positive and negative g during a dive and pullout of an airplane.

tational pull. This effect is called *positive centrifugal acceleration,* and the person is said to be under the influence of +6 g force, or, in other words, a force six times that of gravity.

Negative G. At the beginning of the dive in Figure 21-4, the airplane changes from level flight to a downward direction, which throws the pilot upward against his seat belt. He then is not exerting any force at all against his seat but instead is being held down by the seat belt with a force equal to three times his weight. This effect is called *negative centrifugal acceleration,* and the amount of force exerted is said to be −3 g.

Effect of Centrifugal Acceleration on the Circulatory System. The most important effects of centrifugal acceleration on the body occur in the circulation. Positive g causes the blood in the vascular system to be centrifuged toward the lower part of the body. The veins of the abdomen and legs distend greatly, storing far more blood than usual, so that little or none of it returns uphill to the heart. The cardiac output falls to very low values or even to zero. The arterial pressure falls and the person lapses into coma.

Negative g causes opposite effects; so much blood is centrifuged into the upper part of the body that the cardiac output rises and the arterial pressure rises. Occasionally the person develops such high pressure in his head vessels that brain edema results, and, rarely, vessels rupture in both the brain and eyes,

causing serious mental impairment or visual damage.

Figure 21-5 depicts graphically the time required for blackout (coma) to ensue at various degrees of positive acceleration. A person can usually withstand +4 g almost indefinitely; he might develop a certain amount of dizziness, but, in general, his circulatory control system can maintain sufficient cardiac output to prevent blackout. However, at acceleration values above +4 g, blackout usually occurs quite rapidly, at 5 g in about 8 seconds; at 12 g in about 3 seconds; and at 20 g in about 2½ seconds. Obviously, the amount of time that an aviator can remain conscious at high rates of acceleration limits the rapidity with which he can come out of a dive, and it also limits the sharpness of turns that he can make. Most airplanes are designed to withstand 20 or more positive g. Therefore, the maneuverability of an airplane is dependent more on what the aviator himself can stand than on the strength of the airplane.

Antiblackout Measures. Application of pressure to the outside of the legs and lower abdomen can prevent pooling of blood during positive acceleration.

Another means for partially accomplishing the same effect is for the aviator himself to tighten his abdominal muscles as much as possible during the pullout from a dive. This enables him to withstand a little more acceleration than he otherwise could and allows him several seconds extra exposure time before blacking out.

Different types of "anti-g" suits for accom-

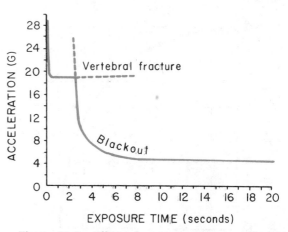

Figure 21-5. Effect of exposure time to different degrees of positive acceleration in producing blackout or vertebral fracture. (From Armstrong: Principles and Practice of Aviation Medicine. The Williams & Wilkins Co.)

plishing this have also been devised. One of these utilizes air bags applied to the abdomen and legs. When the airplane pulls out of the dive, the bags inflate so that positive pressure is applied to the legs and abdomen. In this way the aviator is not uncomfortably subjected to pressure all the time but only when it is needed.

Because of the speed of the modern jet airplane, the tremendous acceleratory forces that can develop make it impossible for an aviator to turn sharply without blacking out. One of the solutions to this problem will probably be for the aviator to lie prone while flying the plane, for the body can withstand +15 g in the horizontal position in comparison with only +4 g in the sitting position.

Effect of Positive Acceleration on the Skeleton. Also illustrated in Figure 21–5 is the effect of positive acceleration in causing vertebral fracture. This graph shows that with the aviator in the sitting position, positive acceleration greater than 20 g is likely to rupture a vertebra because of the intense force of the upper body pressing downward. Only a split second of the intense force can cause the fracture, for the stress that a bone can withstand does not depend on how long the stress is applied, but on how much stress is applied. Similar rupture of other tissues can also occur, such as tearing of the supporting structures of the heart with consequent internal hemorrhage and death.

Deceleration Forces During Parachute Descent. When a person jumps from an airplane, gravity causes him to begin falling toward earth. Figure 21–6 illustrates the velocity of fall after different distances of fall (if the person has not opened his parachute). By the time he has fallen 1500 feet his velocity will have reached approximately 175 feet per second. At this speed, the air resistance exactly opposes the tendency for gravity to increase his velocity even more. Therefore, this velocity of fall, 175 feet per second, is the maximum that will be attained, and it is called the *terminal velocity*. (At very high altitudes the terminal velocity is considerably greater than 175 feet per second because of the rarefied atmosphere, but it will slow down to 175 feet per second when the person falls into the heavier atmosphere.)

One of the deceleratory forces caused by parachute descent is the *opening shock* that occurs when the parachute opens. The force of the shock is roughly proportional to the square of the velocity of fall at the time of opening. Parachutes are designed so that even after the person has reached the terminal velocity of fall the shock will not be sufficient to harm the body.

On landing, the parachutist approaches earth at a velocity of about 15 miles per hour. This is equivalent to jumping without a parachute from a wall approximately 8 feet high. Even this velocity can be dangerous if the parachutist is not ready to cope with the landing jolt. If he lands stiff-legged, he will almost certainly suffer a fracture of a leg, the pelvis, or a vertebra. On the other hand, if he does not tense his muscles when landing, he will pancake on the ground and likely suffer other injuries. Therefore, it is important that he land with his legs flexed but tense.

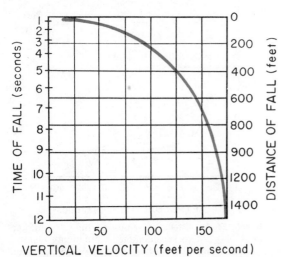

Figure 21–6. Effect of distance of fall on the velocity attained. (From Armstrong: Principles and Practice of Aviation Medicine. The Williams & Wilkins Company.)

SPACE PHYSIOLOGY

The physiologic principles of traveling in spaceships are much the same as those applicable to aviation physiology, except that the importance of some of the factors is intensified. The special problems in space physiology include: (a) weightlessness, (b) intense linear acceleration forces, (c) supply of oxygen and other nutrients, and (d) special environmental hazards, especially radiation.

Weightlessness. Prior to the advent of the spaceship, persons had experienced the phenomenon of weightlessness for not more than a few seconds at a time, this state having been created in the cabins of airplanes undergoing a trajectory path, upward and over a hump, calculated for exact course and velocity to provide zero g forces for a period of 10 to 25 seconds. Yet it was known that once a spaceship was in orbit or was traveling from planet to planet, the spaceship itself would be subjected to exactly the same gravitational forces as the spaceman inside the ship. Therefore, the person would experience the phenomenon of prolonged weightlessness; that is, there would be no force pulling him in any direction toward any part of the spaceship.

Prior to the first space flight, many physiologists were concerned about how a human being would react to weightlessness of long duration. Fortunately, however, now that man has experienced this phenomenon, it has proved to have little significant effect on the physiology of the body, except that the spaceman frequently has difficulty readjusting to the "weightful" state on returning to earth. However, weightlessness does necessitate the use of a few operational devices for flying the spaceship and for other activities within the spaceship. First, since the spaceman "floats" inside the spaceship, he must be either strapped in place or have appropriate hand holds. Second, his food must be in special containers that have closed tops, because the food will literally float off any plate, and fluids will float out of any glass. Often, foods are prepared in tubes of the toothpaste type, and the food is simply squeezed into the mouth. Likewise, water is sucked from a tube. Excreta must also be forced into a container rather than left to float freely in the spaceship. Otherwise, one can imagine what the atmosphere inside the spaceship would be like after a few days of flying. Except for these simple problems, the state of weightlessness is actually very peaceful and pleasant, which is what one might expect from his own experience of floating in water.

RETURN TO EARTH AFTER A PERIOD OF WEIGHTLESSNESS IN SPACE. When a spaceman has remained in the weightlessness state for more than 1 to 3 weeks and then returns to earth, he frequently faints when he first stands. The reason for this is that his circulatory system has become adjusted to the weightlessness state, and it must now become readjusted to the "weightful" state, which requires 2 to 5 days for full readjustment. Two factors seem to be responsible for these adaptive changes in the circulation: First, the blood volume decreases slightly during the space sojourn so that when the person first stands after return to earth, displacement of blood into the lower legs by gravity leads to

reduced cardiac output and a tendency to faint. Second, the blood vessels of the circulation, especially those of the lower body, seem to become more relaxed than normal during a long period of weightlessness. This, too, allows excessive quantities of blood to "pool" in the lower body when the spaceman first stands on returning to earth. It is interesting to note that these same two effects occur in patients who have laid in bed for long periods of time; they, too, often faint after a long period of recumbency.

Linear Acceleration of a Spaceship. At takeoff, a high velocity is reached in only a few minutes while putting the spaceship into orbit. This is achieved using rocket propulsion from one or more boosters. Figure 21–7 shows the intensity of forward acceleration at different times during the 5 minutes required to put a spaceship into orbit, illustrating that linear acceleration increases to 2 g when the first booster begins to fire. As that booster becomes lighter and lighter because of decreasing fuel, the acceleration increases to as much as 9 to 10 g. Then after the first booster has dropped, the second booster goes through a similar pattern of acceleratory changes. Also, the final stage, which carries the space capsule, accelerates for another minute or so before its rocket propulsion completely ceases. Upon cessation of all propulsion at the end of 5 minutes, the spaceman enters the state of weightlessness, which is designated in the figure by zero g acceleration.

In the sitting position the spaceman cannot withstand more than about 4 g, but in the

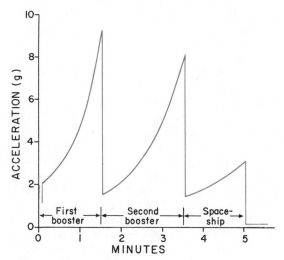

Figure 21–7. Acceleratory forces during takeoff of a spaceship.

reclining position, the position assumed in a reclining easy chair, the spaceman can stand 12 to 15 g, which is greater than any of the forces illustrated in Figure 21–7. Therefore, this is the position that the spaceman assumes during takeoff.

Linear Deceleration When Returning to Earth. One of the greatest problems in all of space physiology is the effect of slowing the spaceship as it reenters the atmosphere. The tremendous amount of kinetic energy stored in the momentum of the moving spaceship must be dissipated as the spaceship decelerates. This energy becomes heat and obviously increases the temperature inside the space capsule to very high levels. Therefore, special precautions are taken to provide a heat dissipation shield at the front of the spaceship which absorbs most of this heat rather than allowing it to be transmitted to the cabin of the capsule.

A spaceship traveling at a speed of Mach 100, a speed 100 times the velocity of sound and about 3 times that required for orbiting the earth, but which might well be used in interplanetary space flights, requires a distance of about 10,000 miles for safe deceleration. Any more rapid deceleration than this would create more g forces than the spaceman could stand.

Replenishment of Oxygen and Other Nutrients. It is a simple matter to provide all the oxygen and nutrients needed by the spaceman if he is to remain in space for not more than a few days to a few weeks. However, interplanetary travel will require that a spaceman remain in space for many months or even years, for which reason it becomes physically impossible to carry sufficient oxygen and other nutrients within the confines allowable in the spaceship. Therefore, many different procedures have been attempted to create a complete *life cycle* within the spaceship that will continually replenish the spaceship's oxygen and also supply adequate nutrients to the spaceman. For instance, algae are capable of living on human excreta, utilizing both carbon dioxide and fecal excreta to form oxygen, carbohydrates, proteins, and fats. Theoretically, therefore, algae can be used as food, and the oxygen can be breathed. Unfortunately, though, the amount of algae necessary to provide this complete life cycle for spacemen is far greater than space limitations in the spaceship will allow. Furthermore, the foods developed from algae have proved to be neither palatable nor totally life

sustaining. Yet future research might provide an adequate means for finally completing this life cycle.

Other methods for completing the life cycle are based on chemical or electrochemical procedures for separating oxygen from carbon dioxide and for resynthesizing certain types of foodstuffs. Here again, the success has not been completely rewarding, so that today's space flights are still limited by the amount of oxygen and nutrients that can be carried from earth at the time of takeoff.

Radiation Hazards in Space Physiology. Early in the space program, scientists began to realize that enough radiation of different types — gamma rays, x-rays, electrons, cosmic rays, and so forth — exists at certain points in space that these rays alone could be lethal. Figure 21–8 illustrates two major belts of such radiation, called *Van Allen radiation belts*, that extend entirely around the earth and that are especially prominent in the equatorial plane of the earth. One of these belts begins at an altitude of about 300 miles and extends to about 3000 miles. An outer belt then begins at about 6000 miles and extends to 20,000 miles.

The intensity of radiation in either one of the two Van Allen belts is great enough that a person in a spaceship orbiting the earth could receive enough radiation in only a few hours to cause death. Indeed, even when the spaceship goes through these two belts on an interplanetary voyage, lack of adequate radiation shielding or going through the two belts slowly could also cause death. However, Figure 21–8 shows that the radiation does not occur to a significant extent at the two poles of the earth, which means that space travel in the future will perhaps take place mainly from one or the other of the two poles. Since significant radiation does not begin until an altitude of about 300 miles is reached, one can also understand why space travel around the earth has thus far been confined to altitudes below 300 miles.

Other Problems of Space Flights. Other hazards at high altitudes include some of the same hazards of aviation but to much severer degree, such as (1) exposure of the spaceman to ultraviolet radiation, which can cause severe sunburn or can blind him, (2) exposure to extreme heat when in the direct pathway of sunrays and to extreme cold when on the opposite side of a spaceship or opposite side of the moon from the sunrays, and (3) exposure to very low barometric pressures in case of

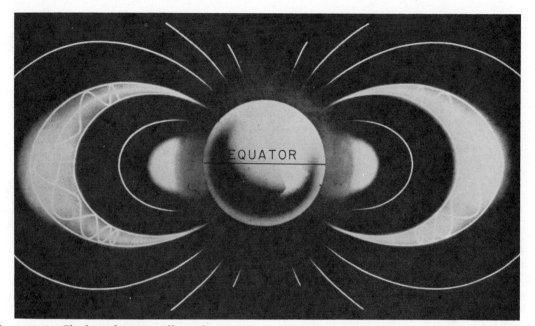

Figure 21-8. The hazardous Van Allen radiation belts around earth. (From Newell, H. E.: *Science,* 131:385, 1960.)

decompression of the spaceship. All of these problems require special engineering such as pressurized suits, protective filters for the eyes, and very intricate heat control systems for the space capsule.

DEEP SEA DIVING AND HIGH PRESSURE PHYSIOLOGY

The human body is occasionally exposed to extremely high barometric pressures such as those which occur when diving deeply beneath the sea or when working in a compression chamber — for instance, in a tunnel beneath a river filled with compressed air to prevent cave-in. The very high barometric pressure increases the amount of gases that pass from the alveoli into the blood and become dissolved in all the body fluids. This extra dissolved gas, unfortunately, often leads to serious physiologic disturbances.

Pressures at Different Depths of the Sea. A person on the surface of the sea is exposed to the pressure of the air above the earth. This pressure is approximately 760 mm. Hg; or it is also stated to be 1 atmosphere pressure. The weight of 33 feet of water is equal to the weight of the atmosphere, so that at a depth of 33 feet beneath the sea the pressure is double that at the surface, or 2 atmos-

pheres of pressure. Thus, the pressure increases as one descends beneath the sea approximately in accordance with the following table:

Feet	Atmosphere(s) pressure
0	1
33	2
67	3
100	4
200	7
300	10

EFFECT OF HIGH GAS PRESSURES ON THE BODY

Figure 21-9 illustrates the pressures of each of the alveolar gases when a diver is breathing (1) air at sea level, (2) compressed air at 33 feet below sea level, and (3) compressed air at 100 feet below sea level. Note that the pressure of water vapor and carbon dioxide remain the same regardless of how far below the surface the person descends. The reason for this is that water vapor pressure is directly dependent on the temperature of the body, as explained in Chapter 19, and the pressure of carbon dioxide is dependent on the rate of carbon dioxide release from the blood

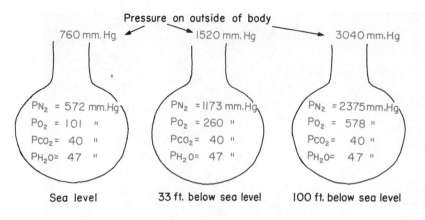

Pressure on outside of body

760 mm. Hg   1520 mm. Hg   3040 mm. Hg

P_{N_2} = 572 mm.Hg
P_{O_2} = 101 "
P_{CO_2} = 40 "
P_{H_2O} = 47 "

P_{N_2} = 1173 mm.Hg
P_{O_2} = 260 "
P_{CO_2} = 40 "
P_{H_2O} = 47 "

P_{N_2} = 2375 mm.Hg
P_{O_2} = 578 "
P_{CO_2} = 40 "
P_{H_2O} = 47 "

Sea level   33 ft. below sea level   100 ft. below sea level

Figure 21–9. Effect of different sea depths on the gas pressures in the alveoli.

into the lungs, which continues to be the same regardless of depth. On the other hand, the pressures of nitrogen and oxygen increase almost directly in proportion to the depth below sea level to which the person descends.

Oxygen Poisoning. Ordinarily a person breathing compressed air can descend to about 200 feet below the sea without danger from excess oxygen absorption. At levels deeper than this, his tissues begin to be damaged very rapidly by the oxygen. The reason for this is the following: Hemoglobin normally releases oxygen to the tissues at pressures ranging between 20 and 45 mm. Hg. In this way the hemoglobin automatically "buffers" the oxygen in the tissues, as was explained in Chapter 19, maintaining a relatively constant oxygen pressure that also allows a relatively constant rate of cellular oxidation. When there are extreme oxygen pressures in the lungs, however, large quantities of oxygen dissolve in the fluids of the blood, and the oxygen transported in this manner, in contrast to the oxygen transported by the hemoglobin, is released to the tissues at pressures that are much greater than 45 mm. Hg, indeed sometimes as high as several thousand mm. Hg, supplying far more than normal amounts of oxygen to the cells. This causes deranged cellular metabolism, often damaging the cells themselves or leading to abnormal cellular function.

The most debilitating effect of oxygen poisoning occurs in the brain, usually manifested by nervous twitchings, convulsions, and coma. Obviously, the person often loses control of himself, which might make him descend too deeply in the sea or fail to operate his breathing apparatus correctly, which will lead to his death.

The dangers of oxygen poisoning can be eliminated by supplying progressively less percentage of oxygen in the diver's breathing mixture as he goes to greater depths.

Nitrogen Poisoning. High pressures of nitrogen, also, can seriously affect a diver's mental functions. Even though this gas does not enter into any metabolic reactions in the body, when it dissolves in the body fluids in high concentrations, it exerts an anesthetic effect on the central nervous system. Ordinarily, a person can descend to approximately 200 feet below sea level before this anesthetic effect becomes serious, which is about the same maximum safe depth as that for the prevention of oxygen poisoning. However, at depths slightly deeper than this the diver will actually fall asleep because of the increasing amounts of nitrogen dissolved in his fluids. Also, in the early stages of this "nitrogen narcosis," the person frequently develops a sense of extreme exhilaration associated with seriously depressed mental acuity, a condition called "raptures of the depths." This state is comparable to severe drunkenness from alcohol, and it can cause the diver to descend much deeper than he should or to perform other dangerous acts that will lead to his death.

To avoid the dangers of nitrogen poisoning, the nitrogen is frequently replaced by helium in the breathing mixture. Helium does not cause an anesthetic effect, and it has an additional advantage of diffusing out of the body fluids more rapidly than nitrogen will when the diver ascends.

The Problem of Carbon Dioxide Wash-out from the Diving Helmet. If high concentrations of carbon dioxide are allowed to collect in a diver's helmet, he will soon develop respiratory acidosis and perhaps even coma. Therefore, carbon dioxide must be washed out of the helmet rapidly enough so that its concentration will never rise to any significant level. To do this, the same *volume* of gas must flow through the helmet at all times. When the diver descends to 200 feet below the sea, his air must be compressed seven times to withstand the pressure. This also reduces the volume seven times, and requires that the compressor pump seven times as much mass of air for carbon dioxide washout as is required at sea level. It can be seen, therefore, that the rate at which air is supplied to the diver must be increased directly in proportion to the depth at which he is working. Indeed, it is this factor of carbon dioxide washout that determines how much air must be pumped to the diver rather than the amount of oxygen that he needs.

SCUBA Diving.   The term SCUBA means "self-contained underwater breathing apparatus." That is, in contrast to the diving helmet or the diving bell, there are no air hoses to an attendant boat on the surface of the sea. The major factor that limits the time a person can remain under water using a SCUBA system is the problem of carbon dioxide washout from the alveoli. The deeper the diver goes, the less becomes the *volume* of each unit mass of gas liberated from his tank. Therefore, the greater must be the *mass* of gas that flows out of his tank to maintain carbon dioxide washout from his lungs. A tank of compressed air thus will last only one-seventh as long at 200 feet as at a few feet below the surface of the sea. For this reason, a SCUBA diver can descend to a 200-foot depth for only 15 to 20 minutes at a time. Other than for this limitation, the physical principles of SCUBA diving are essentially the same as those of helmet diving.

DECOMPRESSION SICKNESS

Solution of Nitrogen in the Body Fluids.   In addition to the anesthetic effect of nitrogen, this gas can also cause very serious damage because of gas bubble formation in the body fluids when the diver returns toward the surface. The amount of nitrogen normally dissolved in the entire body when a person breathes air at sea level pressure is about 1 liter. But if he remains 100 feet below the sea for several hours and breathes compressed air all this time, that is, at 4 atmospheres pressure, the amount of dissolved nitrogen increases to 4 times as much, and at 7 atmospheres pressure to 7 times as much.

Bubble Formation During Decompression.   After a diver has remained at a depth of 200 feet for several hours, he will have dissolved about 7 liters of nitrogen in his body fluids. This amount of dissolved nitrogen exerts a pressure of 3918 mm. Hg. When the diver is still at the 200-foot depth, the pressure of the water is about 5300 mm. Hg, which is far more than enough pressure to keep the nitrogen gas compressed in the body fluids. However, if the diver then suddenly rises to sea level, where the air pressure on the outside of the body is only 760 mm. Hg, a situation has now developed in which the pressure of the 7 liters of dissolved nitrogen in the body fluids is about four times as great as the pressure on the outside of the body. The nitrogen attempts to escape by any route possible, pushing against the inside of the blood vessels, skin, and cells. As a result, small bubbles of nitrogen develop everywhere in the body, and these obviously can cause very severe structural damage to almost any part of the body.

Effect of Bubble Formation on the Body.   The effects of bubble formation on the body are known by many different names: *decompression sickness*, the *bends, diver's paralysis, caisson disease*, and others. The most distressing effects occur when bubbles develop in the central nervous system. They sometimes occur inside neuronal cells, but mostly in the interstitial tissues of the brain and spinal cord, causing mechanical rupture of nerve fiber pathways. The damage can lead to serious mental disorders and very frequently to permanent paralysis because of ruptured fiber pathways to the muscles.

Decompression sickness usually causes severe pain, which may result from bubble damage in the central nervous system, in peripheral nerves, or in pain-sensitive tissues such as the joints and bones; or from sudden distention of the gastrointestinal gases, causing severe bloating of the gut.

In short, many different symptoms are produced by bubble formation and gas expansion in the body, though the most distressing and permanent of these result from damage to the nervous system.

Prevention of Decompression Sickness. The bubbles of decompression sickness can be prevented by allowing the diver to come to the surface very slowly or by decompressing him in a decompression chamber. In either event, the pressure around his body must be decreased so gradually that the excess gases dissolved in his fluids can leave through his lungs before a damaging quantity of bubbles develops.

REFERENCES

Adey, W. R.: The physiology of weightlessness. *Physiologist, 16*:178, 1974.

Bennett, P. B., and Elliott, D. H.: The Physiology and Medicine of Diving and Compressed Air Work. 2nd ed. Baltimore, The Williams & Wilkins Company, 1975.

Burton, R. R., Leverett, S. D., Jr., and Micaelson, E. D.: Man at high sustained +Gz acceleration: a review. *Aerosp. Med., 45*:1115, 1974.

Eccles, J. C.: My scientific odyssey. *Ann. Rev. Physiol., 39*:1, 1977.

Frisancho, A. R.: Functional adaptation to high altitude hypoxia. *Science, 187*:313, 1975.

Gamarra, J. A.: Decompression Sickness. Hagerstown, Harper & Row, Publishers, 1974.

Lahiri, S.: Physiological responses and adaptations to high altitude. *Intern. Rev. Physiol., 15*:217, 1977.

Lanphier, E. H.: Human respiration under increased pressures. *Symp. Soc. Exp. Biol., 26*:379, 1972.

Pace, N.: Respiration at high altitude. *Fed. Proc., 33*:2126, 1974.

QUESTIONS

1. How does low barometric pressure affect oxygen transport to the tissues?
2. What are the respiratory compensations for oxygen deficiency?
3. Approximately how much negative g and positive g can a person stand during flight maneuvers?
4. How much of a hazard is the state of weightlessness in space to the function of the circulation?
5. Explain the problems of linear acceleration as they relate to spaceship travel.
6. What causes the pressure of gases in the lungs to increase deep beneath the sea?
7. What are the toxic effects of high pressures of oxygen? Of high pressures of nitrogen?
8. Explain the role of nitrogen in the development of decompression sickness.

THE CENTRAL VII
NERVOUS SYSTEM

DESIGN OF THE CENTRAL NERVOUS SYSTEM, FUNCTION OF THE SYNAPSE, AND BASIC NEURONAL CIRCUITS

22

The Brain as the Major Control System of the Human Body. The brain is actually a computer that receives literally millions of bits of information from the different sensory organs and then integrates all of these to determine the response to be made by the body. Then it controls the muscles, the glands, and other functional systems of the body.

The purpose of this chapter is to present a general outline of the overall mechanisms by which the nervous system performs its functions. However, before beginning this discussion, the reader should review Chapters 9 and 10, which present the principles of membrane potentials, transmission of nerve impulses, transmission of impulses through neuromuscular junctions, and nervous control of the muscles.

GENERAL DESIGN OF THE NERVOUS SYSTEM

To perform its myriad functions, the nervous system is composed of three major parts: (1) the sensory system, (2) the motor system, and (3) the integrative system, each of which may be described as follows:

The Sensory System. Most activities of the nervous system originate with sensory experience, whether this be visual experience, auditory experience, tactile sensations from the surface of the body, or other sensations. The sensory experience can cause either an immediate reaction, or memory of the experience can be stored in the brain for many weeks or years and can then help to determine bodily actions at some future date.

Figure 22-1 illustrates the general plan of the sensory system, which transmits sensory information from the entire surface and deep structures of the body into the nervous system through the spinal and cranial nerves. This information is conducted into (a) the spinal cord at all levels, (b) the basal regions of the brain, including the medulla and pons, and (c) from these two areas into the higher regions of the brain, including the thalamus and the cerebral cortex. Then signals are relayed secondarily to essentially all other parts of the nervous system to begin analysis and processing of the sensory information.

The Motor System. The most important ultimate role of the nervous system is to control the bodily activities. This is achieved by controlling (a) contraction of skeletal muscles throughout the body, (b) contraction of smooth muscle in the internal organs, and (c) secretion by both exocrine and endocrine glands in many parts of the body. These activities are collectively called *motor functions* of the nervous system, and the portion of the

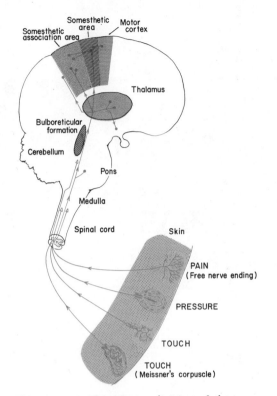

Figure 22-1. The sensory division of the nervous system.

information, which is called *memory,* while others assess sensory information to determine whether it is pleasant or unpleasant, painful or soothing, intense or weak, and so forth. It is in these regions that the appropriate motor responses to incoming sensory information are determined; once the determination is made, signals are then transmitted into the motor centers to cause the motor movements.

THE NEURON AS THE BASIC FUNCTIONAL UNIT OF THE NERVOUS SYSTEM

The basic functional unit of the nervous system is the *neuron*, which is also called a *nerve cell.* The neurons, by way of their filamentous nerve fibers, transmit signals from one to another, in this way providing a widespread communication system. Also, the neurons are responsible for the storage of memories, different patterns of thinking, motor responses, and so forth.

There are 12 billion neurons in the brain

nervous system that is directly concerned with transmitting signals to the muscles and glands is called the *motor division* of the nervous system.

Figure 22-2 illustrates the general plan of the motor axis for controlling skeletal muscle contraction. Signals originate in (a) the motor area of the cerebral cortex, (b) the basal regions of the brain, or (c) the spinal cord and are transmitted through motor nerves to the muscles. Each specific level of the nervous system plays its own special role in controlling bodily movements, the spinal cord and basal regions of the brain being concerned primarily with automatic responses of the body to sensory stimuli and the higher regions with deliberate movements controlled by the thought processes of the cerebrum.

The Integrative System. The term *integrative* means processing of information to determine the correct and appropriate motor action of the body or to provide abstract thinking. Located immediately adjacent to all sensory or motor centers in both the spinal cord and brain are numerous centers concerned almost entirely with integrative processes. Some of these areas are concerned with the storage of

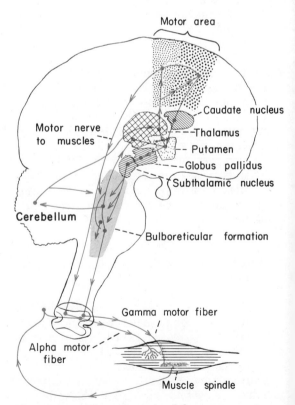

Figure 22-2. The motor axis of the nervous system.

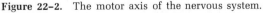

and spinal cord; 9 billion of these are located in the cerebral cortex where almost all storage of memories and other information occurs. Also, there are many different types of neurons, some of which make many connections with other neurons and some of which make relatively few connections. Some neurons are very large and others very small. Some give rise to extremely large nerve fibers which can transmit signals at velocities as great as 100 meters per second. Others give rise to very minute fibers that transmit nerve signals at velocities as slow as 1 meter per second. One of our goals will be to explain how these neurons are organized into different types of nerve networks that perform the myriad functions of the nervous system.

REFLEXES

Many functions of the nervous system are the result of reflexes. A reflex is a motor response that occurs following a sensory stimulus, the response taking place through a *reflex arc* consisting of a *receptor,* a *nerve transmission network,* and an *effector.* A receptor is any type of sensory nerve ending that is capable of detecting one of the usual body sensations such as touch, pressure, smell, sight, and so forth. Once the sensation is detected, a signal is transmitted by the nerve transmission network, which is composed either of a single neuron or of several successive neurons connected in series or parallel with each other or both. Finally, the effector is a skeletal muscle or one of the internal organs such as the heart, gut, or a gland that can be controlled by the nerves.

Let us refer to Figures 22–1 and 22–2 and see how a simple reflex could occur. Assume, for instance, that one of the free nerve endings in the skin is stimulated by a pain stimulus and that the pain signal is transmitted into the spinal cord. On entering the cord it excites other neurons that eventually send signals back to appropriate muscles to cause withdrawal of that part of the body contacting the painful stimulus. Thus, if a person steps on a nail, his foot is automatically withdrawn as soon as a sensory signal can be transmitted to the spinal cord and then back to the muscles of the leg. This is called, very simply, the *withdrawal reflex.* In other words, many of our automatic muscle functions are controlled primarily by reflexes that are processed in the

spinal cord and not by the conscious portions of our brain.

A much more complex reflex would be one in which many different sensory signals pass into the central nervous system from the eyes, ears, skin, and other portions of the sensory nervous system to apprise the person of danger. After a few seconds of integration, an automatic signal is transmitted back to the muscles to make the person run away. This is basically the same type of response as the simple withdrawal reflex except that it involves far more sensory elements, far more integrative elements, and far more motor elements. It also involves memories stored from previous learning, which make one realize the danger.

Thus, if one stretches his imagination far enough, he can explain almost any function of the nervous system on the basis of progressively more complex reflexes. However, most physiologists like to reserve the term reflex for an almost instantaneous automatic motor response to a sensory input signal, and to call the more complex activities of the nervous system the *higher functions.*

THREE MAJOR LEVELS OF NERVOUS SYSTEM FUNCTION

In the development of man through the stages of evolution, the nervous system has progressed from simple nerve fibers connecting different parts of the body, as occur in primitive multicellular animals, up to the very complex nervous system containing 12 billion neurons in the human being. As animalhood reached the multisegmental stage, nerve fibers and neuronal cell bodies aggregated to form a *neural axis* along the body, this neural axis providing the basis for the spinal cord of the human being. Then there developed in the head end of the axis enlarged aggregates of neurons which transmitted control signals down the neural axis to all parts of the body. These portions of the nervous system became highly developed in fishes, reptiles, and birds, and they are comparable to the basal regions of the human brain. Finally, there burst forth an overgrowth of nervous tissue surrounding the basal regions of the brain, forming the cerebral cortex. This occurred primarily in mammals and has reached a tremendous level of development in the human being.

During each of these three stages of development, new functions of the central nervous system were established, and many of these have been inherited all the way from the very lowest animals to the human being. For instance, in the multisegmental worm the most important nervous achievement was reflex response of the worm to noxious stimuli causing the worm's body to withdraw from any damaging influence. The human being has this same type of withdrawal reflex, as previously described, controlled almost entirely by the spinal cord, an inheritance carried all the way from the worm to the human being.

In the more complex animal, such as fishes and reptiles, it became important for the animal to be oriented properly in space, which initiated the process of equilibrium, and the human being has inherited this same ability. Furthermore, it is in the basal regions of the human brain, comparable to the highest levels of the fish's brain, where equilibrium is still controlled.

Finally, mammals have certain mental abilities over and above those of other animals, including especially the ability to store tremendous quantities of information in the form of memories and to utilize this information throughout life. Associated with this vast storage has also come the thinking process, a faculty that has reached its highest level of expression in the human brain. Along with the thinking process has come the art of communication by voice, by sight, or in other ways. All these functions are primarily controlled by the cerebral cortex.

In summary, the three major levels of organization in the central nervous system are (1) the *spinal cord,* which controls many of the basic reflex patterns of the body, (2) the *basal regions of the brain,* which control most of the more complex automatic functions of the body such as equilibrium, eating, gross body movements, walking, and so forth, and finally (3) the cerebral cortex, which harbors our higher thought processes and controls our voluntary discrete motor activities.

TRANSMISSION OF NERVE SIGNALS FROM NEURON TO NEURON: FUNCTION OF THE SYNAPSE

The *synapse* is the junction between two neurons. It is through the synapse that signals are transmitted from one neuron to another. However, the synapse has the capability of transmitting some signals and refusing other signals, thereby making it a valuable tool of the nervous system for choosing which course of events to follow. It is because of this variable transmission of signals that the synapse is perhaps the most important single determinant of central nervous system function.

PHYSIOLOGIC ANATOMY OF THE SYNAPSE

Figure 22–3 illustrates a typical neuron, which is composed of three major parts: the *soma,* or main body of the neuron, and two types of projections, the *dendrites* and the *axon.* Each neuron has only one axon, and signals are transmitted *outward* from the neuron cell body through this axon. The axon may extend only a short distance before it branches into many *terminal fibrils* that excite neighboring neurons, or it may extend for several feet — for instance, all the way from the spinal cord down to the feet to control the foot muscles. Most neurons have many dendrites, sometimes as many as several hun-

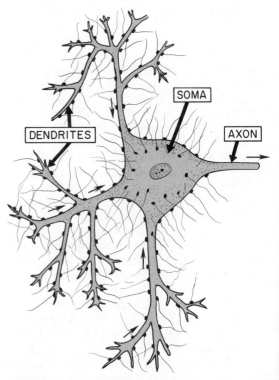

Figure 22–3. A typical motoneuron showing synaptic knobs on the neuronal soma and dendrites. Note also the single axon.

dred, and these transmit signals *into* the neuron cell body. They are usually short projections that extend only a few millimeters in the nervous system where they pick up signals from the axons of other neurons. However, some dendrites, like the axons, are also very long, such as the sensory nerve fibers which are dendrites extending from the peripheral sensory nerve endings to neuronal cell bodies that lie adjacent to the spinal cord and brain stem.

Figure 22–3 also shows hundreds of small fibers leading to the neuron and terminating in small knobs called *synaptic knobs.* These lie on the surfaces of the soma and dendrites. It is these junctions between the synaptic knobs and dendrites or soma that are called the *synapses.* The small fibers are many branches of axons that come from other neurons.

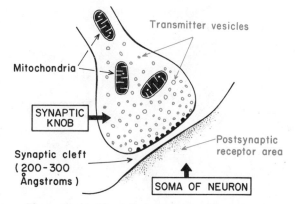

Figure 22–4. Physiologic anatomy of the synapse.

TRANSMISSION OF NERVE SIGNALS THROUGH THE SYNAPSE

Transmission of signals from the synaptic knobs to the dendrites or soma of the neuron occurs in very much the same way that transmission occurs at the neuromuscular junction as explained in Chapter 9. However, there is one major difference: At the neuromuscular junction the nerve endings always secrete acetylcholine which in turn excites the muscle fiber, while at the synapse some synaptic knobs secrete an *excitatory transmitter substance* and others secrete an *inhibitory transmitter substance;* therefore, some of these terminals excite the neuron and some inhibit it.

Excitation of the Neuron — the "Excitatory Transmitter." Figure 22–4 illustrates the structure of a typical synapse, showing a synaptic knob lying adjacent to the membrane of the soma of a neuron. The terminal has many small vesicles containing transmitter substance, and when a nerve impulse reaches the synaptic knob, momentary changes in the membrane structure of the knob allow a few of these vesicles to discharge the transmitter substance into the *synaptic cleft,* a narrow space between the knob and the membrane of the neuron. The transmitter substance then acts on the membrane to cause excitation of the neuron if the transmitter is excitatory, or inhibition if it is inhibitory.

CHEMICAL NATURE OF EXCITATORY TRANSMITTERS. In recent years it has been discovered that several different excitatory transmitters are secreted in different parts of the nervous system. Furthermore, they have varying characteristics of function, some providing prolonged stimulation of neurons while others provide very rapid and brief stimulation.

It should also be noted that all the synaptic knobs derived from a single neuron secrete the same type of transmitter substance. Thus, if a neuron located in one part of the brain sends branches of its axon to two separate brain areas, the transmitter substance secreted in both of these areas will still be the same.

One of the excitatory transmitters in the central nervous system is acetylcholine, the same excitatory transmitter that transmits signals from motor nerves into muscle fibers. However, a more complete listing of the excitatory transmitters that are thus far known to exist includes:

1. Acetylcholine
2. Norepinephrine
3. Serotonin
4. Dopamine (in some areas of the brain)

Excitatory Postsynaptic Potential. The manner in which the excitatory transmitter excites the neuron can be explained by referring to Figure 22–5. The excitatory transmitter increases the permeability of the neuronal membrane immediately beneath the synaptic knob. This allows sodium ions to flow rapidly to the inside of the cell, and since sodium ions carry positive charges, the net result is an increase in positive charges inside the cell, which is called the *excitatory postsynaptic potential.* As a result, an electrical current, illustrated by the arrows in Figure 22–5, is immediately set up through all parts of the

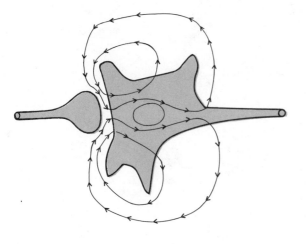

Figure 22–5. Flow of current beneath the synapse and around the neuron, showing that a synaptic discharge causes increased current flow over the entire body of the neuron.

cell body and through the entire membrane surface, including the base of the axon. If the potential becomes great enough, it will initiate an action potential in the axon, which is the nerve fiber leading from the neuron body. However, if the excitatory postsynaptic potential is less than a certain threshold value, nothing will happen — that is, no action potential.

"Summation" of Excitatory Postsynaptic Potentials. Almost invariably, stimulation of a single synaptic knob will not initiate an impulse in the axon. Instead, large numbers of synaptic knobs must become excited at the same time. Referring once again to Figure 22–3, it is readily apparent that hundreds of synaptic knobs could easily be excited simultaneously and that these operating in unison could cause neuronal discharge.

That is, if two knobs release their excitatory transmitter substances simultaneously, twice as much sodium enters the cell body and twice as much excitatory postsynaptic potential develops. This is called *summation.* As more and more knobs fire simultaneously, the current flowing through the cell body becomes greater and greater, until finally it is great enough to excite the axon. And if the postsynaptic potential increases still more, the rate of action potential firing also becomes faster and faster.

SPATIAL AND TEMPORAL SUMMATION. Two different types of summation occur at the synapse in the same way that two types of summation occur in nerve and muscle fibers. These are spatial and temporal summation. *Spatial summation* means that two or more synaptic knobs fire simultaneously, thereby "summing" their individual effects on the excitatory postsynaptic potential. *Temporal*

summation means that the same synaptic knobs fire two or more times in rapid succession, thus adding the effect of the second firing to that of the first before the first effect is over. The effect of the first discharge usually lasts about 15 milliseconds. Therefore, if two successive discharges of the same synaptic knob occur within less than $1/70$ second of each other, temporal summation occurs, and the closer together these discharges, the greater will be the degree of summation.

Repetitive Discharge of the Axon — Threshold for Firing. Once sufficient excitatory transmitter has been secreted by the synaptic knobs to raise the postsynaptic potential above a critical value called the *threshold* of the neuron, the axon will fire repetitively and will continue to do so as long as the potential remains above this threshold. Furthermore, as the postsynaptic potential rises higher and higher above the threshold, the more rapidly will the axon fire. For instance, let us assume that an excitatory postsynaptic potential of 11 millivolts is the threshold value. If the potential rises slightly above this, to 12 millivolts, one would expect the axon to discharge perhaps 5 to 10 times per second. But, if the postsynaptic potential rises to as high as 33 millivolts, 3 times the threshold value, then the axon perhaps would fire as many as 50 to 70 times per second.

Facilitation at the Synapse

When synaptic knobs discharge but fail to cause an action potential in the axon, the neuron still becomes *facilitated;* that is, even though an action potential does not result, the neuron becomes more excitable to impulses from other synaptic knobs. Let us assume, for instance, that 25 synaptic knobs must fire simultaneously to discharge a neuron. If 20 synaptic knobs fire simultaneously, the neuron will not discharge, but it does become sufficiently "facilitated" so that any 5 additional synaptic knobs anywhere on the surface of the cell could now elicit an impulse. This is the situation that develops in a "nervous" person, for large numbers of his neurons become facilitated though not excited; yet very few extraneous impulses can then cause terrific reactions.

Response Characteristics of Different Neurons

The nervous system is composed of hundreds of different types of neurons. Some of the cell bodies are tremendously large in size while others are very minute. Some are capable of transmitting as many as 1000 impulses per second over their axons, and some are capable of transmitting no more than 25 to 50 per second. Also, different neurons have different thresholds. All of these variations are fortunate, because neurons in different parts of the brain perform different functions.

Figure 22–6 illustrates typical *response patterns* of three different neurons. Neuron 1 requires a "threshold" postsynaptic potential of 5 millivolts to discharge it, and it reaches a maximum discharge rate of 100 per second. The threshold of neuron 2 is 8 millivolts, and the maximum frequency of discharge is 35. Finally, neuron 3 has a high threshold of 18 millivolts, but once it becomes excited, its maximum discharge rate is 120 per second.

INHIBITION AT THE SYNAPSE

Some of the synaptic knobs secrete an *inhibitory transmitter* instead of an excitatory transmitter. The inhibitory transmitter depresses the neuron rather than exciting it. Indeed, sometimes when an excitatory transmitter from other synaptic knobs is exciting the neuron, subsequent stimulation of inhibitory synaptic knobs will actually stop all firing of the neuron. The known inhibitory transmitters are:

1. Gamma aminobutyric acid (GABA)
2. Glycine
3. Dopamine (in the basal ganglia)

The inhibitory transmitter has an opposite effect on the synapse to that caused by the excitatory transmitter, usually creating a negative potential called the *inhibitory postsynaptic potential*. An example of the manner in which the inhibitory transmitter operates at the synapse is the following: Let us assume that the threshold for stimulation of a neuron is +10 millivolts and that excitatory synaptic knobs are generating a postsynaptic potential of +20 millivolts. The neuron will be firing repetitively at a rate of perhaps 50 to 200 times per second. At this time, stimulation of inhibitory synaptic knobs releases inhibitory transmitter which creates an inhibitory postsynaptic potential of −15 millivolts. When this summates with the excitatory potential of +20 millivolts, the net value is only +5 millivolts, a level far below the threshold of the neuron. Therefore, the neuron stops discharging.

Excitatory and Inhibitory Neurons. The central nervous system is made up of *excitatory neurons* that secrete excitatory transmit-

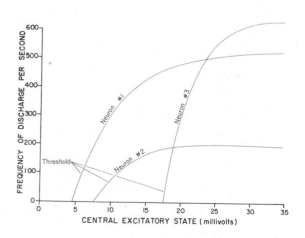

Figure 22–6. Response characteristics of different types of neurons.

ter at their nerve endings, and of *inhibitory neurons* that secrete inhibitory transmitter at their nerve endings. Certain neuronal centers of the central nervous system are composed entirely of excitatory neurons, others entirely of inhibitory neurons, and still others of both excitatory and inhibitory neurons. Therefore, the central nervous system, unlike the peripheral sensory and motor systems, has two modes of activity, either excitation or inhibition, instead of simply excitation alone.

SOME SPECIAL CHARACTERISTICS OF SYNAPTIC TRANSMISSION

One-way Conduction at the Synapse. Impulses traveling over the soma or dendrites of the neuron cannot be transmitted backward through the synapses into the synaptic knobs. Thus, only one-way conduction occurs at the synapse. This is extremely important to the function of the nervous system, for it allows impulses to be channeled in the desired direction.

Fatigue of the Synapse. Transmission of impulses at the synapse is different from transmission in nerve fibers in a very important respect: The synapse often *fatigues* very rapidly while nerve fibers fatigue hardly at all. Figure 22–7 illustrates this effect, showing that at the onset of an input signal, this particular neuron discharges very rapidly at first but then more and more slowly the longer it is stimulated. Some synapses fatigue very rapidly, while others fatigue very slowly.

One might expect this phenomenon of fatigue to be an impediment to the action of the central nervous system, but, on the contrary, it is a necessary feature. Were it not for synaptic fatigue, a person could never stop a thought, a rhythmic muscular activity, or any other prolonged repetitive activity of the nervous system once it had begun. Fatigue of synapses is a means by which the central nervous system allows a nervous reaction to fade away to make way for others. Later in the chapter, specific instances of the importance of fatigue are pointed out.

The "Memory" Function of the Synapse. When large numbers of impulses pass through a synapse, the synapse becomes "permanently" facilitated so that impulses from the same origin can pass through the synapse with greater ease at a later time. It is believed that this is the means by which memory occurs in the central nervous system. For in-

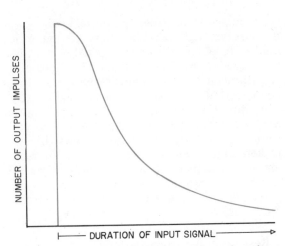

Figure 22–7. Effect of fatigue on the number of output impulses from the neuron after it begins to be stimulated.

stance, a given thought initiated by visual, auditory, or any other type of signal causes impulse transmission through given pathways in the brain. If the same thought is repeated over and over again, such as a thought caused by seeing the same view again and again, then the pathways for the thought become permanently facilitated so that subsequent impulses pass through these pathways with the greatest of ease. Therefore, impulses may later enter the same pathways from some other source besides the visual apparatus, and the person "remembers" the scene rather than actually seeing it. This process of memory will be discussed in greater detail in Chapter 27.

Other Factors That Affect Synaptic Transmission. Figure 22–8 illustrates the effect of several different factors on the number of im-

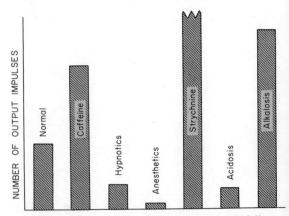

Figure 22–8. Effect of different drugs and different physiologic states on the excitability of a neuron.

pulses transmitted by a synapse in response to a given degree of stimulation of the synaptic knobs. Note that *hypnotics, anesthetics,* and *acidosis* all depress the transmission of impulses at the synapse, while *alkalosis* and *mental stimulants* such as *caffeine, benzedrine,* and *strychnine* all greatly facilitate synaptic transmission. Strychnine facilitates synaptic transmission by inhibiting the inhibitory synaptic knobs, and when given in sufficiently high dosage, it will cause the neurons to discharge spontaneously even in the absence of a synaptic stimulus. It is in this way that strychnine kills an animal, causing so many neuronal impulses to be transmitted throughout the central nervous system and into the motor system that it causes death by spasm of the respiratory muscles.

The Integrative Function of the Neuron. To summarize the overall function of the neuron, one can call it an "integrator," which means a type of calculator that collects information and sums it all together. Signals reach the neuron by way of excitatory and inhibitory synaptic knobs that, in turn, are excited by neurons from other parts of the nervous system. If the resultant sum of all the excitatory and inhibitory effects is above the threshold for excitation, the neuron fires. Some of the neurons fatigue rapidly, others slowly. Some neurons have high thresholds; others have low thresholds. Some fire at rapid rates, others at slow rates. The varying characteristics of the different neurons and their different connections in the nervous system allow the neurons in one portion of the nervous system to control one function of the body while those in another portion control another function. They allow the sorting of signals to determine their meanings, the performance of special skilled motions, the thinking of specific thoughts, and modification of these thoughts by signals arriving from other parts of the nervous system.

BASIC NEURONAL CIRCUITS

THE NEURONAL POOL

The central nervous system is divided into many different anatomic parts, in each of which are located accumulations of neurons called *neuronal pools.* An important feature of the different neuronal pools is that each has a different pattern of organization from all the others. That is, the distribution of nerve

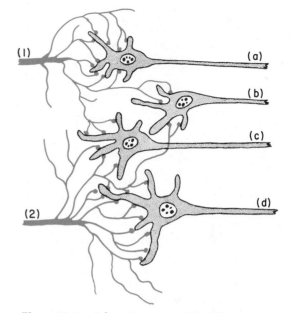

Figure 22–9. Schematic organization of four neurons in a neuronal pool.

fibers within the pool, the number of incoming nerve fibers, the number of outgoing fibers, the types of neurons in the pool, and many other features differ from one pool to another. Each pool is organized to perform a specific function. The purpose of the present section is to discuss, first, the general functions of the neuronal pool and, second, the characteristics of special types of pools.

Synaptic Connections in a Neuronal Pool. A neuronal pool is composed of thousands to millions of neuronal cell bodies. Figure 22–9 shows four typical neurons in a pool stimulated by two incoming nerve fibers. Note that each incoming fiber branches to supply synaptic knobs to several different neurons in the pool. A fiber may deliver one synaptic knob to a given neuron or as many as several hundred. Therefore, when the fiber is stimulated, it will facilitate some neurons and excite others.

Areas of Discharge and Facilitation in the Neuronal Pool. While still observing Figure 22–9, let us expand our imaginations until this pool of four neurons becomes a thousand closely packed neurons with hundreds of thousands of terminal nerve fibrils and synaptic knobs. Near the central point of entry of a fiber to the pool, the fiber usually branches profusely to form many synaptic knobs; this is illustrated in Figure 22–10. But farther away from the center the number of knobs becomes

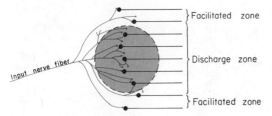

Figure 22–10. "Discharge" and "facilitated" zones of a neuronal pool.

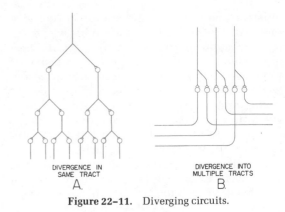

Figure 22–11. Diverging circuits.

less and less. Those neurons that lie in the center of the fiber's "field" are usually supplied with enough synaptic knobs that they discharge each time the input fiber is stimulated. Thus, the dark area in Figure 22–10 is called the *discharge zone*.

In the peripheral portion of the fiber field, the number of terminal fibrils ending on any single neuron is usually too few to cause discharge but, nevertheless, still enough to cause facilitation. Therefore, this area is called the *facilitated zone*. When a neuron is facilitated, a stimulus from some other source can excite it more easily.

SIMPLE CIRCUITS IN NEURONAL POOLS

The simplest circuit in a neuronal pool is that in which one incoming nerve fiber stimulates one outgoing fiber. This type of circuit does not exist in a precise form, though occasionally it is approximated. For instance, in transmitting certain types of sensory signals from peripheral nerves into the brain, a single incoming impulse into a neuronal pool may cause a single outgoing impulse.

Such a circuit obviously acts simply as a relay station, relaying on to additional neurons essentially the same information that enters it.

Diverging Circuits. Figure 22–11 illustrates two types of *diverging circuits*. In the first of these, a single incoming fiber stimulates progressively more fibers farther and farther along the pathway. This can also be called an *amplifying circuit* because an input signal from a single nerve fiber causes an output signal in many different fibers. This type of circuit is exemplified by the system for control of skeletal muscles: Under appropriate conditions, stimulation of a single motor cell in the brain sends a signal down to the spinal cord to stimulate perhaps as many

as a hundred anterior horn cells, and each of these in turn stimulates approximately 150 muscle fibers. Thus, a single motor neuron of the brain sometimes can stimulate as many as 15,000 muscle fibers.

The second diverging circuit in Figure 22–11 allows signals from an incoming pathway to be transmitted into separate pathways, the same information being relayed in different directions at the same time. This type of circuit is common in the sensory nervous system. For instance, when a limb moves, the sensory information from the joints and muscles caused by the movement is transmitted by such a circuit into (a) neuronal pools of the spinal cord, (b) neuronal pools of the cerebellum, (c) neuronal pools of the thalamus, and (d) neuronal pools of the cerebral cortex.

Converging Circuits. Figure 22–12 illustrates two types of *converging circuits*, which are opposites of the diverging circuits. In the figure to the left, different nerve fibers from the same source converge on a single output neuron causing especially strong stimulation.

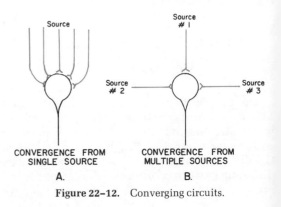

Figure 22–12. Converging circuits.

To the right is another important type of converging circuit in which input fibers from several different sources converge on the same output neuron. This type of circuit allows signals from many different sources to cause the same effect. Thus, the smell of a pigpen might make a person think of a pig; the sight of a pig, the sound of his grunting, touching the pig, or eating pork chops might also lead to the same thought.

Use of Converging Circuits to Perform Complex Functions. Converging circuits can provide functions that are much more complex than can single neurons, for in a converging circuit there may be cells that respond to the incoming signal in several different ways at the same time. For instance, one of the neurons might have a very high threshold but when once stimulated might be so powerful that it causes a very intense reaction. In other parts of the circuit there may be neurons that have low thresholds but when stimulated transmit only weak signals. The output signal from the pool, therefore, might normally be very weak but then suddenly become extremely powerful if a higher threshold of input stimulation should be exceeded. Thus, the converging circuit can be organized to perform almost any type of selective function. The individual neurons of the circuit select which signals shall pass, but the arrangement of the fibers and the combinations of cells determine which incoming signals will have the greater effects, and whether the effects will be excitatory or inhibitory.

Repetitive Circuits. The Reverberating Circuit. Figure 22–13 shows two different types of circuits which, when stimulated only once, will cause the output cell to transmit a series of impulses. The upper circuit of this figure is a *reverberating circuit* which functions as follows: An incoming signal stimulates the first neuron, which then stimulates the second and third. However, branches return to the first neuron and restimulate it. As a result, the signal travels once again through the chain of neurons, this process continuing around and around the circuit indefinitely until one or more of the neurons fail to fire. The usual cause of failure is fatigue, and until it takes place, the output neuron continues to be stimulated every time the signal goes around the circuit, giving a continuous output signal.

The reverberating circuit is the basis of innumerable central nervous system activities,

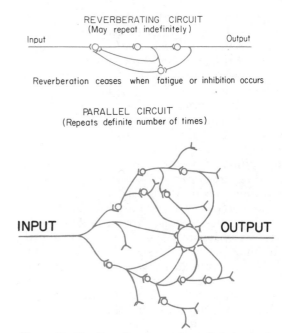

Figure 22–13. Reverberating and parallel circuits for repetitive discharge.

for it allows a single input signal to elicit a response lasting a few seconds, minutes, or hours. Indeed, the lifelong respiratory rhythm is probably caused by a reverberating circuit that continues to reverberate without fatiguing as long as the person remains alive. It is probable, also, that a special circuit in the brain causes a person to awaken when it reverberates, and allows him to relax into a state of sleep when it stops reverberating. Almost all rhythmic muscular activities, including the rhythmic movements of walking, are probably controlled by reverberating circuits.

A few moments of thought about the reverberating type of circuit will emphasize the extreme number of variable functions it can perform. For instance, the number of neurons in the circuit may be great or small. If the number is great, the length of time required for the signal to go around the circuit will be long; if the number is small, the reverberatory period will be short. As a result, the output signal may repeat itself rapidly or slowly. Furthermore, more than one of the neurons in the circuit can give off output signals. One of the cells in the circuit, for instance, might cause an arm to move upward, then another cell a fraction of a second later in the circuit might cause the arm to move to the right, another cell still later cause it to move downward, and another cell cause it to move to the left. As the reverberatory cycle repeats itself once more, the same motions occur again, causing the arm to move around and around continuously until the cycle stops. When the person wishes to perform this motion, all he has to do is set off the reverberatory cycle.

THE PARALLEL CIRCUIT.   The lower part of Figure 22–13 illustrates the *parallel* type of repetitive circuit in which a single input signal stimulates a sequence of neurons that send separate nerve fibers directly to a common output cell. Because a delay of about $1/2000$ second occurs each time an impulse crosses a synapse, the impulse from the first neuron arrives at the output cell $1/2000$ second ahead of the impulse from the second neuron, and impulses continue to arrive at these short intervals until all the neurons have been stimulated. Since there is no feedback mechanism in this circuit, the repetition then ceases entirely.

Parallel circuits in different parts of the central nervous system differ in several ways: First, some circuits are composed of far more neurons than others and give a repetitive output lasting considerably longer. Second, the parallel circuit can be combined with diverging circuits, so that the signal can be amplified to stimulate the output neuron very strongly during part of the repetitive discharge and weakly during other parts of the discharge. In this way the output signal will have a definite amplitude pattern as well as a definite duration.

The parallel type of circuit has certain advantages over the reverberating circuit. For instance, the very variable phenomenon of fatigue is a major factor in determining how long the reverberating circuit will continue to fire, whereas the output duration of the parallel circuit is mainly independent of fatigue. For performing very exact activities such as mathematical calculations the parallel circuit is perhaps very useful, while for rhythmic functions or for greatly prolonged discharges the reverberating circuit is a necessity. Unfortunately, the parallel circuit cannot control functions that last more than a fraction of a second, because one neuron is required in a circuit for each $1/2000$ second. Probably not more than a few dozen neurons are ever organized into parallel circuits, which would limit this type of circuit to activities occurring in less than about $1/50$ second.

Figure 22–14 illustrates the time duration and amplitude characteristics of the output

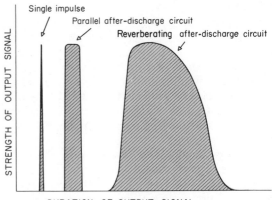

Figure 22–14. Characteristics of signals caused by (1) a single nerve impulse, (2) discharge of a parallel after-discharge circuit, and (3) discharge of a reverberatory after-discharge circuit.

signals from, first, a one-to-one circuit; second, a parallel repetitive circuit; and third, a reverberatory repetitive circuit.

RELATIONSHIP OF THE OUTPUT SIGNAL TO INPUT SIGNAL OF A NEURONAL POOL

Now that the characteristics of some of the important neuronal circuits have been described, we can discuss briefly the overall responses of typical neuronal pools. Almost never is a neuronal pool simply a relay station; that is, almost never do the same number of impulses leave a pool as enter it. Usually incoming signals diverge, converge, or are changed into repetitive signals whose durations last long after the input signal is over. Also, many incoming stimuli do not excite the neuronal pool at all but, instead, inhibit it.

An important characteristic of neuronal pools is that the response to input signals from one source can be altered by input signals coming from secondary sources. For instance, sensory nerve impulses from the skin into the spinal cord ordinarily cause no reflex skeletal muscle effects. However, strong facilitatory impulses transmitted from the brain down the spinal cord to the neurons that control the muscles can make these neurons so excitable that a very light scratch on the skin elicits a strong contraction of the underlying muscle. Thus, a high threshold, low amplification circuit has been changed into a low threshold, high amplification circuit.

It is now up to the imagination of the student to conceive the many possible ways in which individual types of neurons can be organized into different types of neuronal pools and in which different types of neuronal pools can perform an infinite number of reflex and integrative nervous functions. Unfortunately, the precise circuits in most neuronal pools have not yet been worked out in detail. Nevertheless, the types of circuits that are already known can suggest mechanisms by which all the functions of the central nervous system could be performed.

REFERENCES

Axelrod, J.: Neurotransmitters. *Sci. Amer., 230(6)*:58, 1974.
Barker, J. L.: Peptides: roles in neuronal excitability. *Physiol. Rev., 56*:435, 1976.
Bennett, M. V. L. (ed.): Synaptic Transmission and Neuronal Interaction. (Society of General Physiologists Series, Vol. 28.) New York, Raven Press, 1974.
Blumenthal, R., Homsy, Y. M., Katchalsky, A. K., and Rowland, V. (eds.): Dynamic Patterns of Brain Cell Assemblies. Neurosciences Research Program Bulletin. Vol. 12, No. 1. Cambridge, Mass., Massachusetts Institute of Technology, 1974.
Calvin, W. H.: Generations of spike trains in CNS neurons. *Brain Res., 84*:1, 1975.
Martin, A. R.: Synaptic transmissions. *In* MTP International Review of Science: Physiology. Vol. 3. Baltimore. University Park Press, 1974, p. 53.
Nathanson, J. A.: Cyclic nucleotides and nervous system function. *Physiol. Rev., 57*:157, 1977.
Nathanson, J. A., and Greengard, P.: "Second messengers" in the brain. *Sci. Amer., 237(2)*:108, 1977.
Noback, C. R.: The Human Nervous System; Basic Principles of Neurobiology. 2nd ed. New York, McGraw-Hill Book Company, 1975.
Pak, W. L., and Pinto, L. H.: Genetic approach to the study of the nervous system. *Ann. Rev. Biophys. Bioeng., 5*:397, 1976.
Redman, S. J.: A quantitative approach to integrative function of dendrites. *Intern. Rev. Physiol., 10*:1, 1976.
Roberts, E., Chase, T., and Tower, D. B. (eds.): GABA in Nervous System Function. New York, Raven Press, 1975.
Shepherd, G. M.: The Synaptic Organization of the Brain: An Introduction. New York, Oxford University Press, 1974.
Shepherd, G. M.: Microcircuits in the nervous system. *Sci. Amer., 238(2)*:92, 1978.
Snyder, S. H., and Bennett, J. P. Jr.: Neurotransmitter receptors in the brain: biochemical identification. *Ann. Rev. Physiol., 38*:153, 1976.

QUESTIONS

1. What is meant by a reflex?
2. What are the three major levels of nervous system function?
3. Describe the physiologic anatomy of the synapse.
4. How does an excitatory transmitter substance cause an excitatory postsynaptic potential?
5. Describe both the spatial and temporal methods for summation of excitatory postsynaptic potentials.
6. Explain the mechanisms by which inhibitory transmitter substances inhibit synaptic transmission.
7. Explain what is meant by diverging circuits and converging circuits.
8. Explain the function of the reverberating circuit and the factors that determine the duration of reverberation.

23 SOMESTHETIC SENSATIONS AND INTERPRETATION OF SENSORY SIGNALS BY THE BRAIN

The term *somesthetic sensation* means sensation from the body. Also, physiologists frequently speak of subdivisions of the somesthetic sensory system, including *exteroceptive* sensation, *proprioceptive* sensation, and *visceral* sensation. There is much overlap among these different types of sensations.

Exteroceptive sensations are those normally felt from the skin, such as (1) touch, (2) pressure, (3) heat, (4) cold, and (5) pain.

Proprioceptive sensations are those that apprise the brain of the physical state of the body, including such sensations as (1) tension of the muscles, (2) tension of the tendons, (3) angulation of the joints, and (4) deep pressure from the botton of the feet. Note that pressure can be considered to be both an exteroceptive sensation and a proprioceptive sensation.

Visceral sensations are those from the internal organs, including such sensations as (1) pain, (2) fullness, and (3) sometimes the sensation of heat. Thus, the visceral sensations are similar to exteroceptive sensations and are functionally the same except that they originate from inside the body.

GENERAL ORGANIZATION OF THE SOMESTHETIC SENSORY SYSTEM

Figure 23–1 illustrates the general plan for transmission of somesthetic sensory signals

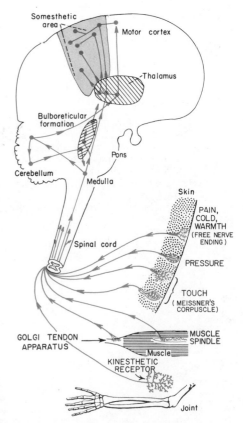

Figure 23–1. Transmission of sensory signals to the brain, showing the sensory receptors and the nerve pathways for transmitting these sensations into the brain.

286

into the brain. The sensations are detected by special nerve endings in the skin, muscles, tendons, or deeper areas of the body, and these emit nerve impulses that are transmitted through nerve trunks into the spinal cord. Upon entering the spinal cord, the sensory nerve fibers branch. Some of the branches end in the spinal cord itself to cause *cord reflexes* which will be discussed in the following chapter, while the others extend to other areas of the cord and brain. The sensory pathways to the brain terminate in several discrete areas as follows: (1) the sensory areas of the brain stem, including the bulboreticular formation and the central gray area, (2) the cerebellum, (3) the thalamus and other closely allied structures, and (4) the cerebral cortex. The signals transmitted into each of these different areas subserve specific functions, which will become clear later in the chapter. Signals transmitted into the cerebellum occur entirely at a subconscious level and are concerned with subconscious control of motor function. Therefore, the cerebellar component of the somesthetic sensory system will be discussed in connection with motor activities of the body in Chapter 25 rather than in the present chapter.

Function of Different Nervous System Levels of Somesthetic Sensation. The sensory nerve fibers that terminate in the spinal cord initiate cord reflexes. These cause immediate and direct motor activities such as contraction of muscles to pull a limb away from a painful object, or perhaps alternate contraction of the limbs to cause walking movements.

The sensory signals that terminate in the lower brain stem cause subconscious motor reactions of a much higher and more complex nature than those caused by the cord reflexes. For instance, it is in the lower brain stem that chewing, control of the body trunk, and control of the muscles that support the body against gravity are all effected.

As the sensory signals travel still farther up the brain and approach the thalamus, they begin to enter the level of consciousness. When sensory signals reach the thalamus their origins are localized crudely in the body, and the types of sensations, which are also called the *modalities of sensation,* begin to be appreciated. Thus, one can then determine whether the sensation is touch, heat, or cold, and so forth. Yet for full appreciation of these qualities, and especially for very discrete localization, the signals must pass on into the cortex. The cortex is a large storehouse of information, of memories of past experiences, and it is this stored information that allows the finer details of interpretation to be achieved.

THE SENSORY RECEPTORS

Some sensory nerve endings in the skin and deeper structures of the body are nothing more than small filamentous branches called *free nerve endings,* while others are special *end-organs* that are designed to respond only to special types of stimuli. Figure 23–1 illustrates some of the more representative sensory receptors, and their functions may be described as follows:

Free Nerve Endings. Free nerve endings detect the sensations of crude touch, pressure, pain, heat, and cold. These sensations occasionally become somewhat confused with each other because the nerve pathways from the free endings interconnect extensively, and the endings themselves also are not always entirely specific for the different types of sensation. For instance, extreme heat or cold is likely to give one a sensation of pain, or very hard pressure might also be confused with pain.

Despite the fact that free nerve endings transmit only crude sensations, they are by far the most common type of nerve ending. They perform most of the general functions of sensation, while the specific functions, such as discrimination of very slight differences between degrees of touch are left to the more specialized receptors.

Special Exteroceptive Receptors. Also shown in Figure 23–1 are several specialized receptors for detection of pressure and touch. The reason why each of these end-organs registers only one type of sensation is that the physiologic organization of the end-organ itself allows each type of sensation to stimulate the nerve ending by some specific physical effect on the end-organ.

Still other specialized sensory receptors are present in the skin. For instance, at the base of each hair is a receptor that allows one to feel even the slightest pressure on any hair. In animals, some of these hair receptors are so well developed that the associated hairs are called *tactile hairs.* A cat actually helps to guide itself in the darkness by its tactile whiskers. Special end-organs have also been found in the lips and snouts of some rooting animals; these presumably aid in the search for food. Finally, in the sexual organs special endings are possibly responsible for the distinctive qualities of sexual sensations.

The Proprioceptor Receptors. Figure 23–1 also illustrates three different types of proprioceptor receptors. One of them, the *joint kinesthetic receptor,* is found in the capsules of joints. These receptors apprise one of the *degree of angulation* of the joints and also of the *rate* at which this degree of angulation changes.

Two special end-organs transmit proprioceptive information from the muscles. These are the *muscle spindle* and the *Golgi tendon apparatus.* The muscle spindle is composed of nerve filaments wrapped around small muscle fibers. The spindle detects the degree of stretch of the muscle. This information is transmitted to the central nervous system to aid in the control of muscle movements. The Golgi tendon apparatus detects the overall tension applied to the tendon, and, therefore, apprises the central nervous system of the effective strength of contraction of the muscle.

ADAPTATION OF THE SENSORY RECEPTORS

When a stimulus is suddenly applied to a sensory receptor, the receptor usually responds very vigorously at first, but progressively less so during the next few seconds or minutes. An example of this effect is the sensation felt when one gets into a tub of hot water, which causes an intense burning sensation at first but after a moment's exposure produces a sensation of pleasing warmth. The same is true to varying degrees for all the other sensations. This loss of sensation during prolonged stimulation is called *adaptation* of sensory receptors.

Adaptation occurs to much greater degrees for some sensations than for others. For instance, the sensation of light touch and that for some types of pressure adapt within a few seconds, in some instances in as little as $1/100$ second. This allows one to feel an object when he first touches it but to lose the sensation very soon thereafter. Were it not for rapid adaptation of the light touch receptors, one would feel intense touch sensations, never stopping, from all the body areas in contact with any object such as the seat of a chair, the shoes, and even the clothing. These sensa-

tions would continue to bombard the brain to such an extent that one could hardly think of anything else. The value of rapid adaptation of certain of the sensations, therefore, is obvious.

The sensations of pain and some types of proprioception usually adapt either very slowly or to only a slight extent. Pain sensations are elicited when tissue damage is occurring. As long as the damage continues to occur, it is important that the person also continue to be apprised of this fact so that he will institute appropriate measures to remove the cause of the damage. The long persistence of some proprioceptive sensations also is desirable because the brain needs to know the physical status of the different parts of the body at all times and not simply immediately after movements have occurred.

DISCRIMINATION OF INTENSITY OF SENSATIONS – THE WEBER-FECHNER LAW

A person can detect the weight of a flea on the tip of his finger or he can detect the weight of a man stepping on the same finger. The difference in weight of these two animals is approximately 70 million times, and yet, from the sensations perceived, a person can estimate the heftiness of the two objects. The reason why one can discriminate such wide differences in intensities is that the number of impulses transmitted by some of the sensory nerves is roughly proportional to the *logarithm* of the intensity of sensation rather than to the actual intensity. This logarithmic response of receptors is illustrated in Figure 23–2, which shows the impulse rate from a muscle spindle as different weights are applied to the muscle. Note that the weight scale doubles at each point, and yet the spindle impulses increase linearly. This is a logarithmic type of response; that is, the number of impulses is approximately proportinal to the logarithm of the weight rather than to the weight itself.

In the case of the flea and the man, the difference between the logarithms of their weights is only about 8 times even though the difference between the actual weights is 70 million times. This logarithmic detection of sensation applies not only to somesthetic sensations, but to some extent also to the sensations of vision, hearing, taste, and smell; it is called the *Weber-Fechner law.*

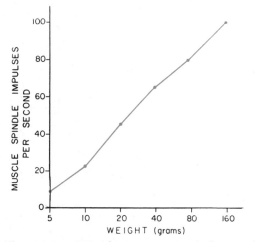

Figure 23-2. Logarithmic response of the muscle spindle. (Drawn from data in Matthews: *J. Physiol. (Lond.)*, 78:1, 1933.)

Another example of the Weber-Fechner law is the following: If one is holding in his hand an object that weighs 1 ounce and he exchanges this object for another that weighs 1.1 ounce, he will be able barely to discriminate the difference in weight of the two objects. The actual difference in weight is 0.1 ounce. Then he picks up an object that weighs 10 pounds, and exchanges this object for another that weighs 10 pounds and 0.1 ounce. He will be completely unable to discriminate the difference in weight. Instead, the second object must now weigh approximately 11 pounds for him to tell the difference. The increase in weight this time must be 1 pound instead of 0.1 ounce. In each instance, the increase is 10 per cent of the original weight. Therefore, another way of expressing the Weber-Fechner law is: Discrimination of intensity level is on a relative basis rather than on an absolute basis.

DETERMINATION OF TYPES OF SENSATION – THE "MODALITY" OF SENSATION

Even though different types of nerve receptors are responsible for detecting different types of sensation — pain, touch, pressure, position, and so forth — nevertheless, it is not the receptor itself that determines the type of sensation that a person feels. Instead, it is the point in the brain to which the signal is transmitted. For instance, if a pain nerve fiber is stimulated by crushing it, by heating it, by bending it, or in any other way, the person will still feel pain regardless of how the pain nerve is stimulated. Likewise, a touch nerve fiber can be stimulated by crushing it, burning it, or bending it, and the person will feel nothing but touch even though the nerve fiber itself is being damaged severely, which one generally considers to cause pain.

Specific Areas of the Brain for Detecting Different Modalities of Sensation. The term "modality" of sensation means the specific quality of sensation felt, that is, whether the feeling be one of pain, touch, pressure, position, vision, hearing, equilibrium, smell, or taste, all of which are different modalities of sensation.

It is frequently stated that the thalamus is the main area of the brain for determining modality of sensation. The reason for stating this is that fiber pathways for almost all of the different modalities terminate at different points in the thalamus, and destruction of the thalamus makes it difficult to distinguish most types of sensory modalities. However, it has now been learned that some of the pain pathways terminate in even lower areas of the brain stem such as the central gray matter of the mesencephalon and in the hypothalamus, and that these areas are principal sites for detection of pain. Furthermore, the cerebral cortex is known to sharpen one's ability to detect the different modalities, even though the lower areas of the brain can provide crude detection.

Thus, it is becoming apparent that widely scattered regions of the brain operate together to determine the different sensory qualities. As the sensory signals enter the brain from the cord, they are picked apart for their different characteristics. One portion of the brain determines whether or not there is a pain element in the sensation, another whether or not there is a touch element, another the area of the body from which the sensation is coming, and so forth.

DUAL SYSTEM FOR TRANSMISSION OF SOMESTHETIC SENSATIONS

On entering the spinal cord, the somesthetic sensory signals may be transmitted up the remainder of the nervous system axis by either one of two pathways called, respectively, the *dorsal column system* and the *spinothalamic system*. These two pathways, illustrated in Figure 23-3, are anatomically distinct and

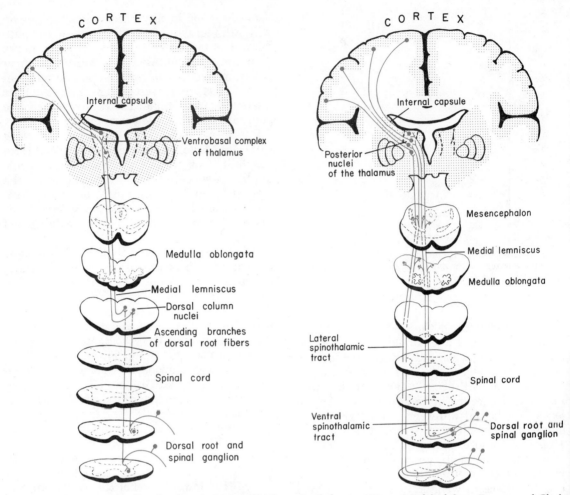

Figure 23–3. (A) The dorsal column system. (B) The spinothalamic system. (Modified from Ranson and Clark: Anatomy of the Nervous System. W. B. Saunders Co.)

also display different characteristics for transmission of signals.

COMPARISON OF THE DORSAL COLUMN AND SPINOTHALAMIC SYSTEMS

The dorsal column system is illustrated in Figure 23–3A. Note that sensory nerve fibers entering this system pass all the way up the same side of the spinal cord until they reach the medulla. Here they terminate on "second order" neurons, which in turn send fibers immediately to the opposite side of the medulla and thence up to the posterior part of the thalamus, terminating in an area called the *ventrobasal complex.* Here these neurons synapse with "third order" neurons which send

fibers thence to the *somesthetic sensory area* of the cerebral cortex. Note especially that there are three separate neurons in the pathway, the first neuron passing all the way from the receptor on the surface of the body to the medulla oblongata, the second from the medulla to the thalamus, and the third from the thalamus to the cortex. Figure 23–4 shows second order neurons synapsing with the third order neurons in the thalamus which then radiate upward to the somesthetic sensory cortex. Note that the leg is represented near the midline of the cortex and the face far to the side.

The spinothalamic system is illustrated in Figure 23–3B. Note that it, too, is composed of three orders of neurons. The first order neurons originate in the receptors in the body and pass to the spinal cord. Here, they ter-

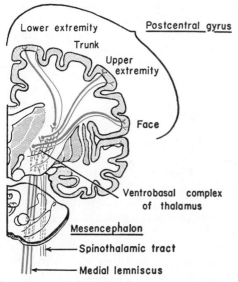

Figure 23–4. Radiation of the sensory pathways from the thalamus to the somesthetic sensory area in the cortex. (Modified from Brodal: Neurological Anatomy in Relation to Clinical Medicine. Oxford University Press.)

minate almost immediately, synapsing with second order neurons located in the posterior horns of the spinal cord gray matter. The fibers from these neurons in turn cross to the opposite side of the cord and travel up to the brain stem through two separate pathways located respectively in the lateral and ventral columns of the cord. These neurons terminate at multiple levels of the brain stem, from the medulla all the way up to the posterior thalamus. In the thalamus, they synapse with third order neurons that transmit signals to the somesthetic cortex.

Discreteness of Signal Transmission in the Dorsal Column Versus the Spinothalamic System. The dorsal column system, as well as the peripheral nerve fibers that connect with it, is composed of very large nerve fibers that transmit signals at velocities of 35 to 80 meters per second. On the other hand, the spinothalamic system and its associated peripheral nerve fibers are composed of very small nerve fibers, some of which are not myelinated at all. These fibers transmit signals at velocities of 1 to 30 meters per second. Since the spinothalamic system conducts signals very slowly, it can be used only for information that the brain can afford to receive after a short delay. On the other hand, the dorsal column system allows transmission of information to the brain within a very small fraction of a second.

Another major difference between the dorsal column and spinothalamic systems is the degree of spatial orientation of the nerve fibers within each of the two tracts. The nerve fibers of the dorsal column system are myelinated so that they are well insulated from each other. Also, there is very little crossover of signals from one part of the tract to other parts where the nerve fibers synapse in the medulla and thalamus. Therefore, when a single receptor is stimulated in the skin or other peripheral area of the body, a very discrete signal is transmitted to a very highly localized point in the thalamus and from there to the cerebral cortex. By contrast, the degree of insulation between the nerve fibers in the spinothalamic system is far less, and there is also far greater diffusion of nerve signals sidewise where the spinothalamic pathways synapse in the spinal cord and thalamus. As a consequence, stimulation of a single nerve receptor exciting the spinothalamic pathway causes excitation of a widely dispersed area in the brain.

Modalities of Sensation Transmitted by the Two Systems. With the foregoing differences in mind, we can now list the types of sensations transmitted in the two systems.

The Dorsal Column System

1. Touch sensations having a high degree of localization of the stimulus and transmitting fine gradations of intensity.
2. Phasic sensations, such as vibratory sensations.
3. Kinesthetic sensations (sensations having to do with body movements).
4. Muscle sensations.
5. Pressure sensations having fine gradations of intensity.

The Spinothalamic System

1. Pain.
2. Thermal sensations, including both warmth and cold sensations.
3. Crude touch sensations capable of gross localization of the stimulus on the surface of the body.
4. Pressure sensations of a cruder nature than those transmitted by the dorsal column system.
5. Tickle and itch sensations.
6. Sexual sensations.

Collateral Signals from the Spinothalamic System to the Basal Regions of the Brain. It is important to note another distinct difference between the spinothalamic and dorsal column systems. That is, a large share of the nerve fibers of the spinothalamic system terminate in the medulla, pons, and mesencephalon even before they reach the thalamus, and many of the fibers that do go all the way to the thalamus give off branches into these lower regions of the brain on the way up. In contrast, the dorsal column system passes through these regions with almost no branches. Thus, the spinothalamic system is much more concerned with sensations that elicit subconscious automatic reactions than is the dorsal column system. On the other hand, the dorsal column system is concerned with very discrete signals that are transmitted primarily into the conscious areas of the brain.

To express this another way, the spinothalamic system is concerned with the older types of sensations, those that occur in even very low forms of animal life, while the dorsal column system is concerned with types of sensation that have appeared recently in the phylogenetic development of man.

MECHANISM FOR LOCALIZING SENSATIONS IN SPECIFIC AREAS OF THE BODY

Among the most important information that the brain must determine about each somesthetic sensation is its point of origin in the body. Once the location has been determined, it is possible to do something about the condition giving rise to the sensation. To provide this localization ability, the nerve fibers are *spatially oriented* in the nerve trunks, spinal cord, hindbrain, and cerebral cortex. This may be explained as follows.

Spatial Orientation of Nerve Fibers. The sensory nerve fibers arising in the leg are separated in the spinal cord and brain from the fibers arising in the arm. The fibers arising from individual fingers are separated from each other, and even the fibers arising from two areas of skin only 1 cm. apart are kept distinct from each other. Also, at each new synaptic relay station along the sensory pathway, the respective sensory signals from adjacent areas of the body are still kept separate from each other all the way to the termination of the signals in the lower brain stem, thalamus, and cerebral cortex.

In general, the thalamus is capable of determining only roughly which part of the body is being stimulated. The thalamus is not organized satisfactorily to localize sensations to very discrete areas of the body; instead, this function is performed primarily in the somesthetic area of the cerebral cortex.

THE SOMESTHETIC CORTEX

Figure 23–5 illustrates the *somesthetic cortex,* which is located immediately posterior to the central sulcus of the brain and extends from the longitudinal fissure at the top of the brain into the Sylvian fissure at the side. This figure also shows the areas of the somesthetic cortex that receive sensations from each of the different parts of the body. Sensations from the foot, for instance, excite the somesthetic cortex where it dips into the longitudinal fissure that runs forward and backward down the middle of the brain. The leg area is approximately at the point where the somesthetic cortex comes out of the longitudinal fissure; then comes the thigh area, followed by the abdomen, thorax, shoulder, arm, hand, fingers, thumb, neck, tongue, palate, and larynx. Thus, the entire body is spatially represented in the somesthetic cortex, each discrete point in the cortex corresponding to a discrete area only a few millimeters in size on the body surface.

A major function of the somesthetic cortex is to localize very exactly the points in the body from which the sensations originate. Though the thalamus is capable of localizing sensations to very general areas, such as to one arm, to a leg, or to the body, it is not capable of localizing sensations to minute areas of the body. Instead, the thalamus re-

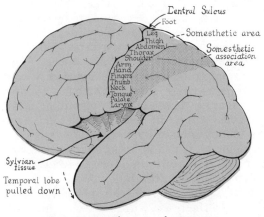

Figure 23–5.   The somesthetic cortex.

lays the necessary signals into the somesthetic cortex where a much better spatial representation is available, and there the job of discrete localization is performed.

Not all types of sensations are localized equally well. The sensations of aching pain, crude touch, warmth, and cold are localized to general areas of the body rather than to discrete areas. It seems that the thalamus performs most of the localization function for these modalities of sensations. On the other hand, the sensations of light touch, pressure, and position are very discretely localized by the somesthetic cortex. These are the sensations normally detected by the special sensory receptors, for instance Meissner's corpuscles for light touch and joint receptors for position sensation.

Function of the Somesthetic Cortex in Analyzing Sensory Signals. Even when the somesthetic cortex is completely removed, a person usually still has the ability to detect the type of sensation that is being received, that is, whether it be pain, touch, heat, cold, or so forth, but his appreciation of these sensations is markedly reduced. Therefore, the cerebral cortex is concerned not so much with detection of type of sensation as with *analysis* of the sensory information after it has already been detected.

INTERPRETATION OF SOMESTHETIC SENSATIONS – THE SOMESTHETIC ASSOCIATION AREA

The area of the cortex a centimeter or so behind the somesthetic cortex is called the *somesthetic association area* (see Fig. 23–5), because it is in this region that some of the more complex qualities of sensation are appreciated. Signals are transmitted directly into this area from several sources: (1) from the thalamus, (2) from other basal regions of the brain, and (3) backward from the somesthetic cortex. In ways not understood, the somesthetic association area puts all this information together and determines the following characteristics of sensations: (1) shape of an object, (2) relative positions of the parts of the body with respect to each other, such as the legs, the hands, and so forth, (3) texture of a surface, such as whether it is rough, smooth, or undulating, and (4) orientation of one object with respect to another object — in other words, the spatial orientation of objects that are felt.

Many of the memories of past sensory experiences are stored in the somesthetic association area, and when new sensations similar to the old ones arrive in the brain, the parallel nature of the two sensations is immediately discerned. It is in this manner that one associates a new sensation with previous ones and thereby recognizes the nature of the sensation. As more and more sensory experiences accumulate, new sensory experiences can be interpreted on the basis of what is remembered from the past.

PAIN

The sensation of pain deserves special comment because it plays an exceedingly important protective role for our body, apprising us of almost any type of damaging process and causing appropriate muscular reactions to remove the body from contact with the damaging stimuli.

The Stimulus That Causes Pain. Pain receptors are stimulated when tissues of the body are *being* damaged. For instance, one feels pain while the skin is *being* cut, but shortly after the cut has been made, the pain generally is gone. Indeed, thousands of soldiers on the battlefield in World War II who had been mortally wounded were asked whether or not they felt pain. In most instances, no pain was actually felt except for a few minutes after the damage had been inflicted or unless the soldier was moved.

Different types of damaging stimuli that can cause pain are *trauma* to the tissues, *ischemia* of the tissues (lack of blood flow), intense *heat* to the tissues, intense *cold* (especially freezing of the tissues), or *chemical irritation* of the tissues. It is believed that as tissues are damaged they release some substance from the cells that stimulates the pain nerve endings. *Bradykinin* has especially been suggested as this substance, though not proved.

Perception of Pain. It is frequently said that one person *perceives* pain more intensely than others. However, experiments in which graded intensities of tissue damage were caused in a large number of different persons showed that all normal persons perceive pain at almost precisely the same degree of tissue damage. For instance, when heat is used to cause tissue damage, almost all persons begin to feel pain when the tissue temperature rises to a level between 44° and 46° Celsius, which is a very narrow range.

Reactivity to Pain. On the other hand, not

all people *react* alike to the same pain, for some react violently to only slight pain while others can withstand tremendous pain before reacting at all. This is determined not by differences in sensitivity of the pain receptors themselves but, instead, by differences in psychic make-up of the individuals. Therefore, when a person is said to be extremely "sensitive" to pain, it is meant that he *reacts* to pain far more than do other persons and not that he perceives far more pain.

THE VISCERAL SENSATIONS AND VISCERAL PAIN

The visceral sensations are those that arise from the internal structures of the body, such as the organs of the abdomen and chest, and they also include sensations from inside the head and from the muscles, bones, and other deep structures.

The usual modalities of visceral sensation are pain, burning sensations, and pressure (which is often manifested as a sensation of fullness). Because these modalities are also exhibited by the exteroceptive sensations, the visceral sensations are sometimes considered to be part of the exteroceptive system. Indeed, the pathways of these two types of sensation are closely related, as will be noted below.

Pathways for Visceral Sensations. Ordinarily one is not at all conscious of his internal organs, but an inflamed organ can transmit pain sensations to the central nervous system through two separate pathways, the *parietal pathway* and the *visceral pathway*, both of which are illustrated for sensations from the appendix in Figure 23–6. The parietal pathway is the same as the pathway for transmission of exteroceptive sensations. In the case of the appendix, the inflamed appendix irritates the peritoneum overlying the appendix, and pain impulses are transmitted from the peritoneum through the same nerve fibers that carry exteroceptive sensations from the outside of the abdominal wall at this point, fibers that enter the spinal cord at the L-1 segment.

The visceral pathway utilizes sensory nerve fibers in the autonomic nervous system. Figure 23–6 illustrates these sensory fibers leaving the appendix to enter the sympathetic chain. After traveling upward a few segments, the fibers leave the sympathetic chain to enter the spinal cord at approximately the T-10 segment in the lower thoracic region. Therefore, the visceral sensations from the

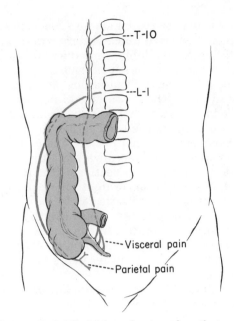

Figure 23–6. Parietal and visceral pathways for transmission of pain from the appendix.

appendix enter the spinal cord at an entirely different point from the parietal sensations.

Referred Visceral Pain. Pain from an internal organ is often felt on a surface area of the body rather than being localized in the organ itself. This is called *referred pain*. It may be referred to the surface immediately above the organ, or often to an area a considerable distance away. The mechanism of referred pain is probably that illustrated in Figure 23–7. This figure shows two visceral nerve fibers as well as two additional nerve fibers from the skin entering the spinal cord. Both of these sets of fibers synapse with the same two neurons in the spinal cord. Therefore, stimu-

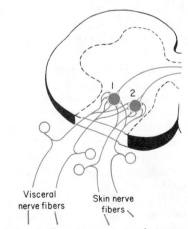

Figure 23–7. The neurogenic mechanism responsible for referred pain.

lating either the visceral fibers or the skin fibers will send impulses up the spinal cord along the same pathway. Because the person has never had reason to know the location of his internal organs but is very familiar with the location of the different surface areas, the impulses from the visceral fibers are usually interpreted as coming from the body surface.

When a visceral sensation is transmitted through a parietal pathway, the pain is usually felt on the surface of the body directly over the respective internal organ, but when the sensation is conducted through a visceral pathway, it is usually referred to an area remote from the organ. The reason for this is that fibers of the visceral pathway usually travel a long distance in the sympathetic chain before entering the spinal cord. Figure 23–8 illustrates the surface areas to which pain is referred from many of the different organs. For instance, though the appendix lies far to the right and in the lower abdomen, its visceral sensations are referred to an area around the umbilicus. Also, referred pain from the kidney and ureter occurs near the midline on the anterior abdominal wall, though these organs actually lie in the posterior part of the abdomen.

Visceral pain from the heart is among the most important of the referred sensations. Figure 23–8 shows the areas to which cardiac pain is often referred, including the upper thorax, the shoulder, and the medial side of the left arm, particularly along the radial artery.

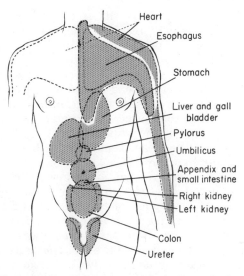

Figure 23–8. Surface areas of the body to which pain from different organs is referred.

The surface area to which visceral pain is referred usually corresponds to that portion of the body from which the organ originated during embryonic development. For example, the heart originates in the neck of the embryo and so does the arm; therefore, heart pain is frequently referred to the arm. The appendix, as another example, originates from the portion of the primitive gut that develops near the umbilicus, which explains the reference of appendiceal pain to the umbilical region. The other areas of referred pain illustrated in Figure 23–8 also correspond to the areas of origin of the different organs.

The Stimulus for Visceral Pain. Cutting through the gut, the heart, the liver, the muscle, or other internal organs with a sharp knife causes almost no pain. Instead, pain from these areas is produced much more easily by (1) stopping the blood flow to the area, (2) application of an irritant chemical over wide areas, (3) stretching the tissues, or (4) spasm of the muscle in the organ.

One of the most important stimuli for visceral sensation is *ischemia,* which means insufficient blood flow. The pain of a heart attack, for example, is caused by poor flow through the coronaries to the heart muscle. Even the skeletal muscles become extremely painful when their blood supply is diminished for a prolonged period of time. The reason for the pain caused by ischemia has never been determined, though it is believed that the lack of blood flow allows metabolic products such as acids, bradykinin, and so forth to build up in the tissues, and that these in turn produce irritant effects on the pain nerve endings.

Stretching the tissues possibly causes pain by stretching the nerve endings or by producing ischemia, for overstretching the tissues occludes the blood vessels. Likewise, spasm of the muscle in an organ can stretch nerve endings as well as compress the blood vessels; spasm also increases the rate of metabolism so that far more than usual quantities of metabolic end-products are dumped into the tissues. These factors together, therefore, could be the reason why spasm of the gut often causes very intense abdominal pain.

HEADACHE

Headache is another type of referred pain, and it is usually caused by irritation or damage occurring in the tissues inside the head.

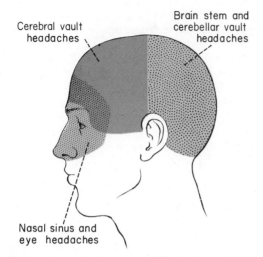

Figure 23-9. Surface areas of the head to which different types of headache are referred.

Figure 23-9 illustrates the surface areas of the head to which pain from the deep structures is referred. Pain fibers from inside the eye and from the nasal sinuses are transmitted through the first and second divisions of the fifth nerve which also supply the skin areas over the lower forehead and around the eye and nose. Therefore, sinus infections or irritation of the eyes caused by their overuse or by intense light can result in dull, aching pain referred diffusely over the frontal and orbital areas of the head.

Irritation occurring inside the skull but above the level of the ears — that is, everywhere around the cerebral cortex — causes headache referred to the frontal and temporal surfaces of the head. The reason for this area of reference is that the third division of the fifth cranial nerve supplies both the upper areas inside the skull and also the surface areas of the skin in the temporal and frontal regions.

Irritative effects in the pocket of the skull where the brain stem and cerebellum reside, that is, beneath the level of the ears, give rise to headache localized over the occipital part of the skull, also shown in Figure 23-9. Both the lower areas inside the cranial vault and the occipital skin areas of the skull are supplied by the upper cervical spinal nerves, which explains the reference of this pain to the occipital regions.

Meningeal Headache. Headache originating inside the skull is often caused by irritation of the *meninges,* which are the membranes surrounding the brain. Some causes of very severe headache of this type are: (1) meningitis, an infection of the meninges, which causes very intense headache; (2) removal of fluid from the spaces around the brain, allowing the brain to rub freely against the meninges and to irritate them, this too causing very intense headache; and (3) an alcoholic binge (the "hangover"), which probably causes headache by irritation of the meninges, for the meninges almost certainly become reddened and inflamed in the same manner as the whites of the eyes the day after an alcoholic bout.

Many physicians believe that headaches do not often result from pain originating in the substance of the brain itself. In fact, a patient who is having a brain operation under local anesthetic — that is, without being asleep — feels little or no pain when the surgeon cuts through the brain tissue. Yet when he cuts the meninges, or especially when he cuts one of the major blood vessels supplying the meninges, intense headache is experienced.

The Everyday Headache. Despite all that we *do* know about the origin of headaches, the modern physician is still completely befuddled by the common everyday headache. It is generally stated that there are two basic types. One, the *migraine* type, supposedly results from spasm of blood vessels supplying some of the intracranial tissues, followed by intense and painful dilatation of these same vessels lasting for many hours. However, this simple explanation is still quite hypothetical.

The second type is the so-called *tension headache,* which results when a person operates under considerable emotional tension. It is associated with tightening of muscles attached to the base of the skull and perhaps with simultaneous vascular spasm or other effects occurring intracranially. It is possible that the muscles themselves are the source of some of the pain and that the pain is referred to the head, though this theory is extremely hypothetical.

It is also well known that constipation is frequently associated with headache, possibly because of absorption of toxic products from the colon. It is possible that diffuse irritation of the brain by these toxic substances can cause headache, despite the fact that cutting through brain tissue in an awake person does not usually cause much pain.

THE INTERPRETATION OF SENSATIONS BY THE BRAIN

CORTICAL LOCALIZATION OF SENSATIONS OTHER THAN SOMESTHETIC

The sensations of vision, hearing, taste, and smell, like the somesthetic sensations, are each transmitted from the receptor organs to small circumscribed areas in the cerebral cortex, each of the areas having a different location. These areas are called the *primary cortical areas* for sensations, and from each primary area the impulses go to an *association area*. The primary area for somesthetic sensation, the somesthetic cortex, and the somesthetic association area were discussed earlier in this chapter, and the peripheral origins of each of the other types of sensation will be discussed in following chapters. However, at this point we will see how all the types of sensation function together to provide conscious interpretation of all sensory experience.

Cortical Areas for Vision. The *primary visual cortex* is located in the posterior part of the brain on the medial side of each hemisphere, as will be discussed in Chapter 28. A small portion of the primary visual cortex extends over the occipital pole, as shown in Figure 23–10, though most of it is hidden inside the longitudinal fissure. This is the area where the signals first arrive in the cerebral cortex from the eyes. However, this primary visual area interprets only the more basic meanings of visual sensations such as whether the object is a line, a square, or a star, and the color of the object.

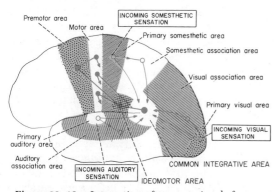

Figure 23–10. Integration of sensory signals from several different sources into a common thought by the common integrative area of the brain, showing also the primary and association areas for vision, for auditory sensations, and for somesthetic sensations.

Accessory visual signals pass from the primary visual cortex and also from the thalamus into adjacent areas of the cortex called the *visual association area,* which also is shown in Figure 23–10. This area interprets the deeper meaning of the visual signals. It interprets the interrelations of the different objects and identifies the objects. Finally, the association area helps to interpret the overall meaning of the scene before the eyes.

Interpretation of written language is one of the most important functions of the visual association area. To accomplish this feat, the visual cortical system must first discern from the light and dark spots the letters themselves, then from the combination of letters the words, and from the sequence of words the thoughts that they express.

Cortical Areas for Hearing. Auditory sensations are transmitted from the ears to a small area called the *primary auditory cortex* in the upper part of the temporal lobe, which will be discussed in Chapter 29. From this area signals pass into the surrounding *auditory association area,* shown in Figure 23–10. The primary auditory cortex interprets the basic characteristics of the sound, such as its pitch and its rhythmicity, while the auditory association area interprets the meaning of the sound. One part of the association area determines whether the sound is noise, music, or speech; then other parts determine the thoughts conveyed by the sound. To interpret the meaning of speech, the auditory association area first combines the various syllables into words, then words into phrases, phrases into sentences, and finally sentences into thoughts.

Cortical Areas for Smell and Taste. The *primary cortex for smell* is located on the bottom surface of the brain, in the *pyriform area* and *uncus,* and the *primary cortex for taste* is located at the bottommost end of the primary somesthetic area, deep in the Sylvian fissure. From these primary areas signals pass into surrounding *smell and taste association areas,* and there the sensations are interpreted in the same manner that somesthetic, visual, and auditory sensations are interpreted by their respective association areas.

THE COMMON INTEGRATIVE AREA OF THE BRAIN – THE GNOSTIC AREA

Figure 23–10 also illustrates the passage of signals from the somesthetic, visual, and auditory association areas into a *common integrative area,* located midway between these three respective association areas. Signals are also transmitted into this area from the taste and smell association areas and directly from the thalamus and other basal areas of the brain. It is here that all the different types of sensations are integrated to determine a common meaning, which is the reason for its name, *common integrative area.* This region of the brain is also called the *gnostic area,* which means the "knowing area."

If a person were in a jungle and heard a noise in the brush, saw the leaves moving, and smelled the scent of an animal, he might not be able to tell from any one of these sensations exactly what was happening, but from all of them together he could quite readily assess his danger. It is in the common integrative area that all the thoughts from the different sensory areas are correlated and weighed against each other for deeper conclusions that can be attained by any one of the association areas alone.

Most of the sensory information arriving in the brain finally is funneled through the common integrative area. For this reason, any damage to this area is likely to leave the person mentally inept; even though the different association areas might still be able to interpret their respective sensations, this information is almost valueless to the brain unless its final meaning can be interpreted. Some signals can be transmitted directly from the association areas to other portions of the brain without going through the common integrative region, but these are so few that the person who loses his entire common integrative area generally becomes seriously confused. This is occasionally the unfortunate result of a brain tumor or a *stroke.* (A stroke is caused by sudden loss of blood supply to an area of the brain because of hemorrhage or thrombosis of a blood vessel.)

Dominance of One Side of the Brain. The common integrative area is located in the *angular gyrus* of the *left* cerebral hemisphere in at least nine-tenths of all people. The angular gyrus of the opposite hemisphere is usually almost nonfunctional, though in one-tenth of the people *it* is the common integrative area and the left angular gyrus is nonfunctional. At birth, both angular gyruses have almost equal functional abilities, but the right angular gyrus region usually becomes supressed as the brain develops, while the left angular gyrus becomes the most important portion of the entire cerebral cortex. Occasionally, though, the relationship between the two is reversed. (This phenomenon of *suppression* occurs very commonly in different parts of the brain when signals from two different sources interfere with each other. For instance, when a person is cross-eyed, so that the visual signals from the two eyes do not correspond satisfactorily, the brain automatically suppresses the signals from one of the eyes. As a result, this eye gradually becomes "functionally" blind until almost all vision is lost, while the person develops excellent vision in the other eye.)

Control of Motor Functions by the Common Integrative Area. Once the common integrative area has integrated all incoming sensations into a common thought, signals are then sent into other portions of the brain to cause appropriate responses. If the integrated thought indicates that muscular activity is needed, the common integrative area sends impulses into the motor portions of the brain to cause muscular contractions.

The basic functions of the cerebral cortex in motor control will be discussed in much more detail in Chapter 25.

ROLE OF THE THALAMUS AND OTHER LOWER BRAIN CENTERS IN THE INTERPRETATION OF SENSATIONS

In the preceding sections of the chapter we have discussed cortical functions as if the cortex were operating almost independently of other parts of the brain. This, however, is entirely untrue. Even when major portions of the sensory cortex are destroyed, the animal is still capable of crude degrees of interpretation of sensation. As has already been pointed out, crude localization and interpretation of many somesthetic sensations can be achieved by the thalamus and other related areas. In the case of vision, the cerebral cortex is not needed in some animals to interpret the overall intensity of light, but it is needed to interpret the shapes and colors of objects. In the case of hearing, an animal can detect the existence of sound and to some extent the direction from which sound is coming without the cerebral cortex, though in general he cannot interpret the finer meanings of sound.

Therefore, whenever it is stated that a particular type of sensation is interpreted in a particular region of the cerebral cortex, it is meant simply that this part of the cortex is responsible for the deeper shades of meaning rather than for total interpretation. Indeed, it would be utterly impossible for the cerebral cortex to operate without preliminary processing of information in the lower regions of the brain.

Activation of Specific Portions of the Cerebral Cortex by the Thalamus. During evolutionary development of the brain, the cerebral cortex originated mainly as an outgrowth of the thalamus, for which reason each area of the cerebral cortex is very closely connected with a corresponding discrete area of the thalamus. Thus, the frontal regions of the cerebral cortex have to-and-fro nerve connections directly with the anterior portions of the thalamus. Likewise, occipital portions of the cerebral cortex are connected with the posterior thalamus, and central portions of the cerebral cortex are connected with lateral and mid areas of the thalamus.

Furthermore, the cortex cannot activate itself but must be activated from lower regions of the brain, an effect that will be discussed in Chapter 27. Especially, signals from specific parts of the thalamus activate specific areas of the cerebral cortex. In this way, function in the thalamus presumably calls forth information stored in the memory pool of the cerebral cortex. Therefore, the thalamus is actually a type of *control center* for the cerebral cortex.

Thus, again it is stressed that even though it is conventional to speak of certain types of sensations being interpreted in specific areas of the cerebral cortex, it is really meant that these cortical areas are the loci of the vast memory information associated with these particular types of sensations. Yet, it is still the lower regions of the brain that are initiating the signals, that are controlling which parts of the cerebral cortex will be activated, that are calling forth stored information from the cerebral cortex, and that are responsible for channeling sensory signals into appropriate parts of the cerebral cortex. Furthermore, many basic aspects of sensations are detected in the lower regions of the brain even before the signals reach the cerebral cortex.

REFERENCES

Bloedel, J. R.: The substrate for integration in the central pain pathways. *Clin. Neurosurg., 21*:194, 1974.

Bonica, J. J. (ed.): International Symposium on Pain. New York, Raven Press, 1974.

Emmers, R., and Tasker, R. R.: The Human Somesthetic Thalamus. New York, Raven Press, 1975.

Fields, H. L., and Bashbaum, A. I.: Brainstem control of spinal pain-transmission neurons. *Ann. Rev. Physiol., 40*:217, 1978.

Halata, Z.: The Mechanoreceptors of the Mammalian Skin. New York, Springer-Verlag New York, Inc., 1975.

Hannington-Kiff, J. G.: Pain Relief. Philadelphia, J. B. Lippincott Company, 1974.

Lynn, B.: Somatosensory receptors and their CNS connections. *Ann. Rev. Physiol., 37*:105, 1975.

Olton, D. S.: Spatial memory. *Sci. Amer., 236(6)*:82, 1977.

Wiederhold, M. L.: Mechanosensory transduction in "sensory" and "motile" cilia. *Ann. Rev. Biophys. Bioeng., 5*:39, 1976.

Wilson, M. E.: The neurological mechanisms of pain. A review. *Anaesthesia, 29*:407, 1974.

Zimmermann, M.: Neurophysiology of nociception. *Intern. Rev. Physiol., 10*:179, 1976.

QUESTIONS

1. Describe the pathways for transmission of sensory signals from the periphery to the cerebral cortex.
2. What are some of the different types of sensory receptors?
3. What is meant by adaptation of sensory receptors?
4. How are different modalities of sensation detected?
5. What types of sensory signals are transmitted through the dorsal column pathway?
6. What types of sensory signals are transmitted through the spinothalamic pathway?
7. Explain the mechanism for localizing sensations in specific areas of the body.
8. Explain the function of the somesthetic association area in the interpretation of sensations.
9. What types of stimuli cause pain?
10. Explain why visceral sensations are frequently referred to areas of the body remote from the viscus.
11. What types of irritation in the cerebral vault will cause headache?
12. Where in the cortex are the primary areas and the association areas for detecting somesthetic sensations, visual sensations, and auditory sensations?
13. What is the role of the common integrative area of the brain in the interpretation of sensory signals?
14. Discuss the role of the thalamus in the interpretation of sensations.

MOTOR FUNCTIONS OF THE SPINAL CORD AND LOWER BRAIN STEM 24

In the last chapter we considered the sensory functions of the central nervous system. To make use of sensory information, the nervous system employs special nervous mechanisms to control the muscles and some of the glands of the body. These are collectively called the *motor functions* of the nervous system. The purpose of the next few chapters will be to discuss these nervous mechanisms, beginning in the present chapter with discussion of the motor functions of the spinal cord and lower brain stem.

Though most of us have an inherent belief that it is only the conscious portion of the brain that causes muscle movements and other motor activities, this is the farthest from the truth, for perhaps the greater proportion of our motor activities is actually controlled by lower regions of the central nervous system, specifically the spinal cord and lower brain stem, which operate primarily at a subconscious level.

PHYSIOLOGIC ANATOMY OF THE SPINAL CORD

Figure 24-1A illustrates a cross-section of the spinal cord, showing that it is composed of two major portions, the *white* and the *gray matter*. The gray matter, which lies deep inside the cord, has the appearance of double horns protruding anteriorly and posteriorly. The cell bodies of the neurons of the cord are located in the gray matter. The white matter, which comprises all other portions of the cord, is composed of fiber tracts. Several long *descending tracts,* shown on the left side of the cord in Figure 24-1A, originate in the brain and pass down the cord to terminate on the neurons in the gray matter; these are all motor tracts. But several other long tracts, the *ascending tracts,* shown on the right side of the cord in the figure, originate in the cord and then pass upward to the brain; these are the sensory tracts, most of which were discussed in the previous chapter.

In addition to the long descending and ascending tracts, many fibers called *propriospinal fibers* pass from one region of the cord to another, as shown in Figure 24-1B.

Figure 24-1B also shows the organization of neurons in the gray matter of the cord. First, let us study the lowermost segment of the cord in this figure. Note that a sensory nerve fiber enters the posterior horn of the gray matter and that it terminates on an *interneuron*. From here signals are transmitted to other interneurons and then eventually to cells in the anterior horn of the gray matter called *anterior motorneurons*. It is these cells that give rise to the motor axons that leave the spinal cord and pass by way of the motor nerves to the muscles.

Note also in Figure 24-1B that the *propriospinal tracts* as well as both the *ascending sensory tracts* and the *descending motor tracts,* which pass all the way to or from the brain, also interconnect with the segmental spinal neurons.

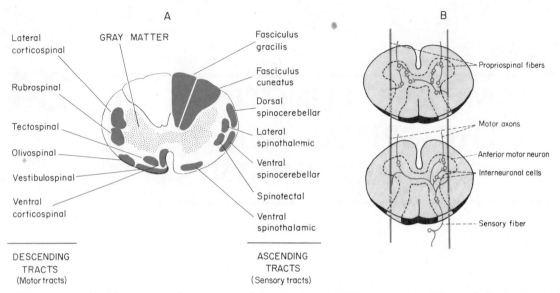

A

GRAY MATTER

Lateral
corticospinal

Rubrospinal

Tectospinal

Olivospinal

Vestibulospinal

Ventral
corticospinal

Fasciculus
gracilis

Fasciculus
cuneatus

Dorsal
spinocerebellar

Lateral
spinothalamic

Ventral
spinocerebellar

Spinotectal

Ventral
spinothalamic

DESCENDING
TRACTS
(Motor tracts)

ASCENDING
TRACTS
(Sensory tracts)

B

Propriospinal fibers

Motor axons

Anterior motor neuron

Interneuronal cells

Sensory fiber

Figure 24–1.   (A) Fiber tracts and gray matter of the spinal cord. (B) Two segments of the cord, illustrating the interneuronal mechanisms that integrate cord reflexes and the propriospinal fibers that interconnect the cord segments.

THE SIMPLE CORD REFLEXES

A nervous reflex is a motor reaction elicited by a sensory signal. For example, a prick of the skin with subsequent passage of the signal through the spinal cord to cause a muscle jerk is a reflex. It was pointed out in Chapter 22 that a large number of the functions of the nervous system are mediated through simple or complex reflexes, some of which are integrated entirely in the spinal cord but some of which utilize signal pathways all the way to the brain and thence back to the muscles or glands to cause motor reactions.

Two essentials must always be present for a reflex to occur: a *receptor* organ and an *effector* organ. All of the sensory nerve receptors are receptor organs for reflexes, and all the muscles of the body, whether they be smooth muscles or skeletal muscles, are effector organs. Glandular cells that can be stimulated by nerve impulses are also effector organs.

THE AXON REFLEX

The simplest nervous reflex is the axon reflex shown to the left in Figure 24–2; this reflex involves only part of a single sensory neuron. Each sensory nerve fiber in a spinal nerve normally has a hundred or more branches, some of which terminate in the skin or other tissues while others terminate on blood vessels or on other structures beneath the skin. It will be recalled that regardless of what point of a neuron is stimulated, the nerve impulse travels over the entire membrane of the fiber. Therefore, stimulating a single sensory receptor causes an impulse to travel backward into all the branches of the fiber as well as upward to the cord. When the impulse reaches the terminals near blood vessels, a hormone, possibly histamine, acetylcholine, or some other similar vasodilatory substance, is secreted, and this causes the blood vessels to dilate.

The axon reflex occurs principally when a tissue is damaged. For example, scratching the skin with a pin elicits an axon reflex,

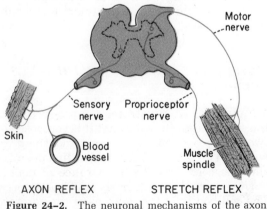

Motor
nerve

Sensory
nerve

Proprioceptor
nerve

Skin

Blood
vessel

Muscle
spindle

AXON REFLEX   STRETCH REFLEX

Figure 24–2.   The neuronal mechanisms of the axon and stretch reflexes.

which in turn causes blood to flow more rapidly than usual in the adjacent vessels. The skin becomes reddened for several millimeters on each side of the scratch mark. In this way, the axon reflex provides increased blood flow to damaged tissues and aids in their repair.

SPINAL CORD REFLEXES THAT HELP TO CONTROL MUSCLE FUNCTION: THE MUSCLE SPINDLE REFLEX AND ITS DAMPING FUNCTION.

Most physical functions of the body are performed by contraction of the skeletal muscles. However, if the muscles are to perform properly, it is essential that the central nervous system know at all times the ongoing state of the muscle, that is, how much it is already contracted and its tension. This information is provided by two different types of receptors, one called *muscle spindles* located throughout the belly of the muscle, and the other called *Golgi tendon apparatuses* located in the tendons. These receptors transmit their information into the spinal cord, and from there information is also relayed to the cerebellum of the brain. This information operates at a subconscious level, but it helps both the brain and the spinal cord control muscle function.

Figure 24–3 illustrates the basic essentials of the muscle spindle. It consists of several small specially adapted muscle fibers called *intrafusal fibers* and nerve endings attached to these fibers. The ends of the intrafusal fibers are connected to the sheaths of the surrounding skeletal muscle fibers. Both ends of the intrafusal fibers can contract in exactly the same manner as other muscle fibers, and

they are excited by a special type of motor nerve fiber called *gamma fibers*. However, the middle portion of each intrafusal fiber does not have the ability to contract. Indeed, when the ends of the fiber contract, the middle portion elongates. Or, when the skeletal muscle fibers surrounding the muscle spindle elongate, so also does this cause the middle of the intrafusal fibers to elongate.

Wrapped around the middle portion of the intrafusal fibers are several nerve endings which are the sensory receptors of the muscle spindle. These transmit signals into the spinal cord to apprise the central nervous system of the degree of elongation of the middle of the spindle.

Function of the Muscle Spindle to "Damp" Muscle Movements — the "Stretch Reflex." When a muscle is suddenly elongated, the middle of the spindle is stretched, and this sends an immediate signal into the spinal cord, as illustrated to the right in Figure 24–2. This signal then excites the motor nerves that control the skeletal muscle fibers immediately surrounding the muscle spindle. Therefore, sudden stretch of the muscle causes an immediate reflex contraction of the same muscle, which automatically opposes further stretch of the muscle. This effect is called the "stretch reflex," and it functions to *damp* changes in muscle length. That is, it prevents the length of the muscle from changing rapidly.

Another feature of the damping process is that it can be turned on or off by stimulation or inhibition of the gamma efferent fibers that supply the intrafusal muscle of the spindle. When gamma efferent signals are exciting the spindle, the two ends of the spindle become taut. Under this condition, the spindle reacts rapidly and strongly to any degree of stretch. On the other hand, when the gamma efferent fibers are silent, the muscle spindle becomes flaccid and does not react to stretch. One can understand very readily the importance of this ability to turn on or off the damping mechanism. For instance, if a person is performing a very discrete function with his fingers, it is highly important that the muscles of the shoulder and elbow be highly damped so that even the slightest force against the hand will initiate an immediate reflex to prevent the hand from being displaced rapidly from its point of fixation. In contrast, when a person wishes to flail his arm in a wide arc, it is essential that the damping mechanism be suppressed, which

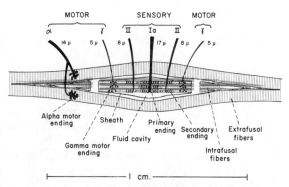

Figure 24–3. The muscle spindle, showing its relationship to the large extrafusal skeletal muscle fibers. Note also both the motor and the sensory innervation of the muscle spindle.

can be achieved by simply turning off the signals that pass through the gamma efferent fibers to the muscle spindles.

The Knee Jerk and Other Muscle Reflexes. One of the best known examples of the stretch reflex is the knee jerk, shown in Figure 24–4. Almost every doctor tests this reflex when he performs a physical examination. A small hammer is used to strike the muscle tendon immediately below the kneecap. The sudden jolt to the tendon stretches the quadriceps muscle in the thigh, which in turn stretches muscle spindles and sends signals to the spinal cord. Then a reflex signal passes back to the same quadriceps muscle, causing a sudden contraction that makes the lower leg jerk forward.

A muscle jerk similar to the knee jerk can be elicited in any muscle of the body by suddenly striking its tendon, or even by striking the muscle itself. The only essential for eliciting such a reflex is to stretch the muscle suddenly.

Muscle reflexes of this type are elicited by physicians for two major purposes: First, if the reflex can be demonstrated, it is certain that both the sensory and motor nerve connections are intact between the muscle and the spinal cord. Second, muscle reflexes can help determine the degree of excitability of the spinal cord. When a large number of facilitatory nerve signals are being transmitted from the brain to the cord, the muscle reflexes will be so active at times that simply tapping the knee tendon with the tip of one's finger might make the leg jump a foot. On the other hand, the cord may be intensely inhibited by inhibitory nerve signals from the brain,

in which case almost no degree of pounding on the muscles or tendons can elicit a response.

The "Servo-Assist" Function of the Muscle Spindle. Experiments have shown that when the sensory nerve roots to the spinal cord have been severed, more nervous energy than normal is required to cause a muscle movement. Therefore, it has been suggested that the muscle spindle plays an important role in helping to provide much of the nervous energy needed to effect motor movements. This function of the muscle spindle is believed to operate similarly to the way that "power steering" works in the automobile, utilizing the principle called *servo-assist*. This may be explained as follows:

When a nerve signal is transmitted from the spinal cord to cause contraction of the large skeletal muscle fibers, a simultaneous signal is transmitted through the gamma motor nerve fibers to cause contraction of the muscle spindles as well. This contraction of the muscle spindles at first stretches the sensory element of the spindles which transmits spindle signals back into the spinal cord. These immediately elicit a muscle spindle reflex that sends large numbers of additional nerve impulses to the large skeletal muscle fibers. Thus, contraction of the skeletal muscles results from two sets of signals: (1) the original signals sent by the spinal cord, and (2) the additional signals transmitted indirectly by way of the muscle spindle. The strength of contraction is, therefore, much greater than it would be in response to the initial signal alone. This is similar to the way that power steering operates; that is, a relatively weak torque on the steering wheel provides a powerful steering effect because of the servo-assistance of the power-steering mechanism.

THE TENDON REFLEX

Figure 24–5 illustrates the tendon apparatus, which is a receptor found in all muscle tendons. It detects the amount of tension in the tendon and transmits this information into the spinal cord and from there also into the cerebellum. The information, in turn, is used by the neural mechanisms to help adjust precisely the tension that is needed to perform the required muscle function.

A second function of the information transmitted into the nervous system from the tendon apparatuses is to protect the muscle it-

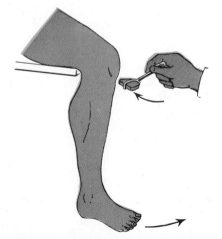

Figure 24–4.   Method for eliciting the knee jerk.

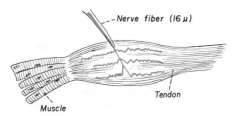

Figure 24–5. Anatomy of the Golgi tendon apparatus.

self. When the tension detected by these apparatuses becomes extremely great, great enough that it might cause tearing of the muscle or rupture of the tendon itself, the neuronal centers in the spinal cord automatically and instantly initiate a reflex to *inhibit* the anterior motor neurons that innervate the muscle. The muscle immediately relaxes, and the excessive stretch is removed from the muscle, a very important protective reflex to prevent damage to either muscle or tendon.

The neuronal circuit in the spinal cord that subserves the tendon reflex is much more complicated than that for the muscle spindle reflex, for it utilizes several interneurons in the spinal cord between the input sensory neuron and the anterior motor neuron, in contrast to the muscle spindle reflex in which the sensory neuron makes at least some direct contacts with the motor neuron.

SPINAL CORD REFLEXES THAT HELP TO SUPPORT THE BODY AGAINST GRAVITY

The Extensor Thrust Reflex. A complex cord reflex that helps support the body against gravity is the *extensor thrust reflex*. Pressure on the pads of the feet causes automatic tightening of the extensor muscles of the legs. This reflex is initiated by pressure receptors in the bottom of the foot. The signal passes to the interneurons (intermediate cells between sensory input and motor output) in the cord. Here it is amplified and diverged into an appropriate pattern of impulses to tighten the extensor muscles, causing the animal or person to keep his leg stiffened automatically when standing.

THE MAGNET REACTION. A reflex closely related to the extensor thrust reflex, but still more complicated, is the so-called *magnet reaction*. In an animal that has had its spinal cord cut so that impulses from the brain will not interfere, one can place the tip of his finger on the pad of the animal's foot, then move his finger in all directions, and the foot will follow the finger. Moving the finger to one side causes appropriate proprioceptor reflexes to make the limb move in the direction of the force. This reaction is an aid to equilibrium of the animal, for excess pressure on one side of the foot indicates that he is falling in that direction, and automatic stiffening of the leg in that direction helps to prevent the fall.

REFLEXES THAT HELP TO PREVENT DAMAGE TO THE BODY

The Flexor Reflex or Withdrawal Reflex. Pain causes automatic withdrawal of any pained portion of the body from the object causing the pain. This is called a *flexor reflex* or sometimes simply a *withdrawal reflex*.

The neuronal mechanism of the flexor reflex is shown to the left in Figure 24–6. On entering the gray matter of the cord, the pain signal stimulates interneurons that transmit a pattern of impulses through appropriate anterior motor neurons to cause withdrawal of the area of the body that is being pained. In Figure 24–6 the pain stimulus is initiated in the right hand, and the biceps muscle, which is a "flexor" muscle because it flexes the arm, becomes excited and pulls the hand away.

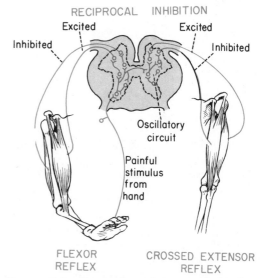

Figure 24–6. Neuronal mechanisms of the flexor reflex, of the crossed extensor reflex, and of reciprocal inhibition.

The Crossed Extensor Reflex. When a flexor reflex occurs in one limb, impulses also pass to the opposite side of the cord, where they stimulate the interneurons controlling the extensor muscles of the opposite limb, thus causing it to extend. This is called a *crossed extensor* reflex. For example, a painful stimulus applied to the right hand causes the left arm to extend as shown in Figure 24–6. Such a reflex pushes the person away from the painful object. That is, at the same time that he withdraws his pained hand on the right, he pushes his entire body away with the left hand.

Withdrawal reflexes are present in all parts of the body, though organized somewhat differently in different areas. For instance, if a needle pricks the small of a person's back, he automatically arches forward, which causes withdrawal. Though this reflex is slightly different from the flexor and crossed extensor reflexes of the limbs, its result is the same.

RECIPROCAL INHIBITION

A very important feature of most reflexes, especially illustrated by the flexor and crossed extensor reflexes, is the phenomenon called *reciprocal inhibition*. That is, when a reflex excites a muscle, it ordinarily inhibits the antagonistic muscle at the same time. For example, when the flexor reflex of Figure 24–6 excites the biceps muscle, it simultaneously inhibits the opposing triceps muscle. Also, the crossed extensor reflex excites the triceps but inhibits the biceps. In all parts of the body where opposing muscles exist, a corresponding reciprocal inhibition circuit is present in the spinal cord. This obviously allows greater ease in the performance of desired activities.

RHYTHMIC REFLEXES OF THE CORD

The reflexes of the spinal cord are not limited to simple reactions; frequently rhythmic reflexes also occur. That is, "patterns of activity" in the cord can make a portion of the body move back and forth in rhythmic motion. Two especially important types of rhythmic reflexes controlled entirely by the spinal cord are the scratch reflex and the walking reflexes.

THE SCRATCH REFLEX

The scratch reflex, though not of importance in the human being, is one of the most important of all protective mechanisms in lower animals. A dog can actually scratch away a flea or other irritating object even after his spinal cord has been completely transected at the neck level.

The scratch reflex depends on two separate integrative abilities of the spinal cord: First, the cord must provide the rhythmic to-and-fro motion of the leg. This is accomplished by reverberating circuits in the interneurons that send impulses first to one group of muscles, then to the opposing muscles, contracting alternately the two sets of muscles to cause the rhythmic motion.

The second cord mechanism utilized in the scratching act pinpoints the area of the body that needs scratching. For instance, a flea moving across the belly of a sleeping dog is followed by the scratching paw. When the flea crawls across the midpoint of the animal's belly this paw stops scratching, but the paw on the opposite side of the body immediately finds the flea and begins to scratch. Thus, the scratch reflex is one of the most complicated of all cord reflexes, for it requires, first, localization of the irritation; second, coordinate movement of the paw into position; and third, reverberating motion of the limb.

WALKING REFLEXES

The spinal cord also is capable of providing the rhythmic to-and-fro movements of the legs that are used in walking. These movements, like the scratching mechanism, are controlled by reverberating circuits in the interneurons that contract alternately the antagonistic pairs of muscles. Reciprocal inhibition also plays a part, for this keeps the muscles on the two sides of the body operating in opposite phase to each other, causing one leg to move forward while the opposite one moves backward. In lower animals, fiber pathways also pass from the hindlimb region of the cord to the forelimb region to cause appropriate phasing of the hind and forelimb movements. This cord control of walking movements with appropriate phasing of the limbs is illustrated in Figure 24–7, which shows a dog with its spinal cord transected in the neck; it is, nevertheless, still exhibiting

Figure 24–7. Walking movements in a dog whose spinal cord has been transected at the neck.

continuous, rhythmic, to-and-fro movements of its limbs.

AUTONOMIC REFLEXES OF THE CORD

In addition to the cord reflexes for controlling skeletal muscular action, the spinal cord also harbors reflex circuits that help to control the visceral functions of the body. These reflexes, called *autonomic reflexes,* are initiated by sensory receptors in the viscera, and the signals are transmitted through sensory nerves to the interneurons of the cord gray matter, where appropriate patterns of reflex responses are determined. Then the signals pass to *autonomic motor neurons* located in the cord gray matter about midway between the anterior and posterior horns. These cells send impulses into the *sympathetic nerves* and back to the viscera to cause autonomic stimulation.

THE PERITONEAL REFLEX

One of the most important of the autonomic reflexes is the peritoneal reflex. Whenever any portion of the peritoneum (the surface lining of the gut and abdominal cavity) is damaged, this reflex slows up or stops all motor activity in the nearby viscera. For example, in appendicitis, which almost always irritates the peritoneum, movement of food through the gastrointestinal tract stops almost completely, preventing further irritation of the inflamed appendix and allowing the reparative processes of the body to func-

tion at optimum efficiency. Without this reflex many would die of appendicitis before aid could be given.

VASCULAR CONTROL BY CORD

The resistance of blood flow in most peripheral blood vessels is normally controlled by vasomotor impulses transmitted from the brain, but after the spinal cord has been transected, these impulses can no longer reach the blood vessels. Yet cord reflexes are still capable of modifying local vascular blood flow in response to such factors as pain, heat, and cold. For instance, when one of the viscera is pained tremendously by overdistention, such as overfilling of the urinary bladder, the vascular resistance often doubles and thereby raises the arterial pressure to as high as twice normal. Though these effects of cord reflexes on the arterial pressure are not important normally, they do illustrate the capability of the cord to help control many of the involuntary reactions of the autonomic nervous system.

THE BLADDER AND RECTAL REFLEXES

Among the most important of the autonomic reflexes of the cord are the bladder and rectal reflexes that cause automatic emptying of the urinary bladder and of the rectum when they become filled. When either the bladder

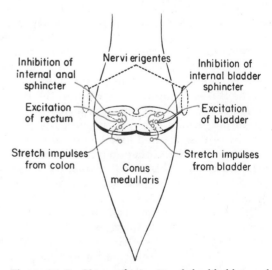

Figure 24–8. Neuronal circuits of the bladder- and rectal-emptying reflexes.

or the rectum becomes excessively full, sensory signals are transmitted into the interneurons of the lower end of the cord, as shown in Figure 24–8. Appropriate signals are then transmitted through parasympathetic nerves back to the bladder or colon. Here they excite the main body of the bladder or colon but at the same time inhibit the internal sphincter of the urethra or anus, thereby causing emptying.

In persons whose spinal cords have been cut, these automatic reflexes are sometimes effective enough to empty the bladder every half-hour or more and the colon one or more times each day. In the normal person, however, these reflexes are inhibited by impulses from the brain until an opportune time arises for emptying. Unfortunately, this inhibition frequently leads to constipation or painful distention of the bladder.

MOTOR FUNCTIONS OF THE LOWER BRAIN STEM

The lower brain stem is composed of the medulla, the pons, and the mesencephalon (Figure 24–9). It is in these areas that many of the centers for arterial pressure control, respiration, and gastrointestinal regulation are located, all of which are discussed in detail in other chapters. The purpose of the present section is not to give an overall discussion of the lower brain stem, but to present its motor functions and their relation to the cord and the forebrain. The motor functions of the lower brain stem can be divided into two major types: first, its function in helping the person support his body against gravity, and, second, its function in the maintenance of equilibrium. But first we need to describe the *bulboreticular formation,* the brain structure that is responsible for these functions.

The Bulboreticular Formation of the Brain Stem. Figure 24–9 shows the bulboreticular formation, which extends through the entire lower brain stem, and also shows some of its connections with other parts of the body. Stimulation of this area transmits signals down several different tracts into the cord to help control the muscles.

The bulboreticular area also receives incoming fibers from many sources: fibers directly from all areas of the body through the spinal cord, fibers from the cerebellum, from the equilibrium apparatus of the ear, from the

motor portion of the cerebral cortex, and from the basal ganglia deep in the cerebrum. Thus, the bulboreticular formation is an integrative area for combining and coordinating (1) sensory information from the body, (2) motor information from the motor cortex, (3) equilibrium information from the vestibular apparatuses, and (4) proprioceptor information (information about body movements) from the cerebellum. With this information available, it controls many of the involuntary muscular activities.

SUPPORT OF THE BODY AGAINST GRAVITY

Widespread stimulation of the middle and upper bulboreticular formation excites mainly the extensor muscles of the body, causing the trunk and the limbs to become stiffened, and making it possible to stand erect. Without this stiffening, the body would immediately crumple because of the pull of gravity.

Suppression of the Bulboreticular Area by the Basal Ganglia; Decerebrate Rigidity When Suppression Is Blocked. When one wants to sit rather than to stand, the excitation of the muscles by the bulboreticular formation must be inhibited. This is accomplished mainly by suppressor impulses from the basal ganglia, special neuronal centers that lie lateral to the thalamus. When the basal ganglia are damaged, or when all the fiber tracts from the cerebrum are cut, the bulboreticular formation automatically becomes so active that the animal becomes rigid all over. This phenomenon, called *decerebrate rigidity,* is illustrated in the dog of Figure 24–10. It shows the necessity of the basal ganglia to inhibit the bulboreticular formation when motor functions other than the support of the body are to be performed. (The animal or person also relaxes when asleep, which also inhibits the bulboreticular area.)

EQUILIBRIUM FUNCTION OF THE LOWER BRAIN STEM

In addition to providing the neurogenic mechanisms for support of the body against gravity, the bulboreticular formation is also capable of maintaining equilibrium by varying the degree of tone in the different muscles. Most of the equilibrium reflexes are ini-

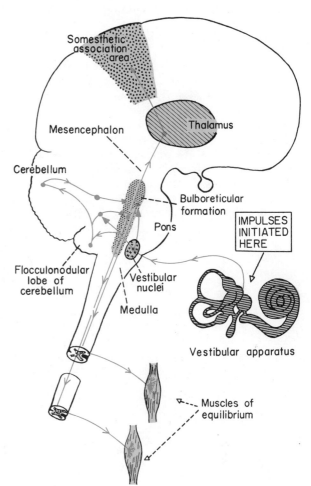

Figure 24–9.   Nervous mechanisms of equilibrium.

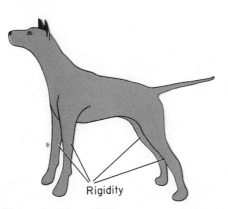
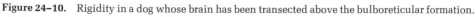

Figure 24–10.   Rigidity in a dog whose brain has been transected above the bulboreticular formation.

tiated by the *vestibular apparatuses* located on each side of the head, adjacent to each internal ear.

Function of the Vestibular Apparatus. The vestibular apparatus is shown in Figure 24–11, and its connections with the equilibrium areas of the central nervous system are shown in Figure 24–9. This apparatus detects the position of the head in space — that is, it determines whether the head is upright with respect to the gravitational pull of the earth or whether it is leaning backward, upside down, or in another position. It also detects sudden changes in movement. To perform these functions the vestibular apparatus is divided into two separate physiologic sec-

tions, the *maculae* of the utricle and the saccule, and the *semicircular canals*.

THE MACULAE.   On the wall of the utricle and the saccule is a structure called the *macula,* illustrated in Figures 24–11A and B. Nerve cells in the base of the macula project "hairs" upward into a gelatinous mass in which are located many minute bonelike calcified granules, like very small granules of sand, called *otoconia*. When the head is bent to one side, the weight of the octoconia pushes the hairs to that side, thereby stimulating the nerves. In this way, the macula of the utricle as well as that of the saccule supplies the equilibrium regions of the central nervous system with the information needed to maintain balance.

The maculae also help the person maintain his balance when he suddenly begins to move forward, to one side, or in any other linear direction. That is, when he begins to move forward, the inertia of the otoconia causes them to lag behind the movement of the rest of the body and, therefore, to bend the hairs backward. This gives the sensation of falling off balance in the backward direction. As a result, the person leans forward to correct this imbalance, which explains why an athlete leans forward automatically when he first starts to run.

On the other hand, when one wishes to decelerate, he must lean backward. Again it is the otoconia of the maculae that initiate automatically this backward leaning, for when one brakes himself the momentum of the otoconia keeps them moving forward while the rest of the body slows. This bends the hairs of the macula in the forward direction, making the person feel as if he were falling head first toward the ground. As a result, the equilibrium mechanism causes the body to lean backward automatically.

THE SEMICIRCULAR CANALS.   The *semicircular canals,* shown in Figure 24–11A, are small circular tubes that contain fluid. The vestibular apparatus on each side of the head has three separate canals located respectively in the three planes of space, one in the horizontal and the other two in the two vertical planes. If one suddenly turns his head in any direction, the fluid in one or more of the semicircular canals lags behind the movement of the head because of the inertia of the fluid. This is the same effect as that observed when one suddenly rotates a glass of water; the glass rotates, but the water remains still. As

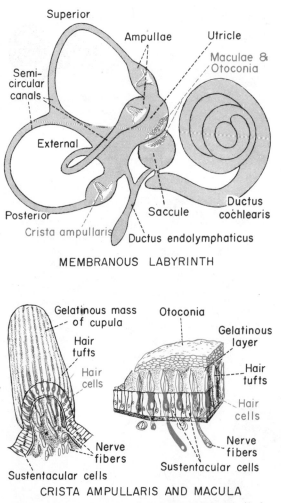

MEMBRANOUS LABYRINTH

CRISTA AMPULLARIS AND MACULA

Figure 24–11.   Functional structures of the vestibular apparatus. (From Goss: Gray's Anatomy of the Human Body. Lea & Febiger; modified from Kolmer by Buchanan: Functional Neuroanatomy, Lea & Febiger).

the fluid moves in the semicircular canal it flows against the *crista ampullaris,* shown in Figures 24–11A and B, which is a valvelike leaflet located at one end of each canal. This structure contains hair tufts (called *cilia*) projecting from hair cells like those in the macula, and bending these cilia to one side or the other gives the person the sensation that his head is *beginning to turn.*

The information transmitted from the semicircular canals apprises the nervous system of sudden *changes in direction of movement.* With this information available, the bulboreticular formation can correct for any imbalance that is likely to occur when running around a corner *even before the imbalance does occur.* This is particularly important when one is changing direction of movement rapidly, as when playing a fast game of almost any type.

Function of the Cerebellum in Equilibrium. In addition to transmitting signals into the bulboreticular formation, the semicircular canals and the maculae also send information to the *flocculonodular lobes* of the cerebellum. Since one of the major functions of the cerebellum is to predict future position of the body in space, which will be discussed in the following chapter, the function of the flocculonodular lobes is probably to predict when a state of imbalance is going to occur. This allows appropriate corrective signals to be given to the bulboreticular formation even before the person falls off balance, and prevents imbalance from occurring rather than necessitating attempts to correct it after it has already occurred. Persons who lose their cerebellum lose this ability to predict, and, as a result, must perform all movements slowly or else fall.

The Neck Proprioceptor Receptors and Their Relationship to the Vestibular Mechanisms. The vestibular apparatuses detect only the position of the *head,* not of the *body,* in relation to the pull of gravity. To translate this information from the head to the whole body, the relationship of the head to the body must be known. This knowledge is provided by proprioceptor receptors (position receptors) in the neck. For instance, when the head bends backward, the vestibular apparatuses send information that the position of the head with respect to gravity is changing, but at the same time the neck proprioceptors send information that the head is angulating backward in relation to the rest of the body. The two sets of impulses allow the brain to determine that the position of the body with respect to gravity has not changed despite the bending of the head. Therefore, the tautness of the postural muscles remains exactly the same. In other words, the proprioceptor reflexes from the neck are as necessary for regulating equilibrium as are the complicated reflexes initiated by the vestibular apparatuses.

Peripheral Proprioceptor and Visual Mechanisms of Equilibrium. If the vestibular apparatuses have been destroyed, a person can still maintain his equilibrium provided he moves slowly. This is accomplished mainly by means of proprioceptor information from the limbs and surfaces of the body and visual information from the eyes. If he begins to fall forward, the pressure on the anterior parts of his feet increases, stimulating the pressure receptors. This information transmitted to his brain helps to correct the imbalance. At the same time, his eyes also detect the lack of equilibrium, and this information too helps to correct the situation.

Unfortunately, the visual and proprioceptor systems for maintaining equilibrium are not organized for rapid action, which explains why a person without his vestibular apparatuses must move slowly.

OVERALL CONTROL OF LOCOMOTION

From the foregoing discussions of the motor functions of the spinal cord, lower brain stem, and cerebrum, it is now possible to construct the overall pattern of locomotion. First, the person must support himself against gravity. This is accomplished partly by the extensor thrust mechanism of the spinal cord that allows stiffening of the limbs when pressure is applied to the pads of the feet; in addition to this, the bulboreticular formation transmits impulses into the extensor muscles to keep the body and limbs stiff. Also, the degree of stiffening of the different parts of the body is varied by the equilibrium system so that when one tends to fall over, appropriate muscles are contracted to bring the body back into the upright position.

Once the body is supported against gravity and maintained in a state of equilibrium, locomotion then depends on rhythmic motion of the limbs. Rhythmic circuits in the spinal cord are capable of providing the to-and-fro movement of the limbs, and the movements of the opposing limbs are kept in opposite phase with each other by the reciprocal inhibition mechanism of the cord. Thus, most of the functions of locomotion can be provided by the cord and brain stem, but the cerebral cortex must control these functions in accord with the desires of the individual. When he wishes to move forward, to stop, or to turn to one side, his motor cortex and basal ganglia simply initiate the action, stop it, or change it by sending *command signals*. The cord and brain stem provide the stereotyped actions required to perform the actual movements. In this way, the energy of the conscious portion of the brain is conserved to perform other mental feats.

REFERENCES

Bizzi, E.: The coordination of eye-head movements. *Sci. Amer., 231(4)*:100, 1974.

Brodal, A. (ed.): Basic Aspects of Central Vestibular Mechanisms. New York, American Elsevier Publishing Company, Inc., 1971.

Creed, R. S., Denny-Brown, D., Eccles, J. C., Liddell, E. G. T., and Sherrington, C. S.: Reflex Activity of the Spinal Cord. New York, Oxford University Press, Inc., 1932.

Goldberg, J. M., and Fernandez, C.: Vestibular mechanisms. *Ann. Rev. Physiol., 37*:129, 1975.

Kawamura, Y.: Neurogenesis of mastication. *Front. Oral Physiol., 1*:77, 1974.

Merton, P. A.: How we control the contraction of our muscles. *Sci. Amer., 226*:30, 1972.

Pearson, K.: The control of walking. *Sci. Amer., 235(6)*: 72, 1976.

Porter, R.: The neurophysiology of movement performance. *In* MTP International Review of Science: Physiology. Vol. 3. Baltimore, University Park Press, 1974, p. 151.

Precht, W.: Vestibular system. *In* MTP International Review of Science: Physiology. Vol. 3. Baltimore, University Park Press, 1974, p. 81.

Purves, D.: Long-term regulation in the vertebrate peripheral nervous system. *Intern. Rev. Physiol., 10*:125, 1976.

Sherrington, C. S.: The Integrative Action of the Nervous System. New Haven, Yale University Press, 1911.

Shik, M. L., and Orlovsky, G. N.: Neurophysiology of locomotor automatism. *Physiol. Rev., 56*:465, 1976.

Stein, R. B.: Peripheral control of movement. *Physiol. Rev., 54*:215, 1974.

Wilson, V. J., and Peterson, B. W.: Peripheral and central substrates of vestibulospinal reflexes. *Physiol. Rev., 58*:80, 1978.

QUESTIONS

1. Describe the physiologic anatomy of the neuronal circuits in the spinal cord.
2. Describe the function of the muscle spindle and of the muscle spindle reflex.
3. Explain the servo-assist function of the muscle spindle.
4. Explain the function of the tendon reflex.
5. How do cord reflexes help to support the body against gravity?
6. What is the functional importance of the flexor reflex and the crossed extensor reflex?
7. Describe the walking reflexes of the spinal cord.
8. What is the role of the bladder and rectal reflexes?
9. Explain the function of the bulboreticular formation in the support of the body against gravity and in the equilibrium process.
10. Describe the vestibular apparatus and its functions in static equilibrium and in equilibrium during motion.
11. Discuss the overall control of locomotion.

CONTROL OF MUSCLE MOVEMENT BY THE CEREBRAL CORTEX, THE BASAL GANGLIA, AND THE CEREBELLUM

25

Though a major share of the motor functions of the body can be performed without involvement of the higher brain centers, one of the distinguishing features of the human being is his ability to carry out extremely complex voluntary muscle activity. He can perform the intricate tasks of talking, writing, using delicate instruments, and achieving specialized patterns of movement required for dance routines, basketball, football, and so forth. All of these activities involve major degrees of control by the higher centers of the brain, especially the cerebral cortex, the basal ganglia, and the cerebellum. It is the goal of this chapter to explain how these higher centers work together to give these special abilities.

BASIC ORGANIZATION OF THE HIGHER CENTERS FOR MUSCLE CONTROL

Figure 25–1 illustrates the interplay among the brain centers for muscle control. These centers are composed of four major parts — the bulboreticular formation, which was discussed in the previous chapter, and the three major parts to be discussed here: (1) the motor cortex, (2) the basal ganglia, and (3) the cerebellum.

The *motor cortex* lies slightly anteriorly in the cerebral cortex. It is this area that plays the greatest role in control of very fine, discrete muscle movements.

The *basal ganglia* lie deep in the cerebral hemispheres and are composed of separate large pools of neurons organized for control of complex semivoluntary movements such as walking, turning, running, and development of bodily postures for performance of specific functions.

The *cerebellum* is located posteriorly and inferiorly in the brain, lying behind the lower brain stem. As we shall see later in the chapter, the cerebellum communicates with the other motor areas of the brain through several large neuronal trunks. The cerebellum helps both the motor cortex and the basal ganglia perform their functions. Mainly, it makes the movements smooth rather than jerky. And it also helps to make groups of muscles operate together in a coordinate manner so that very accurate and very fine degrees of muscle control can be achieved. In other words, unlike both the basal ganglia and the motor cortex, the cerebellum is not directly responsible for control of muscle activity but, instead, operates as a helper to the other two areas.

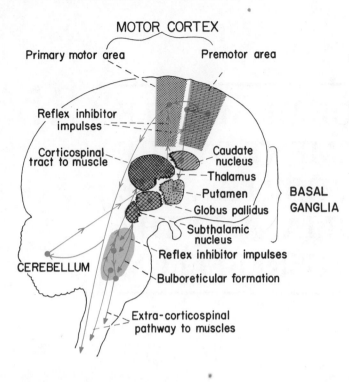

MOTOR CORTEX

Primary motor area   Premotor area

Reflex inhibitor
  impulses

Corticospinal
tract to muscle

Caudate
  nucleus

Thalamus

Putamen

Globus pallidus

BASAL
GANGLIA

Subthalamic
  nucleus

Reflex inhibitor impulses

CEREBELLUM

Bulboreticular formation

Extra-corticospinal
  pathway to muscles

Figure 25–1. Control of muscle activity by the motor cortex, basal ganglia, and cerebellum.

TRANSMISSION OF MOTOR SIGNALS TO THE SPINAL CORD

The Corticospinal Tract. Close study of Figure 25–1 will show that motor signals are transmitted from the brain to the spinal cord through two separate pathways, the *corticospinal tract* and the *extracorticospinal pathway*.

The corticospinal tract is a direct pathway that originates in the motor cortex and passes without synapse all the way to the cord. Figure 25–2 illustrates this pathway in more detail, showing that it is composed of a discrete bundle of nerve fibers coming from each side of the brain. In the lower part of the medulla and first few segments of the spinal cord, the fibers from each side of the brain cross to the opposite side and then proceed downward through the two lateral portions of the cord. Thus, the motor cortex on the left side of the brain controls the muscles of the right side of the body, while the motor cortex of the right side controls those on the left.

The Extracorticospinal Pathway. The extracorticospinal pathway is composed of all other tracts besides the corticospinal tract that transmit motor signals to the cord. These include mainly tracts that originate in the basal ganglia and the bulboreticular formation, as illustrated in Figure 25–1. And Figure 25–3 illustrates the location of these

tracts as well as of the corticospinal tract in a cross-section of the spinal cord. Note in this cross-section that the corticospinal tract is located laterally and that the extracorticospinal tracts lie more anteriorly. These include the *rubrospinal tract,* the *vestibulospinal tract,* the *tectospinal tract,* and two separate *reticulospinal tracts.* It is through these tracts that motor signals for control of most of the stereotyped and subconscious body movements are transmitted.

ROLE OF THE SPINAL SEGMENTS IN MOTOR CONTROL

The nerve fibers of both the corticospinal and the extracorticospinal pathways terminate in a special neuronal network of the spinal cord. And it is this network that then sends direct signals to the muscles.

We saw in the previous chapter that local neuronal mechanisms of the spinal cord are responsible for many cord reflexes that elicit muscle activity even without signals from the brain. When signals also come from the brain, these combine with the signals from the sensory nerves entering each spinal segment, and the combined information is utilized by the cord's neuronal network to control muscle activity. Figure 25–3 illustrates this intrinsic network.

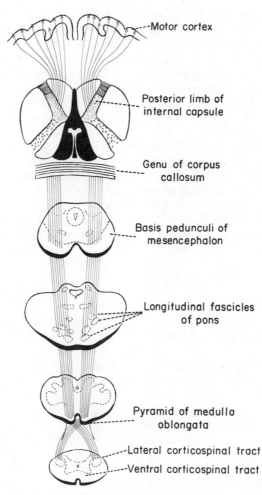

Figure 25–2. The corticospinal tract (also called the pyramidal tract), through which motor signals are transmitted from the motor cortex to the spinal cord. (Modified from Ranson and Clark: Anatomy of the Nervous System, 10th ed. W. B. Saunders Co.)

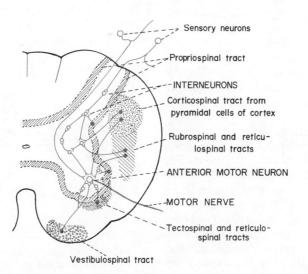
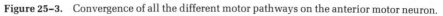

Figure 25–3. Convergence of all the different motor pathways on the anterior motor neuron.

The Anterior Motor Neuron and the Interneurons. Located in the anterior portion of the spinal cord (illustrated in Fig. 25–3) are many very large neurons that send large axons out of the cord into the nerves that supply the muscles. These neurons are called *anterior motor neurons,* and the nerves passing to the muscles are called the *motor nerves.* Note in Figure 25–3 that signals from many different sources impinge on the surface of the anterior motor neuron. Some of the signals arriving at this neuron excite it while others inhibit it. Thus, muscle control is much more complex than simply transmitting a signal that always causes muscle contraction.

Figure 25–3 also illustrates many smaller neurons called *interneurons* located in central portions of the spinal cord gray matter. These interneurons receive signals from incoming sensory nerves, from the corticospinal tract, and from the extracorticospinal pathway, and they in turn transmit signals to the anterior motor neuron. Though a few signals bypass the interneurons and go directly to the anterior motor neuron, by far the larger proportion must first be processed by these cells.

It is in the interneurons that many patterns of muscle contraction are determined. For instance, most of the patterns of the spinal reflexes discussed in the previous chapter are determined here, such as the patterns of the flexor reflex, the crossed extensor reflex, the walking reflexes, the scratch reflex, and different autonomic reflexes.

Control of the Anterior Motor Neurons by Brain Signals. When brain signals reach the interneurons of the spinal cord, they do not necessarily cause a muscle contraction. Instead, the brain signal is first combined with both excitatory and inhibitory signals arriving from other sources. If the preponderance of information is in favor of causing muscle contraction, then contraction occurs. On the other hand, if it is not in favor of contraction, then the muscle might remain as is or even be inhibited.

Also important is the ability of the brain signals to "command" the spinal cord neuronal network to perform specific functions. For instance, a command signal from the brain might direct the muscles to perform walking movements. In response, the interneuronal mechanism for achieving walking movements is set into action. It is this mechanism that actually causes the discrete muscle contractions required for walking; the command signal from the brain, on the other hand, has little to do with controlling the individual muscles, but only commands the cord to perform the appropriate function.

If the reader will now allow his imagination to roam, he will understand how motor signals from the brain can command the spinal cord to achieve many complex motor activities without ever directly exciting an individual muscle. Yet, at times it is also important to make single muscles contract, and this too can be achieved. It is signals from the corticospinal tract rather than the extracorticospinal pathway that have this ability to cause contraction of individual muscles, as will be discussed in more detail later in the chapter.

FUNCTION OF THE BASAL GANGLIA

Now that we have described the general neuronal mechanisms for brain control of muscle function, let us return once again to the higher centers to describe some of the special characteristics of this control. To begin, consider first the functions of the basal ganglia. The basal ganglia are composed of

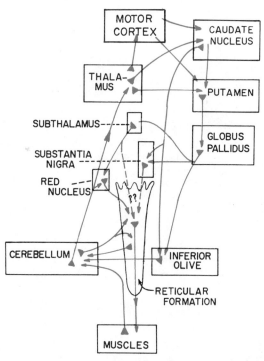

Figure 25–4. Interconnections of the basal ganglia with other parts of the motor system.

large masses of neurons located deep in the substance of the cerebrum and in the upper part of the mesencephalon. These were shown in Figure 25–1, and their multitude of neuronal connections with other portions of the motor control system are illustrated in Figure 25–4. This multitude of connections allows the basal ganglia to control most of the subconscious stereotyped bodily movements, often involving simultaneous contraction of many muscle groups throughout the body. Though it will not be possible to describe the basal ganglia in detail, the most important of these are the *caudate nucleus,* the *putamen,* the *globus pallidus,* the *subthalamic nucleus,* and the *substantia nigra.*

Control of Subconscious Movements by the Basal Ganglia.   Before attempting to discuss the function of the basal ganglia in man, we should speak briefly of their better known functions in lower animals. In birds, for instance, the cerebral cortex is very poorly developed, while the basal ganglia are highly developed. These ganglia perform essentially all the motor functions, even controlling the voluntary movements in much the same manner that the motor cortex of the human being controls voluntary movements. In the cat, and to a less extent in the dog, removal of the cerebral cortex does not interfere with the cat's ability to walk, to eat perfectly well, to fight, to develop rage, to have periodic sleep and wakefulness, and even to participate naturally in sexual activities. However, if a major portion of the basal ganglia is destroyed, only gross stereotype movements remain, which are controlled by the very primitive lower areas in the brain stem — for instance, the bulboreticular area which was described in the previous chapter.

In the human being, many of the potential functions of the basal ganglia are suppressed by the cerebral cortex, but if the cerebral cortex becomes destroyed in a very young human being, many voluntary motor functions do develop. The person will never be able to develop very discrete movements, particularly of the hands, but he can learn to walk, to control his equilibrium, to eat, to rotate his head, to perform almost any type of postural movement, and to carry out most subconscious movements. On the other hand, destruction of a major portion of the caudate nucleus almost totally paralyzes the opposite side of the body except for a few stereotyped reflex movements integrated in the cord or lower brain stem.

FUNCTIONS OF INDIVIDUAL BASAL GANGLIA

Unfortunately, little is known about the functions of most of the individual basal ganglia other than the fact that they operate together in a closely knit unit to perform the subconscious movements. Yet the following are a few of the discrete bits of information that are known about their individual functions.

(1). The *caudate nucleus* controls gross intentional movements of the body, these occurring both subconsciously and consciously and aiding in the overall control of body movements.

(2). The *putamen* operates in conjunction with the caudate nucleus to control gross intentional movements. Both of these nuclei also function in cooperation with the motor cortex to control many of the patterns of movement.

3. The *globus pallidus* probably controls the "background" positioning of the gross parts of the body when a person begins to perform a complex movement pattern. That is, if a person wishes to perform a very exact function with one of his hands, he first positions his body appropriately and then tenses the muscles of the upper arm. These functions are believed to be initiated mainly by the globus pallidus.

4. The *subthalamic nucleus* and associated areas possibly control walking and perhaps other types of gross rhythmic body motions.

ABNORMAL MUSCLE CONTROL ASSOCIATED WITH DAMAGE TO THE BASAL GANGLIA

Even though we do not know all the precise functions of the basal ganglia, we do know many abnormalities that develop when portions of the basal ganglia are destroyed, as follows:

Chorea.   Chorea is random, uncontrolled sequences of motor movement occurring one after the other. Normal progression of movements cannot occur. Instead, the person may perform one pattern of movement for a few seconds and then suddenly jump to an entirely new pattern, such as changing from straight forward movement of the arm to a rotary movement, and then to a flailing movement, this jumping to a new pattern occurring again and again without stopping. This is caused by widespread damage in the *caudate nucleus* and *putamen*.

Athetosis.   Athetosis is characterized by slow, writhing movements of peripheral parts of the body. For instance, the hand or arm may undergo wormlike movements, such as twisting of the arm to one side, then to the other side, then backward, then forward, repeating the same pattern over and over again. The damage is always in the *globus pallidus*.

Hemiballismus.   Hemiballismus is an uncontrollable succession of violent movements of large areas of the body. For instance, a leg may suddenly kick forward, and this may be repeated once every few seconds, or sometimes only once in many minutes. The damage that causes this is in the *subthalamus*.

Parkinson's Disease.   This disease is characterized by *tremor* and *rigidity* of the musculature in either widespread or isolated areas of the body. The typical person with full-blown Parkinson's disease walks in a crouch like an ape, except that his muscles are obviously tense, his face is masklike, and he jerks all over with a violent tremor approximately 6 to 8 times per second. Yet when he attempts to perform voluntary movements, the tremor becomes less or stops temporarily. This disease is caused by destruction of the *substantia nigra,* a less conspicuous basal ganglion that lies anteriorly in the mesencephalon.

FUNCTION OF THE MOTOR CORTEX TO CONTROL MUSCLE MOVEMENTS

As we proceed to a still higher level of motor control, to the cerebral cortex, the types of movements controlled take on several new qualities. First, much more *discrete* movements can be initiated, particularly the very discrete movements by the hands which cannot be controlled by any of the lower centers. Second, the movements are more of the *learned type* rather than of the stereotyped. And, finally, a great majority of the muscle movements controlled by the cerebral cortex are of the *conscious type* rather than of the subconscious type.

THE MOTOR CORTEX

The area of the cerebral cortex most directly concerned with muscle control is the area shown in Figure 25–5 called the *motor cortex.* It is located immediately anterior to the central sulcus of the brain, a deep fissure that separates the front half of the cerebral cortex from the posterior half. This fissure also separates the motor cortex from the somesthetic cortex, which lies immediately behind the sulcus.

Stimulation of a point area in the motor cortex causes a specific muscle or a small closely allied group of muscles on the opposite side of the body to contract. Figures 25–5 and 25–6 show the points of the motor cortex for control of muscles in different parts of the body. Note that the muscles of the lower body are controlled by the cortex near the midline, while the muscles of the upper body are controlled far laterally. Thus, there is point-to-point communication between the motor cortex and specific muscles everywhere in the body. It is by means of signals from discrete parts of the motor cortex that discrete muscle movements are achieved, sometimes involv-

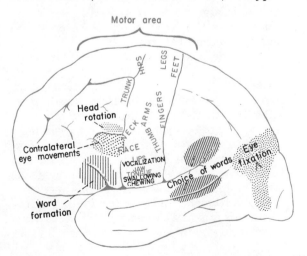

Figure 25–5.   Representation of the different muscles of the body in the motor cortex, and location of other cortical areas responsible for certain types of motor movements.

ing only single muscles, especially in the hands, but more often involving closely allied groups of muscles.

Degree of Representation of Different Muscles in the Motor Cortex. Not all muscles are represented in the motor cortex to the same degree. For instance, stimulation of a single small point in the trunk region of the motor cortex might excite contraction of a large area of back muscles, while stimulation of the same amount of cortical tissue in the finger area might cause nothing more than contraction of a single small finger muscle. Figure 25–6 illustrates diagrammatically the different degrees of representation for different types of muscles, showing that the degree of representation of muscles of the thumb and fingers and also of the mouth and throat regions is as much as 100 times that for the trunk muscles. The special representation of the hands allows the cerebral cortex to control with extreme fidelity essentially all of the fine, learned movements of the hands, which is a special capability of the human being and other higher primates. The high degree of representation in the mouth and throat regions accounts for the ability of the human being to talk, an ability that has not been achieved significantly by lower animals.

CONTROL OF SKILLED MOVEMENTS BY THE MOTOR CORTEX

Essentially all of the skilled movements performed by the human being are learned movements. Furthermore, most skilled movements require function of individual muscles during at least portions of the movement, which involves muscle control by the motor cortex.

Learning Skilled Movements. We all know how awkward a person is when he first begins to learn a highly skilled movement. For instance, a child learning to write letters works meticulously at achieving the appropriate curves and lines. This is a slow and almost painful process that requires a great amount of trial and error. All the while, the sensory nervous system senses the degree of success. The involved sensory signals include those for visual information, somesthetic information, auditory information, and all other types of information that might be useful in determining whether the motor act is successful. After an initial poor degree of success, subsequent attempts to perform the act, combined with sensory correcting signals, fi-

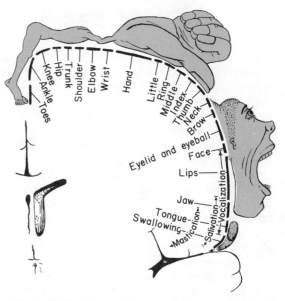

Figure 25–6. Degree of representation of the different body muscles in the primary motor cortex. (From Penfield and Rasmussen: The Cerebral Cortex of Man. The Macmillan Co.)

nally lead to perfection. Thus, the sensory system plays a major role in establishing all skilled, learned motor functions in the brain.

Establishment of Rapid Performance of Stereotyped Skilled Movements — the Concept of a "Premotor Area." As one becomes progressively more skilled in the performance of a particular learned act, he also achieves greater and greater speed in its performance. Sometimes it is important that the act be even more rapid than the time it takes for sensory information to be returned to the brain. For instance, in playing basketball a person performs many motor functions several milliseconds before receiving any feedback information from the sensory system to tell whether or not he is performing the acts properly.

To perform these very rapid acts, the cerebral cortex develops a storehouse of "patterns" of motor activities. These patterns can be called forth by the "thinking" part of the brain, and, once initiated, they will cause a discrete set of muscle movements to occur in an orderly sequence. For instance, when one is writing rapidly, he simply calls forth the pattern of one letter, then of another, still another, and so forth until each word is written. Most of us at times call for the wrong letter pattern and will write the entire letter before we recognize that we have called the wrong letter. Thus, the process of writing a word or sentence is in reality the calling forth of a sequence of patterns of motor activity.

The precise part of the brain that is utilized for storing the skilled patterns of activity is not known. It probably partly involves the motor cortex itself, and also some of the sensory areas, some of the areas of the frontal lobes immediately in front of the motor cortex, and some of the deeper centers for motor control as well, such as the basal ganglia. Of special interest is the area 1 to 3 cm. wide that lies immediately anterior to the motor cortex in the frontal lobes. Electrical stimulation of discrete points in this area frequently elicit skilled patterns of movement, especially skilled movements of the hands. Also located in this area are centers for control of eye movements and for formation of words during speech. Areas for control of some of these functions are indicated in Figure 25–5. These areas are frequently called the *premotor cortex,* and they are generally described as a part of the brain that is capable of producing learned patterns of movement that can be performed extremely rapidly.

Role of the Sensory Nervous System for Control of Very Complex Movements. It is already clear from the discussion of learned motor skills that the sensory portion of the brain plays a major role in establishing such skills. But, in addition to helping establish motor skills, the more complex skills can never be performed satisfactorily in the absence of the sensory system, because these skills require a sequence of successive patterns of movement and the memory loci for control of such sequences is located in the memory bank of the sensory system. For instance, the sensory area in Figure 25–5 labeled "choice of words" lies in the region of the brain known as the *common integrative area,* or the *gnostic area.* It is here that a person puts together sensory information from all sources to make complete meaning. Here, also, the person establishes the thoughts and words that he wishes to express either in written or spoken language. Thus, the sequence of words to express a thought is determined strictly by the sensory portion of the cortex.

COORDINATION OF MOTOR MOVEMENTS BY THE CEREBELLUM

Thus far, we have discussed muscle control without considering the role played also by the cerebellum. However, whether the motor movements be caused by signals generated initially in the bulboreticular area, in the basal ganglia, or in the cerebral cortex, the cerebellum always plays an important role in providing coordinate contraction of the individual muscles. The bases of cerebellar function are the following:

Movements of parts of the body are affected greatly by their inertia and momentum. That is, a limb requires a certain force to start it moving, but once started, it keeps on moving until an opposing force stops the motion. Neither the cerebral cortex nor the basal ganglia are organized to take these physical factors into consideration. Instead, the *cerebellum*

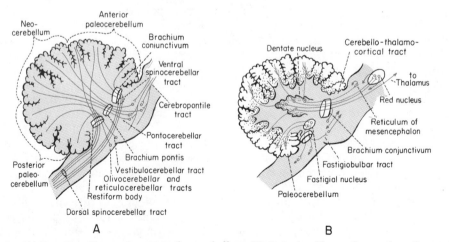

Figure 25–7.   (A) Incoming fiber pathways to the cerebellum. (B) Outgoing fiber pathways from the cerebellum.

makes the automatic adjustments that keep these factors from distorting the patterns of activity. The cerebellum, whose incoming and outgoing fiber pathways are shown in Figure 25–7, is a large structure located posterior to the brain stem. It receives signals from the proprioceptive receptors located in all joints, in all muscles, in the pressure areas of the body, and anywhere else that signals informing of the physical state of the body can be obtained. Signals are transmitted into the cerebellum, too, from the equilibrium apparatus of the ear, and even from the eyes to depict the visual relationship of the body to its surroundings. Finally, the cerebellum receives information directly from the motor cortex as well as from the basal ganglia of all motor signals that are being sent to the muscles. In summary, the cerebellum is a collecting house for all possible information on the instantaneous physical status of the body.

Figure 25–7A illustrates the major fiber tracts that transmit the various types of information into the cerebellum. These include (1) the *spinocerebellar tracts* that carry proprio-ceptor information from the body, (2) the *cortico-ponto-cerebellar tract* that carries motor information from the motor and premotor cortex to the cerebellum, (3) the *olivo-cerebellar tract* that carries signals from the basal ganglia, and (4) the *vestibulocerebellar tract* that carries impulses from the equilibrium apparatus.

Once the cerebellum has collated its information on the physical status of the body, it transmits its analysis to other areas of the brain through the fiber tracts shown in Figure 25–7B, including especially (1) the *fastigiobulbar tracts* that go from the cerebellum into many of the structures of the brain stem and (2) the *cerebello-thalamo-cortical tracts* and related tracts that pass to the thalamus and then to the motor regions of the cerebral cortex and to the basal ganglia.

Feedback Control of Motor Function. The cerebellum acts as a feedback mechanism to help control the muscle movements initiated by the motor cortex and basal ganglia. This is illustrated for the motor cortex in Figure 25–8, and was illustrated for the basal gan-

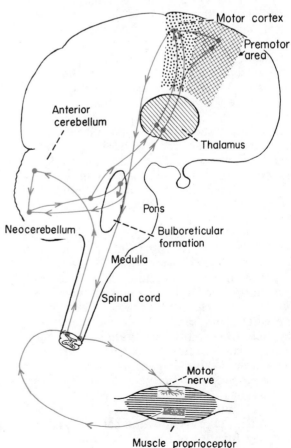

Figure 25–8. Feedback circuits of the cerebellum for damping motor movements.

glia in Figure 25–4. Figure 25–8 shows that the cerebellum receives information from the cortex of the muscular movements that it *intends to perform,* while simultaneously receiving proprioceptive information directly from the body apprising it of the movements *actually performed.* After comparing the intended performance with the actual performance, "corrective" signals are sent back to the motor cortex to bring the actual performance in line with the intended one. Similar effects also occur in the basal ganglia system.

Some of the specific functions of the cerebellum are the following:

Damping Function of the Cerebellum — Tremor in the Absence of the Cerebellum. When one moves his hand rapidly to a new position he can normally stop it exactly at the desired point, but without the cerebellum this cannot be accomplished. If the cerebellum has been removed, the momentum of the arm carries the hand beyond the projected point until some other part of the sensory system besides the cerebellum can detect the fact that the hand has gone too far. The eyes and the proprioceptor mechanisms of the somesthetic cortex finally do detect the overshoot, and they then set into play opposing muscular forces to bring the hand back to the projected point. Again momentum is built up during this return motion, and the hand moves too far in the opposite direction. Once more the sensory mechanisms detect the overshoot and bring the hand back to approach the point. Gradually the successive overshoots become less and less until the hand at last reaches the desired point. However, in the meantime the hand goes through a period of oscillation which is called a "voluntary" tremor because it is initiated by a voluntary muscle movement.

To prevent this effect of momentum, the cerebellum collects information from the moving parts of the body while they are actually moving and determines how much the momentum is affecting the movements. Then, even before each part reaches its destination, the cerebellum sends "feedback" signals to the motor cortex and basal ganglia to initiate appropriate "braking" contractions of opposing muscles to slow up and stop the movement at the proper point. In this way, the hand can be brought to rest at a desired position without the annoying overshoot. This overall mechanism to prevent overshoot is called the *damping* function of the cerebellum.

Ataxia. Lack of the damping function of the cerebellum causes the condition called *ataxia,* which means incoordinate contraction of the different muscles. For example, if a person who has lost his cerebellum tries to run, his feet will overshoot the necessary points on the ground for maintenance of equilibrium and he will fall. Even when walking, his gait will be very severely affected, for placement of the feet can never be precise; he falls first to one side, then overcorrects to the other side, and must correct again and again, giving him a broken gait. Ataxia can also occur in the hands and may be so severe that it becomes impossible for the person to write, to hammer a nail, or to perform any other precise movements without overshooting.

Predictive Function of the Cerebellum. Another function of the cerebellum closely allied to the damping function is its ability to predict the position of the different parts of the body. When the leg is moving very rapidly forward while a person is running, the cerebellum, operating in conjunction with the somesthetic cortex, predicts where the leg will be at each instant during the next few hundredths of a second. It is because of this prediction that the person can send appropriate signals to the leg muscles, directing the exact point on the ground where the foot is to be placed to keep him from falling to one side or the other.

The predictive function also applies to the relationship of the body to surrounding objects, for without a cerebellum a person running toward a wall cannot predict when he will likely reach the wall. Monkeys that have had portions of the cerebellum destroyed have been known to run so rapidly toward walls, being unable to predict how rapidly they are approaching, that they bash their brains.

Equilibrium Function of the Cerebellum. A specific portion of the cerebellum called the *flocculonodular lobes* is concerned with body equilibrium — that is, with maintenance of the body in the upright position against the pull of gravity. Other aspects of the equilibrium mechanism were considered in detail in the previous chapter. Signals are transmitted from the vestibular apparatus into the vestibular nuclei and bulboreticular area of the brain stem and also into the flocculonodular lobes of the cerebellum. The flocculonodular lobes receive equilibrium information especially from the semicircular canals. These organs help the person to anticipate that he will lose his equilibrium when he makes a change in direction of movement; and he corrects his movements ahead of time to prevent this from occurring. When the flocculonodular lobes are destroyed, the person loses this ability to predict that he will lose his equilibrium upon changing direction, so that he must perform all movements much more slowly than usual. A special type of tumor called the *medulloblastoma* frequently occurs in the flocculonodular region of the cerebellum in young children, and the first indication that a child has such a tumor is often a tendency for the child to lose his equilibrium.

THE CONTROL OF SPEECH

The major characteristic of human beings that sets them apart from other animals is their ability to communicate one with the other. This depends on two highly developed functions of the brain: first, the ability to interpret speech and, second, the ability to translate thought into speech. These communicative functions require the highest degree of operative perfection in almost all parts of the brain. Therefore, a description of the act of communication provides an overall review of the integrative functions of the central nervous system.

Interpretation of Communicated Ideas. Ideas are generally communicated from one person to another either by sounds or by written words. In the case of sounds, the information enters the *primary auditory cortex* in the superior part of the temporal lobe, as shown in Figure 25–9. The sounds then are interpreted as words and the words as sentences in the *auditory association areas.* The sentences are interpreted as thoughts in the *common integrative region.* Similarly, combinations of letters seen by the eyes are interpreted as words and the words as sentences by the *visual association areas,* and the sentences become thoughts in the *common integrative region* which was discussed in more detail in Chapter 23.

Motor Functions of Speech. The *common integrative region* of the brain develops thoughts that it wishes to communicate to someone else. It then operates in association with the lateralmost region of the *somesthetic sensory cortex* to initiate a sequence of signals, each signal representing perhaps a syllable or a whole word, and transmits these into the *motor* and *premotor areas* that control the larynx and mouth. The premotor area is called *Broca's area,* or it is sometimes called simply the *speech center.* It is actually no more a speech center than the common inte-

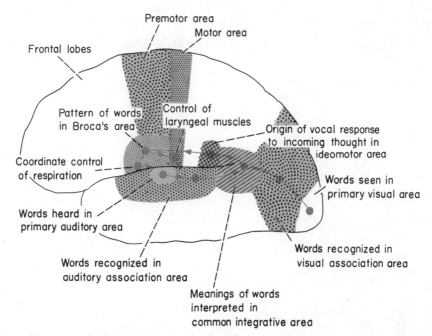

Figure 25–9. Pathways in the brain for communication, showing reception of thoughts by the auditory and visual pathways and control of speech by motor pathways.

grative area, but it is here that the patterns for forming different sounds by the larynx and mouth are controlled. Signals arriving from the common integrative area set off a sequence of patterns in the speech center, and these in turn form the words. In addition to controlling the larynx and mouth, Broca's area also sends signals into an allied region of the secondary motor cortex that controls respiration. Therefore, at the same time that the laryngeal and mouth movements occur, the respiratory muscles are contracted to provide appropriate air flow for the speech process.

Thus, the person's thoughts are formulated into speech.

REFERENCES

Allen, G. I., and Tsukahara, N.: Cerebrocerebellar communication systems. *Physiol. Rev., 54*:957, 1974.

Asanuma, H.: Cerebral cortical control of movement. *Physiologist, 16*:143, 1973.

Asanuma, H.: Recent developments in the study of the columnar arrangement of neurons within the motor cortex. *Physiol. Rev., 55*:143, 1975.

Cooper, I. S., Riklan, M., and Snider, R. S. (eds.): The Cerebellum, Epilepsy, and Behavior. New York, Plenum Publishing Corporation, 1974.

Denny-Brown, D.: The Basal Ganglia, New York, Oxford University Press, Inc., 1962.

Eccles, J. C.: The Understanding of the Brain. New York, McGraw-Hill Book Company, 1973.

Evarts, E. V.: Brain mechanisms in movement. *Sci. Amer., 229*:96, 1973.

Grillner, S.: Locomotion in vertebrates: central mechanisms and reflex interaction. *Physiol. Rev., 55*:247, 1975.

Llinas, R.: Eighteenth Bowditch lecture. Motor aspects of cerebellar control. *Physiologist, 17*:19, 1974.

Porter, R.: Influences of movement detectors on pyramidal tract neurons in primates. *Ann. Rev. Physiol., 38*:121, 1976.

Stein, P. S. G.: Motor systems with specific reference to the control of locomotion. *Ann. Rev. Neurosci., 1*:61, 1978.

QUESTIONS

1. Describe the basic organization of the higher centers for muscle control.
2. Describe the function of the interneurons and anterior motor neurons in the spinal motor segments.
3. How do the basal ganglia control subconscious movements?
4. Give the functions of the individual basal ganglia: the caudate nucleus, the putamen, the globus pallidus, and the subthalamic nucleus.
5. Describe the function of the motor cortex in the control of muscle movements.
6. How does one learn skilled movements, and what is the function of the premotor area?
7. What role does the cerebellum play in the coordination of motor movements?
8. Why does ataxia occur in patients who have major lesions of the cerebellum?
9. Trace the input of speech to the brain, and give the neuronal motor mechanisms for output of speech.

THE AUTONOMIC NERVOUS SYSTEM AND THE HYPOTHALAMUS

26

In our discussion of the nervous system thus far, we have considered mainly the sensory nervous system and the control of muscle movements. However, the nervous system plays still another entirely different and extremely important role: control of many of the involuntary bodily functions such as arterial blood pressure, heart rate, movements of the digestive glands, and even such functions as dilation of the pupils of the eyes.

The portion of the nervous system that is responsible for these internal control functions is called the *autonomic nervous system*. Some of these functions have been described in relation to different organs already discussed, including especially control of arterial pressure. The purpose of the present chapter is to present the overall function of the autonomic nervous system, pointing out its special relationships to other nervous system functions.

SYMPATHETIC AND PARASYMPATHETIC DIVISIONS OF THE AUTONOMIC NERVOUS SYSTEM

The autonomic nervous system has two separate divisions, the *sympathetic* and the *parasympathetic*. The differences between these two are: First, the anatomic distributions of the nerve fibers in the two divisions are distinct from each other. Second, the stimulatory effects of the two divisions on the organs are often antagonistic to each other. Third, the types of hormones secreted at the nerve endings are usually different in the two systems.

ANATOMY OF THE SYMPATHETIC NERVOUS SYSTEM

Figure 26–1 illustrates the anatomy of the sympathetic nervous system, showing that a *sympathetic chain* lies to each side of the spinal cord, and the chain is connected with both the cord and peripheral organs. The sympathetic chain is linked with the spinal cord in a peculiar manner, shown in Figure 26–2. Sympathetic fiber tracts leave the cord through the anterior roots of the spinal nerve. Then, after traveling less than a centimeter, they pass through a small whitish nerve called the *white ramus* into the sympathetic chain. From here, fibers travel in two directions. Some pass into visceral *sympathetic nerves* that innervate the internal organs of the body, while the others return through another small nerve called the *gray ramus* back into the spinal nerve. These latter fibers then travel all through the body along the spinal nerves to supply the blood vessels, the sweat glands, and even the piloerector muscles that cause the hairs to stand on end.

Sympathetic fibers enter the sympathetic chain from the spinal cord only in the thoracic and upper lumbar regions, and none enter the chain in the neck, lower lumbar, or sacral regions. To supply the head with sympathetic innervation, sympathetic fibers from the thoracic chain extend upward into the neck and

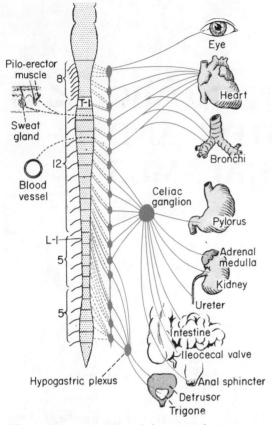

Figure 26–1. Anatomy of the sympathetic nervous system. (The dashed lines represent the gray rami.)

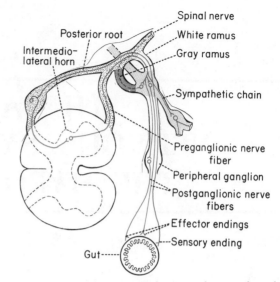

Figure 26–2. Connections between the spinal cord, a spinal nerve, and the sympathetic chain.

then to all the structures of the head. Also, sympathetic fibers pass downward from the chain into the lower abdomen and legs.

Visceral Sensory Fibers in the Sympathetic Nerves.   Sensory fibers also pass through the sympathetic nerves. These arise in the internal organs, then enter the sympathetic nerves, and finally travel by way of the white rami into the spinal nerves. From here they enter the posterior horns of the cord gray matter and either cause sympathetic cord reflexes or transmit sensations to the brain in the same manner that sensations are transmitted from the surface of the body. These sensations are the visceral sensations that were considered in detail in Chapter 23.

Preganglionic and Postganglionic Neurons of the Sympathetic Nervous System.   Figure 26–1 shows bulbous enlargements located periodically along the sympathetic chain and at different points along the sympathetic nerves after they leave the chain. These enlargements are called *ganglia,* and they contain neuronal cell bodies.

Sympathetic signals are transmitted from the spinal cord to the periphery through two successive neurons. The cell body of the first neuron is located in the spinal cord in the lateral gray matter. The fiber from this neuron, called the *preganglionic fiber,* passes into the sympathetic system as illustrated in Figure 26–2. It synapses with a second neuron in either a ganglion of the sympathetic chain or a more peripheral ganglion. The fiber from the second neuron, called the *postganglionic fiber,* passes directly to the organ to be controlled. The first neuron, located in the cord, is called the *preganglionic neuron;* the second, located in the ganglion, is called the *postganglionic neuron.* Thus, the sympathetic motor system is different from the skeletal motor system, for skeletal muscles are stimulated through a single neuron rather than through two.

ANATOMY OF THE PARASYMPATHETIC NERVOUS SYSTEM

Figure 26–3 shows the parasympathetic nervous system. The fibers of this system originate mainly in the tenth cranial nerve, called the *vagus nerve.* However, a few fibers originate in the third, seventh, and ninth cranial nerves, and also in several of the sacral segments of the spinal cord. The vagus nerve supplies parasympathetic fibers to the heart, the lungs, and almost all of the organs of the

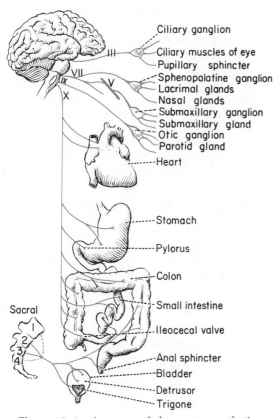

Ciliary ganglion
Ciliary muscles of eye
Pupillary sphincter
Sphenopalatine ganglion
Lacrimal glands
Nasal glands
Submaxillary ganglion
Submaxillary gland
Otic ganglion
Parotid gland
Heart
Stomach
Pylorus
Colon
Small intestine
Ileocecal valve
Anal sphincter
Bladder
Detrusor
Trigone
Sacral

Figure 26–3. Anatomy of the parasympathetic nervous system.

abdomen. The other cranial nerves supply parasympathetic fibers to the head, and the sacral fibers supply the urinary bladder and the lower parts of the colon and the rectum. However, since approximately 90 per cent of all the parasympathetic fibers of the body pass through the vagus nerve, most physiologists, in thinking about the parasympathetic system, think almost automatically of the vagus nerve itself.

Preganglionic and Postganglionic Neurons of the Parasympathetic System. The parasympathetic system is like the sympathetic in that signals must travel through preganglionic and postganglionic neurons before stimulating the various organs. However, there is a different anatomic arrangement of the postganglionic neuron as follows: The cell bodies of the preganglionic neurons are in the brain stem or sacral cord, but their fibers usually pass all the way to the organ to be stimulated instead of to a ganglion as in the sympathetic system. The postganglionic neurons are located in the wall of the organ, and the postganglionic fibers then travel only a few milli-

meters before they reach their final destination on smooth muscle fibers or glandular cells.

ADRENERGIC VERSUS CHOLINERGIC NERVE FIBERS IN THE AUTONOMIC NERVOUS SYSTEM

One of the major differences between parasympathetic and sympathetic nerves is that the postganglionic fibers of the two systems usually secrete different hormones. The postganglionic neurons of the parasympathetic system secrete *acetylcholine*, for which reason these neurons are said to be *cholinergic*. Those of the sympathetic system secrete mainly *norepinephrine;* most postganglionic neurons of the sympathetic system are said to be *adrenergic,* a term derived from noradrenalin, which is the British name for norepinephrine. However, a few are cholinergic like the parasympathetic neurons, as is noted below.

Excitatory and Inhibitory Effects of the Autonomic Hormones. Both acetylcholine and norepinephrine have the capability of exciting some internal organs while inhibiting others. Frequently, if one of these hormones excites an organ, the opposite one inhibits it; but this is not always the case. Therefore, in the following sections we will present the effects of stimulation by both the sympathetic and the parasympathetic nervous systems on most of the important organs and functional systems of the body.

PARASYMPATHETIC AND SYMPATHETIC TONE

Impulses normally are transmitted continuously through all the fibers of both the parasympathetic and sympathetic systems. This allows at least some degree of continuous stimulation of the internal structures, which is called *sympathetic tone* or *parasympathetic tone*. The tone allows nerve stimulation to exert either positive or negative control on a structure. That is, by increasing the number of impulses above the normal value, the stimulation effect can be increased; on the other hand, by decreasing the number of impulses below the normal, the effect can be decreased. As an example, if the sympathetic vasoconstrictor fibers to the blood vessels were normally dormant, it would be possible for sympathetic regulation only to constrict the vessels and never to dilate them; however, because of the normally persistent tone, the blood vessels are always partially constricted so that the sympathetics can either further constrict the vessels by increasing their stimulation or dilate the vessels by decreasing their stimulation. The principles apply throughout the parasympathetic and sympathetic systems and are responsible for a higher degree of effectiveness than would be possible otherwise.

ACTIONS OF THE PARASYMPATHETIC AND SYMPATHETIC NERVOUS SYSTEMS ON DIFFERENT ORGANS

Effect on the Eye.   Table 26–1 lists the effects on different organs caused by sympathetic or parasympathetic stimulation. This table shows that the sympathetic system di-

lates the pupil of the eye, allowing increased quantities of light to enter, and the parasympathetics constrict the pupil, thus decreasing the amount of light. The parasympathetics also control the ciliary muscle that focuses the lens for far and near vision, which will be discussed in more detail in Chapter 28.

Secretion of the Digestive Juices.   The secretion of digestive juices by some of the glands of the gastrointestinal tract is controlled mainly by the parasympathetics, while the sympathetics have very little effect on most of the glands. The salivary glands of the mouth and the fundic glands of the stomach normally are almost entirely controlled by the parasympathetics. On the other hand, the glands in the intestines are controlled only to a slight extent by the parasympathetics, but mainly by local factors in the intestines themselves.

Effects on the Sweat Glands.   The sweat

Table 26–1.   Autonomic Effects on Various Organs of the Body

ORGAN	EFFECT OF SYMPATHETIC STIMULATION	EFFECT OF PARASYMPATHETIC STIMULATION
Eye: Pupil	Dilated	Contracted
Ciliary muscle	None	Excited
Gastrointestinal glands	Vasoconstriction	Stimulation of thin, copious secretion containing many enzymes
Sweat glands	Copious sweating (cholinergic)	None
Heart: Muscle	Increased activity	Decreased activity
Coronaries	Vasodilated	Constricted
Systemic blood vessels:		
Abdominal	Constricted	None
Muscle	Dilated (cholinergic)	None
Skin	Constricted or dilated (cholinergic)	None
Lungs: Bronchi	Dilated	Constricted
Blood vessels	Mildly constricted	None
Gut: Lumen	Decreased peristalsis and tone	Increased peristalsis and tone
Sphincters	Increased tone	Decreased tone
Liver	Glucose released	None
Kidney	Decreased output	None
Bladder: Body	Inhibited	Excited
Sphincter	Excited	Inhibited
Male sexual act	Ejaculation	Erection
Blood glucose	Increased	None
Basal metabolism	Increased up to 50%	None
Mental activity	Increased	None
Adrenal medullary secretion	Increased	None

glands are stimulated by fibers from the sympathetic nervous system. However, these fibers are different from the usual sympathetic fibers, for they are mainly cholinergic rather than adrenergic. Also, they are stimulated by nervous centers in the brain that normally control the parasympathetics, rather than by the centers that control the sympathetics. Therefore, despite the fact that the fibers supplying the sweat glands are anatomically sympathetic, they can be considered physiologically to function like many parasympathetic fibers.

Effects on the Heart. Stimulation of the sympathetic nervous system increases the heart's activity, and also dilates the coronaries so that increased nutrition will be available to the more active heart muscle. On the other hand, parasympathetic stimulation decreases the activity of the heart while constricting the coronaries to conserve blood flow. These effects were discussed in Chapter 13.

Control of the Blood Vessels. Perhaps the most important function of the sympathetic nervous system is to control the blood vessels in the body. Most vessels are constricted by sympathetic stimulation, though a few, the coronaries for instance, are dilated. By controlling the peripheral blood vessels, the sympathetic nervous system is capable of regulating both the cardiac output and arterial pressure; constriction of the veins and venous reservoirs increases the cardiac output, and constriction of the arterioles increases the peripheral resistance, which elevates the arterial pressure.

The parasympathetics, when they affect the blood vessels at all, usually dilate them, but this effect is so slight and occurs in such a few areas of the body that it can be almost totally ignored.

Effect on the Lungs. The bronchi are dilated by sympathetic stimulation. However, the sympathetic system has almost no effect (very slight vasoconstriction) on the blood vessels of the lungs. This is different from the very strong effect on the blood vessels of the remainder of the body.

Control of Gastrointestinal Movements. About 75 per cent of all the parasympathetic nerve fibers are distributed to the gastrointestinal tract, which indicates that by far the most important function of this entire system is regulation of gastrointestinal activities. Parasympathetic stimulation increases peristalsis and at the same time decreases the tone of the gastrointestinal sphincters. Peristalsis propels the food forward while the open sphincters between the different segments of the gastrointestinal tract allow the food to move forward with ease. During extreme parasympathetic stimulation, food can actually pass all the way from the mouth to the anus in approximately 30 minutes, though the normal transmission time is almost 24 hours.

Sympathetic stimulation, on the other hand, inhibits peristalsis and tightens the sphincters. This slows the movement of food through the gastrointestinal tract.

Release of Glucose from the Liver. Sympathetic stimulation causes rapid breakdown of glycogen into glucose in the liver and then release of the glucose into the blood. This increased glucose in the blood provides a quick supply of nutrition for the tissue cells, an effect especially valuable during exercise.

Effect on the Kidneys. Sympathetic stimulation causes intense vasoconstriction of the renal blood vessels and greatly decreases the output of urine. This is a very important mechanism for regulation of blood volume and arterial pressure, for when need be, sympathetic stimulation can cause fluid to be retained in the circulatory system, increasing the blood volume and venous return to the heart as well. These effects, over a period of hours or days, also increase the cardiac output and raise the arterial blood pressure.

Emptying of the Bladder. Emptying of the bladder is caused mainly by parasympathetic stimulation, which excites the muscular wall of the bladder and at the same time inhibits the urethral sphincter that normally holds back the flow of urine. On the other hand, sympathetic stimulation prevents emptying of the bladder. Though the sympathetic effect ordinarily is not important, occasionally when a person has severe peritoneal inflammation in the region of the bladder, a peritoneal reflex excites the sympathetics so greatly that the person becomes unable to urinate.

Control of Sexual Functions. The autonomic nervous system also helps to control the sexual acts of both the male and the female. In the male the parasympathetics cause erection, and the sympathetics cause ejaculation. In the female the parasympathetics cause erection of the erectile tissue around the vaginal opening, which causes tightening, and they also cause the female to secrete large quantities of mucus which facilitates the sexual act. The effect of the sympathetics on the female sexual act is not well understood, but it is believed that these nerves might initiate reverse uterine peristalsis during the female climax.

Metabolic Effects. Generalized sympathetic stimulation increases the metabolism of all cells of the body. The sympathetic nerve fibers are so widespread in all tissues that at least some norepinephrine secreted by the nerve endings seems to reach every functioning cell. The norepinephrine increases the rates of the chemical reactions in all cells, thereby increasing the overall rate of metabolism of the body. In this way, the sympathetics can keep a person warm when he tends to become too cold, and during exercise or other states of activity they can make the body perform greater quantities of work than would be possible otherwise. These metabolic effects can be brought about in only a few seconds, and they can be stopped in another few seconds when the need for increased metabolism is over.

Stimulation of Mental Activities. Another important effect of sympathetic stimulation is an increase in the rate of mental activity. This probably results from the heightened rate of metabolism in the neuronal cells.

Sympathetic Stimulation of the Adrenal Medulla — Secretion of Epinephrine and Norepinephrine. The *adrenal medulla* is the central portion of the adrenal gland; one adrenal gland is located on each side of the body immediately above the kidney. Surrounding the medulla is the adrenal cortex, which is an entirely different gland and which will be discussed in Chapter 35. The secretory cells of the medulla are modified postganglionic sympathetic neurons, and these cells, when stimulated, secrete *epinephrine* and *norepinephrine* into the blood. These two hormones then circulate in the blood and are distributed to all cells of the body. The circulating norepinephrine has the same stimulatory or inhibitory effects as the norepinephrine released directly in the tissues by the sympathetic nerves. And the epinephrine has almost the same effects. Therefore, these two hormones in combination produce almost the same effects on all the organ systems of the body as direct sympathetic nerve stimulation. For instance, they increase the activity of the heart, inhibit peristalsis in the gut, increase the metabolism of all cells, and so forth.

The adrenal medulla mechanism thus provides a second means for causing all or most of the sympathetic activities. During normal sympathetic stimulation, the quantity of epinephrine and norepinephrine secreted by the adrenal medullae is enough to produce between one-quarter and one-half of all the sympathetic effects in the body. In fact, when all the sympathetic nerves besides those to the adrenal medullae are destroyed, the sympathetic system will still function almost normally because the adrenals compensate by secreting even more hormones. Likewise, loss of adrenal medullary secretion is hardly discernible because of compensation by the nerves. To stop the action of the sympathetic nervous system entirely, the functions of both the adrenal medullae and of all the direct sympathetic nerve fibers to the organs must be blocked at the same time.

OVERALL FUNCTION OF THE SYMPATHETIC NERVOUS SYSTEM

Mass Discharge to Prepare the Body for Activity. In general, when the sympathetic centers of the brain become excited they stimulate almost all of the sympathetic nerves at once. As a result, the blood pressure rises, the rate of metabolism increases, the degree of mental activity is enhanced, and the glucose in the blood increases. Considering all of these effects together, it can be seen that mass discharge of the sympathetic nervous system prepares the body for activity. When a person's sympathetic nerves have been destroyed, he cannot "generate steam"; in other words, he simply cannot reach the high levels of vigor often necessary to perform rapid, forceful, and excited activities.

Special Functions of the Sympathetic Nervous System. In addition to the mass discharge mechanism of the sympathetic system, certain physiologic conditions can stimulate localized portions of the system independently of the remainder. Two of these are the following:

HEAT REFLEXES. Heat applied to the skin causes a reflex through the spinal cord and then back again to the skin to dilate the blood vessels in the heated area. Also, heating the blood flowing through the heat control centers of the hypothalamus increases the degree of vasodilatation of all the skin blood vessels without significantly affecting the vessels deep in the tissues.

SHIFT OF BLOOD FLOW TO THE MUSCLES DURING EXERCISE. During exercise, the increased metabolism in the muscles has a loca effect of dilating the muscle blood vessels, but at the same time the sympathetic system constricts the vessels in most other parts of the body. Nerve impulses pass directly from the motor cortex to the sympathetic regions of the brain stem to initiate this effect. The vasodilatation in the muscles allows blood to flow through them with great ease, while vasoconstriction elsewhere, except in the heart and brain, decreases almost all other bodily blood flow. Thus, in exercise the sympathetic nervous system helps to cause massive shift of blood flow to the active muscles.

CONTROL OF THE AUTONOMIC NERVOUS SYSTEM BY NERVE CENTERS IN THE BRAIN — ESPECIALLY THE HYPOTHALAMUS

Not much is known about the exact centers in the brain that regulate the discrete autonomic functions. However, stimulation of widespread areas in the reticular substance of the medulla, pons, and mesencephalon can elicit essentially all of the effects of the sympathetic and parasympathetic system. Also, the hypothalamus can produce most of the functions of the autonomic system. And, finally, stimulation of localized areas of the cerebral cortex can occasionally cause general or sometimes discrete autonomic effects.

IMPORTANT AUTONOMIC CENTERS OF THE BRAIN STEM

Figure 26–4 shows a tentative outline of major centers in the hindbrain and hypothalamus that regulate different functions of the internal organs. It will be noted from this

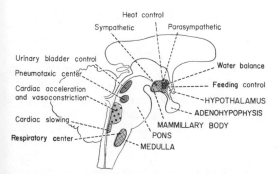

Figure 26–4. The major autonomic control centers of the brain stem and hypothalamus.

figure that the vascular system and respiration are controlled mainly by areas in the reticular substance of the medulla and pons. The urinary bladder is controlled by centers in the mesencephalon, but these centers are usually suppressed by areas in the cerebral cortex that provide conscious regulation of bladder emptying.

AUTONOMIC CENTERS OF THE HYPOTHALAMUS

Figure 26–5 gives much more detailed locations of different control centers in the hypothalamus. Some of the specific functions of the hypothalamus are the following:

Cardiovascular Regulation. In general, stimulation of the *posterior hypothalamus* increases the arterial pressure and heart rate, while stimulation of the preoptic area in the anterior portion of the hypothalamus has exactly opposite effects, causing marked decrease in both heart rate and arterial pressure. These effects are transmitted through the cardiovascular control centers of the lower brain stem and thence through the autonomic nervous system.

Regulation of Body Temperature. Centers in the *anterior hypothalamus* are directly responsive to blood temperature, becoming more active with increasing blood temperature and less active with decreasing temperature. These areas in turn regulate the body temperature, the details of which will be described in Chapter 33. Basically this mechanism (a) controls the amount of blood flow through the skin, thereby controlling the rate of heat loss from the skin; (b) controls sweating, which cools the body; (c) controls shivering, which greatly increases heat production; (d) controls secretion of norepinephrine and epinephrine by the adrenal medullae, which also stimulate heat production; and (e) controls thyroid hormone production, which has a direct effect on heat production in all cells of the body.

Regulation of Body Water. A center located in the *supraoptic nucleus* controls antidiuretic hormone secretion which in turn controls rate of water loss through the kidneys. This mechanism was discussed in detail in Chapter 18 in relation to kidney function.

A closely allied region of the hypothalamus, the *lateral hypothalamus,* controls drinking. It does this by transmitting signals into other parts of the brain to create the conscious feel-

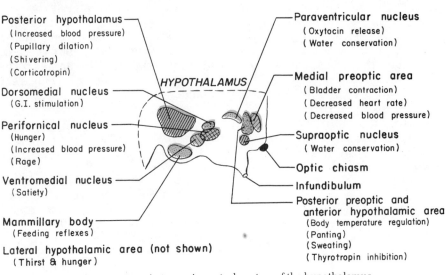

POSTERIOR   ANTERIOR

Posterior hypothalamus ———   ┌── Paraventricular nucleus
 (Increased blood pressure)   │   (Oxytocin release)
 (Pupillary dilation)   │   (Water conservation)
 (Shivering)
 (Corticotropin)   *HYPOTHALAMUS*   ┌── Medial preoptic area
                                               │   (Bladder contraction)
Dorsomedial nucleus ———   │   (Decreased heart rate)
 (G.I. stimulation)   │   (Decreased blood pressure)

Perifornical nucleus ———   ┌── Supraoptic nucleus
 (Hunger)   │   (Water conservation)
 (Increased blood pressure)
 (Rage)   ┌── Optic chiasm

Ventromedial nucleus ———   ─── Infundibulum
 (Satiety)
                                               ┌── Posterior preoptic and
Mammillary body ———   │   anterior hypothalamic area
 (Feeding reflexes)   │   (Body temperature regulation)
                                               │   (Panting)
Lateral hypothalamic area (not shown)   │   (Sweating)
 (Thirst & hunger)   │   (Thyrotropin inhibition)

Figure 25–5.   Autonomic control centers of the hypothalamus.

ing of thirst, thereby making the person seek water.

Regulation of Feeding. Stimulation of areas in the *lateral hypothalamus* also causes an animal to have a voracious appetite and an intense desire to search for food. This area is frequently called the *hunger center*. Located more medially, in the *ventromedial nuclei,* are areas which when stimulated block all feeling of hunger. Therefore, these areas are called the *satiety center*.

Control of Excitement and Rage. In the very middle of the hypothalamus is a nucleus called the *perifornical nucleus* which, when stimulated, causes an animal to become greatly excited and develop elevated arterial blood pressure, dilated pupils, and symptoms of rage such as hissing, arching his back, and assuming a stance ready to attack. Thus, the hypothalamus plays a major role in overall behavior of an animal.

Hypothalamic Control of Endocrine Functions. Some areas of the hypothalamus also secrete so-called *neurosecretory substances.* These are then transported through special veins from the hypothalamus down to the anterior pituitary gland where they promote formation of different anterior pituitary hormones. It is in this way that the hypothalamus controls the secretion of most anterior pituitary hormones and thereby controls many of the metabolic functions of the body. These neurosecretory substances and their

more specific functions will be discussed in Chapter 34.

Summary. A number of discrete areas of the hypothalamus control specific "vegetative" functions of the body. However, these areas overlap so much that the separation of different areas for different hypothalamic functions as presented here is partially artificial. Most important of all, the hypothalamus plays perhaps the key role in setting the basic tenor of bodily function, the basic degree of excitement, the basic level of metabolism.

IMPORTANT CEREBRAL CENTERS

Stimulation of the autonomic nervous system by the cerebral cortex occurs principally during emotional states. Many discrete centers, especially in the prefrontal lobes and temporal regions of the cortex, can increase or decrease the degree of excitation of the hypothalamic centers. Also, the thalamus and closely related structures deep in the cerebrum help to regulate the hypothalamus. Therefore, both conscious and subconscious portions of the cerebrum can cause autonomic effects. Sudden psychic shock, which originates in the cerebral cortex, can cause fainting. This usually results from powerful stimulation of cardiac inhibitory fibers (vagus nerves) which cause the arterial pressure to fall precipitously. On the other hand, extreme degrees of psychic excitement, which can result from either conscious or subconscious stimulation in the cerebrum, can stimulate the vasomotor system to increase the arterial pressure and cardiac output.

Unfortunately, our present knowledge of the relation of the cerebrum to the autonomic nervous system is so scanty that it can be only descriptive rather than explanatory of the actual functional relationships.

REFERENCES

Bennett, M. R.: Autonomic Neuromuscular Transmission. Physiological Society Monograph, 1972.

Black, I. B.: Regulation of autonomic development. *Ann. Rev. Neurosci., 1*:183, 1978.

Haber, E., and Wrenn, S.: Problems in identification of the beta-adrenergic receptor. *Physiol. Rev., 56*:317, 1976.

Hayward, J. N.: Functional and morphological aspects of hypothalamic neurons. *Physiol. Rev., 57*:574, 1977.

Hess, W. R.: The Functional Organization of the Diencephalon. New York, Grune & Stratton, Inc., 1958.

Paton, D. M. (ed.): The Mechanism of Neuronal and Extraneuronal Transport of Catecholamines. New York, Raven Press, 1975.

Patterson, P. H.: Environmental determination of autonomic neurotransmitter functions. *Ann. Rev. Neurosci., 1*:1, 1978.

Reichlin, S., Saperstein, R., Jackson, I. M. D., Boyd, A. E., III, and Patel, Y.: Hypothalamic hormones. *Ann. Rev. Physiol., 38*:389, 1976.

Steer, M. L., Atlas, D., and Levitzki, A.: Interrelations between beta-adrenergic receptors, adenylate cyclase, and calcium. *N. Engl. J. Med., 292*:409, 1975.

Westfall, T. C.: Local regulation of adrenergic neurotransmission. *Physiol. Rev., 57*:659, 1977.

QUESTIONS

1. Describe the sympathetic and parasympathetic divisions of the autonomic nervous system.
2. Explain the preganglionic and postganglionic neurons of the sympathetic and parasympathetic systems.
3. What is the difference between adrenergic and cholinergic nerve fibers?
4. Give the effects of parasympathetic stimulation and of sympathetic stimulation on the following: (1) the eye, (2) secretion of digestive juices, (3) the sweat glands, (4) the heart, (5) the blood vessels, (6) gastrointestinal movements, (7) emptying of the bladder, and (8) body metabolism.
5. Discuss the role of the adrenal medulla in the function of the sympathetic nervous system.
6. What is the role of the lower brain stem in the regulation of autonomic activities?
7. Describe the functions of the hypothalamus for control of the autonomic nervous system.

27 INTELLECTUAL PROCESSES; SLEEP AND WAKEFULNESS; BEHAVIORAL PATTERNS; AND PSYCHOSOMATIC EFFECTS

Another title for this chapter could be "How do the different parts of the brain function together?" That is, what makes us think? What makes our levels of consciousness increase and decrease during sleep and wakefulness? What influences our inner feelings? And, how do our intellectual processes express themselves in the form of bodily function?

Unfortunately, it is these integrative aspects of the brain that are least well understood, for which reason much of what we believe is based on inference from psychological tests rather than from direct experiments in the brain itself.

ROLE OF THE CEREBRAL CORTEX IN THE THINKING PROCESS

At several points we have discussed various aspects of cerebral function and its relation to some phases of the thinking process. To review briefly, the cerebral cortex is in one sense a large outgrowth of the thalamus and related basal areas of the brain, and there are direct communications back and forth between respective portions of the cerebral cortex and thalamus. Whenever a particular part of the thalamus is stimulated, a corresponding part of the cerebral cortex usually

also becomes stimulated. Therefore, we believe that the thalamus has the ability to call forth activity in specific portions of the cerebral cortex. In addition, it is in the cerebral cortex that most of the neurons of the central nervous system are located, 9 billion out of a total of only 12 billion in the entire brain. Therefore, we must ask ourselves the question: What is the purpose of this great mass of neurons in the cerebral cortex? The answer to this seems to be that the cerebral cortex is a vast storehouse for memories, a place where data can be collected and held for days, months, or years until needed at a later time.

Interpretation of Incoming Signals. The cerebral cortex is a large sheet of thin folded tissue lying on the surface of the brain and composed of six separate neuronal cell layers as illustrated in Figure 27–1, each layer performing a different function. Some of these layers send nerve fibers into deeper areas of the brain; others send fibers to adjacent regions of the cortex. It is believed, also, that some of the layers act as *comparators* to compare new incoming signals with memories stored from the past. That is, if the incoming information fits exactly with past memory information that is already stored, one theoretically will immediately identify the new information as something that had been

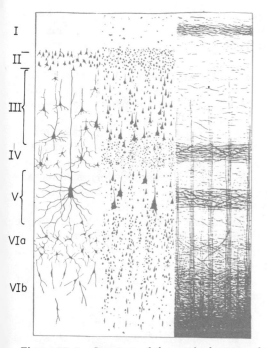

Figure 27-1. Structure of the cerebral cortex, showing to the left the layers of cell bodies and to the right the layers of connecting fibers. (From Ranson and Clark (after Brodmann): Anatomy of the Nervous System. W. B. Saunders Co.)

experienced before. Furthermore, if the previous experience had been associated with some specific quality of sensation such as pain or happiness, displeasure or pleasure, then this too might be remembered.

Nature of Thoughts. We all know intuitively what a thought is, but, even so, a thought is very difficult to define neurophysiologically. To attempt this, let us first describe what happens in the brain when a person sees with his eyes for the first time a new, exciting visual scene. Upon opening his eyes to the scene, nerve signals are transmitted by way of the optic nerve first into many areas at the base of the brain, including the *lateral geniculate body,* the *superior colliculus,* the *lateral thalamus,* and *midregions of the mesencephalon.* From the lateral geniculate body, secondary impulses are transmitted almost immediately to the visual cortex located in the occipital lobe, as described in Chapter 28. These first signals to the cortex terminate in the *primary visual cortex,* but a few milliseconds later signals reach the visual *association cortex,* and soon thereafter they reach the *angular gyrus,* the *hippocampus,* the *prefrontal* cortex, and other parts of the brain. It is the totality of all these signals that gives one the thought expressed by the visual scene.

Therefore, a thought can be described as a specific pattern of signal transmission throughout the brain.

Every thought has specific qualities. It may be pleasant or it may be unpleasant. It may have an element of luminosity, or it may have an element of sound. It may have a sensation of tactile feeling, or it may have a sensation of taste. It may have a characteristic of repetitiveness or of uninterrupted continuity. There is much reason to believe that specific areas in the brain, many of which are located in the basal regions of the brain, interpret these individual qualities. For instance, the feeling of pleasure or displeasure seems to be interpreted by several closely associated but antagonistic areas located in the midportions of the mesencephalon and hypothalamus and surrounding areas. When a visual scene or other type of sensory input to the brain is pleasant, the pleasure areas of these basal regions receive a signal. If they are unpleasant, the displeasure areas receive a signal.

Finally, every thought has its own specific details, such as contrasts between light and dark areas in the visual scene, or contrasts between different frequencies of sound or different intensities of tactile sensations from one second to the next. Memories of these details seem to be stored in the cerebral cortex; therefore, it is presumably the cerebral cortex that is responsible for identification of the fine details of thoughts.

Memories. Memory is the ability to recall thoughts that were originally initiated by incoming sensory signals. Probably most of the memory process occurs in the cerebral cortex, primarily because three-quarters of the neurons of the brain are located here. Yet, we know that essentially every area of the central nervous system can participate in the phenomenon of memory. Indeed, experiments have shown that even the spinal cord can hold crude memories for at least as long as a few minutes to perhaps a few hours.

But what is the nature of a memory? And what causes a memory to persist sometimes for a short time and sometimes for a very long time? There appear to be at least two major types of memory, which can be called *short-term memory* and *long-term memory.*

SHORT-TERM MEMORY. Short-term memory may be defined as persistence of an incoming thought for a few seconds or a few minutes without causing any permanent imprint on the brain. Some short-term memories are possibly caused by continued reverberation of signals within the brain for a short time after the initiating sensation is gone. That is, the incoming thought stimulates neuronal cells connected in reverberating circuits. These cells stimulate secondary cells which then stimulate tertiary cells, and finally the signal gets back to the original cells. Thus, the signal goes around and around the circuit for seconds or minutes after the incoming sensation is over, and as long as these reverberations persist, the person still retains the thought in his mind. In support of this concept is the fact that such reverberating signals can be demonstrated for as long as half an hour back and forth between the cerebral cortex and the thalamus after a strong sensory signal barrages the brain. Furthermore, these reverberating signals are localized to certain areas of the cerebral cortex depending upon the type of sensation that is experienced.

LONG-TERM MEMORY. We all know that some memories last for years even though reverberating signals in the brain cannot possibly persist longer than about an hour at most. These long-term memories almost certainly result from the following mechanism: When a signal passes through a particular set of neuronal synapses, these synapses become *facilitated* for passage of similar signals at a later date. Therefore, when a thought enters the brain, it facilitates those synapses that are used for that particular thought, and this makes it easier for one to recall that same thought at some later date.

Yet, passage of a signal through a synapse only one time usually will not cause sufficient facilitation for the thought to be remembered. Therefore, we must ask the question: How is it that a single sensory experience lasting for only a few seconds can sometimes be remembered for years thereafter? One possible answer to this seems to be that the reverberating signals of the short-term memory mechanism send this same thought through the same synapses many thousands of times for up to an hour after the initial sensory experience is over. A reason for believing this is that if an animal is struck on the head immediately after experiencing a very strong sensory experience, the blow to the head can block the reverberating signals and thereby stop the short-term memory. In such an instance, the long-term memory also fails to develop. Therefore, a persisting short-term memory for a period of at least a few minutes seems to be essential to develop the long-term memory "imprint" or "engram." This develop-

ment of the long-term memory engram is called *consolidation* of the memory. Once this engram has been established, almost any stray signal in the brain can at some later date set off a sequence of signals exactly like those originally initiated by the incoming sensation, whereupon the person experiences the same original thought.

ROLE OF "REHEARSAL" IN TRANSFERRING SHORT-TERM MEMORY INTO LONG-TERM MEMORY. Psychological studies have shown that rehearsal of the same information again and again accelerates and potentiates the degree of transfer of short-term memory into long-term memory. This fits with the above reverberatory theory of short-term memory, because each reverberation is actually a form of continued rehearsal of the same information. Indeed, the brain has a natural tendency to rehearse newly found information, especially that which catches the mind's attention. The importance of rehearsal for establishing long-term memories explains why a person can remember information that is studied in depth far better than he can remember large amounts of information studied only superficially. And it also explains why a person who is wide awake will "consolidate" memories far better than will a person who is in a state of mental fatigue.

Codification of Memories During the Storage Process. While memories are being rehearsed and stored, they are also codified. That is, similar memories that are already stored are called forth and compared with the new memory. And the process involves storing not only the new memory but also its differences from and similarities to the previous memories. Thus, memories are not stored randomly in the mind, but instead are stored in direct association with other memories of the same type, and presumably in unique areas of the cortex. This is obviously necessary if one is to be able to scan the memory store at a later date to find the required information.

Determination of Which Memories to Remember — Role of the Hippocampus and of Pain or Pleasure. Determination of whether or not a memory will be stored for years or is almost immediately forgotten seems to be made by some of the basal regions of the brain and not by the cerebral cortex. The hippocampus, a very old portion of the cerebral cortex located bilaterally on the floor of the lateral ventricle, is essential for storage of many if not most memories. If one has experienced the same thought many times before and, there fore, has become habituated to it, the hippocampus fails to be stimulated. On the other hand, if the thought evinces either pain or pleasure or some other very strong quality, then the hippocampus does become stimulated. In ways not yet understood, this elicitation of signals in the hippocampus is believed to cooperate with other basal regions of the brain to cause the permanent storage of memory.

Because of the importance of the hippocampus for storage of long-term memories, one can easily understand the fact that large brain lesions involving both hippocampi will cause a person to be unable to, or at least find it difficult to, store long-term memories. Therefore, he forgets everything as rapidly as he learns it. This effect is called *anterograde amnesia*.

Recall of Memories After They Have Been Stored. Another great mystery about the memory process is: How do we recall memories once they have been stored? It was pointed out above that memories are codified as they are stored, and they seem to be stored in close association with other memories of the same type. It is also known that the brain has mechanisms whereby its attention can be focused on a succession of stored memories, one after another. This process is called searching or *scanning* of the memory storehouse.

There are several reasons for believing that the *thalamus* is strongly involved in the searching process. First, the thalamus has point-to-point connections with essentially all parts of the cerebral cortex, which would provide easy access to specific memory storage loci. Second, it is known that waves of stimulation travel through the thalamic nuclei and that these cause similar waves of stimulation also to travel through the cerebral cortex; one might imagine that these play some role in the searching process. Third, and most important of all, lesions in the thalamus frequently cause the person to experience difficulty or inability in recalling memories that are known to have been stored at an earlier date. This is called *retrograde amnesia*. It is presumed that these lesions in the thalamus have interfered seriously with the searching process. Sometimes, damage in the thalamic area (or closely associated lower cerebral areas) will cause retrograde amnesia that lasts for days or months and then disappears, presumably because of recovery of the searching process.

Knowledge. Many persons will be surprised to know that the human being is born already having certain types of knowledge. For instance, a newborn baby knows to suck on the breast and even to search for the breast. He knows to cry when pained and to smile, even without being trained to do so, when pleased. Some of the lower animals are born with still other types of knowledge such as the ability to stand and walk and the ability to search out food. Indeed, a major share of the useful knowledge of some lower animals is inherited.

Yet, the human mind is born with much less knowledge than that of many lower animals. Instead, knowledge accumulated in the mind of the human being is generally of an adaptive type based on previous experience, made up almost entirely of memories rather than based on inherited neuronal connections. It is this difference between the lower animal mind and the human mind that gives the human being his great breadth of abilities, one person becoming a mathematical genius, another a vast storehouse of linguistic data, another a depository of jokes, and so forth.

But we also know that the process of forgetfulness allows the character of knowledge in the mind to change with time so that a person's mind may be a great depository of book learning in his school days and yet in later years be filled with practical experience that takes the place of forgotten book learning. Psychological tests show that the quantity of knowledge in a person's mind generally increases during the first 39 years of his life, reaching a peak at this time. Beyond that age the total amount of stored knowledge gradually declines. This does not mean, though, that the amount of stored knowledge of a specific subject might not continue to increase on into very old age.

The Neuronal Cellular Mechanism of Memory — Long-term Facilitation of the Synapse. In the above discussion of memory consolidation, it was pointed out that long-term memories can be stored only when the same sequence of signals passes through the involved synapses many times, probably thousands or even millions of times. Because of this, it is believed that the basic neuronal mechanism of memory is a process of long-term facilitation of the synapses. This concept is also supported by animal experiments. For instance, prolonged repetitive electrical stimulation of nerve fibers entering a neuronal pool will eventually cause that neuronal pool to develop a high degree of sensitivity to subsequent stimulation. This phenomenon is called *posttetanic facilitation,* but one can readily see that it is actually a memory process.

On the other hand, the way in which synapses become increasingly sensitive to successive incoming signals is still a big question. There are two basic theories. The first of these is that prolonged stimulation causes the terminal nerve fibrils in a synaptic pool to sprout forth new terminals, and these in turn provide progressively more synaptic knobs on the surface of the postganglionic neuron. A variant of this theory is that the synaptic knobs themselves enlarge or change their chemical or physical characteristics in such a way that they provide increased quantities of excitatory transmitter substance. This theory that the synaptic knobs increase in number or size is supported by the fact that electronmicroscopic counts of terminal fibrils and synaptic knobs are greater in brain areas that have been purposely stimulated over long periods of time than in brain areas not so stimulated.

The second theory is that the neuronal membrane on the postsynaptic side of the synapse becomes altered, thereby changing the sensitivity of the neuron to incoming signals. One variant of this theory is that synaptic sensitivity is heightened as a result of an increase in ribonucleic acid in the neuronal cell. Here again, experiments have actually demonstrated such an increase in ribonucleic acid, but it is still to be proved that this substance is actually the basis of the memory process rather than some nonspecific effect of greater neuronal activity.

Therefore, as we stand at present, our knowledge of perhaps the most important single function of the brain, memory, is very scanty. When we understand this process more fully, our total understanding of how the mind works will certainly be greatly enhanced.

FUNCTION OF THE COMMON INTEGRATIVE AREA OF THE CEREBRAL CORTEX IN THE THOUGHT PROCESS

It was pointed out in Chapter 23 that most of the sensory input signals to the brain eventually funnel information into the common integrative area located in the posterior portion of the *temporal lobe* and adjacent *angular gyrus region* in the brain's dominant hemisphere, usually the left hemisphere. The primary and association areas for somesthetic, visual, auditory, taste, and smell sensations are all located very near to this common integrative area. Therefore, it is ideally located for integrating the information from all these different sources.

Destruction of the common integrative area of an adult almost completely destroys his intellect and his ability to perform useful functions. For this reason, this portion of the cortex is perhaps the one most important area of all. In a child, destruction of this area is not nearly so serious because the temporal lobe and angular gyrus in the opposite hemisphere of the brain can then develop the same functions, but in the adult this opposite hemisphere has become suppressed so that its functions will not interfere with those of the dominant hemisphere. It is very difficult to develop this region once the process of suppression has taken place.

The fact that the common integrative area is very important to intellectual function of the brain does not mean that it is in this area that all important memories are stored. It merely means that this is one of those key points in the brain where, if the neuronal connections are destroyed, the other parts of the brain cannot function satisfactorily. Other aspects of sensory function in the common integrative area were discussed in Chapter 23. And its function of helping control muscle movements was discussed in Chapter 25.

FUNCTION OF THE PREFRONTAL AREAS OF THE CEREBRAL CORTEX IN THE THOUGHT PROCESS

The prefrontal areas of the cerebral cortex are the most anterior portions of the brain lying in the front 5 to 7 cm. of the frontal lobes, anterior to the motor control areas discussed in Chapter 25. These areas are frequently called "silent" areas because no gross overt effects occur when the prefrontal areas are destroyed. Instead, the person simply loses a major share of what we generally call his intellectual ability, especially his ability for abstract thought.

Function of the Prefrontal Areas in Abstract Thinking. An animal that has lost its prefrontal areas loses the ability to keep small bits of information in its mind for longer than a few seconds at a time. For in stance, if food is placed on one side of the cage and its attention is drawn to the other side, it forgets the locus of the food when it is no longer looking at it. It is believed that the prefrontal areas in the human being perform this same function, holding many small bits of information for short periods of time. In performing mathematical problems, it is necessary to store information temporarily during each stage of the logical process into a small corner of the mind and then to come back to this information a few seconds or a few minutes later. Thus, after storing and processing several hundred bits of information, the problem can be solved. It is believed that the prefrontal areas are the storehouse of these short-lived memories.

Essentially the same mechanisms are used in abstract thinking related to legal processes, diagnosis of diseases by the physician, or analysis of complex business problems.

The Prefrontal Areas as the Locus of Ambition, Conscience, Planning, and Worrying. Obviously, without the ability to think through the consequences of one's actions a person might act in too great haste and perform activities which he would regret in the future. Thus, it is well demonstrated that persons who have lost their prefrontal areas appear also to have lost much of their conscience.

Likewise, planning for the future, worrying about one's activities, or developing ambition are dependent upon his ability to think through interrelationships of all his actions. Here again, the person who loses his prefrontal areas likewise loses all these qualities. On the other hand, he at least is often exempt from prolonged worry for the remainder of his life.

Effect on Behavior Caused by Loss of the Prefrontal Areas. Occasionally the prefrontal areas are completely destroyed by disease or trauma, or sometimes they have been purposely destroyed surgically because of harmful patterns of thought that have developed and cannot be stopped in other ways. When the prefrontal areas are gone, the common integrative area is left by itself in charge of the thought patterns of the brain, without the sobering influences of the prefrontal areas. The person is then likely to exhibit extreme reactions, some of which are very happy in nature while others have the characteristics of raging temper. Often the responses are rapid and are likely to lead to disastrous results. As long as the person is unprovoked he has very much the same personality as a giddy teenager, responding emotionally to everything but without much thought. When provoked, he is likely to fall into a state of rage which cannot be easily quelled at the moment, but which will be totally forgotten a few minutes later.

SLEEP AND WAKEFULNESS

Another of the important mysteries of brain function is its diurnal cycle of sleep and wakefulness. Even when a person remains in total darkness or in total light he still maintains approximately the same sleep-wakefulness cycle, with a periodicity of once every 24 hours. Ordinarily the nervous system shows signs of fatigue shortly before sleep ensues, and it shows signs of having become considerably rested at the termination of sleep. It seems, then, that neuronal fatigue plays an important part in causing sleep, and that sleep, in turn, relieves the fatigue.

Electrical studies of the brain indicate that while a person is awake many nerve impulses pass continuously through the nervous system, never ceasing. However, during most stages of sleep, much fewer impulses are present. Thus, the state of wakefulness seems to be caused by a high degree of activity in the cerebrum, while the state of sleep is caused by a low degree of activity. Therefore, any theory to explain sleep and wakefulness must also explain these changing degrees of cerebral activity during the two states.

Stimulation of the Brain by the "Reticular Activating System." Stimulation of portions of the mesencephalon (midbrain) and thalamus greatly increases the activity of the cortex, an effect shown diagrammatically in Figure 27–2. Stimulation anywhere along the arrows will cause impulses to spread upward and eventually to excite the cortex. This system is called the *reticular activating system,* and it is divided into two separate parts, the *mesencephalic part* and the *thalamic part.*

The mesencephalic part of the reticular activating system is composed mainly of the reticular substance in the mesencephalon and the reticular substance of the upper pons. Stimulation of this region causes very diffuse flow of impulses upward through widespread areas of the thalamus and thence to widespread areas of the cortex, causing generalized increase in cerebral activity.

The thalamic portion of the reticular activating system differs from the mesencepha-

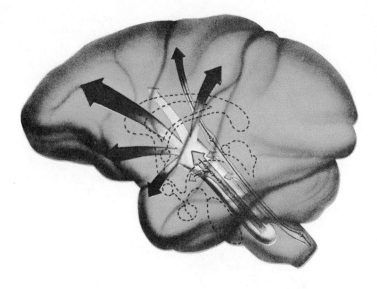

Figure 27–2. The reticular activating system, showing by the arrows passage of impulses from the reticular substance of the mesencephalon upward through the thalamus to all parts of the cortex. (From Lindsley: Reticular Formation of the Brain. Little, Brown and Co.)

lic part in that stimulation here activates *localized* regions of the cerebral cortex. Stimulation posteriorly in the thalamus activates posterior parts of the cerebral cortex while stimulation anteriorly activates anterior parts of the cortex. Thus, signals from specific parts of the thalamus can call forth activity in specific parts of the cortex rather than activating the entire cortex.

It is likely that the function of the thalamic portion of the reticular activating system is to direct one's attention to memories or thoughts stored in specific parts of the cerebral cortex and thereby to allow his consciousness to consider individual thoughts at a time. On the other hand, *activation of the mesencephalic portion of the reticular activating system causes generalized wakefulness.*

THE AROUSAL REACTION. The reticular activating system must itself be stimulated to action by input signals from other sources. When an animal is in very deep sleep, the reticular activating system is in an almost totally dormant state; yet almost any type of sensory signal can immediately activate the system. For instance, pain stimuli from any part of the body, or proprioceptor signals from the vestibular apparatuses, from the joints, and so forth can all cause immediate activation of the reticular activating system. This is called the *arousal reaction,* and it is the means by which sensory stimuli awaken us from deep sleep.

CEREBRAL STIMULATION OF THE RETICULAR ACTIVATING SYSTEM. Signals from the cerebral cortex can also stimulate the reticular activating system and thereby increase its activity. Fiber pathways to both the mesencephalic and thalamic portions of this system are particularly abundant from the *somesthetic cortex,* the *motor cortex,* the *frontal cortex,* and parts of the cortex that deal mainly with emotions, the *cingulate gyrus* and the *hippocampus.* Whenever any of these cortical regions becomes excited, impulses are transmitted into the reticular activating system, thereby increasing the degree of activity of the reticular activating system.

THE FEEDBACK THEORY OF WAKEFULNESS AND SLEEP

From the above discussion we can see that activation of the reticular activating system intensifies the degree of activity of the cerebral cortex, and, in turn, greater activity in the cerebral cortex increases the degree of activity of the reticular activating system. Therefore, it is fairly evident that a so-called "feedback loop" could develop whereby the reticular activating system excites the cortex and the cortex in turn reexcites the reticular activating system, thus setting up a cycle that causes continued intense excitation of both of these regions. Such a cycle is shown in Figure 27–3.

The reticular activating system is also involved in another feedback loop as follows: It sends impulses down the spinal cord to activate the muscles of the body. In turn, the muscle activation excites proprioceptors tha

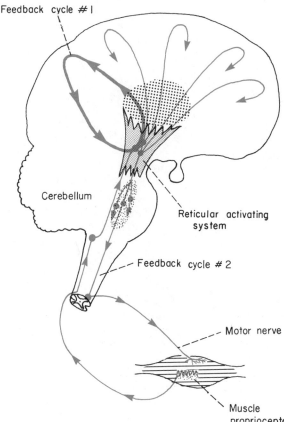

Feedback cycle #1

Cerebellum

Reticular activating system

Feedback cycle #2

Motor nerve

Muscle proprioceptor

Figure 27–3. The feedback theory of wakefulness, showing two feedback cycles passing through the reticular activating system: one to the cerebrum and back (#1), and another to the peripheral muscles and back (#2).

send sensory impulses upward to reexcite the reticular activating system. Thus, here again, activity of the reticular activating system sets off another cycle that in turn further increases the activity in the reticular activating system. This cycle is also shown in Figure 27–3.

One can readily see, therefore, that once either one or both of these feedback loops involving the reticular activating system should become excited, the resulting increased activity in the reticular activating system would stimulate all other parts of the brain, thereby creating a state of wakefulness. This is one of the plausible theories to explain the state of wakefulness.

On the other hand, one might wonder how *sleep* could possibly occur once the different feedback loops should become activated. A possible answer to this is that all synapses of the central nervous system eventually fatigue, so that they either stop or nearly stop transmitting impulses. Therefore, after a prolonged period of wakefulness, one would ex-

pect the synapses in the feedback loops to become fatigued and, consequently, the feedback loops to stop functioning, thus allowing the reticular activating system to become dormant, which is the state of *sleep*.

We also know that a person can exist in all levels of wakefulness and sleep. The feedback theory of wakefulness postulates that the degree of wakefulness depends on the number of feedback loops activated at any one time. Figure 27–3 illustrates a large number of activating pathways passing from the mesencephalic activating system upward through the thalamus to the cortex. If only one of these pathways becomes activated, then obviously the degree of wakefulness would be very slight. If large numbers of the pathways should become activated simultaneously, then the degree of wakefulness would be greatly enhanced. One would expect that after a long period of wakefulness, progressively more and more of the feedback loops would become inactive and a person would fade into a state of sleep. Then, upon arousal, all of the feedback loops might be activated simultaneously, which explains why one could go from a state of sleep to complete wakefulness in a few seconds.

OTHER THEORIES OF WAKEFULNESS AND SLEEP

Though other theories of wakefulness and sleep have not been propounded quite so succinctly as the feedback theory, many neurophysiologists nevertheless, believe that the wakefulness and sleep cycle is based on a cyclic change in some chemical factor secreted into the reticular activating system by specific neurons. Presumably one of the hormonal factors would act as an inhibitory agent to cause sleep while another would act as an excitatory agent to cause wakefulness. Some of the hormones that have been suggested to be involved in such sleep and wakefulness cycles include two well-known synaptic transmitter substances, serotonin for sleep and norepinephrine for wakefulness. Also, a small sleep-producing *polypeptide* has been isolated from the cerebrospinal fluid of animals kept awake for many days. This peptide will put an animal to sleep within minutes after it is injected into the ventricular system of the brain.

In summary, cyclic changes in the hormonal or biochemical status of local neurons in the reticular activating system could well be the cause of the increased and decreased activity of the system during the wakefulness-sleep cycle. Or it is possible that this cycle is caused by some complex combination of the feedback system described above and the hormonal factors.

REM SLEEP, AND THE PROCESS OF DREAMING

Periodically during sleep a person passes through a state of dreaming associated with mild involuntary muscle jerks and rapid eye movements. The rapid eye movements have given this stage of sleep the name *REM sleep*. Electroencephalograms recorded during these periods, strangely enough, show considerable activity in the brain. Yet, muscle tone throughout the body at this time is diminished almost to zero, the heart rate may be as much as 20 beats below normal, and the arterial pressure may be 30 mm. Hg below normal. Thus, the person seems physiologically

to be in very deep sleep rather than light sleep despite the fact that he is dreaming and despite his active electroencephalogram.

REM sleep usually occurs three to four times during each night at intervals of 80 to 120 minutes, each occurrence usually lasting from 5 to 30 minutes. As much as 50 per cent of the infant's sleep cycle is composed of REM sleep; and in the adult approximately 20 per cent is REM sleep.

The cause of REM sleep and its cyclic pattern is still unknown. It has been claimed by some psychologists that in the absence of REM sleep a person develops severe psychic instability. However, this theory seems now to have been disproved. Instead, it seems that serious deficiency of the deep, non-REM sleep is, instead, the culprit in many psychic disorders.

A postulated function of REM sleep is that it represents a periodic test of the degree of excitability of the nervous system. That is, approximately every one and one-half hours this cycle recurs. And if the neurons of the brain are sufficiently rested, the increased neuronal activity during REM sleep can theoretically awaken the person. This theory is supported by the fact that dreams do frequently awaken persons and also by the fact that paradoxical sleep occurs much more frequently and for more prolonged periods of time as morning approaches.

EFFECTS OF WAKEFULNESS

Wakefulness is associated with three major effects: (1) increase in the degree of activity of the cerebrum, (2) transmission of signals directly from the reticular activating system to the muscles, and (3) excitation of the sympathetic nervous system.

The increase in cerebral function is caused by the great number of impulses transmitted from the wakefulness center upward through the thalamus to the cerebral cortex. These impulses continually impinge on the cerebral neurons, facilitating them to activity.

Stimulation of the reticular areas of the brain stem increases the degree of tone in all the muscles of the body. This makes the muscles more excitable than would otherwise be true and prepares them for immediate activity, which explains why a person who is wide awake has the feeling of muscle readiness.

The sympathetic stimulation caused by wakefulness elevates the blood pressure slightly; it increases the rate of metabolism in all the tissues of the body; and, in general, it simply makes the body ready to perform increased amounts of work.

EFFECTS OF SLEEP

Sometimes it is hard to understand why a person needs to sleep at all. Certain parts of the body such as the heart never rest and still are capable of functioning throughout life. One might reason that sleep is a measure to conserve the energies of most parts of the body when they are not needed. Some animals, as a matter of fact, have carried this principle of conservation so far that they pass into a state of very prolonged and deep sleep called *hibernation,* which lasts throughout the entire winter.

A special value of sleep seems to be to reestablish appropriate balance of excitability among the various portions of the nervous system. As a person becomes progressively fatigued, some parts of his central nervous system lose excitability more than others, so that one part may overbalance the others. In fact, extreme fatigue can even precipitate severe psychotic disturbances. Yet after prolonged sleep, all parts of the nervous system usually will have returned once again to appropriate degrees of excitability and to a state of serenity.

Lack of sleep does not *directly* affect the intrinsic functions of the different organs. However, lack of sleep often causes severe autonomic disturbances, and these in turn *indirectly* lead to gastrointestinal upsets, loss of appetite, and other detrimental effects. In this way, loss of sleep can affect the whole body as well as the nervous system itself.

BRAIN WAVES

Electrical impulses called *brain waves* can be recorded from all active parts of the brain and even from the outside surface of the head. The character of these waves is closely related to the level of sleep and wakefulness at the time that the waves are recorded. When a person is awake but not thinking hard, continuous waves at a rate of approximately 10 to 12 per second can be recorded from almost all parts of the cerebral cortex. These are called *alpha waves.* The brain signals that cause them probably originate in the reticu-

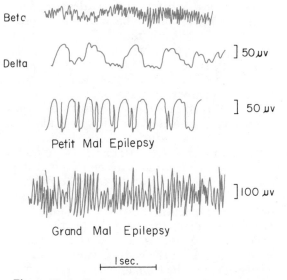

Figure 27–4. Brain wave patterns in the normal person and in two persons with different types of epilepsy.

lar activating system and then spread into the cerebral cortex. These are believed to be the signals that keep the cortex facilitated during wakefulness. The alpha waves are illustrated by the recording at the top of Figure 27–.4.

When any part of the brain becomes very active — for example, the motor region initiating muscular activities — additional waves having a frequency sometimes as high as 50 cycles per second, and intensities often greater than those of the alpha waves, take the place of the normal alpha waves. These, called *beta waves,* are shown by the second recording of Figure 27–4.

During very deep sleep the alpha and beta waves are replaced by a few straggling waves occurring approximately once every 1 to 2 seconds. These are the "sleep waves" or *delta waves,* as shown by the third recording of Figure 27–4.

Abnormal Brain Wave Patterns in Epilepsy. Various abnormalities of the brain can cause strange brain wave patterns. Two of these, caused by different types of *epilepsy,* are shown in the fourth and fifth records of Figure 27–4. The fourth record shows a "spike and dome" picture which occurs in *petit mal* epilepsy. In this disease the person suddenly becomes unconscious for 3 to 10 seconds at a time. Such episodes may occur every few minutes, every few hours, or only once in many months, and when they do occur the person usually continues, while unconscious, whatever physical activity he is already doing even though he might be walking across a crowded street. Petit mal epilepsy seems to result from some abnormality of the reticular activating system of the brain. The transmission of the normal alpha waves to the cerebral cortex is temporarily stopped. Instead, the spike and dome pattern is transmitted, and the person falls asleep for a few seconds until the alpha wave pattern picks up again.

The bottom record of Figure 27–4 shows the brain waves in *grand mal* epilepsy. In this condition the cerebral cortex becomes extremely excited, and many very strong signals spread over the brain at the same time. When these reach the motor cortex they cause rhythmic movements, called *clonic convulsions,* throughout the body. Grand mal epilepsy possibly results from abnormal reverberating cycles developing in the reticular activating system. That is, one portion of the system stimulates another portion, which stimulates a third portion, and this in turn restimulates the first portion, causing a cycle that continues for 2 to 3 minutes until the neurons of the system fatigue so greatly that the reverberation ceases. At the beginning of a grand mal attack a person may experience very violent, abnormal, hallucinatory thoughts; once most of the brain is involved, he no longer has any conscious thoughts at all, because signals are then being transmitted in all directions rather than through discrete thought circuits. Therefore, even though his brain is violently active, he becomes unconscious of his surroundings. Following the attack his brain is so fatigued that he sleeps at least a few minutes and sometimes as long as 24 hours.

BEHAVIORAL FUNCTIONS OF THE BRAIN: THE LIMBIC SYSTEM

Behavior is a function of the entire nervous system, not of any particular portion. However, most involuntary aspects of behavior are controlled by the so-called *limbic system,* which is illustrated in block diagram form in Figure 27–5. This figure shows most of the central structures of the basal brain with a surrounding ring of cerebral cortex called the *limbic cortex.* Most of the cerebral cortex lies still beyond this ring. This ring of limbic cor-

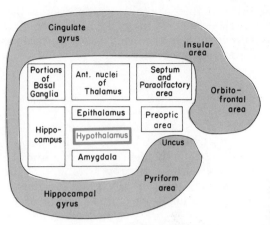

Figure 27–5.   The limbic system.

tex consists of (1) the *uncus,* the *pyriform area,* and the *hippocampal gyrus* on the very bottom of the brain, (2) the *cingulate gyrus* lying deep in the longitudinal fissure of the brain, and (3) the *insular* and *orbital frontal areas* lying on the bottom anterior portion of the brain. All of these areas of the cerebral cortex are phylogenetically old — that is, they were among the earliest portions of the cerebral cortex to evolve in primitive animals.

Perhaps the most important part of the limbic system, from the point of view of behavior, is the *hypothalamus,* the autonomic functions of which were discussed in the previous chapter. Many of the surrounding portions of the limbic system, including especially the hippocampus, the amygdala, and the thalamus, transmit major portions of their output signals into the hypothalamus to cause varied effects in the body, such as to stimulate the autonomic nervous system or to participate in causing such feelings as pain, pleasure, or sensations related to feeding, sex, anger, and so forth.

Some aspects of limbic control are transmitted through the endocrine system, for it will be remembered that the hypothalamus, in addition to controlling the autonomic nervous system, also controls secretion of many of the pituitary hormones. This will be discussed in detail in Chapter 34, but for the time being it should be noted that the limbic system operating through the hypothalamus and anterior pituitary gland can control (1) the rate of secretion of all sex hormones, which together control the various sexual drives of the person, and (2) the rates of secretion of thyroid hormone, growth hormone, and various adrenocortical hormones, which together control most of the person's day-by-day cellular metabolic functions.

Therefore, the limbic system can be said to control the inner being of the person.

With this background in mind, now let us discuss some of the specific mechanisms for control of behavior by the limbic system.

PLEASURE AND PAIN; REWARD AND PUNISHMENT

One of the most important recent discoveries in the field of behavior is the so-called "pleasure and pain" or "reward and punishment" system of the brain. Certain areas in the mesencephalon, hypothalamus, and other closely associated areas, when stimulated, make the animal feel intense punishment as if he were being severely pained. Yet stimulation of other closely related areas causes exactly the opposite effect, making the animal appear to be experiencing extreme pleasure.

One of the experimental methods for studying the reward and punishment centers is to implant an electrode in one or the other of these two areas and then to allow the animal itself to press a lever to control the stimulus, as illustrated in Figure 27–6. If the electrode is implanted in a reward area, the animal will press the lever continually. Indeed, it would rather stimulate its reward center than eat, even though it might be starving. On the other hand, if the electrode is placed in a punishment area, it will avoid stimulation by all means possible.

Relationship of Reward and Punishment to Learning. Experiments on memory have shown that an animal remembers sensory stimuli that cause either reward or punishment but fails to remember sensory stimuli that do not excite either the reward or punishment area. For instance, a food that is very pleasant to the taste is remembered, or, likewise, a food that is exceedingly unpleasant is remembered. On the other hand, food that causes neither pleasure nor displeasure is rapidly forgotten. Similarly, a very painful stimulus, such as touching a hot iron, is remembered well, whereas simply touching a book or stick of wood is forgotten in a few seconds.

Therefore, in the process of memory, two different components must be present for a sensory experience to be remembered. The first component is the sensory experience it-

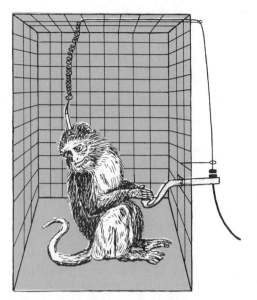

Figure 27–6. Technique for localizing reward and punishment centers in the brain of a monkey.

self, and the second component is an experience of either reward or punishment — that is, an experience of pleasure or pain.

FUNCTION OF THE LIMBIC CORTEX

Even though the limbic cortex is the oldest part of the cerebral cortex, its precise function is least understood of almost all portions of the brain. Electrical stimulation in different parts of the limbic cortex can cause such effects as excitement, depression, increased movement, decreased movement, on rare occasions the phenomenon of rage, on other occasions intense degrees of docility, and so forth. However, these effects cannot be elicited repeatedly from specific points in the limbic cortex. Therefore, it is very difficult to state precise functions for the limbic cortex.

The probable function of the limbic cortex is to act as an *association area* for control of most of the behavioral functions of the body. It presumably stores information about past experiences such as pain, pleasure, appetite, various smells, sexual experiences, and so forth. This store of information is then presumably combined with other information channeled into the limbic areas from surrounding regions of the cerebral cortex, such as from the prefrontal areas and from the sensory areas of the posterior part of the brain. This association of information then presumably provides stimuli for initiating appropriate behavioral responses for each respective occasion, whether this behavior be rage, docility, excitement, lethargy, or so forth. Thus, we believe this to be the part of the cerebral cortex that plays the greatest role in controlling the emotions and other patterns of behavior.

THE DEFENSIVE PATTERN – RAGE

Stimulation of the *perifornical nuclei of the hypothalamus,* located in the very middle of the hypothalamus, gives an animal an extreme sensation of punishment and simultaneously causes it to (1) develop a defense posture, (2) extend its claws, (3) lift its tail, (4) hiss, (5) spit, (6) growl, and (7) develop piloerection, wide-open eyes, and dilated pupils. Furthermore, even the slightest provocation causes an immediate savage attack. This is the pattern of behavior that has been called simply *rage*. It can occur in decorticated animals, illustrating that the basic behavioral patterns for defense and rage are controlled from the lower regions of the brain, especially from the hypothalamus and the mesencephalon.

Exactly opposite emotional behavioral patterns occur when the reward centers are stimulated, namely, docility and tameness. During such stimulation, the animal becomes completely amenable to almost any type of treatment.

FUNCTIONS OF THE AMYGDALA

The amygdala is a complex nucleus located immediately beneath the surface of the cerebral cortex in the anterior pole of each temporal lobe. In lower animals, the amygdala is concerned primarily with smell stimuli, but in human beings it is much larger and operates in very close association with the hypothalamus to control many behavioral patterns. It is believed that the normal function of the various amygdaloid nuclei is to help control the overall pattern of behavior demanded for each social occasion.

Stimulation of various parts of the amygdala can transmit signals through the hypothalamus to cause (1) increase or decrease in arterial pressure, (2) increase or decrease in heart rate, (3) increase or decrease in gastrointestinal activity, (4) defecation or urination, (5) pupillary dilatation or constriction, (6) piloerection, which means hair standing on end, or (7) secretion of various adenohypophyseal hormones.

In addition, the amygdala can transmit signals to areas of the lower brain stem to cause (1) changes in the degree of muscle tone throughout the body, (2) postural movements, such as raising the head or bending the body, (3) circling movements, (4) rhythmic movements, or (5) movements associated with eating, such as licking, chewing, and swallowing.

Finally, excitation of other portions of the amygdala can cause sexual excitement, including erection, copulatory movements, ejaculation, ovulation, uterine activity, and premature labor.

From this foregoing list of functions, one

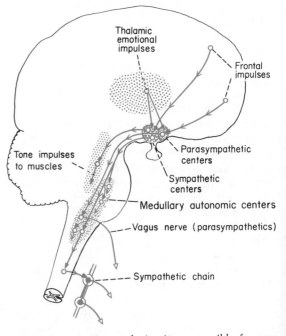

Figure 27-7. Neuronal circuits responsible for psychosomatic effects.

can well understand how the amygdala can play a major role in controlling the body's overall pattern of behavior.

PSYCHOSOMATIC EFFECTS

A psychosomatic effect is a bodily *(somatic)* effect produced by psychological stimulation. The brain can produce such effects in three general ways: (1) by transmission of signals through the autonomic nervous system, (2) by transmission of signals to the muscles through the bulboreticular area, and (3) by control of certain of the endocrine glands. Figure 27–7 illustrates the general neuronal mechanisms believed to be responsible for psychosomatic effects. It shows conscious signals beginning either in the frontal cortex or in the thalamus, then transmitted to the hypothalamus, and finally through the centers of the hindbrain to the cord and thence to the body.

Psychosomatic Effects Transmitted Through the Autonomic Nervous System. Almost any emotion can affect the autonomic nervous system. For example, very intense agitation will increase the excitability of the reticular activating system, and this system in turn excites sympathetic activity throughout the body. Therefore, generalized sympathetic stimulation of the organs is one of the most common of all psychosomatic effects. Many emotions such as excitement, anxiety, or rage often discharge the sympathetics *en masse,* causing marked increase in arterial pressure, palpitation of the heart, and cold chills over the skin.

Psychological effects also often stimulate the parasympathetic centers of the hypothalamus. The emotions of worry, depression, and lethargy, all of which have effects opposite to those that excite the sympathetic system, often stimulate the parasympathetics. On occasion, however, both of the systems may be stimulated simultaneously. Fear, for instance, can cause extreme sympathetic stimulation resulting in elevation of arterial pressure, while at the same time stimulating the parasympathetics to elicit such intense gastrointestinal activity that the person has uncontrolled diarrhea.

Transmission of Psychosomatic Effects Through the Skeletal Muscles. The reticular activating system sends signals downward through the spinal cord and thence directly to the muscles. Therefore, the same emotions that excite the sympathetics and reticular activating system usually increase the tone of the muscles throughout the body. Sometimes the tone becomes so intense that it causes muscle tremor, which explains why certain emotions can culminate in actual shaking.

On the other hand, the emotions that normally stimulate the parasympathetics usually decrease the activity of the bulboreticular formation. As a result, the muscular tone decreases to a very low level, which explains the muscular *asthenia* (muscular weakness) that is characteristic of some psychic states.

Transmission of Psychosomatic Effects Through Glands. The nervous system controls several of the endocrine glands either completely or partially. For instance, the sympathetic nervous system controls the adrenal medulla, and the hypothalamus controls almost all the activities of the pituitary gland. The hypothalamus achieves this control of the anterior pituitary gland by secreting *neurosecretory substances* within the substance of the hypothalamus; these are then absorbed into the local capillaries and carried through minute veins from the hypothalamus to the anterior pituitary gland where they cause secretion of several major hormones: *growth hormone, corticotropin, thyrotropin, prolactin,* and the *gonadotropins.* These hormones, in turn, control growth rate, protein metabolism, overall rate of metabolism, lactation, and most sexual functions. Also, the supraoptic nuclei of the hypothalamus control the secretion of *antidiuretic hormone;* this hormone, in turn, controls the degree of retention of water by the kidneys.

Obviously, therefore, many psychosomatic effects can be mediated through the endocrine glands. For instance, psychic effects that overly stimulate the hypothalamus can produce hyperthyroidism, causing the thyroid gland to secrete excess thyroid hormone and thereby increase the rate of metabolism of all cells of the body. Likewise, psychic signals can affect the output of sex hormones and thus cause failure of ovulation, excess menses, diminished menses, infertility, and other sexual abnormalities.

PSYCHOSOMATIC DISEASES

Perhaps the most common psychosomatic disease is extreme nervous tension associated simultaneously with increased heart rate, elevated arterial pressure, gastrointestinal disturbances, and excess muscle tone. This is often described simply as "nervousness."

Psychosomatic effects can also cause abnormal function of individual organs. For instance, stimulation of the sympathetics can so decrease gastrointestinal activity that constipation results. On the other hand, excessive stimulation of the parasympathetics can increase the degree of gastrointestinal activity so greatly that severe diarrhea results. Another very common psychosomatic disorder is palpitation of the heart caused by excitement, anxiety, or other emotional states.

Occasionally the dysfunction caused by a psychosomatic disorder is so great that tissues are actually destroyed. For instance, stimulation of the parasympathetics to the stomach can cause so much secretion of gastric juices that they eat a hole into the wall of the stomach or upper intestine. This causes the condition called *peptic ulcer*. Such patients can be treated by operative removal of portions of the stomach, by neutralizing the gastric juices with special drugs, or in some instances by psychiatric treatment to alleviate the emotional condition that is initiating the excessive secretion of gastric juices.

Psychosomatic Pain.   Many psychosomatic disorders can lead to pain which is called simply psychosomatic pain. For instance, a stomach ulcer causes intense burning in the pit of the stomach. Also, spasm of the gut caused by excess parasympathetic stimulation can cause cramps in the abdomen. And, occasionally, overstimulation of the heart can even cause cardiac pain. If the psychic condition that causes the functional abnormality can be corrected, then the pain likewise will be corrected.

Myths About Psychosomatic Disease.   Despite the many different ways in which psychosomatic disease can come about, this subject has been greatly overemphasized by newsmen and authors of fiction. It is a common myth that a person can worry so much about the function of one of his organs that he thereby creates disease in that particular organ. Except in a few instances, this is not true. Most psychosomatic diseases exhibit regular patterns such as general states of tension, constipation or diarrhea, ulcer, and a few others of a similar nature.

REFERENCES

Bentley, D., Konishi, M.: Neural control of behavior. *Ann. Rev. Neurosci., 1*:35, 1978.

Block, G. D., and Page, T. L.: Circadian pacemakers in the nervous system. *Ann. Rev. Neurosci., 1*:19, 1978.

Brazier, M. A.: Epilepsy: Its Phenomena in Man. New York, Academic Press, Inc., 1975.

Buser, P.: Higher functions of the nervous system. *Ann. Rev. Physiol., 38*:217, 1978.

de Silva, F. H. L., and Arnolds, D. E. A. T.: Physiology of the hippocampus and related structures. *Ann. Rev. Physiol., 40*:185, 1978.

Harris, A. J.: Inductive functions of the nervous system. *Ann. Rev. Physiol., 36*:251, 1974.

Lester, D.: A Physiological Basis for Personality Traits. Springfield, Ill., Charles C Thomas, Publisher, 1974.

Millon, T.: Medical Behavioral Science. Philadelphia, W. B. Saunders Company, 1975.

Pappenheimer, J. R.: The sleep factor. *Sci. Amer., 235(2)*:24, 1976.

Prince, D. A.: Neurophysiology of epilepsy. *Ann. Rev. Neurosci., 1*:395, 1978.

Rolls, E. T.: The Brain and Reward. Oxford, Pergamon Press, 1975.

Siegel, R. K.: Hallucinations. *Sci. Amer., 237(4)*:132, 1977.

Snyder, S. H.: Opiate receptors and internal opiates. *Sci. Amer., 236(3)*:44, 1977.

Stein, D. G., and Rosen, J. J.: Learning and Memory. New York, The Macmillan Company, 1974.

Thompson, R. F.: Introduction to Physiological Psychology. New York, Harper & Row, Publishers, 1975.

Widroe, H. J. (ed.): Human Behavior and Brain Function. Springfield, Ill., Charles C Thomas, Publisher, 1975.

QUESTIONS

1. What are the differences between short-term memory and long-term memory?
2. What determines which memories will be remembered, and what are the roles of rehearsal and codification in the consolidation of memory?

3. What functions does the thalamus perform in relation to memory, especially recall of memories, and in the thinking process in general?
4. What is the cellular mechanism of memory?
5. What happens to a person when he loses the prefrontal areas of his cerebral cortex?
6. Give the feedback theory of wakefulness and sleep.
7. When recording brain waves, what causes alpha waves, beta waves, and delta waves?
8. What is the function of pleasure and pain in the establishment of memories and in overall brain function?
9. Under what conditions will an animal exhibit rage?
10. Describe the different means by which the brain can cause psychosomatic abnormalities.

THE SPECIAL SENSES VIII

THE EYE 28

The eye is a highly specialized receptor organ of the nervous system. Indeed, the retinal portion of the eye is actually formed from neural tissue that grows outward from the brain during development of the fetus. Furthermore, the retina contains neurons, the same as the brain and spinal cord. The purpose of the present chapter will be, first, to describe the specialized sensory function of the eye, and then to discuss the neurophysiology of vision.

THE EYE AS A CAMERA

Physiologic Anatomy. Figure 28–1 illustrates the general organization of the eye, showing the optical system for focusing the image on the *retina* where the image is projected. The optical system is identical to that of a camera, and the retina corresponds to the photographic film. The retina translates the image into nerve impulses and transmits these into the brain through the optic nerve.

The outer envelope of the eye is a very strong bag composed mainly of a thick fibrous structure called the *sclera*. Anteriorly the sclera connects with the *cornea,* which is the

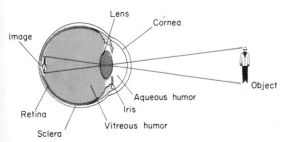

Figure 28–1. General structure of the eye, showing the function of the eye as a camera.

clear part through which light enters the eye. The inside of the eye is filled mainly with fluid, but an ovoid clear body called the *crystalline lens* is located approximately 2 mm. behind the cornea. The fluid anterior to the lens is almost pure extracellular fluid and is called *aqueous humor,* while that behind the lens contains a mucoprotein matrix that forms a gelatinous but clear structure called the *vitreous humor.* Light passes first through the clear cornea, then through the aqueous humor, then the lens, and finally the vitreous humor before impinging on the retina.

The Lens System of the Eye. The lens system of the eye is composed of the cornea and the crystalline lens. Because the cornea is curved on its outside, light rays passing from the air into the cornea are *refracted* (bent) in the same way that any optical lens refracts rays. After the rays pass through the cornea and aqueous humor they strike the curved anterior surface of the crystalline lens, where still more bending occurs, and as they pass through the posterior surface of the lens the rays bend again. Thus, light rays are refracted (bent) at three different major interfaces in the eye. This is analogous to the refraction of light at the different surfaces in a compound lens system of a camera, for in the camera the light rays are also refracted at each interface between the lenses and the air.

Function of a Convex Lens in Forming an Image. Figure 28–2 shows the focusing of light rays from a distant source and also from two points near the lens. It will be noted that the light rays originating from the left-hand side of the figure and striking the outer edges of the convex lens are bent toward the center, while those that strike the lens exactly in the center pass on through without being bent. The reason for this is that the outer rays enter the substance of the lens at an angle, which

353

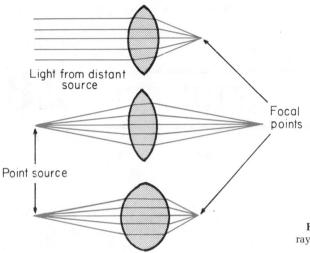

Figure 28–2. Focusing of parallel light rays and light rays from point sources by convex lenses.

causes refraction, while those striking the center enter the lens perpendicular to its surface, which does not cause refraction. The light rays from the edges of the lens angle inward to meet those passing through the center, and they all focus on a common point called the *focal point*.

Figure 28–3A shows the focusing by a single lens of light rays from two different point sources. Note that the rays from the two lights are each focused to focal points on the opposite side of the lens directly in line with the lens center.

Figure 28–3B illustrates the focusing of light rays from different point sources on a human being's body. Those parts of the body that are very bright are point sources of light, whereas those parts that are dark represent the black spaces between the point sources of light. So far as the eyes are concerned, any object is a mosaic of point sources of light. Light from each source is focused by the lens,

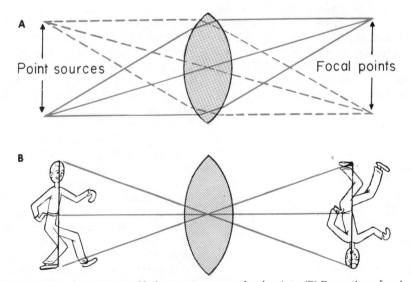

Figure 28–3. (A) Focusing of two points of light to two separate focal points. (B) Formation of an image by a convex lens focusing light rays from an object.

and the focal point of each is always directly in line with the center of the lens and the original source. Consequently, an inverted image is formed by the focal points, as shown in the figure.

ABNORMALITIES OF THE LENS SYSTEM

The normal eye focuses parallel light rays exactly on the retina. This normal focusing of the eye is called *emmetropia,* and it is illustrated at the top of Figure 28–4. However, three different abnormalities frequently occur to prevent focusing of light rays precisely on the retina. These are *hypermetropia, myopia, and astigmatism.*

Hypermetropia. Hypermetropia, or *far-sightedness,* is caused by failure of the lens to bend the light rays enough to bring them to a focal point on the retina, as shown in Figure 28–4. Instead, the light rays are still diffuse when they reach the retina, and thus vision of even distant objects is blurred, and it is even more blurred for near objects. Hypermetropia is called far-sightedness because with this type of vision objects can be seen more clearly at a distance than near at hand.

Myopia. Myopia, which is called *near-sightedness,* is caused by too strong a lens system for the distance of the retina behind the lens. That is, the light rays are focused before they reach the retina, and by the time they do reach the retina they have spread apart again as shown in the figure, causing fuzziness of each point in the image.

Myopia is called near-sightedness because the myope can see objects near him with complete clarity, while not being able to focus any objects that are at a far distance.

Astigmatism. Astigmatism occurs when the lens system becomes ovoid (egg-shaped) rather than spherical. This effect is shown in Figure 28–5. Either the cornea or the crystalline lens becomes elongated in one direction in comparison with the other direction. Because the radius of curvature is greater in the elongated direction than in the short direction, the light rays entering the lens along this lengthened curvature are focused behind the retina, while those entering along the shortened curvature are focused in front of the retina. In other words, the eye is far-sighted for some of the light rays and near-sighted for the remainder. Therefore, the person with astigmatic eyes is unable to focus any object clearly, regardless of how far the object is away from the eyes; for when the near-sighted light rays are in focus, the far-sighted ones are out of focus and vice versa.

Correction of the Optical Abnormalities of the Eye. Glasses with properly prescribed lenses can be used to correct most abnormalities of the lens system of the eye. Glasses bend the light rays before they enter the eye in an appropriate manner to correct for the excess or deficient refractive power of the eye. Figure 28–6 shows the correction of myopia and hypermetropia. In the myopic person the light rays normally focus in front of the retina. To prevent this, a concave lens is placed in front of the eye. This type of lens bends the light rays outward and, therefore, compensates for the excess inward bending of the myopic lens system. By prescribing the appropriate curvature to the concave lens, a myopic person's vision can be made completely normal.

In the hypermetropic eye the lens system normally fails to bend the light rays sufficiently. To correct this abnormality, a convex lens is placed in front of the eye so that the light rays will be partially bent even before they reach the eye. With the aid of this preliminary convergence, the lens system of the eye can then bring the rays to a focal point on the retina.

The lens that must be used to correct astigmatism is somewhat more complicated than those used to correct either myopia or hyper-

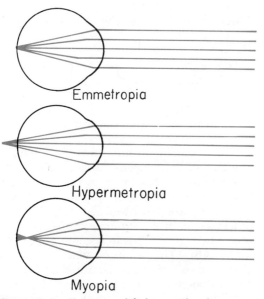

Figure 28–4. Focusing of light rays by the emmetropic eye, the hypermetropic eye, and the myopic eye.

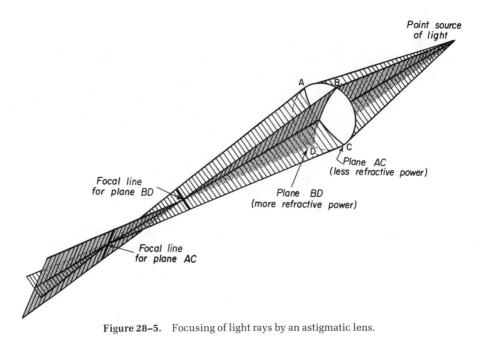

Figure 28-5.   Focusing of light rays by an astigmatic lens.

metropia, for it must be fashioned with more curvature in one direction than the other. However, by prescribing a lens with precisely ground curvatures in exactly the right "axis" in front of the eye, the abnormal refraction of light rays by each portion of the astigmatic eye can be corrected appropriately.

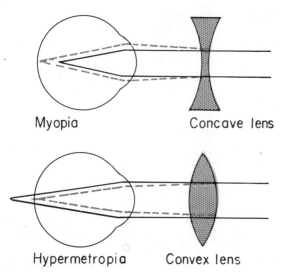

Figure 28-6.   Correction of myopia by a concave lens and of hypermetropia by a convex lens.

FUNCTION OF THE RETINA

Retinal Structure.   Figure 28–7 illustrates the anatomy of the retina, showing that it is composed of several different layers of cells. The light rays enter from the *bottom* of the figure and pass all the way through the retina to the top, finally striking the *rods* and *cones* and the pigment layer. The rods and cones are nerve receptors that are excited by light. These cells change the light energy into neuronal signals that are transmitted into the brain.

The pigment layer of the retina contains large quantities of a very black pigment called *melanin*. The function of the melanin is to absorb the light rays after they have passed through the retina and thereby prevent light reflection throughout the eye. *Albino* persons, who are unable to manufacture melanin in any part of their bodies, have a complete lack of pigment in this layer of the retina. As a result, after passing through the retina the light rays are not absorbed but instead are reflected so intensely in all directions that they cause all images to become bleached out with light. The albino's vision is usually about three times less acute than that of a normal person and he is so blinded by bright sunlight that he must wear dark glasses to see at all.

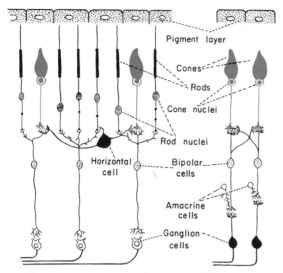

Figure 28-7. Functional anatomy of the retina.

By far the greater number of light receptors in the retina are rods. Light of all colors stimulates the rods, while the cones are stimulated selectively by different colors. Therefore, the cones are responsible for color vision, in contradistinction to the rods, which provide only black and white vision.

THE CHEMISTRY OF VISION

Chemistry of Rod Excitation. Figure 28-8 shows the chemical changes that occur in the rods, both when light strikes the retina and

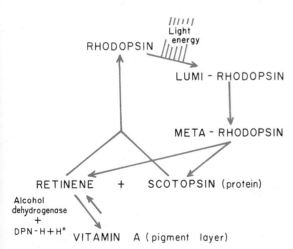

Figure 28-8. The retinene-rhodopsin chemical cycle responsible for light sensitivity of the rods.

during periods between light stimulations. Vitamin A is the basic chemical utilized by both the rods and cones for synthesizing substances sensitive to light. On being absorbed into a rod, vitamin A is converted into a substance called *retinene*. This then combines with a protein in the rods called *scotopsin* to form a light-sensitive chemical, *rhodopsin*. If the eye is not being exposed to light energy, the concentration of rhodopsin builds up to an extremely high level.

When a rod is exposed to light energy, some of the rhodopsin is changed immediately into *lumirhodopsin*. However, lumirhodopsin is a very unstable compound that can last in the retina only about a tenth of a second. It decays almost immediately into another substance called *metarhodopsin*, and this compound, which is also unstable, decays very rapidly into retinene and scotopsin.

Thus, in effect, light energy breaks rhodopsin down into the substances from which the rhodopsin itself had been formed, retinene and scotopsin. In the process of splitting rhodopsin, the rods become excited, probably by ionic charges that develop momentarily on the splitting surfaces of the rhodopsin. These charges last for only a split second. During this slight interval, nerve signals are generated in the rod and transmitted into the optic nerve and thence into the brain.

After rhodopsin has been decomposed by light energy, its decomposition products, retinene and scotopsin, are recombined again during the next few minutes by the metabolic processes of the cell to form new rhodopsin. The new rhodopsin in turn can be utilized again to provide still more excitation of the rods. Thus, a continuous cycle occurs: Rhodopsin is being formed continually, and it is broken down by light energy to excite the rods.

Chemistry of Cone Vision. Almost exactly the same chemical processes occur in the cones as in the rods except that the protein scotopsin of the rods is replaced by one of three similar proteins called *photopsins*. The chemical differences among the photopsins make the three different types of cones selectively sensitive to different colors, which will be discussed later.

Persistence of Images and Fusion of Flickering Images. Following a sudden flash of light that lasts only one-millionth of a second, the eye sees an image of the light that lasts for approximately one-tenth second. The duration of the image is the length of time that the retina remains stimulated following the flash, and this presumably is about as long as the lumirhodopsin remains in the rods.

The persistence of images in the retina allows flickering images to *fuse* when one views a moving picture or television screen. The moving picture flashes 16 to 30 pictures per second, and the television screen provides 60 pictures per second. The image on the retina persists from one picture to the next, which gives one the impression of seeing a continuous picture.

Light and Dark Adaptation. It is common experience to be almost totally blinded when first entering a very bright area from a darkened room and when entering a darkened room from a brightly lighted area. The reason for this is that the sensitivity of the retina is temporarily not attuned to the intensity of the light. To discern the shape, the texture, and other qualities of an object, it is necessary to see both the bright and dark areas of the object at the same time. Fortunately, the retina automatically adjusts its sensitivity in proportion to the degree of light energy available. This phenomenon is called light and dark adaptation.

The mechanism of *light adaptation* can be explained by referring once again to Figure 28–8. When large quantities of light energy strike the rods, large amounts of rhodopsin are broken into retinene and scotopsin; and, because rhodopsin formation is a relatively slow process, requiring several minutes, the concentration of rhodopsin in the rods falls to a very low value as the person remains in the bright light. Essentially the same effects occur in the cones. Therefore, the sensitivity of the retina soon becomes greatly depressed in bright light.

The mechanism of *dark adaptation* is opposite to that of light adaptation. When the person enters a darkened room from a lighted area, the quantity of rhodopsin in his rods (and color-sensitive chemicals in his cones) is at first very slight. As a result, he cannot see anything. Yet, the amount of light energy in the darkened room is also very slight, which means that very little of the rhodopsin being formed in the rods is broken down. Therefore, the concentration of rhodopsin builds up during the ensuing minutes until it finally becomes high enough for even a very minute amount of light to stimulate the rods.

During dark adaptation, the sensitivity of the retina can increase as much as 1000-fold in only a few minutes, and as much as

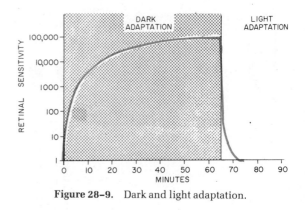

Figure 28–9. Dark and light adaptation.

100,000 times in an hour or more. This effect is illustrated in Figure 28–9, which shows the retinal sensitivity increasing from an arbitrary light-adapted value of 1 up to a dark-adapted value of 100,000 in one hour after the person has left a very bright area and moved into a completely darkened room. Then, on reentering the bright area, light adaptation occurs, and retinal sensitivity decreases from 100,000 back down to 1 in another 10 minutes, which is a more rapid process than dark adaptation.

FUNCTION OF THE CONES – COLOR VISION

The cones are different from the rods in several respects. First, they respond selectively to certain colors, some to one color and others to other colors. Second, cones are considerably less sensitive to light than are rods, for which reason they cannot provide vision in very dim light. Third, many of the cones are connected, one cone to one optic nerve fiber, which provides greater acuity of vision than the rods provide. Ten to 100 rods usually connect with the same optic nerve fiber, which means that impulses transmitted by the rods to the brain do not necessarily originate from one very discrete point on the retina.

Detection of Different Colors by the Cones. The retina contains three different types of cones, each of which responds to a different spectrum of colors. This is illustrated in Figure 28–10, which shows light wave lengths at which the three different types of cones — the blue cone, the green cone, and the red cone — respond. Note that the blue cone responds maximally at a wave length of 430 millimicrons (millimicrometers), which is

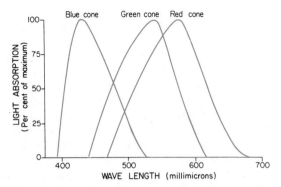

Figure 28-10. Spectral sensitivity curves for blue, green, and red cones. (Drawn from curves recorded by Marks, Dobelle, and MacNichol, Jr.: Science, 143:1181, 1964, and by Brown and Wald: Science, 114:45, 1964.)

a blue color; the green cone at 540 millimicrons, a greenish-yellow color; and the red cone at 575 millimicrons, an orange color. The so-called "red" cone is called the red cone not because its maximal response is in the red range, but because it is the only cone that has any significant response at all above 600 millimicrons (millimicrometers), which is red.

Determination of the Intermediate Colors by the Blue, Green, and Red Cones. It is quite easy to understand how the blue cones determine that an object is blue, how the green cones determine that an object is green, and how the red cones determine that an object is red, but it is more difficult to understand how these cones detect the intermediate colors between the three primary ones. This is accomplished by utilizing a combination of cones. For instance, yellow light stimulates the red and green cones approximately equally. When both of these types of cones are stimulated equally, the brain interprets the color as yellow. Also, when the red cones are stimulated about one and one-half times as strongly as the green cones, which occurs when light with a wave length of 580 millimicrons (millimicrometers) strikes the retina, the brain interprets the color as orange. If both the red and green cones are stimulated, but the green more than the red cones, the color is interpreted as a greenish-yellow. Likewise, when both the green and blue cones are stimulated, the color is interpreted as a bluish-green. Thus, by combining the degrees of stimulation of the different cones, the brain can distinguish not only among the three primary colors but also among the other colors having intermediate wave lengths.

The intensity of colored light is determined by the strength of the signal transmitted to the brain by the cones. For instance, if a yellow color stimulates both the green and red cones and the number of impulses transmitted by the optic nerve fibers from each cone is only 10 per second, the intensity will be relatively weak; but if the number of impulses transmitted is 100 per second, the intensity will be strong. It is especially important to note that a change in intensity does not change the *ratio* between the degree of stimulation of the two types of cones. The brain interprets color on the basis of this ratio and not on the basis of the actual intensity of stimulation of each cone.

Color Blindness. Color blindness also can be understood very readily on the basis of Figure 28-10. Occasionally one of the three primary types of cones is lacking because of failure to inherit the appropriate gene for formation of the cone. The color genes are sex-linked and are found in the female sex chromosome. Since females have two of these chromosomes, they almost never have a deficiency of a color gene, but because males have only one female chromosome, one or more of the color genes is absent in about 4 per cent of all males. For this reason, almost all color-blind people are males.

If a person has complete lack of red cones, he is able to see green, yellow, orange, and orange-red colors by use of his green cones. However, he is not able to distinguish satisfactorily between these colors because he has no red cones to contrast with the green ones. Likewise, if a person has a deficit of green cones, he is able to see all the colors, but he is not able to distinguish between green, yellow, orange, and red colors, because the green cones are not available to contrast with the red. Thus, loss of either the red or the green cones makes it difficult or impossible to distinguish between the colors of the longer wave lengths. This is called *red-green color blindness.*

In very rare instances a person lacks blue cones, in which case he has difficulty distinguishing violet, blue, and green from each other. This type of color blindness is frequently called *blue weakness.*

NEURONAL CONNECTIONS OF THE RETINA WITH THE BRAIN

Figure 28-11 illustrates the connections of the retina with the brain, showing that the

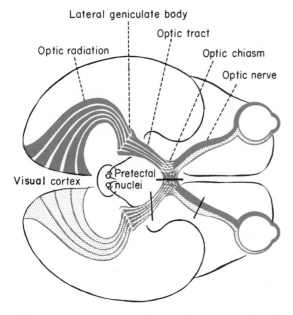

Figure 28–11. Optic pathways for transmission of visual signals from the retinae of the two eyes to the optic cortex. (From Polyak: The Retina. The University of Chicago Press.)

right halves of the two eyes are connected with the right visual cortex and the left halves with the left visual cortex. The *optic nerve* fibers from the nasal half of each retina cross in the *optic chiasm,* located on the bottom surface of the brain, and join the fibers from the temporal half of the opposite retina. Then the combined fibers pass backward through the *optic tract,* synapse in the *lateral geniculate body,* and finally spread through the *optic radiation* into the *visual cortex.*

In addition, fibers pass directly from the optic tract into the *pretectal nuclei;* these fibers carry signals for control of pupillary size in response to light intensity, as will be discussed later in the chapter.

Finally, fibers not shown in Figure 28–11 pass from the *lateral geniculate body* into the lateral thalamus and also into the *superior colliculus* located in the midbrain. These areas help to discriminate certain qualities of the visual scene, as will be discussed in the following section.

Discrimination of the Visual Image at the Retinal Level. Even at the level of the retina the visual image begins to be analyzed, so that the pattern of stimulation in the visual cortex is considerably different from that in the retina. The retina breaks the visual

image into two components. The first component is the level of *luminosity.* That is, some of the optic nerve fibers transmit signals to the brain to indicate the general light intensity of the observed scene. The second component is series of signals to indicate *changes* in light intensity. Indeed, two different types of changes in light intensity are transmitted in the optic fibers. One of these denotes the change in light intensity at any contrast border in the visual image. For instance, the contrast between a white piece of paper and a black line on the paper gives an intense signal that is transmitted in the optic fibers. The second type of change in light intensity is that which occurs when the light entering the eye or the light on any spot in the retina changes rapidly. For example, a person is looking toward a dark wall and suddenly a bright insect flies across his field of vision. At each point where the image of the insect appears on the retina there is a sudden flash of light. And this excites a succession of optic nerve fibers very powerfully.

Thus, the messages that are sent from the retina in the optic nerve fibers do not transmit a true mosaic pattern of the visual image but, instead, transmit one type of signal to denote the general level of illumination and others to denote where in the image there are changes in light intensity. And besides these signals, the respective color cones send additional ones to indicate color contrasts.

Function of the Lateral Geniculate Body in the Analysis of the Visual Image. At the lateral geniculate body, several other aspects of the visual scene begin to be interpreted. One of these is probably *depth perception,* for it is here that signals from the two eyes first come together, and, as we shall see later in the chapter, depth perception depends upon comparing minute differences in shapes of objects as seen by the two separate eyes. The lateral geniculate body is admirably suited for this because it has six layers of neurons, the visual signals from one eye terminating in three of the layers and those from the other eye in the other three layers; thus, the signals from the two eyes are intimately interconnected.

It is also possible that the lateral geniculate body plays a role in *color vision.* A reason for believing this is that one can look at a red light with the left eye and a green light with the right eye and see a yellow color, thus indicating that combination of colors from the two separate eyes occurs at least to some extent in the brain, perhaps at the lateral geniculate level.

Function of the Visual Cortex in Discriminating the Visual Image. Once the visual image reaches the visual cortex, it has by that time been changed to a pattern of stimulation that is considerably different from the actual image on the retina. This is illustrated in Figure 28–12. To the left is the retinal image of a heavy colored cross; to the right is the pattern of stimulation in the visual cortex. Note that stimulation occurs only at the edges of the cross. This is caused by a succession of mechanisms in the retina, the lateral geniculate body, and the visual cortex that allows a cortical point to be stimulated if there is a contrast border between a light area and a dark area. If there is no contrast, the neuron will not be stimulated. Thus, the visual neuronal processing system brings out borders and thereby determines the shapes of images. This explains why a simple line drawing of someone's face can be recognized as a picture of that person. That is, the visual cortex itself actually converts the image of a person to a type of line drawing anyway.

Another feature of discrimination by the visual cortex is that it also determines direction of orientation of the lines and borders in the image. If the cross shown in Figure 28–12 were leaning slightly to the right, an entirely different set of neurons would be stimulated, thus denoting this leaning.

Signals are transmitted into the visual association areas located to the lateral sides of the visual cortex, both directly from the visual cortex and also from the lateral geniculate body. In these areas, the finer meanings of the visual signals are interpreted. For instance, the picture of a letter is believed to be interpreted here as letter A, B, C, or so forth, and a combination of letters is interpreted as a word. Still further away from the primary visual cortex, the combination of words is interpreted as a thought.

Determination of Luminosity in Basal Regions of the Brain. A monkey that has lost its cerebral cortex can still determine light and dark even though it cannot discriminate shapes of objects. This fact indicates that overall light intensity of the visual scene can in at least some animals be interpreted in the basal levels of the brain rather than, or in addition to, the cortex. It has been suggested that the lateral thalamus and superior colliculus play roles in this effect. At any rate,

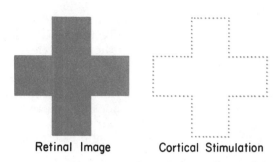

Retinal Image   **Cortical Stimulation**

Figure 28–12. Pattern of stimulation in the visual cortex caused by observing a heavy colored cross.

after this interpretation is made, these lower centers presumably send this information to the visual association areas where it can be used for overall interpretation of the visual image.

FIELDS OF VISION

A means used to determine the extent of a person's normal vision, and also to detect abnormalities of vision, is to plot his *fields of vision*. The person is asked to close one eye and to look straight forward with the other eye. A small spot of light is moved above his central point of vision as far as he can see, then below as far as he can see, then to the right, to the left, and in all directions. In this way, his ability to see in all areas away from his centralmost point of vision is plotted.

Figure 28–13 shows the normal field of vision of the right eye. Far out to the lateral side one can actually see objects at right

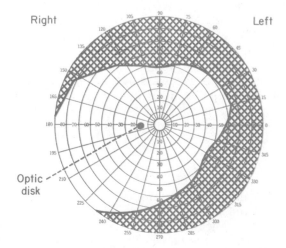

Figure 28–13. The visual field of the right eye.

angles (90 degrees) from the direction in which the eye is looking. To the nasal side, the nose is in the way, and the person can see objects only 50 degrees away from the central point of vision. Likewise, in the upward direction the orbital ridge is in the way, and in the downward direction the cheekbone is in the way. Were it not for these structures around the eye, the field of vision would be considerably greater.

In the field of vision is a *blind spot* caused by the *optic disk,* which is the point where the optic nerve enters the eyeball. At this point no rods or cones are present. The blind spot of each eye is located approximately 15 degrees to the lateral side of the central point of vision. However, the blind spots of the two eyes are on opposite sides in the respective fields of vision so that when the images from the two eyes fuse there is no part of the visual scene that is not covered. Fortunately, also, the blind spots normally are not noticeable in one's vision.

Plotting the fields of vision provides a means for determining and locating damage in the retina or in the neuronal tracts from the retina to the brain. Referring back to Figure 28–11, it will be noted that three dark lines have been placed respectively across the right optic nerve, the optic chiasm, and the right optic tract. If the right optic nerve has been sectioned, the field of vision of the right eye will be zero, or, in other words, the eye will be completely blind. If the optic chiasm has been sectioned, the nasal half of each retina will be blind. This means that the *lateral* half of each visual field will be blind, because the lateral half of the field is registered by the nasal half of the retina. Finally, if the right optic tract is destroyed, the left halves of the visual fields of both eyes will be blind. Obviously, also, damage in any portion of the visual cortex or at any other point in the visual transmission system will cause loss of vision in respective areas in the fields of vision. Thus, the point in the eye or in the brain at which the damage has occurred can often be discerned from the pattern of visual loss.

VISUAL ACUITY

Visual acuity means the degree of detail that the eye can discern in an image. The usual method for expressing visual acuity is to compare the person being tested with the normal person. If at a distance of 20 feet he can barely read letters of exactly the size that the normal person can barely read, he is said to have 20/20 vision, which is normal. If the letters must be as large as those that the normal person can read at a distance of 40 feet, his vision is said to be 20/40, or, in other words, his vision is one-half normal. If he can barely read letters at a 20-foot distance that the normal person can read at 100 feet, his vision is 20/100, or, in other words, his vision is one-fifth normal. An occasional person has better than normal vision, so that he can read at 20 feet what the normal person can read only at 15 feet, in which case his vision is said to be 20/15.

Relationship of the Cones to Visual Acuity. A person normally has very acute vision only in the central area of his visual field. The reason for this is that a small central area of the retina, only a millimeter in diameter, called the *fovea,* is especially adapted for acute vision. No rods are present in the fovea, and the cones there are considerably smaller in diameter than those in the peripheral portions of the retina. Also, the nerve fibers and blood vessels are all pulled to one side, so that light can pass with ease directly to deep layers of the retina where the cones are located. Finally, and especially important, each of the cones of this region connects through an almost direct pathway with the brain, one cone to one optic nerve fiber, so that the impulses from each cone do not become confused with impulses from other cones.

Effect of the Optics of the Eye on Visual Acuity. Figure 28–14 illustrates some of the optical factors that also influence the acuity of vision. It shows the images of point sources of light focused on the retina. If sources are so close together that both of the point images focus on the same cone, then they will be interpreted as a single light rather than as two distinct lights. Likewise, if two lights are focused on two adjacent cones, they still will be interpreted as a single light, because a single light very frequently does stimulate two adjacent cones. Therefore, for the retina to interpret two points of light as separate lights, the images must impinge on cones separated by an unexcited cone between them. And because the diameter of each foveal cone is about 2 to 3 microns (micrometers), the images of point sources of light must be at least 2 to 3 microns apart before they can be seen as two separate points. Thus, in the lower part of Figure 28–14, it is shown that

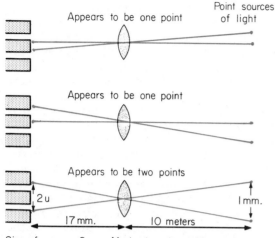

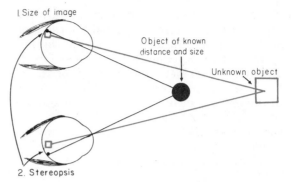

Figure 28-15.   Mechanisms of depth perception.

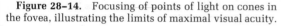

Size of cones = 2 u   Maximal acuity at 10 meters = 1 mm.

Figure 28-14.   Focusing of points of light on cones in the fovea, illustrating the limits of maximal visual acuity.

two points of light 10 meters away from the eye and 1 mm. apart can barely be discerned by the normal eye as two separate lights rather than as a single one. To express this another way, the maximal acuity of vision at 10 meters is 1 mm. separation of point sources of light.

The normal lens system of the eye is sufficiently effective so that visual acuity is determined mainly by the diameters of the cones in the fovea. When the optics of the eye are abnormal, visual acuity then becomes determined instead by the ability of the lens system to focus the image on the retina. Consequently, a person with severe myopia, hypermetropia, or astigmatism may have a visual acuity of 20/40, 20/100, 20/200, or so forth, depending on the degree of optical abnormality.

DEPTH PERCEPTION

The eyes determine the distance of an object by two principal means. These are, first, by the *size of the image* on the retina, and second, by the phenomenon called *parallax*. To determine the distance of an object from the eyes by the image size, the person must have had previous experience with the object and know its actual size. For example, previous experience with other persons makes one remember that their average height is somewhere between 5 and 6 feet. Therefore, even when using one eye, one can determine relatively accurately the distance of a person by

the size of his image on the retina. If he is nearby, his image is very large, but, if he is far away, his image is very small.

The second means for determining the distance of an object, parallax (also called *stereopsis*), depends on slight differences between the shapes and positions of the images on the retinae of the two eyes. This effect, which is shown in Figure 28–15, occurs because the two eyes are set several inches apart. In this figure the two eyes are observing a ball near the eyes and a block at a distance. In the left eye, the image of the block lies to the right of the image of the ball, but in the right eye, the image of the block lies to the left of the image of the ball. In other words, the images of these two objects are actually reversed on the retinae because of the angles from which the two eyes observe them, and the brain interprets their relative distances by the degree to which they are reversed. Obviously, this is an extreme example of parallax, but even when looking at a single object the images in the two eyes are slightly different. It must be noted particularly that this mechanism interprets only the *relative distances* of objects and not the actual distances. However, if the distance of one of the objects is known, that of the second object is also known. For practical considerations, a person always has his hands or other parts of his body to use as objects of known distance, which can be used as reference points for determining the distances of unknown objects.

NEUROGENIC CONTROL OF EYE MOVEMENTS

If the eyes are to function satisfactorily as a camera, their line of sight must be appropriately directed so that the most important portion of the image will fall exactly on the fovea, the area of the retina designed for the most accurate vision. In addition to this, the lens system of the eye must be focused for the distance of the object, and the pupil of the eye must be enlarged or contracted in proportion to the amount of light available, thus aiding in the light and dark adaptation of the eyes. It can be seen, then, that several different sets of eye muscles must be controlled very exactly to attain visual effectiveness.

POSITIONING OF THE EYES

Each eye is positioned by three different pairs of muscles, which are illustrated in Figure 28–16. One pair of these muscles is attached superiorly and inferiorly to the eyeball to move it up or down; another pair is attached horizontally to move the eye from side to side; and another pair is attached around the eyeball respectively on the bottom and top so that it can be rotated in either direction.

To control the eye movements, the visual association areas of the optic cortex must determine first whether or not the eyes are pointing toward the object; if not, impulses are transmitted through *oculomotor centers* in the brain stem to move the eyes in the appropriate direction. The visual association areas also determine whether or not the two eyes

are receiving the same image on the corresponding portions of the two retinae — that is, whether or not the images are *fused* with each other. If not, one or both of the eyes are adjusted up or down, to one side or the other, or rotated so that the images will fuse. These minute adjustments of the eye positions are so important to vision that almost as much of the visual association areas in the brain is concerned with eye movements as with the interpretation of the meaning of visual signals.

Movements of the Eyes for Following Moving Objects. Special mechanisms for allowing the eyes to follow moving objects are also available in the visual association areas, the motor control areas of the frontal lobes, and the cerebellum. To follow successfuly, the eyes move slowly in the same direction that the object is moving, neither falling behind nor getting ahead of the object. If the image does begin to fall behind or to move ahead, the eye position control areas immediately send corrective signals to the eye muscles.

The cerebellum is involved in this mechanism because the object's course of movement must be predicted ahead of time; the predictive ability of the cerebellum allows the eyes to move along with the object rather than having to wait for the object to get to a new position before bringing the eyes along too.

FOCUSING OF THE EYES

The eyes are kept focused on the visual object by changing the curvature of the crystalline lens as the object's distance changes. This is accomplished by the following neuromuscular mechanism:

Figure 28–17 shows the lens of the eye suspended from the sides of the eyeball by the *suspensory ligaments* of which there are about 70. The lens of a young person is an elastic structure having a clear envelope and clear viscous fluid in its center. When the ligaments are not pulling on the lens it assumes a spherical shape, but when the ligaments are tightened it flattens out. Normally, the ligaments are tight and the lens is flattened.

A smooth muscle called the *ciliary muscle* attaches to the suspensory ligaments of the lens where they connect with the eyeball. This muscle is composed of *meridional fibers* and *circular fibers*. The meridional fibers extend from the ends of the suspensory liga-

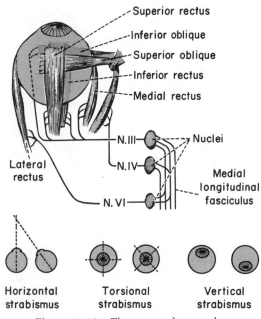

Superior rectus
Inferior oblique
Superior oblique
Inferior rectus
Medial rectus

Lateral rectus

N. III — Nuclei
N. IV
N. VI

Medial longitudinal fasciculus

Horizontal strabismus   Torsional strabismus   Vertical strabismus

Figure 28–16. The extraocular muscles.

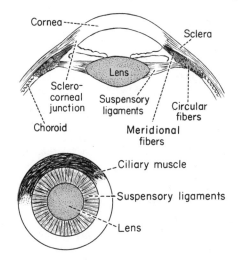

Figure 28–17.   The focusing (accommodation) mechanism of the eye.

ments anteriorly to the sclerocorneal junction; when they contract they pull the ends of the suspensory ligaments forward and loosen them. This allows the flat elastic lens to become thickened and the curvature of the lens to become greater, increasing the focusing power. The circular fibers extend all the way around the eye. When they contract they act as a sphincter, tightening progressively into a smaller and smaller circle, thus again loosening the suspensory ligaments in still another way so that the lens becomes still more convex and develops more focusing power.

Focusing, like positioning of the eyes, is accomplished by signals initiated in the visual association areas. When the image on the retina is out of focus, the fuzziness of the image initiates appropriate reactions to cause a change in the tension in the ciliary muscle. As the focus becomes better and better, it finally reaches a point at which the image is seen with greatest acuity. At this point, the focusing mechanism "locks in" and holds until the distance of the observed object changes. Then the focusing mechanism proceeds to the new focus in about one-half second.

CONTROL OF THE PUPIL

The pupil of the eye is the round opening of the iris through which light passes to the interior. The iris can constrict until the pupillary diameter is no greater than 1.5 mm., or it can relax until the diameter becomes 8 to 9 mm. Constriction of the pupil is caused by contraction of the *pupillary sphincter,* a circular muscle around the pupillary opening that is controlled by the parasympathetic nerves to the eye. Dilation of the pupil is caused by relaxation of the sphincter or by contraction of *radial muscle fibers* that extend from the edge of the pupil to the outer border of the iris; this muscle is controlled by the sympathetic nerves to the eye.

Because the amount of light that enters the eye is proportional to the area of the pupil and not to the diameter, the amount of light entering the eye changes with the square of the diameter, and can be varied between its two extremes approximately 30-fold. This provides a mechanism for light and dark adaptation in addition to the retinal mechanism for light and dark adaptation. Changes in pupillary size can occur in less than 1 second, in contrast with retinal adaptation which requires several minutes.

The size of the pupil is regulated by a *pupillary light reflex.* When the retina is stimulated by light, some of the optic nerve inpulses, instead of passing all the way to the visual cortex, leave the optic tract as shown in Figure 28–11 and pass to the pretectal nuclei of the brain stem. From here, signals are transmitted to the oculomotor center of the brain stem that controls the muscles of the eye and finally back to the iris. When the light intensity in the eyes is great, the size of the pupillary opening diminishes. On the other hand, when the light intensity becomes slight, the opening of the pupil increases.

The pupillary light reflex is frequently absent in a person with syphilis of the central nervous system, for this disease has a special predilection for destroying the pretectal nuclei.

REFERENCES

Campbell, C. J., Koester, C. J., Rittler, M. C., and Tackaberry, R. B.: Physiological Optics. Hagerstown, Md., Harper & Row, Publishers, 1974.

Dunn-Rankin, P.: The visual characteristics of words. *Sci. Amer., 238(1)*:122, 1978.

Glezer, V. D., Leushina, L. I., Nevskaya, A. A., and Prazdnikova, N. V.: Studies on visual pattern recognition in man and animals. *Vision Res., 14*:555, 1974.

Glickstein, M., and Gibson, A. R.: Visual cells in the pons of the brain. *Sci. Amer., 235(5)*:90, 1976.

Hagen, M. A.: Picture perception: toward a theoretical model. *Psychol. Bull., 81*:471, 1974.

Horridge, G. A.: The compound eye of insects. *237(1)*:108, 1977.

Janisse, M. P. (ed.): Pupillary Dynamics and Behavior. New York, Plenum Publishing Corporation, 1974.

Johansson, G.: Visual motion perception. *Sci. Amer., 232(6)*:1975.

McIlwain, J. T.: Large receptive fields and spatial transformations in the visual system. *Intern. Rev. Physiol., 10*:223, 1976.

Michael, C. R.: Color vision. *N. Engl. J. Med., 288*:724, 1973.

Raphan, T., and Cohen, B.: Brainstem mechanisms for rapid and slow eye movements. *Ann. Rev. Physiol., 40*:527, 1978.

Rushton, W. A. H.: Visual pigments and color blindness. *Sci. Amer., 232(3)*:64, 1975.

Singer, W.: Control of thalamic transmission by corticofugal and ascending reticular pathways in the visual system. *Physiol. Rev., 57*:386, 1977.

QUESTIONS

1. Describe the optical system of the eye and the formation of an image on the retina.
2. What causes hypermetropia, myopia, and astigmatism?
3. Describe the rhodopsin cycle and its relationship to vision.
4. Explain light and dark adaptation.
5. Give the mechanism for color vision.
6. Describe the neuronal connections of the retina with the brain.
7. In what forms are the visual signals transmitted from the retina back to the cerebral cortex?
8. What factors determine visual acuity?
9. Explain depth perception.
10. Explain the mechanism for focusing of the eyes and control of this by the brain.
11. Give the mechanism of the pupillary light reflex.

HEARING, TASTE, AND SMELL 29

The function of the ear is to convert sound into nerve impulses. Figure 29–1 illustrates the general organization of the ear, showing the *external ear,* the *auditory canal,* the *tympanic membrane,* and the *ossicular system,* composed of the *malleus, incus,* and *stapes,* that transmits sound into the *cochlea.* The cochlea is also called the inner ear, and it is here that sound is converted into nerve impulses.

TRANSMISSION OF SOUND TO THE INNER EAR

Characteristics of Sound and Sound Waves. Sound is caused by compression waves traveling through the air. The transmitter of the sound, whether it be another person's voice, a radio speaker, or some noise-making device, creates the sound by alternately compressing the air and then relaxing the compression. For instance, a vibrating violin string creates sound by moving back and forth. When the string moves foward it com-

presses the air, and when it moves backward it decreases the amount of compression even below normal. This alternate compression and decompression of the air produces sound.

Sound waves travel through the air in very much the same way that waves travel over the surface of water. Thus, compression of the air adjacent to a violin string builds up extra pressure in this region, and this in turn causes the air a little farther away to become compressed. The pressure in this second region then compresses the air still farther away, and this process continues on and on until the wave finally reaches the ear.

FUNCTION OF THE TYMPANIC MEMBRANE (TYMPANUM) AND OSSICLES

When sound waves strike the tympanic membrane, the alternate compression and decompression of the air adjacent to the membrane cause the membrane to move backward and forward. The center of this membrane is connected to the handle of the *malleus;* this in turn is connected to the *incus,* and the incus to the *stapes.* These bones are called the *ossicles.* They are suspended in the *middle ear* by ligaments, so that they can rock back and forth. Movement of the handle of the malleus, therefore, causes the stapes also to move back and forth against the *oval window* of the cochlea, thus transmitting the sound into the cochlear fluid.

Transformation of the Pressure of the Sound Waves by the Ossicular Lever System. If sound waves were applied directly to

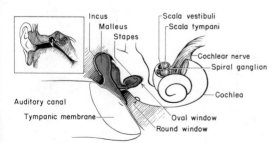

Figure 29–1. General organization of the ear, showing the external ear, the ossicular system, and the cochlea.

367

the oval window, they would not have enough pressure to move the fluid in the cochlea backward and forward to produce adequate hearing, because fluid has many times as much inertia as air, and a correspondingly greater amount of pressure is required to cause movement of fluid. The tympanic membrane and the ossicular system transform the pressure of the sound waves into a usable form by the following means:

The sound waves are collected by a very large membrane, the tympanic membrane, the area of which is approximately 70 sq. mm., or 22 times that of the oval window, which has an area of only 3.2 sq. mm. Therefore, 22 times as much sound energy is collected as could be collected by the oval window alone, and all of this is transmitted through the ossicles to the oval window. Thus, the pressure of movement of the foot of the stapes is increased to about 22 times that which could be effected by applying sound waves directly to the oval window. This pressure is now sufficient to move the fluid of the cochlea backward and forward.

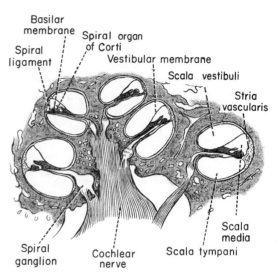

Figure 29-2. Cross-section of the cochlea, showing its coiled tubes composed of the scala vestibuli, the scala tympani, and the scala media. The cochlear nerve leading from the basilar membrane is also illustrated. (From Goss: Gray's Anatomy of the Human Body. Lea & Febiger.)

TRANSMISSION OF SOUND INTO THE COCHLEA BY THE BONES OF THE HEAD

The cochlea lies in a bony chamber inside the temporal bone of the skull. Therefore, any vibration of the skull also can cause vibration of the fluid in the cochlea. Ordinarily, sound waves in the air cause almost no vibration in the skull bones, but clicking of the teeth or holding vibrating devices such as a tuning fork or special sound vibrators against the skull can cause bone vibrations. In this way, instead of the fluid vibrating inside the cochlea, the cochlea vibrates around the fluid, allowing the cochlea to react to the sound as if the sound had entered via the tympanic membrane and ossicles.

FUNCTION OF THE COCHLEA

Physiologic Anatomy of the Cochlea. The cochlea is a membranous device formed of coiled tubes. The outside appearance of the cochlea is illustrated in Figure 29–1, and a cross-section of its structure is shown in Figure 29–2. And, finally, a functional diagram of the cochlea is illustrated in Figure 29–3, showing the coiled tubes of the cochlea

stretched out linearly. The total length of this uncoiled cochlea is about 3.5 cm.

The cross-sectional diagram of Figure 29–2 shows that the cochlea is actually composed of three separate tubes lying side by side, called the *scala vestibuli,* the *scala tympani,* and the *scala media.* All of these tubes are filled with fluid, and they are separated from each other by membranes. The membrane between the scala vestibuli and scala media is so thin that it never obstructs the passage of sound waves. Its function is simply to separate the fluid of the scala media from that of the scala vestibuli. The two fluids have separate origins, and the chemical differences between them are important for proper operation of the sound receptor cells. On the other hand, the membrane separating the scala media from

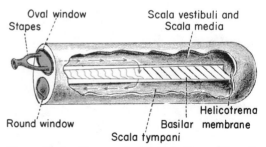

Figure 29-3. Movement of fluid in the cochlea when the stapes is thrust forward.

the scala tympani, called the *basilar membrane,* is a very strong structure that does impede the sound waves. It is supported by about 25,000 reedlike, thin spines that project into the membrane from one side. These spines, called *basilar fibers,* protrude most of the distance across the membrane. And, finally, located on the surface of the membrane are the sound receptor cells called the *hair cells.*

In the uncoiled cochlea depicted in Figure 29–3, the membrane between the scala vestibuli and the scala media has been eliminated because it does not affect sound transmission in the cochlea in any way. So far as the transmission of sound is concerned, the cochlea is composed of two separate tubes rather than three, and the two tubes are separated by the basilar membrane. The reedlike basilar fibers near the oval window at the base of the cochlea are short, but they become progressively longer, like the reeds of a harmonica, farther and farther up the cochlea, until at the tip of the cochlea the fibers are about two and one-half times as long as those near the base.

RESONANCE OF SOUND IN THE COCHLEA

Conduction of Sound in the Cochlea. As each sound vibration enters the cochlea, the oval window at first moves inward and pushes the fluid of the scala vestibuli deeper into the cochlea, as indicated by the arrows in Figure 29–3. And the increased pressure in the scala vestibuli bulges the basilar membrane into the scala tympani; this pushes fluid in this chamber toward the round window and causes it to bulge outward. Then, as the sound vibration causes the stapes to move backward, the procedure is reversed, with the fluid moving in the opposite direction along the same pathway and the basilar membrane bulging now into the scala vestibuli.

Resonance in the Cochlea. A phenomenon called *resonance* occures in the cochlea and causes each sound frequency to vibrate a different section of the basilar membrane. These vibrations are similar to those that occur in many musical instruments and can be explained as follows: When the string of a violin is pulled to one side it becomes stretched a little more than usual, and this stretch makes the string then move back in the other direction. However, as it moves, it builds up momentum and, therefore, does not stop moving when it reaches its normal straight position. The momentum causes the string to become stretched once again but this time in the opposite direction; and the string then moves back in the first direction. The cycle continues over and over again so that once the string starts vibrating it continues for a while.

Two factors determine the frequeny at which a string will vibrate. First, the greater the *tension* developed by the string, the more rapidly it turns around and moves in the opposite direction, and the higher is the frequency of vibration. The second factor is the *mass* of the string. The greater its mass, the greater is the momentum developed during vibration, and the longer it will take for the string to change direction of movement; therefore, the lower will be the sound frequency. These same two factors, mass and elastic tension, apply to resonance in the cochlea. The vibrating mass in the cochlea that vibrates back and forth is the fluid between the oval and round windows, while the elastic tension is mainly the tension developed by bending the basilar fibers.

When *high frequency* sound enters the oval window, the sound wave travels along the basilar membrane only a short distance, as shown in Figure 29–4A, before a point of resonance is reached. The basilar membrane at this point vibrates "in tune" or "in resonance" with the frequency of the sound. As a result, the membrane moves back and forth forcibly at this point, while the vibrational movement is very slight elsewhere in the membrane. When a *medium frequency* sound enters the oval window, the wave travels much farther along the basilar membrane before the area of resonance is reached, as illustrated in Figure 29–4B. And, finally, a *low frequency* sound wave travels almost all the way to the end of the membrane before it reaches its resonance point, as illustrated in Figure 29–4C.

High frequency sound waves resonate near the base of the cochlea for two reasons: First, the mass of fluid from the oval and round windows to the basal regions of the basilar membrane is very slight. Second, the basilar fibers, which provide the elastic tension for the vibrating system, have far greater rigidity at the basal end of the membrane than toward its tip. The combined effect of these two factors, the low mass and the greater rigidity, causes the basilar membrane to resonate at very high frequencies near its base.

On the other hand, when a sound wave

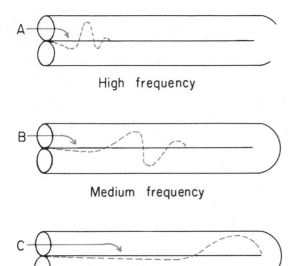

High frequency

Medium frequency

Low frequency

Figure 29-4. Diagrammatic representation of sound waves traveling along the basilar membrane and resonating at different points of high, medium, and low frequency sounds.

travels all the way to the tip of the basilar membrane, the mass of fluid that must move is very great. Also, the basilar fibers are longer and less rigid near the tip of the cochlea. The combined effect of these two factors,

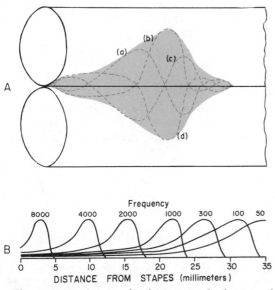

Figure 29-5. (A) Amplitude pattern of vibration of the basilar membrane for a medium frequency sound. (B) Amplitude patterns for sounds of frequencies between 50 and 8000 per second, showing resonance at different points for the different frequencies.

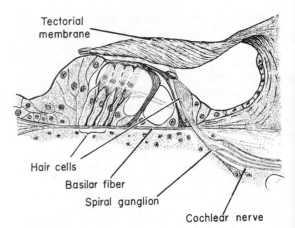

Figure 29-6. The organ of Corti, showing the hair cells and the tectorial membrane pressing against the projecting hairs. (Modified from Bloom and Fawcett: A Textbook of Histology, 10th ed. W. B. Saunders Co.)

the great mass and the low rigidity of the elastic component, makes the basilar membrane near this end resonate at a very low frequency. Similarly, intermediate frequency sounds resonate at intermediate points along the membrane. Figure 29-5A illustrates the amplitude of vibration of different parts of the basilar membrane for a medium frequency sound wave, and Figure 29-5B represents the degree of vibration of the membrane for several different sound frequencies from very low to very high.

Determination of Pitch by the Cochlea. Located on the surface of all the basilar fibers (Fig. 29-6) are special nerve receptors, called *hair cells* because of hairy projections (cilia) at one end. These cells are stimulated when the basilar fibers vibrate back and forth. When the hair cells near the base of the cochlea are stimulated, the brain interprets the pitch of the sound as that of a very high frequency. When the hair cells in mid-cochlea are stimulated, the brain interprets the sound as one of intermediate pitch, and stimulation of those at the tip of the cochlea is interpreted as low pitch. It can be seen, therefore, that the pitch of a sound is determined by the point in the cochlea at which the basilar membrane vibrates.

Determination of Loudness of a Sound. The loudness of a sound is determined by the intensity of movement of the basilar fibers. The greater the displacement back and forth, the more intensely are the hair cells stimulated, and the greater is the number of impulses transmitted into the brain to indicate the degree of loudness. For instance, if a hair cell near the base of the cochlea transmits only one impulse per second, the frequency of the sound will still be interpreted as one of very high pitch, but the loudness of the sound will be almost zero. If the same hair cell is stimulated 1000 times per second, the pitch of the sound will remain the same, but the loudness will be extreme.

CONVERSION OF SOUND ENERGY INTO NERVE IMPULSES

The hair cells of the organ of Corti are almost identical with the hair cells of the vestibular apparatus for maintaining equilibrium, which were described in Chapter 24. It is known that the hair cells in the latter apparatus are stimulated by bending the cilia that extend from the ends of the cells. Therefore, it can be presumed that the hair cells of the cochlea are also stimulated by bending their cilia when sound vibrates the basilar membrane. A suggested mechanism by which this could occur involves the *tectorial membrane,* which is also shown in Figure 29–6. Vibration of the basilar membrane causes the hair cells lying on its upper surface to *rock* back and forth; this bends the projecting cilia where they abut against the tectorial membrane. Bending of the cilia excites the cells and generates impulses in the small filamentous cochlear nerve endings that entwine the hair cells.

TRANSMISSION OF SOUND INTO THE CENTRAL NERVOUS SYSTEM

Figure 29–7 shows the pathway for transmission of sound impulses from the cochlea into the central nervous system. After passing through the cochlear nerve, the impulses are transmitted through at least five different levels in the brain, which may be listed: (1) the *dorsal* and *ventral cochlear nuclei,* (2) the *trapezoid body* and *superior olivary nucleus,* (3) the *inferior colliculus,* (4) the *medial geniculate body,* and (5) the *auditory cortex.*

Function of Lower Auditory Centers — Determination of Sound Direction. The function of the lower auditory centers is not well understood, but several of them are especially involved in the localization of the direction from which sound is coming. For instance, if sound comes from the left, it reaches the left

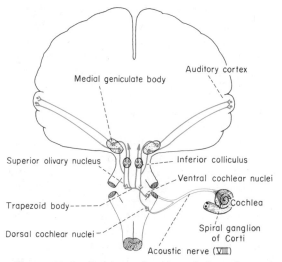

Figure 29–7. Pathways for transmission of sound impulses from the cochlea into the central nervous system.

ear before it reaches the right. If the sound is coming from directly in front, it reaches both ears simultaneously. That is, as the sound direction changes from directly ahead to one side, the time lag becomes progressively greater. This time lag is interpreted in the trapezoid body and superior olivary nucleus to determine the direction from which the sound is coming.

Another function of the auditory centers in the brain stem is reflex production of rapid motions of the head, of the eyes, or even of the entire body in response to auditory signals. That is, auditory signals pass directly into the medulla, pons, and cerebellum to alter a person's equilibrium, to make his head jerk to one side, or to make his eyes roll in one direction. Most of these rapid responses can occur even though the cerebral portions of the auditory system have been destroyed.

Figure 29–7 shows that the auditory signals from each ear are also transmitted approximately equally into the auditory pathways on both sides of the brain stem and cerebral cortex. Therefore, destruction of one of the pathways will not greatly affect one's hearing ability in either ear.

Function of the Auditory Cortex. The function of the auditory cortex has already been discussed in Chaper 23. In brief, the *primary auditory cortex* is located in the middle of the superior gyrus of the temporal lobe. This area receives the sound signals and interprets them as different sounds. However, the signals must also be transmitted into surrounding *auditory association areas* before their meanings become clear. Finally, the signals are transmitted into the *common integrative center* of the cortex, where the overall meaning of all combined auditory, visual, and other types of sensation is determined.

Destruction of one auditory cortex reduces hardly at all one's ability to hear, but destruction of both cortices greatly depresses the hearing. It is especially interesting, though, that even with both auditory cortices destroyed, an animal is still capable of hearing sounds of extreme intensity. It seems that the medial geniculate bodies (which are in reality extended protrusions of the thalamus) are the regions that receive these sounds, for destruction of these areas then causes total deafness. This illustrates again that many sensory functions of the brain can be performed by the thalamus and its related structures independently of the cerebral cortex.

DEAFNESS AND HEARING TESTS

Any damage to the ossicular system, the cochlea, the cochlear nerve, or to the pathways for transmission of sound signals to the auditory cortex can cause partial or total deafness. The types of deafness are generally separated into two categories, *conduction deafness* and *nerve deafness*. The term conduction deafness means deafness caused by failure of sound waves to be conducted from the tympanic membrane through the ossicular system into the cochlea. Nerve deafness, on the other hand, means failure of auditory nerve signals to reach the auditory cortex because of damage to the cochlea itself or to any portion of the neurogenic transmission system for sound.

One of the most common causes of *conduction deafness* is repeated blockage of the *eustachian tube.* This tube connects the middle ear with the nasopharynx, and it opens temporarily every time one swallows. Its function is to keep the pressure inside the middle ear equal to the pressure of the air, so that no pressure difference will exist between the two sides of the tympanic membrane. If this tube becomes plugged because of a cold, because of allergic swelling of the nasal membranes, or for some other reason, then the air in the middle ear becomes absorbed, and in its place a serous fluid collects. And this also causes the tympanic membrane to be pulled inward because of lowered pressure in the middle ear. Fibro-

blasts tend to grow into the serous fluid and cause fibrous tissue between the ossicles and the walls of the middle ear. If this process continues long enough, the ossicles finally become so firmly bound to the walls of the middle ear that sound conduction by way of the ossicles into the cochlea becomes almost nil.

Nerve deafness is characteristic of old age; almost all older people normally develop at least some degree of this type of deafness, especially for very high frequency sounds. This is probably caused by the aging process in the cochlea itself.

HEARING TESTS

Almost any type of sound instrument can be used to check a person's hearing; the most common of these for many years has been the *tuning fork* or the tick of a watch. After striking a tuning fork, the sound can usually be heard for about 30 seconds if the fork is held near the normal ear. If the person has conduction deafness, the ear is unable to hear the sound, but placing the butt of the fork against the skull will allow transmission of vibrations to the cochlea via the bones of the skull. If the cochlea and the neural transmission system are both still functioning properly, the sound is then heard even though it could not be heard by air conduction. On the other hand, if the person has nerve deafness, he will still be unable to hear the tuning fork even by bone conduction.

The Audiometer. In audiology clinics a special sound emitter, called an audiometer, is used to measure the degree of deafness. This is an electronic apparatus capable of generating sound of all frequencies in an earphone or in a vibrator placed against the bone of the skull. The apparatus is calibrated so that the zero mark corresponds to the intensity of sound required for a normal person barely to hear it. If the person is deaf for sound of a particular frequency, the sound intensity must be increased far above the zero level before he can hear it, and the amount of extra sound energy that must be added is said to be the *hearing loss* for that particular frequency.

Figure 29–8 shows an *audiogram* from a person who has conduction deafness. Approximately 40 to 60 decibels of extra sound energy had to be transmitted into the earphone at each frequency for the person to hear the sound. However, when the skull vibrator,

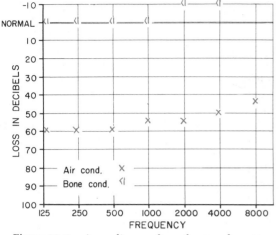

Figure 29–8. An audiogram from the ear of a person with conduction deafness.

which causes bone conduction, was used instead of the earphone, no extra energy was required; indeed, at the high frequencies the hearing was even better than normal for bone conduction. This illustrates that bone conduction was at least normal, which means that the cochlea and auditory pathway were also normal. However, conduction of sound through the ossicular system was greatly impaired.

The decibel system for expressing sound energy is a logarithmic scale rather than a linear one. Thus, a change of 10 decibels is a 10-fold change in sound energy, 20 decibels is a 100-fold change, 30 decibels is 1000-fold, and 60 decibels is 1,000,000-fold. For example, in the audiogram of Figure 29–8, the hearing loss at most frequencies is about 60 decibels, which means that this deaf person actually had a hearing loss of about 1,000,000-fold.

Sounds that the normal ear hears frequently vary in intensity a hundred million-fold or more. For instance, the intensity of sound in a very noisy factory is about one million times as great as that of a quiet whisper. Therefore, a person with 60 decibels hearing loss — a one million-fold loss — can still hear sounds of very strong intensity.

TASTE

The senses of taste and smell are called the *chemical senses* because their receptors are excited by chemical stimuli. The taste receptors are excited by chemical substances in the food that we eat, while the smell receptors are excited by chemical substances in the air.

THE TASTE BUD

The sensory receptor for taste is the *taste bud,* shown in Figure 29–9. It is composed of epithelioid *taste receptor cells* arranged around a central pore in the mucous membrane of the mouth. Several very thin hair-like projections called *microvilli* protrude from the surface of each taste cell and thence through the pore into the mouth. These microvilli provide the receptor surface for taste.

Interweaving among the taste cells of each taste bud is a branching network of two or three taste nerve fibers that are stimulated by the taste cells.

Before a substance can be tasted, it must be dissolved in the fluid of the mouth, and it must diffuse into the taste pore around the microvilli. Therefore, highly soluble and highly diffusible substances such as salts or other small molecular compounds generally cause greater degrees of taste than do less soluble and less diffusible substances such as proteins or others of very large molecular size.

The Four Primary Sensations of Taste. Psychologically, we can detect four major types of taste called the *primary* taste sensations. They are: (1) salty, (2) sweet, (3) bitter, and (4) sour.

Until recent years it was believed that four entirely different types of taste buds existed, each type detecting one particular primary taste sensation. It has now been learned that every taste bud has some degree of sensitivity

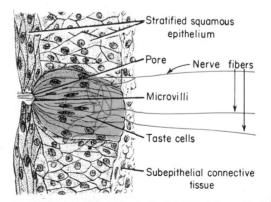

Figure 29–9. The taste bud. (Modified from Bloom and Fawcett: A Textbook of Histology, 8th ed. W. B. Saunders Co.)

for all the primary taste sensations. However, each bud usually has a greater degree of sensitivity to one or two of the taste sensations than to the others. The brain detects the type of taste by the ratio of stimulation of the different taste buds. That is, if a bud that detects mainly saltiness is stimulated to a higher intensity than buds that respond more to the other tastes, the brain interprets this as a sensation of saltiness even though the other buds are stimulated to a lesser extent at the same time.

Almost any ion will stimulate the salty buds. These buds in general determine the concentration of salts and other ionic substances present in food. The most familiar salt that stimulates these buds is sodium chloride, common table salt. Therefore, we normally associate the function of these buds almost entirely with the content of table salt in food.

The sweet taste buds detect the amount of sugars in food. This is one of the means by which animals determine whether or not fruits are ripe and whether or not unknown foods are nutritious. Essentially all wild foods that are sweet are safe to eat and contain a considerable amount of nutrition.

The bitter taste buds provide a protective function, for they detect principally the poisons in wild plants. The alkaloidal compounds of wild herbs, for instance, are among the poisons detected by the bitter taste buds. These same alkaloids, when given in appropriate quantities, are frequently very valuable drugs. Many drugs such as quinine, though of very bitter taste, have a beneficial effect when used properly, and yet can be lethal when used in too great a quantity.

The sour taste buds detect the degree of acidity of food — that is, they detect the concentration of hydrogen ion in the mouth. If the degree of acidity is slight, such as that of dilute vinegar, the food is usually very palatable; if the acidity is extremely strong, the taste is very unpleasant, and the food is rejected.

REGULATION OF THE DIET BY THE TASTE SENSATIONS

The taste sensations obviously help to regulate the diet. For example, the sweet taste is usually pleasant, causing an animal to choose foods that are sweet. On the other hand, the bitter sensation is always unpleasant, caus-

ing bitter foods, which are often poisonous, to be rejected. The sour taste is sometimes pleasant and sometimes unpleasant, and, likewise, the salty taste is sometimes unpleasant. The pleasantness of these types of taste is often determined by the momentary state of nutrition of the body. If a person has been long without salt, then, for reasons which are not yet understood, the salty sensation becomes extremely pleasant. If a person has been eating an excess of salt, the salty taste is very unpleasant. The same is true for the sour taste and to a lesser extent for the sweet taste. In this way the quality of the diet is automatically varied in accord with the needs of the body. That is, lack of a particular type of nutrient often intensifies one or more of the taste sensations and causes the person to choose foods having a taste characteristic of the deficient nutrient.

Importance of Smell to the Sensation of Taste.   Much of what we call taste is actually smell, because foods entering the mouth give off odors that spread into the nose. Often, a person who has a cold states that he has lost his sense of taste, but on testing for the four primary sensations of taste they are all still present.

The smell sensations, which are discussed in following sections of the chapter, function along with taste sensations to help control the appetite and the intake of food.

TRANSMISSION OF TASTE SIGNALS TO THE CENTRAL NERVOUS SYSTEM

Figure 29–10 shows the pathways for transmission of taste signals into the brain stem and then into the cerebral cortex. The signals pass from the taste buds in the mouth to the *tractus solitarius* located in the medulla. Then the signals are transmitted to the *thalamus*, and from there to the *primary taste cortex* of the opercularinsular region, as well as into surrounding taste association areas and into the common integrative area, which integrates all sensations.

Taste Reflexes   One of the functions of the taste apparatus is to provide reflexes to the salivary glands of the mouth. To do this, impulses are transmitted from the tractus solitarius in the brain stem into nearby nuclei that control secretion by the parotid, submaxillary, and other salivary glands. When food is eaten, the quality of the taste sensations, operating through these reflexes, helps to de

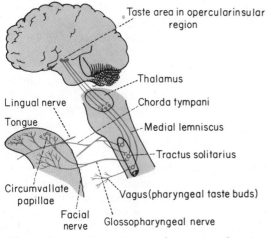

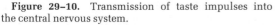

Figure 29–10. Transmission of taste impulses into the central nervous system.

termine whether the output of saliva will be great or slight.

SMELL

The sense of smell is vested in the *olfactory epithelium,* located on each side in the upper reaches of the nasal cavity. The location of these areas is illustrated in Figure 29–11, which shows a cross-section of the air passages of the nose as well as connections of the

olfactory epithelium with the nervous system.

THE OLFACTORY RECEPTORS

The olfactory epithelium contains numerous nerve receptors, called *olfactory cells,* which are shown in Figure 29–12. These are special types of nerve cells that project small microvilli called *olfactory hairs* or *cilia* outward from the nasal epithelium into the overlying mucus. It is the olfactory hairs that detect the different odors.

The means by which odors excite the olfactory hairs are not well understood. However, those odors most easily smelled are, first, the very highly volatile substances, and, second, substances that are highly soluble in fats. The necessity for volatility is easily understood, for the only means by which an odor can reach the high spaces of the nose is by air transport to this region. The reason for the fat solubility factor seems to be that the olfactory hairs themselves are outcroppings of the cell membrane of the olfactory cell, and all cell membranes are composed mainly of fatty substances. Presumably, then, the odoriferous substance becomes dissolved in the olfactory hairs, and this produces nerve impulses in the olfactory cells.

The Primary Sensations of Smell. It has been very difficult to study individual olfactory cells, for which reason we are not yet sure

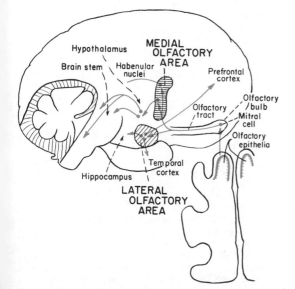

Figure 29–11. Organization of the olfactory system, and pathways for transmission of olfactory impulses into the central nervous system.

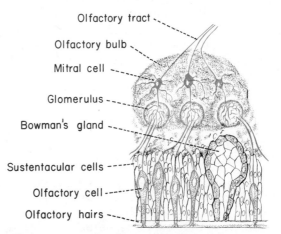

Figure 29–12. The olfactory membrane, showing especially the olfactory cells with their cilia protruding into the overlying mucus of the nose. (Modified from Bloom and Fawcett: A Textbook of Histology, 8th ed. W. B. Saunders Co.)

what primary chemical stimuli excite the various types of olfactory cells. Yet on the basis of crude experiments, the following primary sensations have been postulated:

1. Camphoraceous
2. Musky
3. Floral
4. Pepperminty
5. Ethereal
6. Pungent
7. Putrid

Thus, smell, like all the other sensations, is probably subserved by a few discrete types of cells that give rise to specific primary olfactory sensations. However, the preceding listing is mainly conjecture and probably is in error.

Adaptation of Smell. Smell, like vision, can adapt tremendously. On first exposure to a very strong odor, the smell may be very strong, but after a minute or more the odor will hardly be noticeable. The olfactory receptors apprise the person of the presence of an odor, but do not keep belaboring him with its presence. This is especially valuable when one must work in pungent surroundings.

Masking of Odors. Unlike the eye's ability to see a number of different colors at the same time, the olfactory system detects the sensation of only a single odor at a time. However, the odor may be a combination of many different odors. If both a putrid odor and a sweet odor are present, the one that dominates the other is the one that has the greater intensity, or, if both are of about equal intensity, the sensation of smell is between that of sweetness and putridness. The ability of a high intensity odor to dominate is called *masking*. This effect is used in hospitals, toilets, and other areas to make pungent surroundings pleasant; incense may be burned or some odoriferous but pleasant-smelling substance may be evaporated into the air to mask the less desirable odors.

TRANSMISSION OF SMELL SIGNALS INTO THE NERVOUS SYSTEM

Because smell is a subjective phenomenon that can be studied satisfactorily only in human beings, very little is known about transmission of smell signals into the brain. Smell pathways terminate in two major areas of the brain called the *medial olfactory area* and the *lateral olfactory area* respectively,

both illustrated in Figure 29–11. The medial olfactory area lies in the very middle of the brain, while the lateral olfactory area lies laterally on the undersurface of the brain covered over by the temporal lobe.

The medial olfactory area is responsible primarily for the primitive functions of the olfactory system, such as eliciting salivation in response to smell, licking the lips, and causing an animal to stalk a juicy meal.

On the other hand, the lateral olfactory area includes portions of the amygdala, and it is very closely associated with higher functions of the nervous system; direct pathways pass from this area into the temporal cortex, the hippocampus, and the prefrontal cortex, all important regions of cortical function. The lateral olfactory area is concerned with complicated responses that depend on olfactory stimuli. Thus, recognition of a certain type of smell as belonging to a particular animal is a function believed to be performed by this area. And recognition of various delectable or detestable foods presumably is also a function of this area. In human beings, tumors located in this region frequently cause the person to perceive very abnormal smells which may be of any type, pleasant or unpleasant, and continuing sometimes for months on end.

REFERENCES

Ainsworth, W. A.: The perception of speech signals. *Sci. Prog. 62*:33, 1975.

Aitkin, L. M.: Tonotopic organization at higher levels of the auditory pathway. *Intern. Rev. Physiol., 10*:249, 1976.

Alberts, J. R.: Producing and interpreting experimental olfactory deficits. *Physiol. Behav., 12*:657, 1974.

Brugge, J. F., and Geisler, C. D.: Auditory mechanisms of the lower brainstem. *Ann. Rev. Neurosci., 1*:363, 1978.

Douek, E.: The Sense of Smell and Its Abnormalities. New York, Churchill Livingstone, Div. of Longman, Inc., 1974.

Gerber, S. E.: Introductory Hearing Science. Philadelphia, W. B. Saunders Company, 1975.

Gulick, W. L.: Hearing: Physiology and Psychophysics. New York, Oxford University Press, 1971.

Moulton, D. G.: Spatial patterning of response to odors in the peripheral olfactory system. *Physiol. Rev., 56*:578, 1976.

Moushegian, G., and Rupert, A. L.: Relations between the psychophysics and the neurophysiology of sound localization. *Fed. Proc., 33*:1924, 1974.

Norsiek, F. W.: The sweet tooth. *Am. Sci., 60*:41, 1972.

Ohloff, G., and Thomas, A. F. (eds.): Gustation and Olfaction. New York, Academic Press, Inc., 1971.

Starr, A.: Neurophysiological mechanisms of sound localization. *Fed. Proc., 33*:1911, 1974.

QUESTIONS

1. Describe the ossicular system of the ear and its transmission of sound.
2. Describe the physiologic anatomy of the cochlea.
3. How does resonance of sound occur in the cochlea?
4. Explain how the cochlea determines the loudness of a sound.
5. How do the hair cells convert sound energy into nerve impulses?
6. Trace the nerve signals from the ear to the cortex.
7. Explain how the taste bud converts taste sensations into nerve signals.
8. What are the four primary sensations of taste, and what chemical substances stimulate each?
9. Give the characteristics of function by the olfactory receptors.
10. Trace the transmission of smell signals into the nervous system.

THE GASTROINTESTINAL AND METABOLIC SYSTEMS

IX

GASTROINTESTINAL MOVEMENTS AND SECRETION, AND THEIR REGULATION

30

The function of the gastrointestinal system, illustrated in Figure 30–1, is to provide nutrients for the body. Food, after entering the mouth, is propelled through the esophagus into the stomach, and then through the small and large intestines before finally emptying out the anus. While the food passes through the gastrointestinal tract, digestive enzymes secreted by the gastrointestinal glands act on the food, breaking it into simple chemical

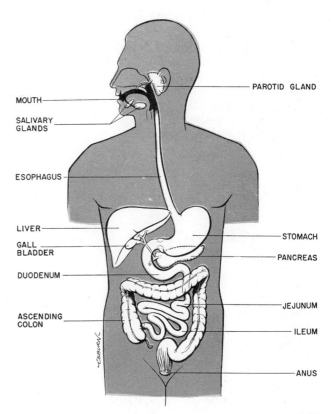

MOUTH

SALIVARY GLANDS

ESOPHAGUS

LIVER

GALL BLADDER

DUODENUM

ASCENDING COLON

PAROTID GLAND

STOMACH

PANCREAS

JEJUNUM

ILEUM

ANUS

Figure 30–1 The gastrointestinal tract from the mouth to the anus.

substances that can be absorbed through the intestinal wall into the circulating blood. The general functions of the gastrointestinal tract, therefore, can be divided into (1) propulsion and mixing of the gastrointestinal contents, (2) secretion of digestive juices, (3) digestion of food, and (4) absorption of food. The first two of these functions are discussed in the present chapter, and the remaining two in the following chapter.

PHYSIOLOGIC ANATOMY OF THE GASTROINTESTINAL TRACT

The gastrointestinal tract is essentially a long muscular tube with an inner lining that secretes digestive juices and absorbs nutrients. Figure 30–2 illustrates a typical cross-section of the gut, showing that most of the outer part is *smooth muscle* arranged in two layers, a *longitudinal layer* and a *circular layer*. Contraction of the longitudinal muscle shortens the gut, and contraction of the circular layer constricts it. The inner lining of the gut is called the *mucosa,* and it is covered on the interior by an *epithelium*. Small glands called *mucosal glands* penetrate into the deeper layers of the mucosa. These glands secrete digestive juices.

The Intramural Nerve Plexus. One of the primay controllers of gastrointestinal function is a plexus of nerves called the *intramural plexus* which is present in the wall of the gut all the way from the esophagus to the anus, forming an intertwining web of nerve fibers and nerve cell bodies. It controls muscular contraction in the gut and controls secretion by many of the glands.

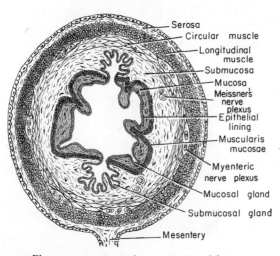

Figure 30–2.   Typical cross-section of the gut.

Parasympathetic nerve fibers from the brain to the gut, carried mainly in the vagus nerve, terminate in the intramural plexus and, when stimulated, increase the degree of activity of the nerve network, and therefore, of the gut itself. *Sympathetic nerve fibers* from the spinal cord to the gut also terminate in the intramural plexus or directly in the gut wall itself and, when stimulated, have exactly the opposite effect on activity of the plexus, decreasing its level of activity.

GASTROINTESTINAL MOVEMENTS

Two basic types of movement occur in the gastrointestinal tract, *propulsive movements* and *mixing movements*. The propulsive movements keep food moving along the gut, and the mixing movements mix the food with the gastrointestinal secretions. The movements in different parts of the gastrointestinal tract exhibit differences that need to be described separately for certain portions of the gut. However, let us first consider the general characteristics of the two types of movements.

PROPULSIVE MOVEMENTS OF THE GASTROINTESTINAL TRACT – PERISTALSIS

Food is moved along the gastrointestinal tract by *peristalsis* which is caused by slow advancement of a circular constriction, as illustrated in Figure 30–3. This has very much the same effect as encircling one's fingers tightly around a thin tube full of paste and then pulling the fingers along the tube. Any material in front of the fingers will be squeezed forward.

Mechanism of Peristalsis. Peristalsis is caused by nerve impulses that move along the intramural nerve plexus. Stimulation of any single point of the plexus causes impulses to travel around the gut and also lengthwise in both directions. The impulses traveling around the gut constrict it, and those traveling lengthwise cause the constriction to move forward. The usual rate of movement of this constriction is a few centimeters per second.

The usual stimulus that initiates peristalsis is *distention* of the gut, for this excites the local nerve plexus, causing a circular constriction to begin and to advance along the gut.

Law of the Gut. Even though peristalsis can move in both directions along the gut, it

Labels on Figure 30–2: Serosa; Circular muscle; Longitudinal muscle; Submucosa; Mucosa; Meissner's nerve plexus; Epithelial lining; Muscularis mucosae; Myenteric nerve plexus; Mucosal gland; Submucosal gland; Mesentery

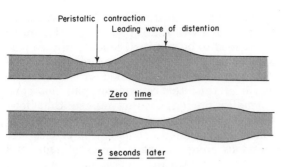

Peristaltic contraction
Leading wave of distention

Zero time

5 seconds later

Figure 30–3. Peristalsis.

most frequently moves toward the anus. The probable reason for this is that the intramural nerve plexus itself is "polarized" in this direction. When a portion of the gut becomes distended, it causes constriction of the gut on the headward side of the distention and relaxation on the anal side. The constriction pushes the food forward and the relaxation allows easy forward movement; the movement, therefore, is normally analward rather than backward. This is called the *law of the gut.*

MIXING MOVEMENTS IN THE GASTROINTESTINAL TRACT

The mixing movements consist of two basic types, (1) *weak peristaltic movements* that fail to move the food forward but nevertheless do succeed in mixing the intestinal contents adjacent to the wall of the gut, and (2) *segmental movements,* which are isolated constrictions that occur at many points along the gut at the same time. The segmental movements occur rapidly, several times each minute, and each time "chop" the food into new segments. The precise characteristics of the mixing movements are quite different in the different parts of the gut, for which reason they will be described specifically for each part of the gastrointestinal tract.

Let us now begin at the upper end of the gastrointestinal tract and describe both the propulsive and mixing movements as the food passes analward.

SWALLOWING

Swallowing is initiated when a bolus of food is pushed backward by the tongue into the pharynx. The bolus stimulates *swallowing re-ceptor areas* located all around the opening of the pharynx, and impulses pass to the brain stem to initiate a series of automatic muscular contractions as follows:

1. The soft palate is pulled upward to close the posterior part of the nose from the mouth.

2. The vocal cords in the larynx close strongly, and the epiglottis swings backward over the glottis, these two effects combining to prevent food from going into the trachea.

3. The muscular esophageal sphincter that normally keeps the upper end of the esophagus closed becomes relaxed, followed immediately by upward movement of the larynx that pulls the esophagus open.

4. The pharyngeal muscles then constrict to force the bolus of food from the pharynx downward into the esophagus.

NERVOUS CONTROL OF SWALLOWING. Figure 30–4 illustrates the nervous pathways involved in the swallowing mechanism, showing that the swallowing receptors transmit impulses from the posterior mouth and throat mainly through the *trigeminal nerve* into the reticular substance of the *medulla oblongata* where the swallowing center is located. Once this center has been activated, the sequence of muscular reactions listed above occurs automatically and usually cannot be stopped. Nerve signals go by way of the *glossopharyngeal* and *vagus nerves* to the pharyngeal and laryngeal muscles, and contraction of these move the bolus of food into the upper esopha-

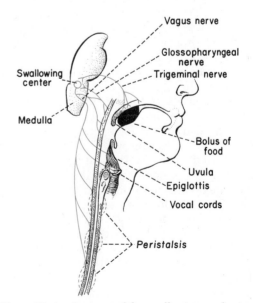

Vagus nerve

Glossopharyngeal nerve

Trigeminal nerve

Swallowing center

Medulla

Bolus of food

Uvula

Epiglottis

Vocal cords

Peristalsis

Figure 30–4. Anatomy of the swallowing mechanism.

gus. Then impulses from the vagi activate the proximal portion of the esophagus to push the food on toward the stomach.

Esophageal Stage of Swallowing. The musculature of the pharynx and of the upper one-third of the esophagus is different from that in the rest of the gastrointestinal tract, for this muscle is skeletal muscle, controlled directly by nerves from the brain; in contrast, the remainder of the muscle in the esophagus and gastrointestinal tract is smooth muscle and is only indirectly controlled by the central nervous system through the effects of the autonomic nervous system on the intramural plexus. Therefore, all contractions of the pharynx and upper one-third of the esophagus are initiated directly by vagal and glossopharnygeal nerve impulses, and without these nerves the act of swallowing becomes paralyzed.

However, once food has reached the middle third of the esophagus, the distention elicits a typical peristaltic wave called the *secondary peristaltic wave* of the esophagus that pushes the food the rest of the way into the stomach.

The entire time required for food to pass from the pharynx to the stomach is 5 to 10 seconds.

Function of the Gastroesophageal Constrictor. Approximately 5 cm. above the point where the esophagus empties into the stomach, the wall of the esophagus is thickened, and the muscle coat is considerably stronger than elsewhere in the esophagus. Furthermore, this portion of the esophagus remains mildly constricted under normal conditions, in contrast to the remainder of the esophagus which is normally relaxed except when peristalsis is occurring. This thickened and normally constricted area of the esophagus is called the *gastroesophageal constrictor,* and its function is to help prevent reflux of gastric contents into the esophagus.

When swallowed food passes down the esophagus, a "leading wave" of relaxation is transmitted through the intramural nerve plexus of the esophageal wall to the gastroesophageal constrictor and causes it to relax. This allows the food to pass on into the stomach. On the other hand, when food in the stomach attempts to move backward up the esophagus, the constrictor normally does not relax but, instead, prevents any backward flux.

However, the gastroesophageal constrictor operates differently in two abnormal conditions. First, in vomiting, reverse peristalsis in the stomach will then open the constrictor to allow food to be vomited. Second, in rare persons, the intramural nerve plexus of the esophagus is poorly formed or absent. Because of this, the constrictor will not open normally during swallowing, so that food tends to collect above the constrictor. This condition is called *achalasia;* it causes tremendous enlargement of the esophagus, and the collected food in the esophagus often becomes infected with bacteria. And the infection often then spreads to the esophageal wall, causing ulcers to develop. The esophagus sometimes holds as much as a full liter of swallowed food for many hours. It also causes pain in the chest or in the deep throat.

MOTOR FUNCTIONS OF THE STOMACH

The motor functions of the stomach are threefold: (1) storage of large quantities of food immediately after a meal, (2) mixing of this food with gastric secretions, and (3) emptying of the food from the stomach into the small intestine. The basic functional parts of the stomach are illustrated in Figure 30–5. Physiologically the stomach can be divided into two major parts, the *corpus* and the *antrum.*

Storage Function of the Stomach. Food, on emptying from the esophagus into the stomach, first enters the corpus, which is an extremely elastic bag that can store very large quantities of food. Furthermore, the

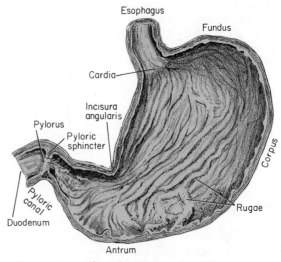

Figure 30–5. Physiologic anatomy of the stomach.

tone of the corpus is normally very slight, so that even extreme amounts of food do not increase the pressure in the stomach greatly. Thus, the corpus is mainly a receptive organ for holding food until it can be utilized by the remainder of the gastrointestinal tract.

Mixing in the Stomach and Formation of Chyme. The gastric glands, which cover most of the mucosa of the corpus, secrete large quantities of digestive juices which come into contact with the stored food. Weak, rippling peristaltic waves, called *tonus waves* or *mixing waves,* pass along the stomach wall approximately once every 20 seconds. These begin in any part of the corpus and spread for variable distances toward the antrum, and they become more intense when food is present in the stomach than when it is not present.

The mixing waves mix the gastric secretions with the outermost layer of food, gradually eating the stored food away, and these waves also gradually move the mixture toward the antral part of the stomach. On entering the antrum, the waves become stronger, and the food and gastric secretions become mixed to a greater and greater degree of fluidity. As the food becomes thoroughly mixed with the gastric secretions, the mixture takes on a milky-white sludge appearance and is then called *chyme.*

Propulsion of Chyme Through the Pylorus, and Emptying of the Stomach. The opening from the stomach into the duodenum is known as the *pylorus.* At this point, the muscular coat is greatly hypertrophied, forming a very strong muscular sphincter called the *pyloric sphincter.* The tonus or mixing waves are rarely strong enough to push chyme through the pylorus into the duodenum. However, occasional very powerful peristaltic contractions occur, beginning either in the corpus or antrum, and these generate as much as 50 mm. Hg pressure in the prepyloric portion of the antrum. This is usually enough pressure to push open the pyloric sphinter and propel the chyme on into the duodenum.

REGULATION OF STOMACH EMPTYING. Emptying of the stomach is controlled mainly by the intensity of the strong peristaltic waves. The pylorus itself usually remains at least slightly constricted all the time. Therefore, a certain degree of pressure is required before chyme can pass through. A weak peristaltic wave fails to move the chyme, but a strong peristaltic wave succeeds.

Among the different factors that determine whether or not the peristaltic wave will succeed are the following:

Degree of fluidity of the chyme. Obviously, the better the food has become mixed with gastric secretions the more easily it can flow through the narrow passageway of the pylorus. Therefore, ordinarily, food will not pass out of the stomach until it has been thoroughly mixed.

Quantity of chyme already present in the small intestine. When a large amount of chyme has already emptied into the small intestine, particularly when a large portion of this is still present in the duodenum, a reflex called an *enterogastric reflex* spreads backward through the intramural nerve plexus from the duodenum to the stomach to inhibit peristalsis. In this way, the duodenum keeps itself from becoming overfilled.

Presence of acids and irritants in the small intestine. The gastric secretions, as will be discussed later in the chapter, are highly acidic, but the acid in the chyme entering the duodenum is ordinarily neutralized by pancreatic secretions that also empty into the duodenum. Until the acid becomes neutralized by pancreatic juice, irritation of the wall of the duodenum elicits an enterogastric reflex similar to that elicited by distention; this too inhibits the peristaltic waves in the stomach, thus stopping gastric emptying. In this way, the duodenum protects itself from too much acidity. Likewise, any other irritant also causes an enterogastric reflex that will do the same thing.

Presence of fats in the small intestine. When fats enter the small intestine from the stomach, they extract from the mucosa of the duodenum and jejunum several hormones, including cholecystokinin, small amounts of secretin, and probably others. These are absorbed into the blood and carried to the stomach. Here they inhibit stomach peristalsis and slow stomach emptying. This mechanism allows adequate time for fat digestion to occur in the small intestine. Proteins and carbohydrates, on the other hand, have much less inhibitory effect on stomach emptying. Fortunately, both of these foods are digested much more easily in the intestinte than is fat.

In summary, gastric emptying is determined mainly by fluidity of the contents in the stomach and the state of the duodenum. If the duodenum is already filled, has irritant substances in it, or has fat in it, emptying will proceed very slowly, but if the duodenum is empty and the contents of the stomach are very fluid, emptying will proceed rapidly.

MOVEMENTS OF THE SMALL INTESTINE

Propulsive Movements.   It is in the small intestine that the most typical peristalsis occurs, for distention of any portion of the small intestine with chyme initiates a peristaltic wave. Peristalsis is far more intense when the parasympathetic nerves are stimulated, and sympathetic stimulation can inhibit greatly or totally block peristalsis.

Mixing Contractions – Segmentation.   The presence of chyme in the small intestine initiates a type of contraction called *segmentation,* which is illustrated in Figure 30–6. When the small intestine becomes distended, many constrictions occur either regularly or irregularly along the distended area. As shown in Figure 30–6, the intestine becomes "chopped" into small sausagelike vesicles. The constrictions then relax, but others occur at different points a few seconds later. Thus, repetitive "chopping" of the chyme keeps it mixed continually while it is in the small intestine.

Both the segmentation and peristaltic movements of the small intestine are controlled by the intramural nerve plexus.

Emptying of Intestinal Contents at the Ileocecal Valve.   Figure 30–7 shows the *ileocecal valve* where the small intestine empties into the large intestine. Note that the small intestine actually protrudes forward into the colon. This projection of the valve prevents the contents of the colon from regurgitating into the small intestine; instead the lips of the valve

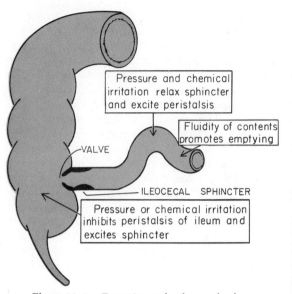

Figure 30–7.   Emptying at the ileocecal valve.

simply close on themselves when pressure builds up in the colon.

Emptying of the small intestine at the ileocecal valve occurs in very much the same way that the stomach empties; that is, peristaltic waves in the small intestine build up pressure behind the valve and push chyme forward into the colon, but if the colon has become too full, intramural nerve reflexes can inhibit peristalsis and thereby slow or stop the emptying.

MOVEMENTS OF THE COLON

The functions of the colon, illustrated in Figure 30–1, are (1) absorption of water and electrolytes from chyme and (2) storage of fecal matter until it can be expelled. The first half of the colon is concerned mainly with absorption and the distal half with storage. And, except when the bowels are to be emptied, the movements of the colon are usually very sluggish.

Mixing Movements.   The mixing movements of the colon are similar to the segmentation movements of the small intestine, but they occur much more slowly. Concentric contractions of the colon divide the colon into large pockets called *haustrations,* which are illustrated in Figures 30–1, 30–7, and 30–8. The circular constrictions last for about 30 seconds, and after another few minutes occur

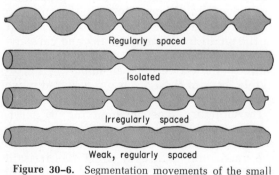

Figure 30–6.   Segmentation movements of the small intestine.

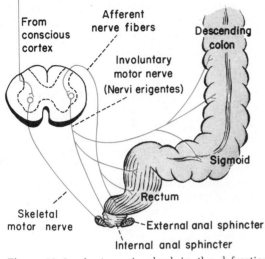

Figure 30-8. Anatomy involved in the defecation reflex.

again in nearby but not the same areas. Thus, fecal material is slowly "dug" into and rolled over in much the same manner that one spades the earth.

Ordinarily, 500 to 800 ml. of chyme is emptied into the colon each day, and of this, most of the water and electrolytes are reabsorbed before defecation takes place, leaving an average volume of feces of 100 to 200 ml. each day.

Propulsive Movements. The typical peristaltic movements that occur in the small intestine do not occur in the colon. In fact, the colon has no peristaltic movements at all 95 to 99 per cent of the time. Yet when the colon becomes overfilled, several strong peristaltic movements, called *mass movements,* occur one after another. These propel the fecal material long distances, sometimes all the way from the ascending colon to the descending colon. After a few minutes the mass movements cease, but reappear again many hours later when some part of the colon becomes overfilled again.

Defecation. When the mass movements of the colon have succeeded in moving fecal material into the rectum, a special reflex called the *defecation reflex* occurs. This causes emptying of the rectum and lower parts of the colon. Figure 30-8 shows that filling the rectum excites nerve endings that send signals into the lower part of the spinal cord. These cause reflex signals to be transmitted through the sacral parasympathetic nerves to the descending colon, sigmoid flexure, rectum, and *internal anal sphincter,* causing contraction of

the gut wall but relaxation of the sphincter. This reflex thus causes emptying of the bowels if the *external anal sphincter* is also relaxed. However, the external anal sphincter is a skeletal muscle that guards the outer opening of the anus, controlled by voluntary skeletal nerves that can be relaxed or tightened according to the will of the person. If the time is not propitious for emptying the bowels, tightening of the external anal sphincter prevents defecation despite the defecation reflex. On the other hand, relaxation of the sphincter allows defecation to take place.

If a person prevents defecation when the defecation reflex occurs, the reflex usually dies out after a few minutes but returns a few hours later. Also, a person can frequently initiate a defecation reflex at will by tightening his abdominal muscles, which compresses the rectal wall and elicits the typical reflex. Unfortunately, this elicited reflex is usually much weaker than the natural reflex, so that defecation is less efficacious under these conditions than when the reflex has been elicited naturally.

SPECIAL TYPES OF GASTROINTESTINAL MOVEMENTS

Antiperistalsis and Vomiting. Occasionally, some intensely irritating substance enters the gastrointestinal tract. An immediate effect of this is an increase in the rate of local secretion of mucus, which helps to protect the interior of the gut. Simultaneously, the local gut wall contracts intensely. For reasons not too well understood, these intense local contractions elicit *antiperistalsis,* meaning peristalsis backward toward the mouth, rather than forward peristalsis. Food usually cannot move backward from the colon into the small intestine because of the ileocecal valve, but it can move all the way from the tip of the small intestine back to the stomach.

On reaching the stomach, the irritative material is then rapidly expelled by the vomiting process as follows: The intense irritation of the gut initiates signals transmitted through the visceral sensory nerves into the brain. These cause a sensation of *nausea,* and, if the signals are strong enough, they will also cause an automatic reflex integrated in the medulla oblongata called the *vomiting reflex.* This reflex first closes the airway into the trachea, then causes relaxation of the gastroesophageal constrictor and very tight contraction of both the diaphragm and the abdominal muscles. The squeezing action of these muscles on the stomach pushes food out of the stomach upward through the esophagus and mouth. This is the vomiting process.

Gastrocolic and Duodenocolic Reflexes. Almost everyone is familiar with the natural desire to defecate following either a heavy meal or the first meal of the day. The cause of this is the *gastrocolic* and *duodenocolic reflexes,* mainly the latter. These reflexes are elicited by increased filling of the stomach and duodenum, which in turn transmit signals downward along the intramural nerve plexus to the colon to cause increased excitability of the entire colon, initiating both mass movements and defecation reflexes.

The Peritoneal Reflex. Irritation of the peritoneum in any way, whether caused by cutting the peritoneum during an abdominal operation, by infection of the peritoneum, or even by a severe blow to the abdomen that causes trauma of the peritoneum, will elicit a *peritoneal reflex* that strongly excites the sympathetic nerves to the gut. These nerves in turn *inhibit* gastrointestinal activity and thereby stop or slow movement of chyme along the intestinal tract. Obviously, lack of movement aids in the repair of the peritoneal damage.

Mucosal Reflexes. Irritation inside the gut or distention of the gut ordinarily excites the intramural plexus rather than inhibiting it. Reflexes occur mainly locally, causing increased local secretion and increased local motor activity. The secretion dilutes the irritating factor, and if the irritation is not strong enough to initiate antiperistalsis and vomiting, then instead the motor activity moves the irritant on through the gut. When the irritation is great it causes *diarrhea.* And if the degree of irritation increases still more, it can cause diarrhea from the lower part of the intestinal tract while causing vomiting from the upper part. Obviously, these reflexes are protective in nature and prevent irritating substances or infectious processes inside the intestinal tract from causing severe permanent damage.

GASTROINTESTINAL SECRETIONS

Glands are present throughout the gastrointestinal tract to secrete chemicals that mix with the food and digest it. These secretions are of two types: first, mucus which protects the wall of the gastrointestinal tract, and, second, enzymes and allied substances that break the large chemical compounds of the food into simple compounds.

Mucus. Mucus is secreted by every portion of the gastrointestinal tract. It contains a large amount of mucoprotein that is resistant to almost all digestive juices. Mucus also lubricates the passage of food along the mucosa, and it forms a thin film everywhere to prevent the food from excoriating the mucosa. And it is *amphoteric,* which means it is capable of neutralizing either acids or bases. All these properties of mucus make it an excellent substance to protect the mucosa from physical damage, and to prevent the digestion of the wall of the gut by the digestive juices.

SALIVARY SECRETION

Saliva is secreted by the *parotid, submaxillary, sublingual,* and smaller glands in the mouth. Salvia is about half *mucus* and half a solution of the enzyme *ptyalin.* The function of the mucus is to provide lubrication for swallowing. Without mucus one can hardly swallow. If one mixes food with water to take the place of the mucus, approximately ten times as much water as mucus is necessary to provide the same degree of lubrication.

The function of the ptyalin in the saliva is to begin the digestion of starches and other carbohydrates in the food. Ordinarily the food is not exposed to saliva in the mouth long enough for more than 5 to 10 per cent of the starches to become digested. However, the mixed saliva and food is usually stored in the corpus of the stomach for 30 minutes to several hours before it is mixed with stomach secretions. During this time the saliva may digest more than 50 per cent of the starches.

Regulation of Salivary Secretion. The *superior and inferior salivatory nuclei* located in the brain stem, shown in Figure 30–9, control secretion by the salivary glands. These nuclei in turn are controlled mainly by taste impulses and tactile sensory impulses from the mouth. Foods that have a pleasant taste ordinarily cause the secretion of large quantities of saliva, while some unpleasant foods may decrease salivary secretion so greatly that swallowing is made very difficult. Also, the tactile sensation of smooth-textured foods inside the mouth increases salivation, while the sensation of roughness decreases salivation. This effect presumably allows those foods that will not abrade the mucosa to be

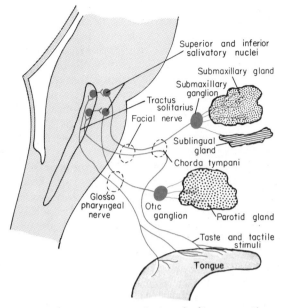

Figure 30-9. Nervous regulation of salivary secretion.

swallowed with ease, and causes the rejection of abrasive foods.

The Phases of Salivary Secretion. In addition to the salivation that occurs while food is actually in the mouth, salivation frequently occurs even before food enters the mouth — that is, when a person is thinking about or smelling pleasant food — and it continues to occur even after the food has been swallowed. Therefore, salivary secretion can be divided into three phases, the *psychic phase,* the *gustatory phase,* and the *gastrointestinal phase.* The psychic phase presumably makes the mouth ready for food and aids in the secretion of saliva as the food is presented to the mouth. The gustatory phase supplies the saliva that mixes with the food while one is chewing, and the gastrointestinal phase continues the secretion of saliva even after the food has passed for storage into the stomach. Secretion during the gastrointestinal phase is especially likely to be abundant when one has swallowed irritant foods because of nerve reflexes from the stomach wall. The saliva, on being swallowed, helps to neutralize the irritant substance, thereby relieving the irritation of the stomach.

ESOPHAGEAL SECRETIONS

The esophagus secretes only *mucus.* Normally, food passes from the mouth through the esophagus and into the stomach in about 7 seconds. This food has not been subjected to the mixing movements of the gastrointestinal tract and, therefore, is in its most abrasive state. Fortunately the esophaus is supplied with a great abundance of mucous glands that secrete mucus to protect the mucosa from excoriation.

GASTRIC SECRETIONS

Mucus. The primary function of the gastric secretions is to begin the digestion of proteins. Unfortunately, though, the wall of the stomach is itself constructed mainly of smooth muscle which itself is mainly protein. Therefore, the surface of the stomach must be exceptionally well protected at all times against its own digestion. This function is performed mainly by mucus that is secreted in great abundance in all parts of the stomach. The entire surface of the stomach is covered by a layer of very small *mucous cells* which themselves are composed almost entirely of mucus; this mucus prevents gastric secretions from ever touching the deeper layers of the stomach wall. In addition, the gastric glands that secrete the stomach digestive enzymes also secrete a major amount of mucus at the same time. And, in the antral region of the stomach where the powerful peristaltic movements occur and where excoriation of the stomach wall is particularly likely to occur, mucus is secreted by special large mucous glands that extend deep into the mucosa and that pour large amounts of viscid mucus onto the mucosal surface of the atrium. In the absence of mucus secretion, holes are eaten into the wall of the stomach in only a few hours. These holes are called *stomach ulcers.*

Digestive Substances. The major digestive substances secreted by the stomach are *hydrochloric acid* and *pepsinogen.* The hydrochloric acid activates the pepsinogen to form *pepsin,* which is an enzyme that begins the digestion of proteins.

Less abundant enzymes secreted by the stomach are *gastric lipase* for beginning the digestion of fats, and *rennin* for aiding in the digestion of *casein,* one of the proteins in milk. These enzymes are secreted in such minor quantities that they are of almost no importance.

The total quantity of stomach secretion each day is about 2000 ml.

Regulation of Gastric Secretion. NEURO-GENIC MECHANISMS. Stomach secretion is regulated by both neurogenic and hormonal mechanisms, as shown in Figure 30–10. Some of the neurogenic mechanisms are quite similar to those regulating salivary secretion. For instance, food in the stomach can cause local nervous reflexes (called submucosal reflexes) that occur entirely in the wall of the stomach itself to cause local secretion. Also, signals from the stomach mucosa to the medulla of the brain can cause reflexes back to the stomach through the vagus nerves to cause secretion. In addition, secretory signals from the medulla to the stomach can be excited by impulses originating in various other areas of the brain, particularly in the cerebral cortex.

THE "GASTRIN" MECHANISM. Gastric secretion is also regulated by a hormone called *gastrin.* When meats and certain other foods reach the antral portion of the stomach, they cause the hormone *gastrin,* a large polypeptide, to be extracted from the antral mucosa and to be absorbed into the blood stream. This hormone then passes by way of the blood to the fundic glands of the stomach, and causes them to secrete a strongly *acidic* gastric juice. The acid, in turn, greatly aids in the digestion of the meats that first initiated the gastrin

mechanism. In this way the stomach helps to tailor-make the secretion to fit the particular type of food that is eaten.

THE PHASES OF GASTRIC SECRETION. Large amounts of stomach juices are often secreted when pleasant food is simply thought of or particularly when it is smelled. This is called the *cephalic phase* of stomach secretion. It prepares the stomach for food that is to be eaten. The second phase of gastric secretion is the *gastric phase,* which is the secretion that occurs while the food is in the stomach itself. This is caused mainly by reflexes initiated by food in the stomach and by the gastrin mechanism. Finally, even after the food has left the stomach, gastric secretions continue for several hours. This is called the *intestinal phase* of gastric secretion. It is probably caused by hormones that pass to the stomach in the blood after being extracted from the intestinal mucosa by the food. Ordinarily, the amount of secretion during the intestinal phase is only about 10 per cent of the total during the other two phases.

PANCREATIC SECRETIONS

The pancreas, shown in Figure 30–11, is a large gland located immediately beneath the

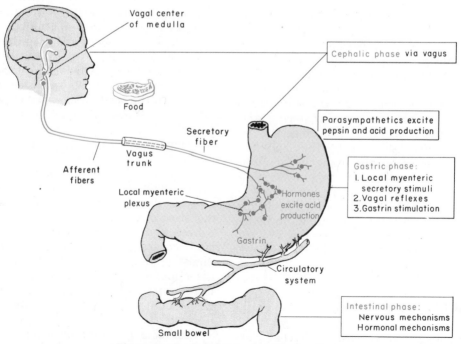

Figure 30–10.   Regulation of gastric secretion.

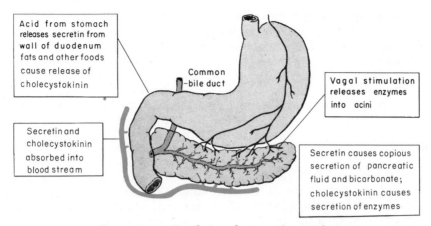

Figure 30–11.   Regulation of pancreatic secretion.

stomach. It empties about 1200 ml. of secretions each day into the upper portion of the small intestine 4 cm. beyond the pylorus. These secretions contain large quantities of *amylase* for digesting carbohydrates, *trypsin* and *chymotrypsin* for digesting proteins, *pancreatic lipase* for digesting fats, and other less important enzymes. It is obvious from this list that the pancreatic secretions are as important for digesting the food as any others of the entire gastrointestinal tract.

In addition to the digestive enzymes, pancreatic secretions contain large amounts of *sodium bicarbonate,* which react with the hydrochloric acid emptied into the duodenum in the chyme from the stomach to form sodium chloride and carbonic acid. The carbonic acid then is absorbed into the blood, becomes water and carbon dioxide, and the carbon dioxide is expired through the lungs. The net result is an increase in the quantity of sodium chloride, a neutral salt, in the intestine. Thus, pancreatic secretions neutralize the acidity of the chyme coming from the stomach. This is one of the most important functions of pancreatic secretion.

Regulation of Pancreatic Secretion.   THE "SECRETIN" MECHANISM AND NEUTRALIZATION OF CHYME.   When chyme enters the upper small intestine it causes a hormone called *secretin* to be released from the intestinal mucosa; the quantity of secretin released is especially abundant when the chyme is highly acidic. The secretin in turn is absorbed into the blood and then carried to the glandular cells of the pancreas. There it causes the cells to secrete large quantities of fluids containing extra large amounts of sodium bicarbonate. The bicarbonate then reacts with the acid of

the chyme to neutralize it. Thus, the secretin mechanism is an automatic process to prevent excess acid in the upper small intestine.

When satisfactory neutralization does not occur, the acidic chyme, containing also large quantities of the protein-digesting enzyme pepsin, is likely to erode through the wall of the duodenum, the uppermost part of the small intestine, causing duodenal ulcers. In fact, ulcer of the duodenum is about four times as common as ulcer of the stomach, because this area is not as well protected by mucous glands as is the stomach.

THE "CHOLECYSTOKININ" MECHANISM TO CAUSE ENZYME SECRETION.   At the same time that secretin is extracted from the intestinal mucosa, another hormone, *cholecystokinin,* also is extracted mainly in response to fats, but to a lesser extent in response to proteins and carbohydrates. Cholecystokinin, like secretin, passes by way of the blood to the pancreas, but, unlike secretin, it causes the secretory cells to secrete large quantities of digestive enzymes instead of sodium bicarbonate. These enzymes, on entering the duodenum, begin digesting the foods.

VAGAL REGULATION OF THE PANCREAS. Stimulation of the vagus nerve also causes the secretory cells of the pancreas to secrete highly concentrated enzymes. The quantity of fluid secreted, however, is usually so small that the enzymes remain in the ducts of the pancreas and later are floated into the intestinal tract by the copious secretion of fluid that follows secretin stimulation.

Vagal stimulation of pancreatic secretion seems to be a by-product of the vagal reflexes to the stomach. That is, some of the reflex impulses initiated by food in the stomach return to the pancreas rather than to the stomach. This allows preliminary formation of pancreatic enzymes even before the food enters the intestine. However, vagal stimulation of pancreatic secretion is probably unimportant in comparison with the hormonal stimulation by secretin and cholecystokinin.

LIVER SECRETION

The liver, shown in Figure 30–12, secretes a solution called *bile* that contains a large quantity of *bile salts,* a moderate quantity of *cholesterol,* a small quantity of the green pigment *bilirubin* which is a waste product of red blood cell destruction, and a number of other less important substances. The only substance in bile that is of importance to the digestive functions of the gastrointestinal tract is the bile salts. The remaining contents are actually waste products being *excreted* from the body fluids by this route.

The bile salts are not enzymes for digesting foods, but they act as a powerful *detergent* (a substance that lowers the surface tension at the surface between water and fats). This helps the mixing movements of the intestine break the large fat globules of the food into small globules, thus allowing the lipases of the intestinal tract, which are water soluble, to attack larger surface areas of the fat, and to digest it. Without this action of bile almost none of the fats in the food would be digested. Bile salts also help in the absorption of the fat digestion products, as will be discussed more fully in the following chapter.

Regulation of Bile Secretion.   Liver secre-

tion, unlike secretion by other gastrointestinal glands, continues steadily throughout the day and does not increase and decrease significantly in response to food in the intestine. The only hormone known to affect it is secretin, which can increase the output of bile some 10 to 20 per cent but does not have the powerful effect on the liver that it has on the pancreas.

STORAGE OF BILE IN THE GALLBLADDER.   Even though bile secretion is a continuous process, the flow of bile into the gastrointestinal tract is not continuous. As illustrated in Figure 30–12, a circular muscle around the outlet where the common bile duct empties into the duodenum, called the *sphincter of Oddi,* normally blocks the flow of bile into the gut. Instead, the bile flows into the *gallbladder,* which is attached to the side of the common bile duct. Much of the fluid and electrolytes of the bile are then reabsorbed into the blood by the gallbladder mucosa. This concentrates by as much as 12-fold the bile acids, cholesterol, and bilirubin, all of which cannot be reabsorbed, and allows the gallbladder, even though it has a maximum volume of only 50 ml., to accommodate the active components (bile acids) of an entire day's liver secretion of bile (600 ml.).

EMPTYING OF THE GALLBLADDER — THE "CHOLECYSTOKININ" MECHANISM.   When food enters the small intestine, two mechanisms simultaneously cause the gallbladder to empty into the small intestine. First, the *cholecystokinin* extracted from the wall of the duodenum by fats and other foods in the chyme passes through the blood to the gallbladder and causes the muscular wall to con-

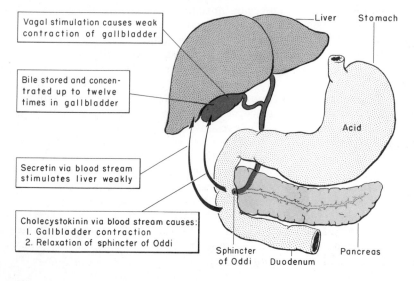

Vagal stimulation causes weak contraction of gallbladder

Bile stored and concentrated up to twelve times in gallbladder

Secretin via blood stream stimulates liver weakly

Cholecystokinin via blood stream causes:
1. Gallbladder contraction
2. Relaxation of sphincter of Oddi

Liver   Stomach

Acid

Sphincter of Oddi   Duodenum   Pancreas

Figure 30–12. Bile secretion, bile storage in the gallbladder, and the cholecystokinin mechanism for promoting gallbladder emptying.

tract. This is the same cholecystokinin that causes the pancreas to secrete large quantities of enzymes. Second, the presence of food in the duodenum causes duodenal peristalsis, and the peristaltic waves send periodic inhibitory nerve signals to the sphincter of Oddi through the intramural plexus to open it. This combination of gallbladder contraction and opening of the sphincter of Oddi allows the stored bile to empty into the intestine, and the bile salts immediately begin their emulsifying action on the fats.

GALLSTONES. Gallstones are caused mainly by the fatty waste product *cholesterol* which is excreted in the bile. Cholesterol is relatively insoluble, but it is normally held in solution in the bile by physical attraction to the bile salts. Often, though, when bile becomes concentrated in the gallbladder, too much water is removed from the bile, and the cholesterol then becomes too concentrated to remain in solution. Crystals of cholesterol begin to precipitate, and these sometimes grow to fill the entire bladder, thus forming gallstones.

A means for preventing the formation of gallstones is to eat a diet low in fat, for cholesterol is formed by the liver in great abundance in response to a high fat diet. Once gallstones have been formed, however, the only treatment is removal of the stones, or, preferably, removal of the gallbladder itself along with the stones. Absence of the gallbladder does not greatly affect the digestion of fats, because bile continues to be excreted into the intestine, though now it enters the gut almost all the time rather than periodically.

SECRETION IN THE SMALL INTESTINE

The small intestine secretes the enzymes *sucrase, maltase,* and *lactase* for splitting disaccharides into monosaccharides, the final digestion products of carbohydrates. Also secreted are large quantities of *peptidases* for performing the final steps in protein digestion, and small quantities of *lipases* for splitting fats.

However, secretion in the small intestine does not occur in the usual manner. Instead, the digestive enzymes are formed in the epithelial cells lining the intestinal wall, and much of the digestive process occurs either inside these cells or in close proximity to them. Also some of the cells slough off into the intestinal lumen, then dissolve and release small amounts of the enzymes that act directly on the food of the chyme.

Mucus Secretion in the Small Intestine. The small intestine also secretes along its entire surface large quantities of mucus, which provide the same protective function in this part of the gastrointestinal tract as in the stomach, the esophagus, and elsewhere. In the first few centimeters of the duodenum, an especially abundant amount of mucus is secreted by large mucous glands, called *Brunner's glands,* lying deep in the mucosa. The function of this secretion is to protect this portion of the intestinal tract from the powerful digestive action of pepsin and hydrochloric acid in the chyme newly arrived from the stomach. Once the chyme has been neutralized by pancreatic juice, however, it no longer has such a strong tendency to digest the wall of the intestine, which explains why Brunner's glands are needed only in this uppermost region of the intestinal tract.

Quantity of Secretion. The total amount of secretion of the small intestine is about 3000 ml. per day, which compares with about 500 ml. of saliva, 2000 ml. of gastric juice, 1200 ml. of pancreatic juice, and 600 ml. of bile. In other words, the total quantity of intestinal secretion is almost as much as that of all the remaining gastrointestinal glands put together.

Absorption of Fluid from the Small Intestine. Almost all the secretions that enter the small intestine, including saliva, gastric secretion, pancreatic secretion, bile, and intestinal secretion, are absorbed before entering the large intestine. Any fluid that is ingested is absorbed as well. That is, 8 liters or more of fluid is absorbed by the mucosa of the small intestine each day. A remaining 500 to 800 ml. passes along with the chyme into the large intestine, and most of this is absorbed there before the feces are expelled. Thus, fluid circulation occurs continually between the body fluids and the gastrointestinal tract. Because the fluid utilized in forming the gastrointestinal secretions is mainly extracellular fluid, any loss of fluid from the gastrointestinal tract, such as by vomiting or diarrhea, is actually a loss of extracellular fluid and can cause one to become extremely dehydrated.

Regulation of Secretion in the Small Intestine. Most secretion in the small intestine is probably regulated by local nervous reflexes. That is, food distending the intestinal tract or irritating the intestinal mucosa initiates reflexes in the intramural plexus to stimulate secretion by the intestinal mucosa.

However, there probably is also a hormonal mechanism for regulating intestinal secretion. Though not as important quantitatively as the nervous reflex mechanism, the hormonal system is believed to help determine the types of enzymes secreted. One theory is that food in the small intestine extracts a mixture of hormones from the mucosa, and these hormones then stimulate the intestinal glands to secrete the appropriate enzymes for digesting the food.

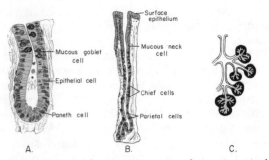

Figure 30–13. The anatomic types of gastrointestinal glands: (A) A simple tubular gland represented by a crypt of Lieberkühn. (B) A more elongated tubular gland represented by a gastric gland. (C) A compound acinous gland represented by the pancreas.

SECRETIONS OF THE LARGE INTESTINE

The large intestine, like the esophagus, performs no digestive functions. Therefore, its only significant secretion is mucus. The entire mucosa is coated with mucous cells that provide lubrication for the passage of feces from the ileocecal valve to the anus and that also protect the large intestine from digestion by the enzymes emptied from the small intestine. The portion of the large intestine near the ileocecal valve is protected against the digestive hormones better than the distal portion. Consequently, during severe diarrhea the rapid flow of digestive enzymes from the small intestine into the distal colon is very likely to cause extreme irritation. Prolonged and severe diarrhea is sometimes associated with a condition called *ulcerative colitis,* which occasionally leads to holes (ulcers) in the colon that cause death. The mucus secreted in the large intestine normally protects against this.

from the stomach through the colon, are large numbers of single glandular cells called *mucous goblet cells.* These form mucus that is continually secreted from the surface of the cell onto the lining of the gastrointestinal tract. The purpose of this mucus is to protect the gut wall from excoriation by the gastrointestinal contents.

Tubular Glands.   The most abundant type of gland in the gastrointestinal tract is the tubular gland. A simple example of this type of gland is illustrated in Figure 30–13A, and a more elongated type is shown in Figure 30–13B. The first of these is actually nothing more than a depression in the wall of the intestine, called a *crypt of Lieberkühn.* Within this gland are a number of mucous goblet cells. In addition, however, the epithelial cells lining the crypt also form small amounts of intestinal digestive enzymes.

CELLULAR MECHANISMS OF SECRETION

Thus far, we have discussed the gastrointestinal glands and their secretions without stating the cellular mechanisms involved in glandular secretion. Let us now explain some of these mechanisms.

ANATOMIC TYPES OF GASTROINTESTINAL GLANDS

Single Cell Mucous Glands.   On the gastrointestinal epithelial surface, all the way

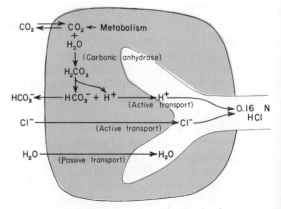

Figure 30–14.   Postulated mechanism for the secretion of hydrochloric acid.

The more elongated gland in Figure 30–13B is a gastric gland of the type found in the main body of the stomach. It contains a large number of *mucous cells* that secrete mucus, *chief cells* that secrete gastric enzymes (mainly pepsin), and *parietal cells* that secrete hydrochloric acid.

Compound Glands. Some glands, such as the salivary glands, the pancreas, and the liver, have very complex structures and are called compound glands. One of the simpler of these is the *compound acinous gland* illustrated in Figure 30–13C, which is represented by the pancreas. The acini are lined with secreting glandular cells, and the secretions then are transported through a duct system into the gastrointestinal tract.

the cell called the *canaliculi*. At the same time, chloride ions that have diffused into the cell from the blood are also actively transported into the canaliculi. Thus, in the canaliculi the hydrogen and chloride ions come together to form hydrochloric acid. This then creates an osmotic force that pulls water also into the canaliculi. And this movement of hydrogen ions, chloride ions, and water into the canaliculi creates a flow of secretions outward through openings from the canaliculi into the lumen of the gastric gland.

The other glands of the gastrointestinal tract synthesize and secrete their products in a similar manner, though of course the details of the processes are different.

MECHANISM OF SECRETION BY A GLANDULAR CELL

Each glandular cell has its own peculiar mechanism for secretion depending on the substance that is to be secreted. In general, protein enzymes are formed inside the cell by the ribosomes attached to the surface of the endoplasmic reticulum, in the manner described in Chapters 2 and 3. Then the proteins are transported through the endoplasmic reticulum to the Golgi apparatus where they are packaged into secretory granules. Finally, the secretory granules are extruded through the surface of the cell into the lumen of the gland. Simultaneously, active transport of ions through the glandular cell membrane causes secretion of sodium, chloride, bicarbonate, and other ions. And secretion of the ions causes osmosis of water, giving rise to the fluid volume of the secretion.

To give an example of a special secretion, Figure 30–14 illustrates a postulated mechanism for the secretion of hydrochloric acid by the parietal cells of the gastric glands. This secretion begins with diffusion of carbon dioxide into the parietal cell from the blood or formation of carbon dioxide within the parietal cell as a result of cellular metabolism. The carbon dioxide then combines with water to form carbonic acid, and the carbonic acid in turn dissociates into bicarbonate ion and hydrogen ions. The hydrogen ions in turn are actively transported through the wall of a specialized endoplasmic ductal system within

REFERENCES

Atanassova, E., and Papasova, M.: Gastrointestinal motility. *Intern. Rev. Physiol.*, 12:35, 1977.

Bortoff, A.: Myogenic control of intestinal mobility. *Physiol. Rev.*, 56:418, 1976.

Forker, E. L.: Mechanisms of hepatic bile formation. *Ann. Rev. Physiol.*, 39:323, 1977.

Hendrix, T. R., and Paulk, H. T.: Intestinal secretion. *Intern. Rev. Physiol.*, 12:257, 1977.

Johnson, L. R.: Gastrointestinal hormones and their functions. *Ann. Rev. Physiol.*, 39:135, 1977.

Makhlouf, G. M.: The neuroendocrine design of the gut. The play of chemicals in a chemical playground. *Gastroenterology*, 67:159, 1974.

Mason, D. K.: Salivary Glands in Health and Disease. Philadelphia, W. B. Saunders Company, 1975.

Moller, E.: Action of the muscles of mastication. *Front Oral Physiol.*, 1:121, 1974.

Rothman, S. S.: The digestive enzymes of the pancreas: a mixture of inconstant proportions. *Ann. Rev. Physiol.*, 39:373, 1977.

Sachs, G., Spenney, J. G., and Rehm, W. S.: Gastric secretion. *Intern. Rev. Physiol.*, 12:127, 1977.

Sachs, G., Spenney, J. G., and Lewin, M.: H^+ transport: regulation and mechanism in gastric mucosa and membrane vesicles. *Physiol. Rev.*, 58:106, 1078.

Sarles, H.: The exocrine pancreas. *Intern. Rev. Physiol.*, 12:173, 1977.

Snook, J. T.: Adaptive and nonadaptive changes in digestive enzyme capacity influencing digestive function. *Fed. Proc.*, 33:88, 1974.

Stroud, R. M., Kossiakoff, A. A., and Chambers, J. L.: Mechanisms of zymogen activation. *Ann. Rev. Biophys. Bioeng.*, 6:177, 1977.

Walsh, J. H.: Circulating gastrin. *Ann. Rev. Physiol.*, 37:81, 1975.

Wood, J. D.: Neurophysiology of Auerbach's plexus and control of intestinal motility. *Physiol. Rev.*, 55:307, 1975.

Young, J. A., and van Lennep, E. W.: Morphology and physiology of salivary myoepithelial cells. *Intern. Rev. Physiol.*, 12:105, 1977.

QUESTIONS

1. Give the physiologic anatomy of the gastrointestinal tract.
2. Explain the function of both the propulsive movements and the mixing movements of the gastrointestinal tract.
3. Explain the sequence of events during the swallowing process.
4. What are the motor functions of the stomach, including especially the control of stomach emptying?
5. How do the movements of the small intestine and of the colon differ from each other and also from movements in other portions of the GI tract?
6. How does vomiting occur?
7. How are the different phases of salivary secretion controlled?
8. What are the different secretions of the stomach, and how are they regulated?
9. What are the components of pancreatic secretion, and how is each of these controlled?
10. Give the characteristics of bile, and explain how its release into the intestinal tract is regulated.
11. How do the secretions of the small intestine differ from those of the large intestine, and how are both controlled?
12. Explain the cellular mechanisms of secretion.

DIGESTION AND ASSIMILATION OF CARBOHYDRATES, FATS, AND PROTEINS

31

The term *digestion* means the splitting of large chemical compounds in the foods into simpler substances that can be used by the body. The term *assimilation* includes several functions that may be listed: (1) absorption of the digestive end-products into the body fluids, (2) transport of these to the cells where they will be used, and (3) chemical change of some of them into other substances that are specially needed for various purposes. The function of the digestive and assimilative processes is to provide nutrients for the chemical reactions of metabolism.

DIGESTION, ABSORPTION, AND DISTRIBUTION OF CARBOHYDRATES

Carbohydrates are composed of carbon, hydrogen, and oxygen. The basic unit of a carbohydrate is a *monosaccharide,* the most common of which in the food is *glucose,* and which has the following chemical formula:

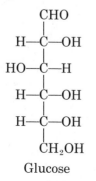

Glucose

Two other monosaccharides very frequently present in food are *fructose* and *galactose,* the formulas for which are the same as that of glucose except that some of the "H" and "OH" radicals are transposed.

Glucose and other monosaccharides are usually *polymerized* (combined) into larger chemical compounds such as *starches, glycogens, pectins,* and *dextrins.* By far the most common carbohydrate of the diet is starch, which is a polymer of glucose. The glucose molecules in starches are joined together in the manner illustrated on the following page.

It will be noted from this formula that the successive molecules of glucose are combined with each other by a *condensation* process, which means that one glucose molecule loses a hydrogen ion and the next loses a hydroxyl ion. The hydrogen and hydroxyl ions combine to form water, and the two glucose molecules connect together at the points where the ions were removed.

In addition to starches, another common source of carbohydrates is the disaccharides, which are combinations of only two molecules of monosaccharides. The common disaccharides in the diet are maltose, sucrose, and lactose. *Maltose* is a combination of two glucose molecules, and it is derived mainly by splitting starches into their disaccharide components, all of which are maltose. *Sucrose* is a combination of one molecule of glucose and one molecule of fructose. It is the same as cane sugar which is common table sugar. *Lac-*

397

tose is a combination of one molecule of glucose and one molecule of galactose. It is the sugar present in milk.

The intestinal epithelial cells contain the enzymes *maltase, lactase,* and *sucrase* which split maltose, lactose, and sucrose into their

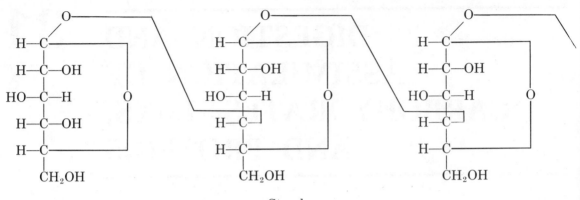

Starch

BASIC MECHANISM OF CARBOHYDRATE DIGESTION – THE PROCESS OF HYDROLYSIS

Carbohydrate digestion breaks the starch or other carbohydrate polymers into their component monasaccharides. To do this, one molecule of water must be added to the compound at each point where two successive monosaccharides are joined. This is a process of *hydrolysis,* which is opposite to the condensation process by which the successive monosaccharides are combined with each other. The secretions of the digestive tract contain enzymes that catalyze this hydrolysis process.

Scheme of Digestion of Carbohydrates. The schema in Figure 31–1 shows digestion of the three most common carbohydrates in the diet, the starches, lactose, and sucrose. Starches and other large carbohydrates are digested principally by *ptyalin* in the saliva and *amylase* in the pancreatic juice, but perhaps to a slight extent also by *hydrochloric acid* in the stomach and *intestinal amylase* in the small intestine. The resulting product of these reactions is the disaccharide maltose.

respective monosaccharides as these are absorbed through the intestinal wall into the blood. The final products of carbohydrate digestion are *glucose, galactose,* and *fructose,* as shown by the schema. Because all the monosaccharides derived from maltose are glucose and half of those derived from the other two carbohydrates are glucose, it is evident that this substance is by far the most abundant end-product of carbohydrate digestion. On the average, about 80 per cent of the monosaccharides formed by digestion is glucose, 10 per cent galactose, and 10 per cent fructose.

ABSORPTION OF MONOSACCHARIDES

The Absorptive Epithelium of the Intestine. Figure 31–2A illustrates a longitudinal section through a segment of the intestine. It shows the mucosa protruding into the lumen of the intestine in the form of large folds called *valvulae conniventes.* Also, on the surface of the mucosa everywhere are literally millions of small *villi* only a millimeter or so in length. The structure of a villus is shown in Figure 31–2B. On the surface of the villus is an epithelial lining and within its sub-

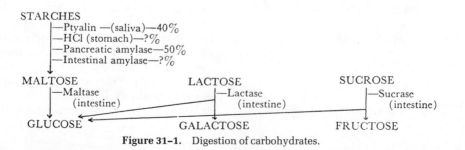

Figure 31–1.   Digestion of carbohydrates.

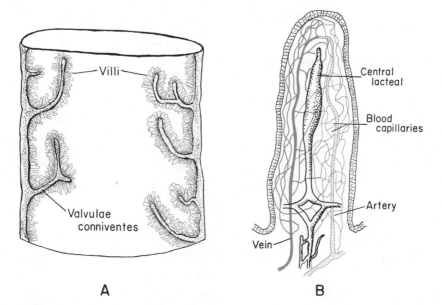

A   **B**

Figure 31–2. (A) Distribution of villi and mucosal folds on the inner surface of the small intestine. (B) Structure of a villus, showing the blood vessel system and the central lacteal.

stance is an arteriole, a blood capillary, a vein, and a large central lymphatic capillary called the central *lacteal*. It is into the blood capillaries and this central lacteal that substances are absorbed.

The epithelial cells on the lumenal surface of each villus have a "brush border" (Fig. 31–3) which comprises thousands of minute *microvilli* only 0.1 micron (micrometer) in diameter. These microvilli provide a tremendous surface area through which substances can be absorbed into the interior of the cell.

Because of (a) the mucosal folds, (b) the villi, and (c) the microvilli on the surface of the epithelial cells, the total absorptive area of the intestine is about 600 times as great as

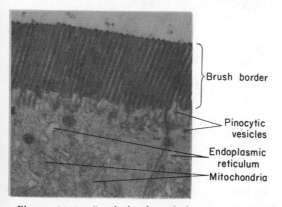

Figure 31–3. Brush border of the gastrointestinal epithelial cell, showing also pinocytic vesicles, mitochondria, and endoplasmic reticulum lying immediately beneath the brush border. (Courtesy of Dr. Wm. Lockwood.)

it would be without these structures, giving a total absorptive area of about 550 square meters for the entire small intestine.

Route of Absorption of Monosaccharides. The monosaccharides are absorbed through the intestinal epithelium into the blood of the villus capillaries. The blood then empties into the portal venous system and finally flows through the liver into the general circulation. As the monosaccharides pass through the liver, they are partially processed prior to reaching the peripheral cells for metabolism.

Mechanism of Absorption of Monosaccharides. Monosaccharides are absorbed from the gastrointestinal tract by active absorption. The mechanism of this process was discussed in Chapter 8, and is also essentially the same as that for absorption from the tubules of the kidney, which was discussed in detail in Chapter 17. Briefly, the monosaccharides combine with a carrier substance in the epithelial cells and are transmitted in this combined form from the intestinal lumen to the opposite side of the cells; there they are released from the carrier into the tissue fluids and thence into the capillary blood. For this process to occur, energy must be expended by the epithelial cells, which is the reason why the process is called "active" absorption.

Active absorption of the monosaccharides is very important because it allows absorption to occur even when monosaccharides are present in the intestine in extremely small concentrations, concentrations even smaller than those in the blood itself.

FATE OF THE MONOSACCHARIDES IN THE BODY

Glucose in the Blood and Extracellular Fluid — Conversion of Fructose and Galactose to Glucose. When a person eats the usual diet that is high in carbohydrates, approximately 80 per cent of the monosaccharides absorbed from the gut is glucose, and essentially all of the remainder is fructose and galactose. However, almost immediately, both these monosaccharides are also converted into glucose. The fructose is mainly converted as it is absorbed through the intestinal epithelial cells because of a metabolic interconversion that takes place in these cells. The galactose is absorbed very rapidly by the liver cells, converted into glucose, and then returned to the blood. Thus, for practical purposes, all carbohydrates finally reach the individual tissue cells in the form of glucose, which supplies a major share of the cellular energy.

The concentration of glucose in the blood and extracellular fluid is approximately 90 mg. in each 100 ml., while the concentrations of fructose and galactose are usually very slight because of their rapid conversion to glucose.

Transport of Glucose Through the Cell Membrane — Effect of Insulin. Before glucose can be used by the cells it must be transported through the cell membrane. Unfortunately, the pores of the cell membrane are too small to allow glucose to enter by the process of simple diffusion. Here again it must be transported by a chemical process, called *facilitated diffusion,* the general principles of which are shown in Figure 31–4. Glucose first combines with a carrier, believed to be a protein, in the cell membrane. Then it is transported to the inside of the cell where it breaks away from the carrier.

In some way that has not yet been completely explained, the hormone *insulin* greatly enhances this facilitated transport of glucose through the cell membrane. Some possible ways in which insulin might do this are: (1) by catalyzing the reaction between glucose and the carrier, (2) by removing glucose from the carrier on the inside of the cell, or (3) by forming one of the components of the carrier mechanism itself. Regardless of which of these might be correct, the rate at which glucose can be transported through the cell

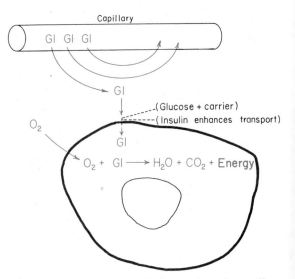

Figure 31–4. Transport of glucose from the capillary into the extracellular fluid and then from the extracellular fluid into the cell to be utilized for energy.

membrane is determined mainly by the amount of insulin available. When the pancreas fails to secrete insulin, as much as five times less glucose enters than the cell needs. When insulin is secreted in excessive abundance, glucose enters so rapidly that glucose metabolism becomes much greater than normal. It is obvious, therefore, that the rate of carbohydrate metabolism is regulated in accordance with the rate of insulin secretion by the pancreas.

Regulation of Blood Glucose Concentration. BUFFER EFFECT OF THE LIVER. After a meal, large quantities of monosaccharides are absorbed into the blood, and the glucose in the portal blood coming from the intestines rises from the normal concentration of 90 mg. per 100 ml. to as high as double this. However, this portal blood flows through the liver before it reaches the general circulation, and the liver removes about two-thirds of the excess glucose. In this way the liver keeps the general blood concentration of glucose from rising above 120 to 130 mg. per 100 ml., despite very rapid absorption from the intestines.

The mechanism by which the liver removes the glucose from the portal blood is the following: Glucose is first absorbed through the cellular membranes into the liver cells. Then it is converted to *glycogen* a polymer of glucose, and stored until a later time. When the blood glucose level falls to lower values several hours after a meal, the glycogen is split back into glucose, which is transferred out of the liver into the blood.

In essence, then, the liver is a "buffer" organ for blood glucose regulation, for it keeps the blood glucose level from rising too high and from falling too low.

INSULIN PRODUCTION BY THE PANCREAS AS A MEANS FOR CONTROLLING BLOOD GLUCOSE CONCENTRATION.   After a person eats a large meal, the rise in blood glucose concentration stimulates the pancreas to produce large quantities of insulin. The insulin in turn promotes rapid transport of glucose into the cells, thus decreasing the blood glucose level back toward normal. Therefore, in addition to the liver buffer mechanism, this pancreatic production of increased quantities of insulin also aids in preventing excessive rises in blood glucose concentration.

EFFECT OF EPINEPHRINE, SYMPATHETIC STIMULATION, AND GLUCAGON IN PREVENTING LOW BLOOD GLUCOSE CONCENTRATION.   A low blood glucose level stimulates the sympathetic centers of the brain, causing secretion of norepinephrine and epinephrine by the adrenal glands and excitation of all the sympathetic nerves throughout the body. Also, the low glucose concentration directly stimulates the pancreas to secrete the hormone *glucagon*. This glucagon, the norepinephrine, epinephrine, and sympathetic stimulation all cause liver glycogen to split into glucose which is then emptied into the blood. This returns the blood glucose concentration back toward normal, acting as a protective mechanism against low blood glucose levels.

Gluconeogenesis.   Another effect that occurs when the blood glucose level falls too low is formation of glucose mainly from proteins. This phenomenon is called *gluconeogenesis*. The importance of gluconeogenesis is that it provides glucose to the blood even during periods of starvation. Glucose, unfortunately, is not stored to a major extent in the body, for only 300 grams at most is stored in the form of glycogen in both the liver and all the remainder of the body, mainly the muscles. Ordinarily this amount is not sufficient by itself to maintain the blood glucose concentration at normal values for more than 24 hours. However, as the blood glucose level falls below normal, gluconeogenesis begins and continues until an adequate supply of glucose is available again.

Later in this chapter it will be noted that most of the cells of the body can utilize fats for energy when glucose is not available. However, the neurons of the brain are unable to utilize fats, and without an adequate supply of glucose these cells begin to die. This is the major reason why it is very important that the blood glucose concentration remain essentially normal even during long periods of starvation.

ENERGY FROM GLUCOSE

The major function of glucose in the body is to provide energy, though some glucose molecules are used as building stones for synthesis of other needed compounds. Energy is derived from glucose by two means: first, by splitting the molecules of glucose into smaller compounds and oxidizing these to form water, which liberates an extremely large amount of energy. These mechanisms of energy release are discussed in the following chapter.

DIGESTION, ABSORPTION, AND DISTRIBUTION OF NEUTRAL FATS

Neutral fats, like carbohydrates, are composed of carbon, hydrogen, and oxygen, though the relative abundance of oxygen in fats is considerably less than in carbohydrates. A representative molecule of neutral fat and its digestive end-products is the following:

$$CH_3-(CH_2)_{16}-\overset{\overset{\displaystyle O}{\|}}{C}-O-CH_2$$

$$CH_3-(CH_2)_{16}-\overset{\overset{\displaystyle O}{\|}}{C}-O-CH \xrightarrow{\ lipase\ } 3CH_3-(CH_2)_{16}-\overset{\overset{\displaystyle O}{\|}}{C}-O-H + \begin{array}{l} HO-CH_2 \\ | \\ HO-CH \\ | \\ HO-CH_2 \end{array}$$

$$CH_3-(CH_2)_{16}-\overset{\overset{\displaystyle O}{\|}}{C}-O-CH_2$$

Tristearin   Stearic acid   Glycerol

It is evident from these formulas that a fat molecule contains two major components: first, a glycerol nucleus and, second, three fatty acid radicals. Each fatty acid radical is combined with the glycerol by a condensation process, that is, removal of a hydroxyl radical from the glycerol and a hydrogen ion from the fatty acid, with formation of a water molecule and bonding of the fatty acid and the glycerol at the points of removal. This mechanism was noted previously as the means by which monosaccharides also combine with each other to form complicated carbohydrates.

The differences among various fats lie in the composition of the fatty acids in the molecule. Most fats in the human body have fatty acids with 16 or 18 carbon atoms in their chains. The fats containing the longer chain fatty acids are more solid than those containing the shorter fatty acids. Other than this, the chemical and physical properties of most fats do not vary greatly from one to the other.

Some of the fatty acids in the body and in the diet are *unsaturated,* which means that at various points in the carbon chain the atoms are bonded together by double bonds rather than single bonds, and that there is a corresponding lack of two hydrogen atoms. The unsaturated fats are needed to form a few special structures of the cells, but otherwise even these perform the same principal function as the saturated fats, which is to provide energy for the metabolic processes.

DIGESTION OF FATS

The digestion of fats, like that of carbohydrates, is a *hydrolysis* process. It is catalyzed by enzymes called *lipases,* secreted in the stomach, pancreatic, and intestinal juices. The diagram in Figure 31–5 shows the complete schema of fat digestion. Though a minute quantity of fat is digested in the stomach, most of its is digested by the pancreatic and intestinal lipases after it enters the small intestine.

The end-products of fat digestion are *fatty acids, glycerol,* and *glycerides.* Glycerides are composed of a glycerol nucleus with one or two of the fatty acid chains still attached. Though the end result of complete fat digestion is the splitting of the fats entirely into fatty acids and glycerol, the process goes to completion in only about 40 per cent of the fat molecules, leaving many glycerides (glycerol still attached to one or two fatty acids) still among the digestive products.

Because glycerides pass through the intestinal membrane with almost the same ease as glycerol and fatty acids, the digestive process is quite adequate for absorption to occur. In fact, a very small amount of finely emulsified neutral fat that has not been digested at all can even be absorbed through the membrane.

Role of Bile Salts in Fat Digestion. The bile salts secreted by the liver are essential for complete digestion of fat in the intestine even though they perform no digestive enzyme function. In the absence of bile salts, as much as 50 per cent of the fat passes undigested all the way through the gastrointestinal tract and is expelled in the feces.

The bile salts play two roles in fat digestion. The first of these is its effect of acting as a detergent; that is, it greatly decreases the surface tension of the fatty globules in the food. This allows the mixing movements of the intestines to break the fat globules into very finely emulsified particles, thereby providing greatly increased surface area on which the water-soluble digestive enzymes, the lipases, can act.

The second way in which bile salts increase fat digestion is to transport the end-products of digestion, the fatty acids and the glycerides, away from the fat globules as the digestive process proceeds. The bile salt molecules themselves aggregate to form colloidal particles called *micelles.* These have a fatty core, but they are still stable in the fluids of the intestines because the surfaces of the micelles are ionized, which is a property that promotes water solubility. The fatty acids and the glycerides become absorbed in the fatty portions of these micelles as they are split away from the fat globules, and in this form they are "ferried" from the fat globules to the

Fat $\xrightarrow{\text{(Bile + Agitation)}}$ Emulsified fat

Emulsified fat $\xrightarrow{\textit{Pancreatic lipase}}$ $\begin{cases} \text{Fatty acids} \\ \text{Glycerol} \end{cases} 40\% \ (?) \\ \text{Glycerides } 60\% \ (?)$

Figure 31–5. Digestion of fats.

intestinal epithelium where absorption occurs.

ABSORPTION OF THE END-PRODUCTS OF FAT DIGESTION

Once the micelles have ferried the products of fat digestion to the intestinal epithelium, the fatty acids and glycerides are released and, like the monosaccharides, are also absorbed by the villi of the intestinal mucosa; but, unlike the monosaccharides, they are absorbed into the central lacteal, a lymph vessel in the center of the villus, as shown in Figure 31–2B, instead of into the blood. The mechanism by which this fat absorption occurs is the following:

The fatty acids and glyceride molecules are very soluble in the brush border of the epithelial cells lining the surfaces of the villi. Therefore, they diffuse readily from the intestinal lumen into the interior of these cells. Then, the endoplasmic reticulum inside the cell resynthesizes new molecules of neutral fat and expels the newly formed fat into the interstitial fluid of the villi in the form of small fat globules called *chylomicrons;* these are immediately picked up by the central lacteal.

Transport of Fat Through the Lymphatics. Lymph is "milked" from the central lacteals into the abdominal lymphatics by rhythmic contraction of the villi. This contraction is caused by a hormone, *villikinin,* which is released from the intestinal mucosa when fats are in the chyme. After leaving the central lacteals, the neutral fat is eventually transported upward through the *thoracic duct,* the major lymphatic channel of the body, to empty into the blood circulation at the juncture of the internal jugular and subclavian veins.

CHYLOMICRONS. The fat absorbed into the central lacteal is in the form of small fatty droplets about 1 micron (micrometer) in dimeter. These are the *chylomicrons.* They immediately adsorb proteins to their surfaces, which keeps them suspended in the lymph and prevents them from sticking to each other or to the walls of the lymphatics or blood vessels. It is in this form that fats are transported through the lymphatics and finally into the blood.

After a fatty meal the level of chylomicrons in the circulating blood reaches a maximum in approximately 2 to 4 hours, sometimes becoming as much as 1 to 2 per cent of the blood, but within 2 to 3 hours almost all of them will have been deposited in the fat tissue of the body or in the liver.

FAT TISSUE

Fat tissue is a special type of connective tissue that has been modified to allow storage of neutral fat. It is found beneath the skin, between the muscles, between the various organs, and in almost all spaces not filled by other portions of the body. The cytoplasm of fat cells sometimes contains as much as 95 per cent neutral fat. These cells store fat until it is needed to provide energy elsewhere in the body.

Fat tissue provides a *buffer* function for fat in the circulating fluids. After a fatty meal the high concentration of fat in the blood is very soon lowered by deposition of the extra fat in the fat tissue. Then, when the body needs fat for energy or other purposes, it can be mobilized from the fat tissues and returned to the circulating blood, as will be explained below.

Because of this buffer function of the fat tissues, the fat in the fat cells is in a constant state of flux. Ordinarily, half of it is removed every 8 days, and new fat is deposited in its place.

TRANSPORT OF FATS IN THE BODY FLUIDS

Free Fatty Acids. Most fat is transported in the blood from one part of the body to another in the form of *free fatty acids.* These usually exist in the blood as a loose combination of free fatty acids with albumin, one of the plasma proteins. Every fat cell contains large quantities of the fat digestive enzyme *lipase.* However, this remains in an inactive form except when there is need for release of fat from the fat tissue. Several hormones, especially some of the adrenocortical hormones, can activate the lipase. This then digests the neutral fat in the fat cell into glycerol and fatty acids. The fatty acids diffuse out of the cell and immediately combine with albumin in the blood and are then transported to other tissues of the body where they are released from the albumin. Some combine with glycerol in fat tissue areas to form new neutral fat. Others enter other tissue cells where they are split into smaller molecules that are used to supply energy, as is explained below.

CONCENTRATION OF FREE FATTY ACIDS. Even though almost all fat transport in the body is in the form of free fatty acids, these are normally present in the blood in a concentration of only about 10 mg. per 100 ml. of blood, or one-tenth the concentration of glucose. However, the fatty acids are transported to their destination within a few seconds to a few minutes, remaining in the blood only a short time, and can therefore account for tremendous amounts of available energy to the cells. When fats are being used in great quantities by the cells, the blood concentration of fatty acids increases as much as fourfold or more.

Lipoproteins. Lipoproteins are minute fatty particles covered by a layer of adsorbed protein. These are suspended in a colloid form in the plasma and to a lesser extent in other extracellular fluids. The *chylomicrons* are a type of lipoprotein because they are composed of lipid substances (neutral fat, phospholipids, and cholesterol) and a layer of adsorbed protein. However, in addition to the chylomicrons, large numbers of much smaller lipoprotein particles are also present in the blood. These are formed almost entirely in the liver, and their function is probably to transport neutral fat, phospholipids, and cholesterol from the liver to the different cells of the body.

SYNTHESIS OF FAT FROM GLUCOSE AND PROTEINS

Much of the fat in the body is not derived directly from the diet but instead is synthesized in the body. The fat cells themselves are capable of synthesizing small amounts of fat, but most fat is synthesized in the liver and then transported to the fat cells. Both glucose and amino acids derived from proteins can be converted into fat, but by far the most important source is glucose.

When an excess of glucose is in the diet and sufficient insulin is secreted by the pancreas to cause all the glucose to enter the cells, almost all the extra glucose not used immediately for energy passes into the fat cells and the cells of the liver to be converted into fat. Thus, the fat tissues provide a means for storing energy derived from carbohydrates as well as from fats. This conversion of other foods to fats explains why eating any type of food, whether it be fat, carbohydrate, or protein, can increase the amount of fat tissue.

FUNCTIONS OF THE LIVER IN THE UTILIZATION OF FAT

The liver is undoubtedly the most important organ of the body for controlling fat utilization by the body. The liver, in addition to converting much of the excess glucose into fat, converts fat into substances that can be used elsewhere in the body for special purposes. For example, some of the fats must be desaturated to provide the unsaturated fats required by all cells of the body for their metabolic processes; some must be converted into the fatty substances, *cholesterol* and *phospholipids,* needed for cellular structures; and others are broken into smaller molecules that can be used easily by the cells for energy. The liver performs all these functions.

When the body is depending mainly on fats instead of glucose for energy, the quantity of fat in the liver gradually increases. This is enhanced by adrenocortical hormones, for they cause the fat tissue cells to mobilize their fat.

ENERGY FROM FATS

Fatty acids can be utilized for energy by almost all cells of the body with the exception of the neuronal cells of the brain. However, about 40 per cent of the fatty acids used for energy are first split in the liver into *acetoacetic acid* and then transported to the cells for use as explained in the following chapter.

The first stage in the utilization of fats for energy is to split the neutral fat into glycerol and fatty acids. The glycerol, being very similar to some of the breakdown products of glucose, can then be used for energy in very much the same manner as glucose. However, by far the major amount of energy in the fat molecule is in the fatty acid chains, and before these can be used for energy they must be split into still smaller chemical compounds. Ordinarily, this is accomplished by a chemical process called *alternate oxidation* of the carbon chain, which is illustrated below.

The net result is the formation of many

$$CH_3—CH_2—CH_2—CH_2—CH_2—CH_2—CH_2—CH_2—CH_2—CH_2—CH_2—C—OH \longrightarrow 6CH_3—C—OH$$

Fatty acid   Acetic acid

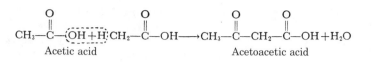

$$CH_3-\overset{\overset{O}{\|}}{C}-\overset{\cdots}{OH+H}CH_2-\overset{\overset{O}{\|}}{C}-OH \longrightarrow CH_3-\overset{\overset{O}{\|}}{C}-CH_2-\overset{\overset{O}{\|}}{C}-OH+H_2O$$

Acetic acid   Acetoacetic acid

molecules of *acetic acid*. This is then oxidized in the cells in a manner almost identical with the oxidation of glucose, giving tremendous amounts of energy to the cells, as will be explained in the following chapter.

In the liver, most of the acetic acid molecules formed from the breakdown of fatty acids condense two at a time to form *acetoacetic acid. (See above.)*

The acetoacetic acid is called a *keto acid,* and it in turn can change into several other closely related forms of keto acids, or even into acetone.

The keto acids are highly diffusible through cellular membranes. Therefore, as shown in Figure 31–6, the keto acids formed in the liver diffuse immediately into the blood and are transported to all cells throughout the body. There the keto acids are oxidized in the same manner as glucose to provide energy for cellular functions, as will be explained further in the following chapter.

Fat-Sparing Effect of Carbohydrates. As long as sufficient glucose is available to supply the energy needs of the cells, it is burned in preference to fatty acids and keto acids, and, if more than enough glucose is available, the excess is converted into fat. Thus, when glucose is available, the burning of fat stops; for this reason glucose is said to be a *fat sparer.*

Conversely, whenever the available glucose is very slight, the body automatically shifts its metabolic system to derive energy from fat instead of carbohydrates. This conversion to fat utilization is caused by two major hormonal changes: First, the decrease in blood glucose concentration causes the pancreas to decrease its rate of insulin secretion, and this in turn causes the fat cells to release greatly increased quantities of fatty acids into the circulating blood. Second, lack of sufficient carbohydrates indirectly causes the adrenal glands to secrete increased quantities of cortisol, one of the adrenocortical hormones. This has a direct effect on fat cells of activating the cellular lipase and causing the release of fatty acids into the circulating blood, therefore increasing the utilization of free fatty acids for energy.

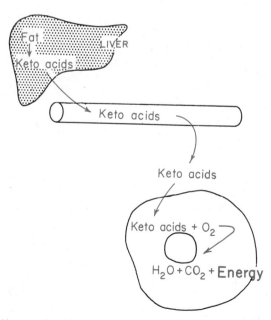

Figure 31–6. Formation of keto acids in the liver, their transport to the cells, and their utilization for energy.

PHOSPHOLIPIDS AND CHOLESTEROL

Two additional substances, *phospholipids* and *cholesterol,* which have physical properties similar to those of neutral fats, are present in large quantities throughout the body, especially as structural components of cellular and intracellular membranes. The phospholipids are composed of glycerol, fatty acids, and a phosphate side chain. Cholesterol is composed mainly of a sterol nucelus that is synthesized from acetic acid. The chemical structures of these substances are the following:

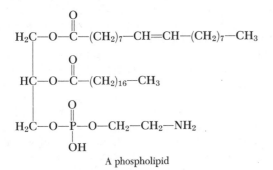

A phospholipid

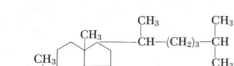

Cholesterol

Both phospholipids and cholesterol are fat-soluble and are slightly water-soluble. They are synthesized in all cells of the body, though to a much greater extent in the liver cells than in other cells. Both are transported in lipoproteins from the liver to other parts of the body.

Both phospholipids and cholesterol are major constituents of cell membranes and membranes of intracellular structures such as the nuclear membrane, the membranes of the endoplasmic reticulum, the membranes of mitochondria, of lysosomes, and so forth.

The precise functions of the phospholipids and cholesterol in the membranes are unknown, though they possibly play roles in determining membrane permeability and its transport properties.

DIGESTION, ABSORPTION, AND DISTRIBUTION OF PROTEINS

Proteins are large molecules made up of many *amino acids* joined together. Amino acids, in turn, are small organic compounds that have an amino radical, —NH₂, and an acidic radical, —COOH, both on the same molecule. Twenty-one important amino acids are known to be present in the body proteins; the formulas of these, plus two special amino acids, diiodotyrosine and thyroxine (the thyroid hormones), are shown in Figure 31–7.

Some of the amino acids can be synthesized in the body from other amino acids, but ten of them cannot. These ten are called *essential amino acids,* for they must be provided in the diet in order for the human body to form the proteins necessary for life.

Amino acids combine with each other to form proteins by means of *peptide linkages,* an example of which is illustrated by the equation shown at the bottom of this page. It will be noted that the product of the two combined acids, which is called a *peptide,* still has an amino radical and an acid radical, both of which can provide reactive points for combinations with still additional amino acids. Most proteins contain several hundred to several thousand amino acids combined in this manner. The nature of the protein is determined by the types of amino acids in the protein and also by the pattern in which they are joined.

DIGESTION OF PROTEINS TO FORM AMINO ACIDS

Referring to the preceding equation, it is evident that a peptide linkage is another example of *condensation,* which is the same

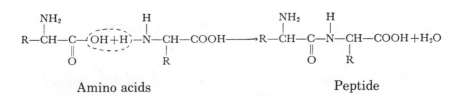

Amino acids   Peptide

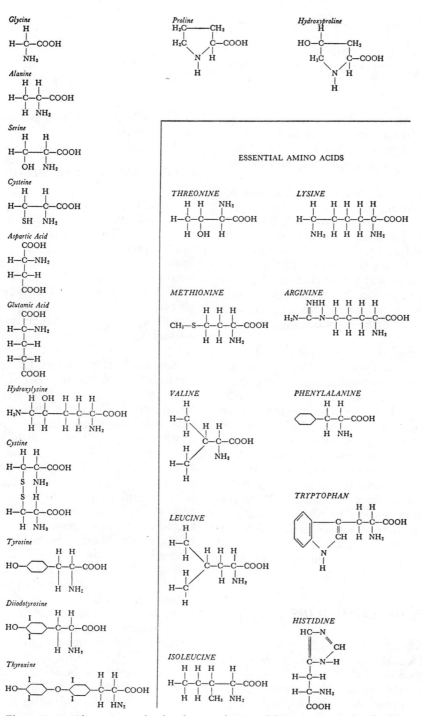

Figure 31–7. The amino acids, also showing that ten of these are essential in the diet.

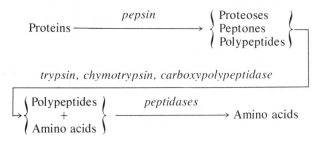

Figure 31–8. Digestion of proteins.

means by which the component parts of fats and carbohydrates are combined. Therefore, the digestion of protein, like that of carbohydrates and fats, is accomplished by a process of *hydrolysis*.

The schema in Figure 31–8 shows the sequence of protein digestion, which begins with *pepsin* action in the stomach. Pepsin is secreted into the stomach in the form of *pepsinogen,* a substance that has no digestive properties, but once it comes in contact with the hydrochloric acid of the stomach it is activated to form pepsin. The hydrochloric acid also provides an appropriate reactive medium for pepsin, for it can split proteins only in acid surroundiings.

Protein is digested in the stomach into *proteoses, peptones,* and *polypeptides,* all of which are smaller combinations of amino acids than proteins — the proteoses are nearly as large as proteins, the peptones are intermediate in size, and the polypeptides are combinations of only a few amino acids. After entering the small intestine, these substances are further split by *trypsin, chymotrypsin,* and *carboxypolypeptidase* of the pancreatic juice into small polypeptides and some amino acids. Then the small polypeptides are finally split by *peptidases* of the pancreatic and intestinal juices into amino acids. Thus, the final products of protein digestion are the basic components of proteins, the *amino acids*.

ABSORPTION OF AMINO ACIDS

Amino acids are absorbed from the gastrointestinal tract in almost exactly the same manner as monosaccharides, that is, by active transport into the blood of the intestinal villi. The transport is carrier-mediated, and energy expenditure is required for it to occur; these are also characteristics of monosaccharide transport.

After absorption through the intestinal mucosa, the amino acids pass into the capillaries of the villi and thence into the portal blood, flowing through the liver before entering the general circulation.

AMINO ACIDS IN THE BLOOD

All of the different amino acids circulate in the blood and extracellular fluid in small quantities. However, their total concentration is only about 30 mg. in each 100 ml. of fluid; or, to express this another way, the total concentration of all the 21 different amino acids together is only about one-third that of glucose. The reason for this small concentration is that the amino acids, on coming in contact with cells, are absorbed very rapidly.

BUFFER ACTION OF THE LIVER AND TISSUE CELLS FOR REGULATING BLOOD AMINO ACID CONCENTRATION. The liver acts as a buffer for amino acids in the same manner that it acts as a buffer for glucose. When the blood concentration of amino acids rises high, a large proportion of them is absorbed into the liver cells, where they can be stored, probably combined with each other to form small protein molecules. When the amino acid concentration in the blood falls below normal, the stored amino acids pass back out of the liver cells into the blood to be used as needed elsewhere in the body.

Most other cells of the body also have this ability to store amino acids to at least some extent and to release these into the blood when the blood content of amino acids falls. As a result, amino acids are in a state of continual flux from one part of the body to another. If the amount of amino acids in the cells of one tissue falls too low, then amino acids will enter these cells from the blood and will be replaced by animo acids released from other cells. This continual flux of the amino acids among the various cells is illustrated by the diagram of Figure 31–9.

EFFECT OF CORTISOL ON AMINO ACID FLUX. Recent research has shown that cortisol, one of the adrenocortical hormones, aids

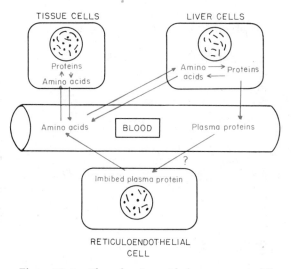

Figure 31-9. Flux of amino acids from one part of the body to another, and reversible equilibrium between the proteins of the different areas of the body.

in the movement of amino acids from one area of the body to another. Though the precise nature of this mobilization function of cortisol is yet unknown, it is believed that it increases the rate of active transport of amino acids through the cellular membranes and thereby promotes their rapid transfer from tissue to tissue. When an area of the body is damaged and is in need of amino acids for repair of its cells, the rate of cortisol secretion becomes greatly increased, and the resultant mobilization of amino acids supplies the needed materials.

THE TISSUE PROTEINS AND THEIR SYNTHESIS

The proteins of the cells perform two major functions. First, they provide most of the structural elements of the cells, and, second, they are the enzymes that control the different cellular chemical reactions. Therefore, the types of proteins in each cell determine its functions. Each cell is capable of synthesizing its own proteins, and this synthesis is controlled by the genes of the cell's nucleus in the manner described in Chapter 4. Basically, this process is the following:

Regulation of Protein Synthesis by the Genes. The nucleus of each cell of the human body contains 46 chromosomes arranged in 23 pairs. Each of these chromosomes contains several thousand *deoxyribose nucleic acid* molecules. Each of these mole-

cules is a separate *gene,* and its function is determined by its intrisic chemical structure and also by its position in the chromosome thread.

Each gene of the nucleus controls the formation of a corresponding type of *ribose nucleic acid* that is transported to the cytoplasm of the cell. This has a slightly different chemical composition from the deoxyribose nucleic acid of the nuclear gene, and it in turn acts as a "template" to control the formation of a protein by the ribosomes in the cytoplasm. Since it is the proteins that perform the structural and enzymatic functions of the cell, the nuclear genes regulate, in a roundabout way, the entire function of the cell.

Formation of Plasma Proteins. The proteins in the plasma are of three different types: *albumin,* which provides the colloid osmotic pressure in the plasma; *globulins,* which provide the antibodies; and *fibrinogen,* which is used in the process of blood clotting. Almost all these are formed in the liver and then released into the blood, though a small portion, of the globulins in particular, are formed by reticuloendothelial cells, plasma cells, and large lymphocytes. The plasma proteins are the same as many of the intracellular proteins except that they have been extruded into the circulating blood. The method of extrusion from the liver cells and plasma cells is unknown, but in the case of the large lymphocytes, it has been observed that the cellular membrane actually ruptures and empties its cytoplasmic contents into the lymph. Presumably some similar method is utlized by the liver and plasma cells.

Whenever the concentration of proteins in the plasma falls too low to maintain normal colloid osmotic pressure, the production of plasma proteins by the liver increases markedly. Though the means by which this control function works is unknown, it obviously is of great value in maintaining normal circulatory dynamics, for, if the colloid osmotic pressure of the blood should vary either above or below normal, the transfer of fluids through the capillary membranes into and out of the interstitial spaces would become abnormal.

Conversion of Proteins into Amino Acids. Most of the body's cells synthesize far more proteins than are absolutely necessary to maintain life of the cells. Therefore, if amino acids are needed elsewhere in the body, some of the cellular proteins can be reconverted into amino acids and then transported in this form. The reconversion process is catalyzed by enzymes, called *cathepsins,* that are in all cells, stored normally in the lysosomes.

The quantity of proteins in a cell is determined by a balance between their rate of synthesis and their rate of destruction. Even the plasma proteins circulating in the blood are subject to reconversion into amino acids, for they can be imbibed by reticuloendothelial cells and other cells and then split into amino acids by the intracellular enzymes.

The constant balance between amino acids and proteins in the cells, and between amino acids and the plasma proteins, is shown in Figure 31–9. By this constant interchange of amino acids, the proteins in all parts of the body are maintained in reasonable equilibrium with each other. If one tissue suffers loss of proteins or if the blood suffers loss of plasma proteins, many of the proteins in the remainder of the body will soon be converted into amino acids, which are transported to the appropriate point to form new protein. For example, in widespread cancer that is using extreme quantities of amino acids for formation of new cancer cells, the amino acids are derived continually from the tissue proteins, leading to serious debility. Also, when large quantities of blood are lost, the plasma proteins are replenished to normal within approximately 5 to 7 days by the transfer of amino acids from the tissue proteins to the liver, where new plasma proteins are formed.

USE OF AMINO ACIDS TO SYNTHESIZE NEEDED CHEMICAL SUBSTANCES

Most metabolic reactions in the cells require special chemical substances to keep them operating, and most of these chemicals are synthesized from amino acids. For instance, the muscles require large quantities of *adenine* and *creatine* to cause muscular con-

traction, the blood cells require large amounts of *heme* to form hemoglobin, and the kidneys require large quantities of *glutamine* to be used in forming ammonia. All of these substances are synthesized from amino acids.

Also, many of the hormones secreted by the endocrine glands are synthesized from amino acids. These include *norepinephrine, epinephrine* and *thyroxine,* which are synthesized from *tyrosine; histamine,* which is synthesized from histidine; and several *pituitary hormones, parathyroid hormone,* and *insulin,* which are all small proteins.

DERIVATION OF ENERGY FROM AMINO ACIDS

In addition to the use of amino acids for synthesizing new proteins or other chemical substances, some of them are also used for energy, as shown in Figure 31–10. The first step in using proteins for energy is to remove the amino radical. This process is called *deamination,* and it *occurs in the liver.* In the process, the removed amino radical is converted into ammonia, which in turn combines with carbon dioxide to form urea, all of these reactions also occurring in the liver. The urea is excreted by the kidneys into the urine. Thus, once again the liver is extremely important for one of the metabolic processes.

Referring back to the formulas of the amino acids, it will be evident that removal of the amino radical from certain of these acids still leaves very complicated chemical compounds. A few of these cannot then be utilized by the body because of their nature, and, therefore, are excreted directly into the gastrointestinal tract and lost in the feces. But most of the deaminated amino acids have sufficiently simple chemical structures so that they can enter into the same cellular reactions as glucose and keto acids. These are often directly

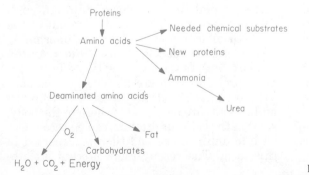

Figure 31–10. Complete schema for utilization of proteins in the body.

oxidized to form water and carbon dioxide, liberating energy in the process; or, if energy is not needed at the moment, they can be converted into fat or carbohydrate and later utilized for energy in the form of keto acids or glucose.

Conversion of Proteins to Fats or Carbohydrates. Ordinarily the body's cells must synthesize about 45 gm. of new proteins each day to replace the proteins being destroyed by the natural processes of wear and tear. If extra quantities of amino acids above the amount needed for this purpose are eaten, these normally are deaminated and converted to fats or carbohydrates, or are used for energy. In other words, even normally, excess proteins are not stored in the body in the form of proteins.

Protein-Sparing Effect of Carbohydrates and Fats During Starvation. When an insufficient quantity of food is eaten, the major portion of the energy needed for the chemical processes of the cells is derived from carbohydrates and fats as long as these are available, and the proteins are spared. This is called the *protein-sparing* effect of these substances. However, when the stores of the carbohydrates and fats are finally depleted, amino acids then begin to be mobilized and deaminated to be used for energy. One can live for another few days on this energy derived from the proteins, but this final process rapidly depletes the cells of their functional elements and soon leads to death.

ABSORPTION OF IONS AND WATER

Absorption of Ions. Ions are absorbed from the gastrointestinal tract in almost exactly the same manner as from the tubules in the kidneys as described in Chapter 17. Sodium, for instance, is *actively absorbed;* that is, it combines with a carrier in the epithelial cells and is transported through the intestinal membrane in this form to be released on the opposite side into the blood.

Though less definitive experiments are available for absorption of other electrolytes from the gastrointestinal tract, it is known that potassium, calcium, magnesium, chloride, phosphates, and iron are all also actively absorbed in a similar manner.

Absorption of Water. Water is absorbed by *diffusion* and *osmosis*. Diffusion means simply that random motion of the water molecules eventually carries them through the ep-

ithelial pores into the extracellular fluid, but the diffusion process can be affected greatly by osmotic forces, as was discussed in Chapter 8.

Water absorption from the gastrointestinal tract is controlled almost entirely by osmotic forces that operate as follows: When the monosaccharides, amino acids, and ions are absorbed from the small intestine by active absorption, the osmotic pressure of the intestinal fluids becomes very slight because of loss of the solutes. On the other hand, the osmotic pressure of the interstitial fluid on the opposite side of the epithelial membrane becomes increased. As a result, an osmotic pressure gradient develops across the intestinal membrane which causes water to be absorbed by osmosis from the intestinal lumen into the extracellular fluids. It is in this way that 8 or more liters of gastrointestinal fluid are normally absorbed from the gastrointestinal tract each day.

REFERENCES

Benditt, E. P.: The origin of atherosclerosis. *Sci. Amer., 236(2)*:74, 1977.

Bockus, H. L., Berk, J. E., Haubrich, W. S., Kalser, M., Roth, J. L. A., and Vilardell, F. (eds.): Gastroenterology. 3rd ed. Philadelphia, W. B. Saunders Company, 1975.

Brown, H. (ed.): Protein Nutrition. Springfield, Ill., Charles C Thomas, Publisher, 1974.

Csaky, T. Z. (ed.): Intestinal Absorption and Malabsorption. New York. Raven Press, 1975.

Gerolami, A., and Sarles, J. C.: Biliary secretion and motility. *Intern. Rev. Physiol., 12*:223, 1977.

Homsher E., and Kean, C. J.: Skeletal muscle energetics and metabolism. *Ann. Rev. Physiol., 40*:93, 1978.

Huijing, F.: Glycogen metabolism and glycogen-storage diseases. *Physiol. Rev., 55*:609, 1975.

Jackson, R. L., Morrisett, J. D., and Gotto, A. M., Jr.: Lipoprotein structure and metabolism. *Physiol. Rev., 56*:259, 1976.

Kappas, A., and Alvares, A. P.: How the liver metabolizes foreign substances. *Sci. Amer., 232(6)*:22, 1975.

Lieber, C. S.: The metabolism of alcohol. *Sci. Amer., 234(3)*:25, 1976.

Masoro, E. J.: Lipids and lipid metabolism. *Ann. Rev. Physiol., 39*:301, 1977.

Matthews, D. M.: Intestinal absorption of peptides. *Physiol. Rev., 55*:537, 1975.

Racker, E. (ed.): Energy Transducing Mechanisms. Baltimore, University Park Press, 1975.

Smyth, D. H. (ed.): Intestinal Absorption. Vols. 4A and 4B. New York. Plenum Publishing Corporation, 1974.

Verger, R., and Haas, G. H.: Interfacial enzyme kinetics of lipolysis. *Ann. Rev. Biophys. Bioeng., 5*:77, 1976.

Volpe, J. J., and Vagelos, P. R.: Mechanisms and regulation of biosynthesis of saturated fatty acids. *Physiol. Rev., 56*:339, 1976.

QUESTIONS

1. Explain the role that hydrolysis plays in the digestion of carbohydrates, fats, and proteins.
2. Give the mechanisms for absorption of monosaccharides and proteins.
3. How does the absorption of lipids differ from the absorption of monosaccharides?
4. Explain why glucose is the most important of all the monosaccharides in bodily metabolism.
5. How is blood glucose concentration controlled, and what is the role of insulin in this?
6. Explain the role of bile salts in fat digestion and absorption.
7. What is the relative importance of free fatty acids, chylomicrons, and lipoproteins for transport of fats in the body fluids?
8. What functions do phospholipids and cholesterol subserve in the body?
9. Explain how amino acids can be converted into glucose or fatty acid or can be used for energy.
10. Explain the absorption of ions and of water.

RELEASE OF ENERGY FROM FOODS; AND NUTRITION 32

The major function of all the digestive and metabolic processes of the body is to provide energy for performing the various bodily functions. Energy is required to lift an arm, to move a leg, or to do any activity employing muscular contraction. It is needed for the secretion of digestive juices, the development of membrane potentials in nerves and other cells, the synthesis of new chemical compounds, and active absorption of substances from the gastrointestinal tract or kidney tubules. In short, almost all functions performed by the body require energy that in turn must be supplied by the ingested food. The final steps for release of energy from the foods — that is, the end stages of metabolism — are described in the present chapter.

ADENOSINE TRIPHOSPHATE AS THE COMMON PATHWAY OF ALMOST ALL ENERGY

The cells do not use the actual foods for their immediate supply of energy. Instead, they use almost entirely a chemical compound called *adenosine triphosphate* (ATP) for this energy. The foods, in turn, are then used to synthesize more adenosine triphosphate after it has been used. The importance of this compound to the function of the cell was pointed out in Chapter 3. The present chapter explains the role of adenosine triphosphate in the overall utilization of energy by the body.

The formula for adenosine triphosphate is shown at the bottom of the page. Extremely large amounts of energy are stored in this molecule at the *bonds* where the last two phosphate radicals join with the remainder of the molecule. These bonds (~) are called *high energy phosphate bonds*. Every time a cell needs energy a phosphate radical is broken away from adenosine triphosphate at the high energy bond, and this liberates the needed energy. Each mole of adenosine triphosphate releases 8000 calories of energy for each of the high energy bonds that is broken. In short, there is a storehouse of adenosine triphos-

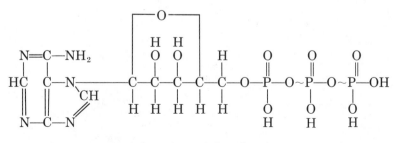

Adenosine triphosphate

phate in each cell that provides the necessary energy for muscular contraction, for development of membrane potentials, for active absorption, and so forth, but this adenosine triphosphate must be replenished continually.

FORMATION OF ADENOSINE TRIPHOSPHATE

Use of Energy from Carbohydrates to Form Adenosine Triphosphate. In the preceding chapter it was noted that carbohydrates are digested to form glucose, or are changed into glucose after absorption. Then the glucose is used by the cells for energy. Part of the energy is released from glucose by a process called *glycolysis* that does not require oxygen, but by far the major amount of energy is released when the glucose is oxidized.

GLYCOLYSIS. In glycolysis, the glucose molecule, which has six carbon atoms, is split by a series of cellular enzymes into two smaller molecules having only three carbon atoms. Then the three-carbon-chain molecules are modified to form *pyruvic acid,* which has the following formula:

$$CH_3-\overset{\overset{\displaystyle O}{\|}}{C}-COOH$$

During glycolysis a small amount of energy is released from the glucose molecule, and this energy is used to form adenosine triphosphate. By splitting glucose to form two pyruvic acid molecules, energy is liberated without the expenditure of any oxygen. This is called energy liberation by *anaerobic metabo-*

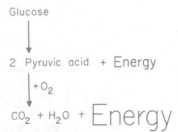

Figure 32–1. Derivation of energy from glucose by glycolysis and by oxidation.

lism, and it is illustrated by the first stage of the reaction in Figure 32–1.

OXIDATIVE RELEASE OF ENERGY FROM CARBOHYDRATES. After the glucose has been split into pyruvic acid molecules, these can then be metabolized with oxygen to form carbon dioxide and water. This reaction is shown by the second stage in Figure 32–1. The oxidative metabolism of pyruvic acid provides about 18 times as much energy as the glycolytic breakdown of glucose to form pyruvic acid. Therefore, by far the major amount of energy liberated from carbohydrates for the performance of cellular function is derived from *oxidative metabolism.*

The chemical reactions by which pyruvic acid is oxidized to supply energy have been worked out in great detail, the general principles of which are shown in Figure 32–2. The reactions of the first stage, called the *citric acid cycle* or *Krebs cycle,* split the pyruvic acid molecule into carbon dioxide and hydrogen; the carbon dioxide is removed by enzymes called *decarboxylases,* and hydrogen atoms are removed by *dehydrogenases.* In the second

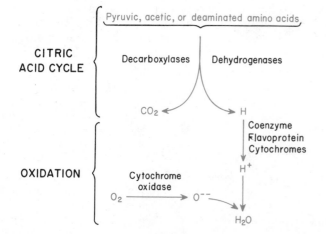

Figure 32–2. Splitting of pyruvic acid, acetic acid, or deaminated amino acids into carbon dioxide and hydrogen in the tricarboxylic acid cycle, and oxidation of the released hydrogen atoms by the cellular oxidative enzymes.

stage, called *oxidation,* the hydrogen reacts with oxygen to form water.

When hydrogen atoms are split away from pyruvic acid by the dehydrogenases, they immediately combine with a substance called *coenzyme.* Then, under the influence of other enzymes, hydrogen atoms are passed to *flavoprotein molecules* and finally to *cytochrome molecules.* At this point the hydrogen atoms are released into the surrounding fluid as hydrogen ions. Simultaneously, dissolved oxygen that has been carried to the tissues by hemoglobin is changed into oxygen ions by *cytochrome oxidase.* The presence of ionic hydrogen and ionic oxygen in the same solution provides two highly reactive substances that immediately form water molecules. Thus, the hydrogen atoms removed from the pyruvic acid become oxidized with oxygen to form water.

FORMATION OF ADENOSINE TRIPHOSPHATE DURING GLUCOSE METABOLISM One might ask why it is necessary for the hydrogen and oxygen to go through the complicated stages of the above reactions, for it is well known that hydrogen and oxygen can combine with each other very rapidly simply by being burned together. The answer to this question is that the indirect procedure is required to channel the released energy in the proper direction to form new adenosine triphosphate.

The quantity of adenosine triphosphate formed in the different stages of glucose metabolism is the following: For each molecule of glucose metabolized, 2 molecules of adenosine triphosphate are formed during glycolysis, 2 are formed in the citric acid cycle, and 34 are formed during oxidation of hydrogen, making a total of 38 molecules of adenosine triphosphate for each molecule of glucose metabolized.

The total amount of energy in each mole of glucose is 686,000 calories. Of this amount, 266,000 become stored in the form of ATP. The remainder is lost as heat caused by the chemical reactions. Thus, the overall *efficiency* of energy transfer from glucose to ATP is 39 per cent, the remaining 61 per cent of the energy becoming heat, which represents wasted energy.

Use of Energy from Fats and Proteins to Form Adenosine Triphosphate. It was pointed out in the preceding chapter that fatty acids are split into acetic acid, and that amino acids derived from proteins are deaminated to form deaminated amino acids. The same decarboxylases and dehydrogenases that remove carbon dioxide and hydrogen from pyruvic acid do the same for the acetic and most of the deaminated amino acids, and the hydrogen atoms are oxidized as explained above for the carbohydrates. Large amounts of energy are released, especially during oxidation of the hydrogen atoms, to synthesize adenosine triphosphate.

The quantity of energy derived in this manner from fats and proteins represents about 80 per cent of the total energy derived by the cells in contrast to only 20 per cent from oxidation of glucose. The reason for this is that half or more of the carbohydrate eaten by a person is first stored in the body as fat and then later used for energy in the form of fatty acids.

Regulation of Food Metabolism by Adenosine Diphosphate (ADP). The reactions that result in the formation of adenosine triphosphate cannot occur unless *adenosine diphosphate* is available from which the adenosine triphosphate can be formed. Therefore, the breakdown of foodstuffs is controlled by the presence or absence of adenosine diphosphate. Every time adenosine triphosphate is used by the cells for energy it loses one phosphate radical and therefore becomes adenosine diphosphate. The newly formed adenosine diphosphate immediately initiates reactions with the foodstuffs to cause release of new energy that is used to convert the adenosine diphosphate back into adenosine triphosphate. These interrelationships are shown in Figure 32–3. Then, when all the adenosine diphosphate has been resynthesized into adenosine triphosphate, the metabolism of the food ceases. It is in this way that the amount of food used for energy is automatically geared to the needs of the body.

INTERACTION OF ADENOSINE TRIPHOSPHATE WITH CREATINE PHOSPHATE

Another substance that contains high energy phosphate bonds, creatine phosphate, is present in the cells in quantities several times as great as those of adenosine triphosphate. When adequate amounts of adenosine triphosphate are available, much of it is used to form creatine phosphate in accordance with the reactions shown in Figure 32–3, and as rapidly as it is used more adenosine triphosphate is formed. This results in the buildup of large quantities of creatine phosphate. Then

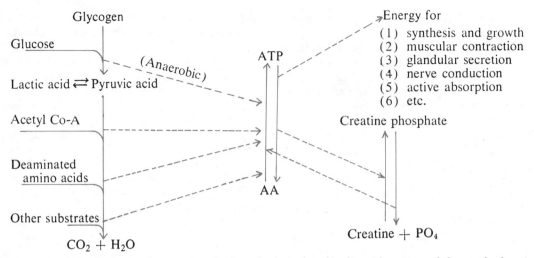

Figure 32–3. Overall schema of energy transfer from foods to the adenylic acid system and then to the functional elements of the cells. (Modified from Soskin and Levine: Carbohydrate Metabolism. University of Chicago Press.)

when the cell demands energy rapidly and in large amounts, energy is released from adenosine triphosphate directly to the functional elements of the cells. And the energy stored in the creatine phosphate is used immediately to reform new adenosine triphosphate. This reaction occurs in a fraction of a second, much more rapidly than the oxidative reactions, and it provides a very rapid source of extra energy that can be used to keep the cell functioning at an extremely high rate of metabolism for a short period of time, even though the oxidative methods for reconstituting ATP are much slower to respond.

An Overall Schema for the Energy Mechanisms of the Cell. Figure 32–3 also illustrates an overall schema for the chemical reactions that provide energy for cellular functions. This schema shows that the breakdown of adenosine triphosphate to adenosine diphosphate releases the needed energy for muscular contraction, glandular secretion, neuronal activity, intermediary metabolism, and other energy functions of the cell. Within a few seconds, creatine phosphate also breaks down, providing energy for resynthesis of much of the adenosine triphosphate. During the ensuing minutes, the adenosine triphosphate and creatine phosphate are both resynthesized by energy from the foodstuffs, part of it by glycolysis and other anaerobic procedures, but 90 per cent or more by the oxidation of pyruvic acid, acetic acid, deaminated amino acids, and a few other substances such as alcohol, glycerol, and lactic acid.

The Oxygen Debt. For a few seconds at a time a person can perform very strenuous feats of exercise requiring energy release many times that which can be sustained over a long period of time. This immediate burst of energy is provided to a great extent by the adenosine triphosphate and creatine phosphate stored in the cells. During the next few minutes, while the adenosine triphosphate and creatine phosphate are being resynthesized by the oxidative processes of metabolism, an extra quantity of oxygen above the usual amount must be utilized. Also, some of the stores of oxygen in the hemoglobin of the blood, in the myoglobin of muscle, and in the dissolved state in the body fluids will have been used during the rapid burst of energy release, and this, too, must be replenished after the exercise is over. The extra oxygen that must be used to restore completely normal conditions after exercise is over is called the *oxygen debt*.

In essence, our ability to develop an oxygen debt explains why an athlete continues to breathe very hard for many minutes after running a race, but even more important, it explains why we can perform feats involving tremendous amounts of activity for a few seconds, even though this activity cannot be sustained for long.

NUTRITION

The term *nutrition* means the supplying of foods that are required to keep one alive and healthy. These foods include carbohydrates and fats, which supply most of the body's energy; and proteins, vitamins, and minerals, which are required for synthesis of special structures and special chemical compounds needed by the body.

FOODS SUPPLYING ENERGY

Peoples in various parts of the world have widely differing diets, and some, in comparison with others, even obtain different proportions of their energy from the various types of foodstuffs. Approximately 40 per cent of the energy in the food of the average American diet is in the form of carbohydrate, 45 per cent fat, and 15 per cent protein. In other less prosperous parts of the world, the abundance of fat and protein in the diets is often less than half these values, while the energy derived from carbohydrate sometimes rises to as much as 80 per cent of the total.

The Calorie As a Measure of Energy. The energy of a food is measured in terms of the amount of heat liberated by complete breakdown of the food into its metabolic end-products, and this is expressed in *Calories,* a unit for measuring heat. One Calorie (spelled with a capital "C") is the amount of heat required to raise the temperature of 1 kilogram of water 1 degree Centigrade.

The Calorie is a very good measure of food energy, for most of the energy released in the body eventually becomes heat anyway. For example, the chemical reactions for extracting energy from the foods are so inefficient that 61 per cent of the energy becomes heat as ATP is formed, and by the time the ATP is used to perform cellular functions, 75 to 90 per cent of the energy becomes heat. The remaining 10 to 25 per cent is converted into muscular action or other functional activities of the body. Then as these functions are performed, almost all of this remaining energy still becomes heat. As an example, a large amount of energy is used to pump blood around the circulatory system. As the blood flows through the blood vessels, the energy that has been imparted to the blood is converted into heat because of friction between the blood and the walls of the vessels. In this way, all the energy expended by the heart eventually becomes heat. Likewise, almost all the energy expended by the skeletal muscles of the body eventually becomes heat, because most of it is used to overcome the friction of the joints and the viscosity of the tissues, and these two effects in turn convert the energy into heat.

Energy Content of the Different Types of Foods. The amount of energy released to the body by oxidation of 1 gram of each of the three types of foods is the following:

	Calories
Carbohydrate	4.1
Fat	9.3
Protein	4.1

From these figures it is evident that 1 gram of fat supplies more than twice as much energy as 1 gram of either carbohydrate or protein. For this reason, fats in the diet are very deceptive. Often one thinks that he is deriving very little of his energy from fat and yet may be obtaining as much from fat as from carbohydrate. A second reason why fat is deceptive is that it often occurs in a pure form in foods, while carbohydrates and proteins are generally diluted several times over with water. When one eats potatoes with butter, the fat of the butter generally contains almost as much energy as the entire potato, for two reasons: First, the butter supplies two and one-quarter times as much energy per gram as the starch of the potato, and, second, the starch comprises only about one-sixth the bulk of the potato, because most of the potato is water.

DAILY ENERGY REQUIREMENTS

An average man of 70 kilograms who lies in bed all day long and does nothing else except eat and exist usually requires about 1650 Calories of energy each day. If he simply sits in a chair, another 200 or more Calories are required. Therefore, about 1800 Calories per day is the normal basal amount of energy required simply for living. In addition, any type of exercise requires still more energy, which is shown by the values in Table 32–1 for different types of activity. From this table it is evident that walking upstairs requires approximately 17 times as much energy as lying in bed asleep. However, this tremendous rate of energy utilization can be continued only for short periods of time. Over long periods, a well-conditioned worker can average as much as 6000 to 8000 Calories of energy expenditure each 24 hours, or, in other words, as much as four times the basal rate.

Table 32–1. Energy Expenditure per Hour During Various Types of Activity for a 70 Kg. Man*

FORM OF ACTIVITY	CALORIES PER HOUR
Sleeping	65
Awake lying still	77
Sitting at rest	100
Standing relaxed	105
Dressing and undressing	118
Tailoring	135
Typewriting rapidly	140
"Light exercise"	170
Walking slowly (2.6 miles per hour)	200
Carpentry, metal working, industrial painting	240
"Active exercise"	290
"Severe exercise"	450
Sawing wood	480
Swimming	500
Running (5.3 miles per hour)	570
"Very severe exercise"	600
Walking very fast (5.3 miles per hour)	650
Walking up stairs	1100

*Extracted from data compiled by Professor M. S. Rose.

PROTEIN REQUIREMENTS OF THE BODY

However low the concentration of amino acids might fall to in the blood, the liver still deaminates some of them all the time, and these then are used mainly for energy. Because of this continual loss of amino acids, the proteins of the body diminish constantly unless they are replenished by proteins in the diet. Normally, even when more than enough carbohydrates and fats are present in the diet to provide their protein-sparing effect, approximately 45 grams of protein still are required each day to replace this continual loss.

Partial Proteins. Certain types of proteins do not contain all the essential amino acids in the proportion in which they are needed in the body. When one eats such a protein his requirement will be considerably above 45 grams per day. For instance, the proteins from vegetables or grains have ratios of the various amino acids different from those found in the human body. On the other hand, the proteins of animal origin, in general, have almost exactly the same amino acid compositions as those of the human being. Therefore, the person who requires 45 grams of animal protein in his diet each day might require as much as 65 grams of vegetable protein, the exact amount depending on the type of vegetable eaten. The proteins that cannot supply the right proportions of the different amino acids are called *partial proteins,* because they supply only part of the needed variety of amino acids.

An especially serious example of nutritional deficiency that occurs because of too much partial protein and too little complete protein in the diet is the disease condition called *kwashiorkor,* occurring mostly in young African children. The major diet of these children is a cornmeal mush. The corn contains a moderate amount of protein, but this protein is very deficient in one of the essential amino acids, tryptophan. Therefore, animal proteins cannot be synthesized from the amino acids derived from this diet; instead, by far the greater majority of the amino acids are simply degraded and used for energy. As a result, the child becomes very protein-deficient, which causes greatly stunted growth and, in many instances, death. Those children who survive are frequently mentally deficient for the remainder of their lives because their brains have not developed properly.

SPECIAL NEED FOR HIGHLY UNSATURATED FAT IN THE DIET

A small amount of highly unsaturated fat is essential for nutrition of animals. The body cannot desaturate fat sufficiently by its own metabolic processes to supply this. The types of unsaturated fatty acids usually available in the diet are *arachidonic, linoleic,* and *linolenic* acids. These are believed to be needed by the cells to form cellular structures. Without them animals develop skin sores, mental changes, and other evidence of general cellular debility. Whether or not these same effects would occur in the human being is not known, because such a small amount of these substances is required in the diet that no human being has been proved to be suffering from a deficiency of them.

VITAMIN REQUIREMENTS OF THE BODY

The vitamins are chemical compounds needed in only minute quantities by the body to perform special functions. The daily requirements of each of the vitamins is given in Table 32–2, and the vitamin content of dif-

Table 32–2. **Daily Requirements of the Vitamins**

Vitamin	Daily Requirement
A	5000 units
Thiamin	1.8 mgm.
Riboflavin	1.8 mgm.
Niacin	18.0 mgm.
Ascorbic acid	75.0 mgm.
D (children and during pregnancy)	400 units
E	unknown
K	none
P	unknown
Folic acid	0.5 mgm. (?)
B_{12}	2.0 μgm. (?)
Inositol	unknown
Pyridoxine	2.0 mgm. (?)
Pantothenic acid	unknown
Biotin	unknown
Para-aminobenzoic acid	unknown

ferent foods is given in Table 32–3. In general, eating a balanced diet will provide an adequate quantity of all the different vitamins. Occasionally, though, an abnormality makes it impossible to utilize one of the vitamins, in which case a vitamin deficiency disease can occur even in the presence of a normally satisfactory diet.

The precise chemical functions of many of the vitamins in the body are only partially known, but from the physiologic effects caused by lack of them in the diet, their functions can at least be speculated upon as follows.

Vitamin A. Vitamin A is used by the eyes to synthesize the light-sensitive retinal pigments used by the rods and cones for vision. This was discussed in detail in Chapter 28. Also, lack of vitamin A causes the epithelial structures, such as the skin, the intestinal mucosa, and the germinal epithelium of the ovaries and testes, to become highly *keratinized,* or horny. This leads to scaliness of the skin, failure of growth of young animals, failure of reproduction, and even hardening of the cornea, occasionally causing corneal opacity and blindness.

Thiamine — Beriberi. Thiamine forms a compound in the cells called *thiamine pyrophosphate,* which is part of a decarboxylase that removes carbon dioxide from pyruvic acid and other substances. Without thiamine the oxidative processes for metabolizing especially carbohydrates and fats become defi-

Table 32–3. **Vitamin Content of Various Foods***

Vitamin A Units/100 gm.	Thiamine mgm./100 gm.	Riboflavin mgm./100 gm.	Niacin mgm./100 gm.	Vitamin C mgm./100 gm.
Apricots 5000	Barley 0.450	Dried beans 0.30	Asparagus 1.2	Asparagus 45
Broccoli 4000	Dried beans 0.540	Almonds 0.50	Barley 4.7	Brussel sprouts 95
Butter 3500	Buckwheat 0.450	Beef 0.20	Soybeans 4.0	Cabbage 65
Carrots 6000	Dried cowpeas 0.900	Cheese 0.55	Dried beans 3.0	Butter 70
Collards 8000	Egg yolk 0.320	Chicken 0.30	Beef 5.4	Cauliflower 82
Chard 9000	Whole wheat 0.585	Collards 0.25	Chicken 8.5	Lemon juice 53
Endive 15,000	Pork meat 1.0	Eggs 0.40	Collards 2.3	
Kale 20,000	Lamb meat 0.330	Lamb 0.24	Corn 1.3	Orange juice 45
Liver 70,000	Beef meat 0.120	Liver 2.60	Lamb 8.0	Fish 20
Mustard greens 10,000	Liver 0.400	Milk 0.18	Liver 17.0	Potatoes 10
Sweet potatoes 3000	Millet 0.700	Peanuts 0.45	Mackerel 2.1	Lean meats 5
Pumpkin 7000	Brown rice 0.370	Pork 0.24	Pork 7.0	Liver 25
Spinach 25,000	Wheat germ 2.0	Grains 0.10	Salmon 6.2	
			Wheat 4.3	

	Pantothenic Acid mgm./100 gm.	Pyridoxine mgm./100 gm.	Inositol mgm./100 gm.	Biotin mgm./100 gm.	Folic Acid mgm./100 gm.
Meats	0.8	0.1	60	0.007	0.15
Eggs	1.4	0.02	33	0.009	0.90
Milk	0.3	0.13	18	0.005	0.005
Cereals	0.4	0.05	65	0.005	0.10
Fruits	0.2	0.05	95	0.003	0.05
Vegetables	0.5	0.15	70	0.007	0.08

*Modified from W. H. Eddy and G. Dalldorf.

cient, and this can cause almost any type of abnormality in the body. It especially affects the functions of the nervous system, the heart, and the gastrointestinal system.

Lack of thiamine causes pathologic changes in neurons and in the myelin sheaths of nerve fibers, often resulting in actual destruction of the cells or irritation or degeneration of peripheral nerves. Frequently the fibers of peripheral nerves become so irritable that a condition called *polyneuritis* results, which manifests itself by excruciating pain along the course of the nerves.

In the heart, thiamine deficiency decreases the strength of the muscle. The heart becomes greatly dilated and pumps with little force, resulting in *congestive heart failure.*

In the gastrointestinal tract, thiamine deficiency causes weakness of the intestinal muscle, poor secretion of digestive juices, and poor maintenance of the intestinal mucosa. Severe indigestion, constipation, lack of appetite, and other symptoms very often develop.

A thiamine-deficient person who has neuritis, enlargement of the heart, and gastrointestinal symptoms all at the same time is said to have beriberi.

Niacin and Riboflavin — Pellagra. Niacin and riboflavin are utilized in the body to form respectively *coenzyme* and *flavoprotein.* Earlier in the chapter it was pointed out that these two substances perform necessary functions in the oxidation of hydrogen atoms after they are removed from pyruvic acid and other substrates. Without niacin and riboflavin, therefore, the oxidation of foods becomes deficient, and the cells fail to receive adequate quantities of energy.

Niacin deficiency leads especially to discoloration of the skin, which becomes darkened on exposure to sunlight. Along with this, severe muscular weakness, diarrhea, and one or more psychoses often occur. This is the clinical picture of the condition called *pellagra.*

Deficiency of riboflavin is not as likely to lead to such severe difficulties as deficiency of niacin, because other types of flavoproteins that can perform almost the same functions as the protein derived from riboflavin are present in the body. Nevertheless, riboflavin deficiency, when added to deficiencies of niacin or thiamine, can greatly intensify the symptoms observed in pellagra or beriberi. A common result of riboflavin deficiency alone is cracking at the angles of the mouth, called *cheilosis.*

Vitamin B$_{12}$ and Folic Acid — Pernicious Anemia. These two substances are needed by the bone marrow to form red blood cells and are also needed in all the other tissues of the body for adequate growth. When they are lacking, the red cells released into the blood are few in number, and those that are released are usually larger than normal, poorly formed, and very fragile. Therefore, the person develops very severe anemia called *pernicious anemia.*

On studying the bone marrow, one finds that the new red cells being formed have abnormal structures, for which reason it is clear that vitamin B$_{12}$ and folic acid are necessary for the formation of the structural elements of the cells. However, lack of these vitamins does not affect the formation of hemoglobin.

Red cells are affected more than other cells of the body by lack of these two substances probably because these cells are produced much more rapidly than most other cells. This is supported by the fact that cancer cells, which are also rapidly reproduced cells, fail to reproduce in the absence of these vitamins. In fact, purposeful deficiency of these two vitamins can actually be used as a means for slowing up the growth of most cancers.

Less Important Vitamin B Compounds. For approximately the first 20 years after beriberi was discovered to be a nutritional disease, it was believed that the vitamin extract used to treat this disease contained only one single vitamin, and this was called *vitamin B.* Later, many different vitamins were discovered in the extract, and these have come to be known as the *vitamin B complex.* The vitamins included in this complex are thiamine (vitamin B$_1$), riboflavin (vitamin B$_2$), niacin, vitamin B$_{12}$, folic acid, *pyridoxine, pantothenic acid, biotin, inositol, choline,* and *para-aminobenzoic acid.* The functions of some of these vitamins in the human being are not well known, but some of the functions of pyridoxine and pantothenic acid, the two most important of them, are the following:

PYRIDOXINE. Pyridoxine is needed for synthesis of some of the amino acids and perhaps other compounds. Dietary lack of this vitamin in lower animals can cause *dermatitis, decreased rate of growth,* development of a *fatty liver, anemia,* and different types of *mental symptoms.* The effects of lack in the human being are not too well known, because pyridoxine deficiency almost never occurs without simultaneous deficiency of other vitamins of the B complex.

Pantothenic Acid. Pantothenic acid is used in the body to form a special chemical called *coenzyme A,* which catalyzes the acetylation of many substances in the cells. For example, acetic acid must be acetylated before it can be oxidized, and acetylcholine is formed from choline by the process of acetylation. Pantothenic acid deficiency has never been known to develop in the human being, because this substance is widespread in almost all foods of the diet. In animals, deficiency can be created artificially, and the results are retarded growth, failure of reproduction, graying of the hair, dermatitis, fatty liver, and many other effects. All these effects testify to the importance of pantothenic acid to the metabolism of the body.

Ascorbic Acid (Vitamin C) — Scurvy. The major function of ascorbic acid is to maintain normal intercellular substances throughout the body. These include the connective tissue fibers that hold the cells together, the intercellular cement substance between the cells, the matrix of bone, the dentin of the teeth, and other substances excreted by the cells into the intercellular spaces.

Deficiency of ascorbic acid in the diet causes failure of wounds to heal because new fibers and new cement substance are not deposited. Also, it causes bone growth to cease and blood vessels to become so fragile that they bleed on the slightest provocation. In short, lack of vitamin C leads to loss of integrity of many of the tissues of the body, and the resulting picture, characterized especially by bleeding of the gums, splotchy hemorrhages beneath the skin, and many other internal abnormalities, is the disease called *scurvy.*

Vitamin D — Rickets. Vitamin D promotes calcium absorption from the gastrointestinal tract. Without adequate quantities of vitamin D the bones become decalcified, a process called *rickets,* and in very severe vitamin D deficiency the level of ionic calcium in the blood may fall so low that muscular tetany develops. All these effects of vitamin D deficiency will be discussed in relation to calcium metabolism and parathyroid hormone in Chapter 36.

Vitamin E. In lower animals, lack of vitamin E can cause the male germinal epithelium in the testes to degenerate, causing sterility; in the female it can cause reabsorption of a fetus even after conception. For these reasons vitamin E is sometimes called the *antisterility* vitamin. Severe lack of the vitamin in animals can also cause muscle degeneration and paralysis. However, deficiency of this vitamin in the human being has never been proved, and the precise chemical means by which vitamin E functions also has not been determined.

Vitamin K. Vitamin K is required for the formation of prothrombin by the liver. Therefore, deficiency of vitamin K depresses blood coagulation so that there is excessive bleeding. This subject was discussed in greater detail in Chapter 7. Vitamin K usually is not in the diet in large quantities. Yet the normal person does not experience vitamin K deficiency because a large amount of this substance is synthesized in the colon by bacteria and is then absorbed. If the bacteria of the colon are destroyed by administration of antibiotic drugs, vitamin K deficiency usually develops within the next few days.

MINERAL REQUIREMENTS OF THE BODY

Table 32–4 gives the amounts of different minerals and some other substances in the entire body, and Table 32–5 gives the daily requirements of minerals. Some of these minerals have already been discussed in relation to other phases of the physiology of man. For instance, sodium, chloride, and calcium are major constituents of the extracellular fluid, and potassium, phosphate, and magnesium are major constituents of the intracellular fluid. These minerals are responsible for development of electrical potentials at the cell membrane and for the maintenance of proper osmotic equilibria between the extracellular and intracellular fluids. In addition, calcium and phosphate are major constituents of bone, and phosphate forms a great number of different chemicals used for a myriad of functions inside all cells, some of which were discussed earlier in this chapter. The remaining

Table 32–4. **Content in Grams of a 70 Kg. Adult Man**

Water	41,400	Mg	21
Fat	12,600	Cl	85
Protein	12,600	P	670
Carbohydrate	300	S	112
Na	63	Fe	3
K	150	I	0.014
Ca	1,160		

Table 32–5.   **Daily Requirements
of Minerals**

Na	3.0 gm.	I	250.0 μgm.
K	2.5 gm.	Mg	unknown
Cl	2.5 gm.	Co	trace
Ca	1.0 gm.	Cu	trace
PO$_4$	1.5 gm.	Zn	trace
Fe	12.0 mgm.	F	trace

minerals that need special comment are iron, iodine, cobalt, copper, zinc, and fluorine.

Iron.   About two-thirds of the iron of the body is in the hemoglobin in the blood; most of the remainder is stored in the liver in the form of *ferritin.* The iron in ferritin can be mobilized when needed, and carried in the blood to the bone marrow, where it is used to form hemoglobin.

Iron is also present in some of the enzymes of the cells — especially in the cytochromes. Therefore, a second function of iron is to aid in the oxidation of food in the tissue cells.

Iodine.   Iodine is used by the thyroid gland in the synthesis of thyroxine, a hormone that increases the rate of metabolism of the body. The functions of thyroxine and its relation to iodine metabolism will be discussed in Chapter 34.

Copper and Cobalt.   Copper and cobalt both affect the formation of red blood cells. Copper, in some way yet unknown, helps to catalyze the formation of hemoglobin. Very rarely does copper deficiency exist, however, so that this is almost never a cause of hemoglobin deficiency. Cobalt is an essential element in vitamin B$_{12}$, and in this way is essential for maturation of red blood cells. When cobalt in other forms besides vitamin B$_{12}$ is in the diet in excess, the bone marrow, for reasons not yet understood, produces far too many red blood cells, causing polycythemia.

Zinc.   Zinc forms part of the structure of the enzyme *carbonic anhydrase,* which is present in many parts of the body, especially in the red blood cells and in the epithelium of the kidney tubules. This substance catalyzes the reaction of carbon dioxide and water to form carbonic acid and also catalyzes the same reaction in the reverse direction. It causes carbon dioxide to combine with water about 210 times as rapidly as it would otherwise. This allows the blood to transport carbon dioxide far more easily than could be possible in the absence of this enzyme.

In the kidney tubules, carbonic anhydrase catalyzes the reactions that cause the secretion of hydrogen ions into the tubular fluid, thereby helping to regulate the acid-base balance of the body fluids.

In addition to being utilized in carbonic anhydrase, zinc is also a component of many other enzymes, including especially many of the peptidases that are required for digestion of proteins in the intestines.

Fluorine.   Fluorine in the diet protects against *carious* teeth. It does not make the teeth any stronger, but is believed to inactivate bacterial secretions that can cause tooth decay. Only a small trace of this substance in the drinking water of the growing child usually provides a major degree of protection against tooth decay throughout life, which is the reason why city water supplies are now usually fluorinated.

REGULATION OF FOOD INTAKE

Food intake is regulated by the sensations of *hunger* and *appetite.* Hunger means craving for food, and the term appetite is often used in the same sense as hunger except that it usually does not imply actual discomfort as is frequently the case with hunger. Also, appetite often means a desire for specific food instead of for food in general.

The term *satiety* means the opposite of hunger, a feeling of complete fulfillment in the quest for food. Satiety results from a filling meal.

Neural Centers for Regulation of Food Intake.   Stimulation of the *lateral hypothalamus* causes an animal to eat voraciously, while stimulation of the *medial nuclei of the hypothalamus* causes complete satiety even in the presence of highly appetizing food. Therefore, we can call the lateral hypothalamus the *hunger center* or the *feeding center,* while we can call the medial hypothalamus a *satiety center.*

In addition to the hypothalamic centers, however, the conscious centers of the cerebral cortex and the amygdaloid nuclei also enter into regulation of food intake — not so much into the regulation of total quantity of intake, but into the specific choice of food, for it is in these areas that memories of previously eaten pleasant or unpleasant foods are stored, and these memories adjust the appetite for different foods accordingly.

Long-Term and Short-Term Regulation of Food Intake. There are two entirely different types of food intake regulation, called respectively "long-term regulation" and "short-term" regulation.

Long-term regulation means regulation of food intake in relation to the amount of nutritive stores in the body. For instance, a person who has been underfed for many weeks has intensified hunger until his normal nutritive stores have been replenished. Conversely, an animal that has been force-fed until it is greatly overweight has almost no hunger, a condition that can last for weeks until its normal body weight has returned.

The precise mechanism by which the nutritive stores affect hunger is not known, though it is known that a decreased level of glucose in the body fluids, which usually goes along with decreased stores of other nutrients in the body, increases a person's hunger. Decreased quantities of amino acids and fatty acids in the body fluids seem also to cause the same effect, in this way controlling the degree of hunger.

Short-term regulation means regulation of dietary intake in relation to the amount of food that can be processed by the gastrointestinal system in a given period of time. For instance, if a person overeats, he can so overload his gastrointestinal tract that he will become sick from this alone. Therefore, during the process of eating, two principal mechanisms prevent such overeating. These are (1) "metering" of food as it passes through the mouth and (2) reflexes caused by distention of the upper gastrointestinal tract. Metering of food means that sensory receptors in the mouth and pharynx detect the amount of chewing, salivation, swallowing, and tasting and thereby quantitate the amount of food that passes through the mouth. In some way that is not understood, this information passes to the hypothalamic feeding center to inhibit hunger for up to 30 minutes or an hour, but no longer. Likewise, as food fills the stomach and other regions of the upper gastrointestinal tract, visceral sensory impulses, caused mainly by distention of the gut, are transmitted to the feeding center and inhibit it. In this way, overfilling of the gastrointestinal tract is prevented until the food that is already there has had time to be digested.

OBESITY

Much obesity is caused simply by overeating as a result of poor eating habits. For instance, many people eat three meals a day simply because of habit rather than because of hunger at the time of the meal.

Many instances of obesity also result from *inherited* imbalance between the hunger and satiety centers of the hypothalamus. For instance, when an exceedingly obese person forces himself to diet until he reduces to many pounds below his obese weight, he develops voracious hunger and, if left to his own means, will regain weight almost exactly to his original obese level. Once he reaches this level, his hunger becomes essentially the same as that of a normal person. Thereafter, he eats merely enough to maintain his weight rather than to gain still more. This is analogous to setting the thermostat in a house to a high level. In other words, the "hungerstat" is set at a higher level in these persons than in others, causing excessive eating until the person becomes very obese.

STARVATION

During starvation, essentially all the stored carbohydrates in the body, which amounts to about 300 grams of glycogen in the liver and muscles, becomes used up within the first 12 to 24 hours. Thereafter, the person exists on his stored fats and proteins. For the first 2 to 4 weeks, almost all of the energy used by the body is derived from the stored fats. But eventually even these are almost depleted, so that finally the proteins must also be used. Most tissues can give up as much as one-half of their proteins before cellular death begins. Therefore, for at least another few days to a week, the body can derive its energy from proteins. But, finally, death ensues because proteins are the necessary chemical elements for performance of cellular functions. This usually occurs 4 to 7 weeks after starvation begins.

REFERENCES

Bennett, A. F.: Activity metabolism of the lower vertebrates. *Ann. Rev. Physiol., 40*:447, 1978.

Bray, G. A.: Endocrine factors in the control of food intake. *Fed. Proc. 33*:1140, 1974.

Bray, G. A., and Campfield, L. A.: Metabolic factors in the control of energy stores. *Metabolism, 24*:99, 1975.

Buskirk, F. R.: Obesity: a brief overview with emphasis on exercise. *Fed Proc., 33*:1948, 1975.

Dickie, R. S.: Diet in Health and Disease. Springfield, Ill., Charles C Thomas, Publisher, 1974.

Mayer, J.: Human Nutrition. Springfield, Ill., Charles C Thomas, Publisher, 1974.

Novin, D., Wyrwicka, W., and Bray, G. A. (eds.): Hunger: Basic Mechanisms and Clinical Implications. New York, Raven Press, 1975.

Olson, R. E. (ed.): Protein-Calorie Malnutrition. New York, Academic Press, Inc., 1975.

Panksepp, J.: Hypothalamic regulation of energy balance and feeding behavior. *Fed. Proc., 33*:1150, 1974.

Williams, R. J.: Physician's Handbook of Nutrition. Springfield, Ill., Charles C Thomas, Publisher, 1974.

QUESTIONS

1. Give reasons why adenosine triphosphate is so important to cellular metabolism.
2. Explain how carbohydrates and fats are utilized in the formation of adenosine triphosphate.
3. Explain the role of creatine phosphate in cellular metabolism.
4. What is the energy equivalent of carbohydrate, fat, and protein?
5. What are the approximate daily energy requirements at rest and during work?
6. What are the daily requirements for proteins and fats?
7. What are the effects of deficiency of each of the different vitamins?
8. Review the different mineral requirements of the body, and give the function of each of the minerals.
9. Explain the mechanisms of and the differences between long-term and short-term regulation of food intake.

BODY HEAT, AND TEMPERATURE REGULATION

33

HEAT PRODUCTION IN THE BODY

Metabolic Rate. In the preceding chapter it was pointed out that essentially all of the energy released from foods eventually becomes heat. Therefore, the rate of heat production by the body is a measure of the rate at which energy is released from foods. And this is called the *metabolic rate*.

The metabolic rate is measured in Calories, which is the same term used to express the amount of energy in foods. When a normal individual is in a very quiet state, the metabolic rate may be as low as 60 to 70 Calories per hour. On the other hand, it may be as high as 1000 to 2000 Calories per hour for a few minutes at a time, or as high as 200 to 300 Calories per hour for several hours at a time.

FACTORS THAT AFFECT THE METABOLIC RATE

Any factor that increases the rate of energy release from foods also increases the metabolic rate. Some of the more important of these are the following:

Exercise. Exercise is perhaps the most powerful stimulus for increasing the metabolic rate. When muscles contract, a tremendous quantity of adenosine triphosphate is degraded to adenosine diphosphate in the muscle cells, and this then enhances the rate of oxidation of the foodstuffs. During very strenu-

ous exercise lasting only a moment or more the metabolic rate can actually increase to as much as 40 times that during rest.

Effect of Sympathetic Stimulation and Norepinephrine on Metabolic Rate. When the sympathetic nervous system is stimulated, norepinephrine is released directly into the tissues by the sympathetic nerve endings. Also, large quantities of this hormone and of epinephrine are released into the blood by the adrenal medullae. These two hormones then exert a direct effect on all cells to raise their metabolic rates. They enhance the breakdown of glycogen into glucose and also increase the rates of some of the enzymatic reactions that promote oxidation of foods.

Strong sympathetic stimulation can increase the metabolic rate as much as 100 per cent, but the metabolic rate remains elevated for only a few minutes after the sympathetic stimulation ceases. By controlling the activity of the sympathetics, the central nervous system has a means for regulating the rates of activity of all the cells of the body, increasing their activity when this is required and decreasing their activity when the need no longer exists.

Effect of Thyroid Hormone on Metabolic Rate. Thyroid hormone has an effect on all cells of the body similar to that of norepinephrine and epinephrine, except that the thyroid hormone requires several days to begin acting but then continues for as long as 4 to 8 weeks after its release from the thyroid gland, rather than for only a few minutes. Thyroid

425

hormone secreted in very large amounts can, like norepinephrine and epinephrine, increase the metabolic rate as much as 100 per cent. On the other hand, complete lack of secretion by the thyroid gland causes the metabolic rate to fall to as low as 50 per cent below normal. In other words, the overall span from total lack of thyroid hormone to great excess can increase the metabolic rate as much as fourfold.

The precise mechanism by which thyroid hormone affects the cells is not known, but it is known to increase the quantities of most of the cellular enzymes, which perhaps explains its metabolic effects. The actions of thyroid hormone will be discussed at further length in Chapter 34.

Most of the other endocrine hormones, except norepinephrine, epinephrine, and thyroxine, have only minor effects on the overall metabolic rate, though insulin from the pancreas, growth hormone from the anterior pituitary gland, testosterone from the testes, and adrenocortical hormones all can increase the general rate of metabolism as much as 5 to 15 per cent.

Effect of Body Temperature on Metabolic Rate.   The higher the temperature of a chemically reactive medium, the more rapid are the chemical reactions. This effect is also observed in the chemical reactions of the cells in the body. Each degree Centigrade increase in temperature raises the metabolic rate approximately 10 per cent. With very high fever the metabolic rate may be as much as twice normal because of the fever itself.

Specific Dynamic Action of Foods on Metabolic Rate.   After a meal, the metabolic rate usually rises and remains elevated for the ensuing 2 to 10 hours. In general, a meal containing large quantities of fats and carbohydrates will increase the metabolic rate only 4 to 15 per cent. In contrast, a meal containing large quantities of proteins can increase the metabolic rate as much as 30 to 60 per cent. This effect is called the specific dynamic action of foods.

The specific dynamic action of foods is believed to be caused at least partially by the greater metabolism required for digestion, absorption, and assimilation of the foods. But in addition to this effect, proteins probably increase the metabolic rate an extra amount because of direct stimulatory effects of some of the amino acids or of some other breakdown products of proteins.

BASAL METABOLIC RATE

Because of the many different factors that can affect the metabolic rate, it is extremely difficult to compare metabolic rates from one person to another. To have any valid comparison at all, a person's rate of energy utilization must be measured when he is in a so-called *basal* state. This means that (1) he is not exercising and has not been exercising for at least 30 minutes to an hour, (2) he is at complete mental rest so that his sympathetic nervous system is not overactive, (3) the temperature of the air is completely comfortable so that his sympathetic nervous system is not unduly stimulated by this factor, (4) he has not eaten any food within the last 12 hours that could cause a specific dynamic action on the metabolic rate, and (5) his body temperature is normal to avoid the effect of fever on metabolism.

Under these *basal conditions* most of the factors that affect the metabolic rate are controlled, but this basal state does not remove the effect of the constantly secreted thyroid hormone. Therefore, the *basal metabolic rate* in essence is determined by two major factors: first, the inherent rates of chemical reactions of the cells, and, second, the amount of thyroid hormone acting on the cells. Because the inherent rate of cellular activity is relatively constant from one person to another, an abnormal basal metabolic rate is usually caused by an abnormal secretory rate of thyroid hormone. For this reason basal metabolic rates are often measured to assess the degree of thyroid activity.

Direct Method for Measuring Basal Metabolic Rate.   Since the basal metabolic rate is actually a measure of the amount of heat produced by the body under basal conditions, it can be determined by measuring the heat given off from the body in a known period of time. To do this, human subjects are placed in a large chamber called a *human calorimeter*. This chamber is cooled by water flowing through a radiator system. The heat given off from the body is picked up by the cooling system and then measured by appropriate physical apparatus. This is called the *direct* method for measuring the basal metabolic rate because it measures the heat output directly. Obviously, it is very cumbersome, but it has been a very important tool in experimental studies.

Indirect Method for Measuring Basal Metabolic Rate.   An indirect method for measuring the basal metabolic rate is based on the amount of oxygen burned by the body in a given period of time. From this the rate of energy release can be calculated.

The amount of energy released when 1 liter of oxygen burns with carbohydrates is 5.05 Calories. When 1 liter of oxygen burns with fats the amount released is 4.70 Calories. And when 1 liter of oxygen burns with proteins the amount is 4.60 Calories. It is obvious from these figures that every time 1 liter of oxygen is burned in the metabolic fires of the body, essentially the same amount of energy is released regardless of which one of the three different foods is being used for energy. Therefore, it is reasonable to use an approximate average of these values, 4.825 Calories, as the amount of energy released in the body every time 1 liter of oxygen is burned. When calculating the basal metabolic rate in this way, the value obtained is never more than 4 per cent in error, even though a great excess of one type of food or the other might be used for metabolism momentarily, and 4 per cent is far less than the error of measurement anyway. Therefore, to determine the amount of energy being generated in the body, one needs only to determine the amount of oxygen being used. This is done using a respirometer.

THE RESPIROMETER.   Figure 33–1 shows a respirometer. The subject places a mouthpiece in his mouth and breathes in and out of an inverted, counterbalanced can that rides up and down in a water bath. The respirometer contains pure oxygen that is breathed back and forth into the lungs. The oxygen is gradually absorbed into the blood, and in its place carbon dioxide is expired. Then soda lime in the respirometer reacts chemically with the carbon dioxide to remove it from the respiratory gases. The net result, therefore, is a continual loss of oxygen from the respirometer; this loss causes the inverted can to sink deeper and deeper into the water. The recording pen also falls lower and lower, giving a record of the rate of oxygen utilization.

Method for Expressing the Basal Metabolic Rate.   The basal metabolic rate is generally expressed in terms of Calories per square meter of body surface area per hour. The reason for expressing it in terms of body surface area is to allow comparisons between individuals of different sizes. Experimental studies have shown that the basal metabolic rate varies from one normal person to another approximately in proportion to the body surface area and not in proportion to the weight. Thus, someone who weighs 200 pounds does not have a basal metabolic rate twice that of someone weighing 100 pounds, but only 30 per cent greater. His surface area also is about 30 per cent greater.

A representative calculation of basal metabolic rate might be the following: A person is found to use 1.8 liters of oxygen in 6 minutes. Therefore, in 1 hour he uses 18 liters of oxygen, and his total basal metabolic rate for this period would be 18 times 4.825 or 86.85 Calories per hour. To express this in Calories per square meter per hour one uses the chart in Figure 33–2A, which shows the square meters

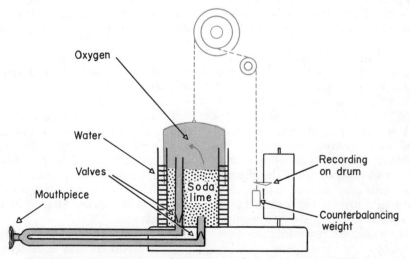

Figure 33–1.   A respirometer for measuring the rate of oxygen utilization.

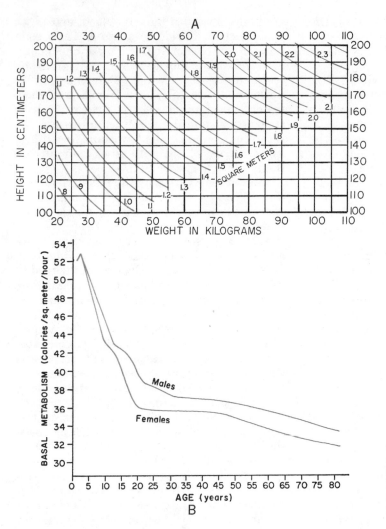

Figure 33–2. (A) Diagram for determining the body surface area when the weight and height are known. (From DuBois: Basal Metabolism in Health and Disease. Lea & Febiger.) (B) Normal basal metabolic rates for males and females at different ages.

of surface area in relation to weight and height. If the person weighs 70 kg. and is 180 cm. tall, his surface area is 1.9 square meters. Therefore, his basal metabolic rate is 86.85 Calories per hour divided by 1.9, or 45.7 Calories per square meter per hour.

The basal metabolic rate is often then expressed in per cent of normal. To do this one refers to the chart in Figure 33–2B, which shows the normal basal metabolic rate for males and females at different ages. If this person is a boy aged 18, his normal basal metabolic rate is 40 Calories per square meter per hour. However, his actual basal metabolic rate is 5.7 Calories greater than the normal value, or 14 per cent greater than normal. His basal metabolic rate, therefore, is stated to be +14. If his basal metabolic rate had been 25 per cent less than normal, it would have been expressed as −25, rather than plus.

When the thyroid gland is secreting an ex-treme quantity of thyroxine, the basal metabolic rate can at times go as high as +100. On the other hand, when the thyroid gland is secreting almost no thyroxine, the basal metabolic rate falls to as low as −40 to −50.

MAJOR LOCI OF HEAT PRODUCTION IN THE BODY

Though all the tissues of the body produce heat, those that have rapid chemical reactions produce the major amounts. In the resting state, the liver, heart, brain, and most of the endocrine glands produce large amounts of heat. This causes their temperatures to be a degree or so higher than that of most of the other tissues.

During rest the amount of heat produced by each skeletal muscle is not very great, but, because half of the entire mass of the body is composed of muscles, heat production of all the skeletal muscles together nevertheless accounts for about 30 per cent of all the body's heat production even at rest. Therefore, a slight increase or decrease in the degree of muscular tone can have considerable effect on the amount of heat produced. During severe exercise the amount of heat produced by the muscles can rise for about a minute to as great as 50 times that produced by all the remaining tissues together. For this reason, changes in the degree of muscular activity constitute one of the most important means by which the body regulates its temperature. This is discussed in detail later in this chapter.

HEAT LOSS FROM THE BODY

The heat produced in the body must be removed continually or otherwise the body temperature would continue rising indefinitely. The body loses heat to its surroundings in three ways: radiation, conduction, and evaporation.

HEAT LOSS BY RADIATION

About 60 per cent of the heat loss from a nude person sitting in a room at 70° F. (21° C.) is by radiation, as shown in Figure 33–3. Heat loss in this manner is based on the principle that objects facing each other are always radiating heat toward one another. The human being radiates heat toward the walls, and the walls radiate heat toward him. However, since his body temperature is usually greater than that of the walls, he radiates more heat to walls than he receives from them.

The principle of radiation is used in radiant heating systems in commercial buildings and in many homes. The floors, ceilings, and walls of the rooms are heated to temperatures between 70 and 85° F. (21 and 29.4° C.). These are temperatures approximately comfortable for the human being, because they permit him to radiate heat at a rate that allows his body temperature to remain constant. A person is often comfortable in a radiantly heated room even though the air temperature may be in the 50's. This fact illustrates very forcefully how important radiation can be as a means of heat exchange.

HEAT LOSS BY CONDUCTION

Approximately 18 per cent of the heat lost by the nude person in Figure 33–3 leaves his body by conduction, 15 per cent by conduction to the air and 3 per cent to the floor and stool. These values are only average, for the colder the air and adjacent solid objects, the greater is the conduction of heat into them.

Effect of Convection Air Currents on Loss of Heat by Conduction. Even though the temperature of the air remains constant at 70° F. (21° C.), the rate at which heat can be conducted into the air depends on how much the air is moving. If it is flowing past the body, every time the air adjacent to the body becomes warmed it is carried away and replaced by cooler air. The more rapidly the air moves, the greater is the amount of heat conducted from the body. For this reason it is sometimes said that large quantities of heat are lost from the body by *convection*. This is not actually true; the heat is still being lost by conduction, though the convection currents carry the heated air away.

HEAT LOSS BY EVAPORATION

A small amount of water continually diffuses through the skin and evaporates, and evaporation of each gram of water removes approximately one-half Calorie of heat from the body. Therefore, even normally, about 22 per cent of the heat formed in the body is removed by evaporation. Evaporation of only 150 ml. of water each hour would remove all the heat produced in the body under basal conditions. This shows how important evaporation can be as a cooling mechanism.

The Sweat Mechanism. In addition to the continual diffusion of water through the skin, the sweat glands produce large quantities of

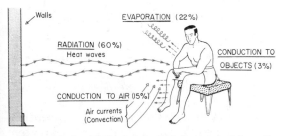

Figure 33–3. Methods by which heat is lost from the nude body. The percentage values are those for heat loss by each method when the person is in a room at 70°F.

sweat when the body becomes overheated. The sweat obviously increases the amount of heat that can be lost by evaporation. Under extreme conditions about 1.5 liters of sweat can be secreted each hour, which, if it all evaporates, will remove as much as 800 Calories of heat from the body, or about 12 times the basal level of heat production.

Necessity for Evaporation in Tropical Climates. Evaporation as a means of heat loss is extremely important in tropical climates, for when the temperature of the air and of the surroundings rises above the temperature of the body, heat cannot be lost by either radiation or conduction. Instead, the body gains heat by these means. However, evaporation of sweat can still keep the body temperature below that of the surroundings. Therefore, one's only means for maintaining normal body temperature in the tropics is by evaporation. Those few unfortunate persons who are born without sweat glands must forever live in temperate or cold climates.

Effect of Air Convection on Evaporation. Air currents have much the same effect on the removal of heat by evaporation as by conduction, for water evaporating from the skin quickly saturates the air immediately adjacent to the skin. If this air does not move away rapidly, the evaporation process will cease. However, if new air continually replaces the old, the air next to the skin never becomes saturated with moisture, and evaporation can continue unabated. This explains why in hot climates a fan is an important means for keeping cool; it also explains why one is usually much cooler outdoors under the shade of a tree, where the air is moving, than he is inside a house where the air is not moving, even though the air temperature may be the same.

EFFECT OF CLOTHING ON HEAT LOSS

Clothing is a barrier to the transfer of heat from the body to the surroundings. Heat radiated or conducted from the skin is absorbed by the inner surface of the clothing, and before it can be radiated or conducted to the surroundings it must pass through the mesh of the cloth. The inside of the clothing becomes warm in comparison with the outside; the inside warmth decreases the rate of conduction of heat from the body to the clothing and even radiates much of the heat back to the body.

The insulation properties of most clothing result mainly from air entrapped within the clothing mesh and not from the actual material composing the clothing. A furred animal utilizes this principle in the winter time, for his hair grows long and stands on end, entrapping large amounts of air that become warm and act as a body insulator.

When clothing becomes wet, the spaces are filled with water, and this conducts heat many hundred times as rapidly as does air. As a result, the clothing is no longer an adequate heat insulator; instead, heat is conducted almost as if the clothing did not exist. Therefore, wet clothing has almost no value for keeping the body warm in cold climates, and one of the most important lessons to be learned for arctic survival is always to keep the clothing dry whatever the circumstances.

Effect of Clothing in the Tropics. In the tropics, the effect of clothing on conduction and radiation usually is not important, because the temperature of the surroundings is close to that of the body. Instead, clothing must be designed for two other purposes: to protect one from direct exposure to the sun's heat rays and to allow maximum evaporation of sweat. Almost any clothing that blocks light rays also blocks the heat rays of the sun, but black clothing absorbs light rays and converts these into heat, whereas white clothing reflects the rays. Also different types of clothing have vastly different effects on evaporation. Those types of clothing that rapidly absorb water, such as the cottons and the linens, usually do not depress the evaporative cooling of the body, because sweat is absorbed into the clothing and evaporated from its surface in the same manner as from the body. This is not true of wool and especially of plastic materials, for they are not absorptive enough to allow satisfactory evaporative cooling.

THE BODY TEMPERATURE

The body temperature is determined by the balance between heat production and heat loss, as shown by Figure 33–4. If these two are exactly equal, the body temperature neither rises nor falls. When heat production is greater than heat loss, the body temperature rises; conversely, when heat loss is greater than heat production, the body temperature falls. Appropriate regulatory systems are always at work in the body to keep heat pro-

1.Basal metabolism

2.Muscular activity
— Shivering —

3.Thyroxine effect on cells

4.Epinephrine effect on cells

5.Temperature effect on cells

1.Radiation

2.Evaporation
—Convection—

3.Conduction
—Convection—

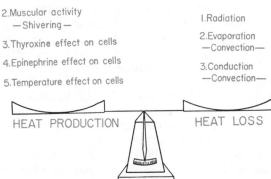

HEAT PRODUCTION   HEAT LOSS

Figure 33–4. Balance between heat production and heat loss, illustrating that the body temperature remains constant as long as these two are equal.

duction and heat loss approximately equal, thereby maintaining a normal body temperature.

Normal Values of Body Temperature. The normal body temperature varies no more than a degree or so from one person to another. However, when a person is exposed to extremely cold or hot weather, his overall body temperature may vary as much as 1° F. (0.6° C.). Also, when he experiences extreme emotions that cause excessive stimulation of the sympathetic nervous system, the amount of heat production may become great enough to raise the body temperature a degree or so. And, finally, extremely hard exercise can sometimes increase the body temperature as much as 5 to 6° F. (2 to 3° C.), though usually within 10 to 20 minutes after the exercise is over the temperature will have fallen back to that of the basal state.

Figure 33–5 depicts the normal temperature under different conditions as measured both in the mouth and rectally. The average normal oral temperature is usually considered to be between 98.0 and 98.6° F. (36.6 and 37° C.), while the average normal rectal temperature is generally about 1° F. (0.6° C.) higher. The reason for the higher rectal temperature is that the mouth is continually cooled by the facial surfaces and by evaporation in the mouth and nose.

REGULATION OF BODY TEMPERATURE

HYPOTHALAMIC REGULATION OF TEMPERATURE

The Preoptic Temperature-Sensitive Center. Located in the anteriormost portion of the hypothalamus, in the *preoptic area,* is a group of neurons that respond directly to temperature. When the temperature of the blood increases, the rates of discharge of these cells likewise increase. When the temperature decreases, the rates of discharge decrease.

From this preoptic temperature-sensitive area, signals radiate to various other portions of the hypothalamus to control either heat production or heat loss. In general, the hypothalamus can be divided into two major heat control divisions: an anterior *heat-losing center* which, when stimulated, reduces the body heat and a posterior *heat-promoting center* which, when stimulated, increases the body heat. These areas are shown in Figure 33–6. The anterior center is composed mainly of hypothalamic parasympathetic centers,

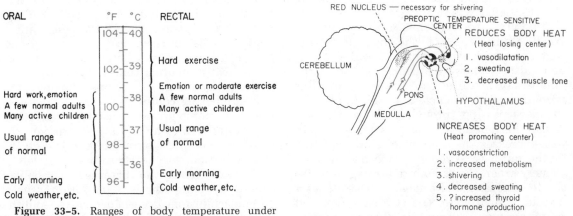

ORAL   °F °C   RECTAL

104–40

102–39   Hard exercise

Emotion or moderate exercise

Hard work, emotion   100–38   A few normal adults
A few normal adults   Many active children
Many active children

37   Usual range
Usual range   98–   of normal
of normal

36

Early morning   96–   Early morning
Cold weather, etc.   Cold weather, etc.

Figure 33–5. Ranges of body temperature under different normal conditions. (From DuBois: Fever and the Regulation of Body Temperature. Charles C Thomas.)

RED NUCLEUS — necessary for shivering

PREOPTIC TEMPERATURE SENSITIVE
CENTER
REDUCES BODY HEAT
(Heat losing center)
1. vasodilatation
2. sweating
3. decreased muscle tone

CEREBELLUM

PONS   HYPOTHALAMUS

MEDULLA

INCREASES BODY HEAT
(Heat promoting center)
1. vasoconstriction
2. increased metabolism
3. shivering
4. decreased sweating
5. ? increased thyroid
hormone production

Figure 33–6. The hypothalamic and brain stem mechanisms for regulating body temperature.

while the posterior center is principally sympathetic in nature.

Hypothalamic Mechanisms for Increasing the Body Temperature. Whenever blood colder than normal passes into the preoptic region of the hypothalamus, the preoptic heat-sensitive cells are inhibited, and this in turn strongly activates the *hypothalamic heat-promoting center* in the posterior hypothalamus. Stimulation of this center automatically activates several different mechanisms to increase the body heat. These are the following:

VASOCONSTRICTION OF THE SKIN TO INCREASE BODY TEMPERATURE. Stimulation of the heat-promoting center excites the sympathetics to cause the blood vessels of the skin to constrict powerfully. This lessens the flow of warm blood from the internal structures to the skin, decreasing the transfer of heat to the body surface from those organs producing most of the heat. Very little heat is transported directly from the internal structures to the body surface by means other than through the flowing blood because the fat beneath the skin is a very adequate heat insulator. When vasoconstriction occurs, the temperature of the skin falls to approach the temperature of the surroundings; this causes heat loss to be greatly diminished, allowing the internal body to retain its heat and the temperature to rise.

SYMPATHETIC STIMULATION OF METABOLISM TO INCREASE BODY TEMPERATURE. Stimulation of the sympathetic nerves releases norepinephrine throughout the body tissues and also causes both epinephrine and norepinephrine to be secreted into the blood by the adrenal medullae. These hormones in turn raise the rates of metabolism of all cells, thereby increasing heat production. This, too, tends to elevate the body temperature.

SHIVERING TO INCREASE BODY TEMPERATURE. Stimulation of the heat-promoting center increases the degree of wakefulness, and also causes transmission of large numbers of impulses into the bulboreticular formation and red nucleus of the hindbrain. Impulses passing through these regions increase the tone of all the muscles, which in turn increases the amount of heat produced by the muscles. When the tone becomes extremely great, the stretch reflex begins to oscillate; that is, a slight movement stretches a muscle, which elicits the stretch reflex and causes the muscle to contract. The contraction stretches the antagonistic muscle, which then develops a stretch reflex itself. Its contraction then stretches the first muscle, and the cycle becomes repetitive so that continual shaking develops. All this muscular activity can increase the rate of heat production by several hundred per cent. Consequently, when the body is exposed to extreme cold, shivering is a very powerful force to maintain normal body temperature.

PILOERECTION TO INCREASE BODY TEMPERATURE. Piloerection means that hairs stand on end. This occurs when the sympathetic centers are stimulated, for the sympathetic nerves excite small piloerector muscles located at the bases of the hairs. In the human being this mechanism does not protect against heat loss because of the scarcity of hairs, but in lower animals piloerection entraps large quantities of air in the zone adjacent to the body and provides increased insulation against cold.

INCREASED THYROID HORMONE PRODUCTION TO INCREASE BODY TEMPERATURE. If the body is exposed to cold for several weeks, as at the beginning of winter, the thyroid gland begins to produce greater quantities of thyroid hormone. This is caused by formation of a neurosecretory hormone in the preoptic region of the hypothalamus (see Chapter 34) that passes in the blood to the anterior pituitary gland to augment its production of thyrotropic hormone. This hormone in turn excites the thyroid gland. The increase in thyroid hormone production over a period of weeks steps up the rate of heat production up to 20 to 40 per cent and allows one to withstand the prolonged cold.

Hypothalamic Mechanisms for Decreasing the Body Temperature. When the temperature of the preoptic temperature-sensitive region rises too high, the neurons of this area become overly excited, which also stimulates the *anterior hypothalamic heat-losing* center. And because of reciprocal innervation between this center and the posterior hypothalamic heat-promoting center, the latter is inhibited. As a result, all the mechanisms of the heat-promoting center that tend to increase the body temperature become inoperative. For example, instead of vasoconstriction, the blood vessels to the skin dilate, allowing the skin to become very warm so that heat can be lost rapidly. Also, the increased metabolism caused by sympathetic stimulation ceases, and muscular tone greatly decreases. And the production of thyroid hormone diminishes gradually. The reversal of all of these effects allows the rate of heat loss to increase and the rate of heat production to decrease, thus making the body temperature fall.

In addition to reversing the heat promoting effects, stimulation of the heat-losing center causes two effects of its own to decrease the temperature. These are sweating and panting.

SWEATING TO DECREASE BODY TEMPERATURE. If the reversal of the heat-promoting effects is not sufficient to bring the body temperature back to normal, the anterior hypothalamus initiates the sweating process by sending sweat-promoting signals to all the sweat glands of the body through the sympathetic nerves. As much as a gallon of sweat can be poured onto the surface of the skin in an hour, and under favorable conditions a large proportion of this will evaporate. By this means, a refrigeration mechanism is initiated to reduce the body temperature when it tends to rise too high.

PANTING. In lower animals, though not in the human being, excessive stimulation of the hypothalamus by heat initiates a neurogenic mechanism in the pons of the hindbrain to cause panting. The animal breathes very rapidly but very shallowly, so that a tremendous quantity of air passes into and out of the respiratory passageways. Evaporation of water from the respiratory passageways is greatly increased, allowing a major amount of heat loss. Many lower animals — dogs, for instance — do not have well developed sweat mechanisms, and panting is often their only means of regulating body temperature in hot climates. Because of the shallowness of breathing during panting, most of the air entering the alveoli is dead space air, so that the alveolar ventilation of the animal remains essentially normal. This prevents overventilation despite the great turnover of air in the upper respiratory passageways.

Summary of Hypothalamic Functions in Body Temperature Regulation. The hypothalamus, when exposed to a temperature above normal, decreases heat production and increases heat loss. Conversely, when exposed to a temperature below normal, it increases heat production and decreases heat loss. Figure 33–7 illustrates these effects, showing an experiment in which the internal body temperature was progressively increased while heat production and evaporative heat loss were measured. Note that when body temperature was below normal, heat production in-

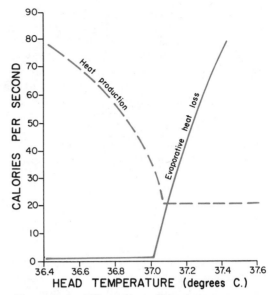

Figure 33–7. Effect of hypothalamic temperature on: (1) evaporative heat loss from the body and (2) heat production caused primarily by muscular activity and shivering. This figure demonstrates the extremely critical temperature level at which increased heat loss begins and increased heat production stops. (Drawn from data in Benzinger, Kitzinger, and Pratt: Temperature, Part III, ed. by Hardy, p. 637. Reinhold Publishing Corp.)

creased markedly; when above normal, heat loss by evaporation increased markedly.

Thus, the hypothalamus acts as a thermostat, maintaining the internal body temperature usually within one-half degree of the normal average value. The efficiency of the hypothalamus as a thermostat is indicated by the fact that the nude body can be exposed for several hours to dry air temperatures as low as 50° F. (10° C.) or as high as 150° F. (65.5° C.) without changing the internal body temperature more than 1 to 2° F. (0.6 to 1° C.).

Effect of Skin Thermoreceptors on Body Temperature Control. Though it is the hypothalamus that is the basic temperature controller of the body, signals transmitted to the hypothalamus from thermoreceptors in all areas of the skin can modify the "thermostatic setting" of the hypothalamus. That is, when the air temperature falls very low, which excites the cold receptors of the skin, the level of the hypothalamic thermostat is automatically set to a temperature level several tenths of a degree above normal body temperature. Conversely, exposure of the skin to hot air reduces the "set" to slightly below normal.

These effects play an important role in helping the body adapt to extremes of air temperature, for the following reason: If the body becomes exposed to an extreme of temperature and no readjustment is made in the control system, the internal body temperature will have to become abnormal before the hypothalamus will react. Fortunately, the thermal signals from the skin initiate the hypothalamic reaction *before* the internal body temperature becomes abnormal, thus helping to maintain a very even internal temperature.

FEVER

Fever means a body temperature that is elevated beyond the normal range. It occurs in many disease states and, therefore, is of extreme importance in assessing the severity of a person's affliction. The most frequent cause of fever is severe bacterial or viral infection such as occurs in pneumonia, typhoid fever, tuberculosis, diphtheria, measles, yellow fever, mumps, poliomyelitis, and so forth. A less frequent cause of fever is the destruction of body tissues by some means other than infection. For instance, a person after a severe heart attack usually develops fever, and a patient exposed to destruction of his tissues by x-ray or nuclear radiation may have fever for the ensuing few days.

RESETTING OF THE HYPOTHALAMIC THERMOSTAT BY PROTEINS OR POLYSACCHARIDES AS A CAUSE OF FEVER

Fever is usually caused by abnormal proteins or polysaccharides released into the body fluids during disease processes. These substances have either direct or indirect effects on the hypothalamic thermostat to reset its normal operating range at a higher temperature level. For example, only 0.001 gram of a polysaccharide derived from the bodies of typhoid bacilli can reset the level of the thermostat from normal up to as high as 110° F. (43.3° C.). When this happens, all the heat-promoting mechanisms for increasing the body temperature become active and remain active until the body temperature reaches 110° F. (43.3° C.). Then the mechanisms for heat loss and heat production become equalized again, and the temperature continues to

be regulated at this elevated level as long as the abnormal polysaccharide is present. In other words, the body temperature is still regulated during fever, but it is regulated at a level considerably above normal because of resetting of the thermostat.

Chills and Sweating in Feverish Conditions. Figure 33–8 shows the effect of disease on the course of the body temperature. At the beginning of this chart, the oral temperature is 98.6° F. (37° C.), but after a few hours the disease suddenly sets the thermostat to a level of 103° F. (39.4° C.). During the ensuing hours, all the heat-promoting mechanisms for increasing body temperature operate at full force. These include skin vasoconstriction, increased metabolism caused by norepinephrine and epinephrine secretion, and shivering. Even when the body temperature reaches 101° F. (38.3° C.), it still has not reached the setting of the thermostat, and the heat-promoting phenomena continue to take place. Therefore, even though the temperature is high, the skin remains cold, and shivering occurs. This is called a *chill;* when someone is having a chill it is quite certain that his body temperature is actively rising.

After a number of hours the body temperature reaches the setting of the thermostat, and the chills disappear, but the temperature remains regulated at the high thermostatic setting until some factor breaks the disease process.

In Figure 33–8 the disease process is corrected after another few hours, and the thermostat is suddenly set back to its normal level of 98.6° F. (37° C.). However, at first the body temperature still has not fallen. This causes

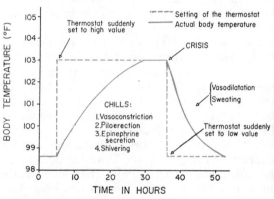

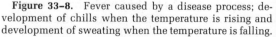

Figure 33–8. Fever caused by a disease process; development of chills when the temperature is rising and development of sweating when the temperature is falling.

the heat-losing mechanisms, including especially skin vasodilatation and sweating, to decrease the body temperature. The person's skin suddenly becomes warm, and he begins to sweat, which effect is known as the *crisis*. When this happens one knows that the patient's temperature is beginning to fall. In the days before the "miracle" drugs (the antibiotics), many, many persons with bacterial diseases died, and the duty of the doctor was mainly to keep the sick person comfortable until he either died or a crisis appeared. If the crisis did appear, the doctor left the sickroom and announced to the family that all was now going to be well. The family in turn revered the doctor for his miraculous achievement. Unfortunately for the reputation of the present-day doctor, he too frequently cures the patient long before a crisis can develop.

Possible Value of Fever in Disease. One wonders what the value of fever in disease might be. At present no clear-cut answer is available, but suggested values are: (1) Many bacterial and viral agents do not survive as well at high temperatures as at normal body temperatures. Therefore, elevation of the temperature might well be a means for combating these infections. For example, high temperature has been shown to be especially lethal to gonococcal and syphilitic bacteria. (2) Because high rates of chemical reactions occur in the cells at high temperatures, it is possible that these increased rates allow the cells to repair damage much more rapidly than they could otherwise.

EFFECT OF TEMPERATURE ITSELF ON THE BODY

When the body temperature rises to between 108° and 110° F. (42.2 and 43.3° C.) it becomes very difficult, and often impossible, for the temperature regulatory mechanisms to return the body temperature to normal again. The main reason for this is that at these high temperatures the rates of cellular metabolism become so greatly increased because of the temperature itself that often no amount of regulation can overcome this very rapid rate of heat production. As a result, the body temperature rises still higher, the rate of heat production rises higher, and a vicious cycle develops, causing the body temperature to keep on rising.

Also, when the body temperature rises above 108° F. (42.2° C.), the metabolic rates of the cells become so great that the cells begin to "burn themselves out"; and when the body temperature rises to 112 to 114° F. (44.4 to 45.5° C.), death almost always ensues simply because of the heat itself. The most damaging effect of a very high body temperature occurs in the neuronal cells of the brain, sometimes causing permanent destruction of these cells even though the person recovers.

When the body temperature falls far below normal, below about 92° F. (33.3° C.), heat regulation again becomes impaired, but for opposite reasons. The low body temperature causes such slow rates of chemical reactions in the cells that no amount of regulation can bring the rate of heat production up high enough to return the body temperature to normal. The cold temperature decreases the rate of cellular metabolism, and this allows the temperature to fall still lower, which decreases the rate of metabolism more, again creating a vicious cycle until the body temperature falls so low that the person dies. Usually death occurs when the body temperature reaches about 75° F. (23.9° C.).

Body temperatures down to about 85° F. (29.4° C.) do not cause significant damage to the body, though the bodily functions become so greatly slowed that the person remains in a state of suspended animation until he is rewarmed.

REFERENCES

Cabanac, M.: Temperature regulation. *Ann. Rev. Physiol.*, 37:415, 1975.

Elizondo, R.: Temperature regulation in primates. *Intern. Rev. Physiol.*, 15:71, 1977.

Hales, J. R. S.: Physiological responses to heat. *In* MTP International Review of Science: Physiology. Vol. 7. Baltimore, University Park Press, 1974, p. 107.

Hegsted, D. M.: Energy needs and energy utilization. *Nutr. Rev.*, 32:33, 1974.

Hensel, H.: Thermoreceptors. *Ann. Rev. Physiol.*, 36:233, 1974.

Himms-Hagen, J.: Cellular thermogenesis. *Ann. Rev. Physiol.*, 38:315, 1976.

Mildvan, A. S.: Mechanism of enzyme action. *Ann. Rev. Biochem.*, 43:357, 1974.

Mitchell, D.: Physical basis of thermoregulation. *In* MTP International Review of Science: Physiology. Vol. 7. Baltimore, University Park Press, 1974, p. 1.

Reeves, R. B.: The interaction of body temperature and acid-base balance in ectothermic vertebrates. *Ann. Rev. Physiol.*, 39:559, 1977.

Robertshaw, D.: Role of the adrenal medulla in thermoregulation. *Intern. Rev. Physiol.*, 15:189, 1977.

Schmidt-Nielsen, K.: Animal Physiology: Adaptation and Environment. London, Cambridge University Press, 1975.

Thompson, G. E.: Physiological effects of cold exposure. *Intern. Rev. Physiol.,* 15:29, 1977.

Toporek, M., and Maurer, P. H.: Metabolic biochemistry. *In* Sodeman, W. A., Jr., and Sodeman, W. A. (eds.): Pathologic Physiology: Mechanisms of Disease. 5th ed. Philadelphia, W. B. Saunders Company, 1974, p. 3.

QUESTIONS

1. Explain how each of the following factors affects metabolic rate: exercise, sympathetic stimulation, thyroid hormone, body temperature, and foods.
2. What is meant by basal metabolic rate, and how is it determined?
3. Explain how heat is lost by radiation, conduction, and evaporation.
4. What is the role of the hypothalamus in body temperature regulation?
5. Explain how the body increases heat production when the temperature falls too low.
6. How does the temperature-controlling mechanism reduce heat production and increase heat loss when the body temperature becomes too great?
7. Explain how resetting of the hypothalamic thermostat causes fever.
8. What are the effects on the body of either too high a body temperature or too low a body temperature?

ENDOCRINOLOGY AND REPRODUCTION

X

INTRODUCTION TO ENDOCRINOLOGY: THE PITUITARY HORMONES AND THYROXINE

34

THE NATURE OF HORMONES AND THEIR FUNCTION

A hormone is a chemical substance elaborated by one part of the body that controls or helps to control some function elsewhere in the body. In general, hormones are divided into two types: first, the *local hormones,* which affect cells in the vicinity of the organ secreting the hormone, including such hormones as acetylcholine, histamine, and the gastrointestinal hormones, all of which have been discussed at different points in this text; and, second, the *general hormones,* which are emptied into the blood by endocrine glands and then flow throughout the entire circulation to affect cells and organs in far distant parts of the body. Some general hormones affect all cells almost equally; others affect specific cells far more than others. For example, growth hormone secreted by the thyroid gland affect all cells of the body. On the other hand, the pituitary gland produces gonadotropic hormones that affect the sex organs much more than other tissues, even though these hormones are secreted into the general circulation.

Regulatory Functions of Hormones. Some examples of the regulatory functions of the general hormones are: (1) The anterior pituitary gland secretes six different hormones that regulate respectively the rate of growth of all tissues of the body, the rate of secretion of thyroid hormone by the thyroid gland, the rate of secretion of adrenocortical hormones by the adrenal gland, the rate of formation of milk by the mammary glands, and the rates of secretion of several different sex hormones. (2) Thyroxine secreted by the thyroid gland controls the rate of metabolism of all cells. (3) The adrenal medullary hormones epinephrine and norepinephrine, which have been discussed in connection with the sympathetic nervous system, cause the same reactions throughout the body as stimulation of the general sympathetic nervous system. (4) Hormones secreted by the adrenal cortex regulate reabsorption of sodium by the kidneys, and also regulate some aspects of carbohydrate, fat, and protein metabolism. (5) Insulin secreted by the pancreas regulates the utilization of glucose throughout the body. (6) Hormones secreted by the testes control the sexual and reproductive functions of the male. (7) Hormones secreted by the ovaries and the placenta control the sexual and reproductive functions of the female. (8) Hormones secreted by the parathyroid glands regulate the concentration of calcium in the body fluids.

In the present chapter and the following few chapters, the functions of the general hormones are considered in detail. The study of these and their functions is called *endocrinology,* and the glands that secrete the general hormones are called *endocrine glands.* The word *endocrine* means secretion to the interior of the body as opposed to *exocrine,* which means secretion to the exterior as is true of the intestinal glands, the sweat glands, and others.

MECHANISMS OF HORMONAL ACTION

The way in which the different hormones control activity levels of target tissues differs from one hormone to another. However, two important general mechanisms by which hormones function are: (1) activation of the *cyclic AMP system* of cells, which in turn elicits specific cellular functions, or (2) *activation of the genes* of the cell, which then cause formation of intracellular proteins that in turn initiate specific cellular functions. These two mechanisms may be described as follows:

CYCLIC AMP, AN INTRACELLULAR HORMONE MEDIATOR

Many hormones exert their effects on cells by first causing cyclic AMP *(cyclic 3',5'-adenosine monophosphate)* to be formed in the cell. Once formed, the cyclic AMP initiates the desired activities inside the cell. Thus, cyclic AMP is an *intracellular hormonal mediator*. It is also frequently called the "second

messenger" for hormone mediation — the "first messenger" being the original stimulating hormone.

Figure 34–1 illustrates the function of the cyclic AMP system in more detail. The first event in this mechanism of hormonal control is the action of the stimulating hormone on a specific receptor in the cell membrane. In fact, whether or not the hormone will affect a particular cell is determined by the presence or absence of such a specific receptor for that hormone.

Second, once the hormone binds with the receptor, an enzyme in the cell membrane called *adenylcyclase* is activated.

Third, the adenylcyclase then converts some of the adenosine triphosphate (ATP) inside the cell cytoplasm into cyclic AMP. The cyclic AMP persists in the cell for a few seconds to many minutes before it is reconverted into ATP. However, as long as the hormone remains active at the receptor, still more cyclic AMP will continue to be formed.

Fourth, once generated inside the cell, the cyclic AMP in turn performs any number of different physiologic functions, including:

1. Activating enzymes
2. Altering cell permeability
3. Altering the degree of smooth muscle contraction

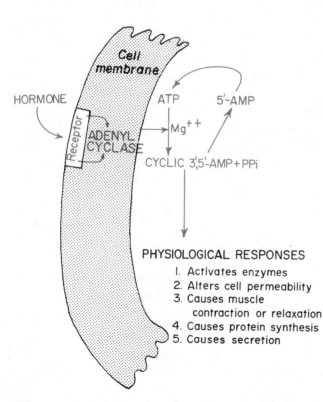

PHYSIOLOGICAL RESPONSES

1. Activates enzymes
2. Alters cell permeability
3. Causes muscle contraction or relaxation
4. Causes protein synthesis
5. Causes secretion

Figure 34–1. The cyclic AMP mechanism by which many hormones exert their control of cell function.

4. Activating protein synthesis

5. Causing secretion by the cell

The specific effect that occurs in each individual cell is determined by the characteristics of the cell. Thus, if the cell is a glandular cell, it will form its specific secretion; or if the cell is a smooth muscle cell, it will contract or relax depending upon whether the cyclic AMP is excitatory or inhibitory in that particular cell.

The cyclic AMP mechanism has been shown to act as an intracellular hormonal mediator system for at least some of the functions of each of the following hormones:

1. Corticotropin (ACTH)

2. Thyroid-stimulating hormone

3. The gonadotropic hormones that stimulate the sex glands

4. Antidiuretic hormone

5. Parathyroid hormone that controls extracellular fluid calcium

6. Glucagon that helps to control blood glucose concentration

7. Epinephrine that controls contraction of many smooth muscles

8. Secretin that controls secretion by the pancreas

9. The hypothalamic releasing factors that control secretion of most of the anterior pituitary hormones

HORMONAL CONTROL BY ACTIVATION OF GENES

A second important way in which hormones act is to activate one or more genes in the nucleus; these in turn cause synthesis of proteins in the target cells. This occurs especially in response to steroid hormones secreted by the adrenal cortex, the ovaries, and the testes.

The general mechanism of steroid hormone function is the following: First, the hormone enters the cytoplasm of the cell where it combines with a specific receptor protein. Second, after several additional steps that occur in the cytoplasm and nucleus, one or more specific genes are activated in the nucleus. Third, the genes cause the formation of specific messenger RNA molecules. Fourth, the RNA diffuses to the cytoplasm and causes the formation of specific proteins. Fifth, the proteins then increase specific activities of the cells. For instance, in the renal tubules, protein enzymes are formed in response to aldos-

terone stimulation to cause sodium absorption and potassium secretion.

THE ANTERIOR PITUITARY HORMONES

The *pituitary gland,* shown in Figure 34–2 and known also as the *hypophysis,* is about the size of the tip of the little finger, and it lies in a small bony cavity beneath the base of the brain. It is divided into two completely separate parts, the *posterior pituitary gland,* known also as the *neurohypophysis,* which is connected by a stalk with the hypothalamus of the brain, and the *anterior pituitary gland,* called also the *adenohypophysis,* which lies anterior to the posterior pituitary gland and has functions not related to those of the posterior pituitary gland.

The anterior pituitary gland secretes at least six different hormones, all of which are small proteins or large polypeptides, and various research workers have postulated many more that might be produced by this gland. However, only the six have thus far been proved to have major significance. These six hormones are the following:

1. Growth hormone

2. Thyrotropin

3. Corticotropin

4. Prolactin

5. Follicle-stimulating hormone

6. Luteinizing hormone

The last two are called *gonadotropic hormones* because they regulate the functions of the sex glands.

The general functions of these six hormones are illustrated in Figure 34–3.

GROWTH HORMONE

Growth hormone is secreted by the anterior pituitary gland in large quantities throughout life even though most growth in the body stops at adolescence. At this time the production of growth hormone diminishes but does not stop. The function of growth hormone during the growing phase of an animal's life is to promote development and enlargement of all bodily tissues. It increases the sizes of all bodily cells and also their number. Consequently, each tissue and each organ becomes larger under the influence of growth hormone. The bones enlarge and lengthen, the skin thick-

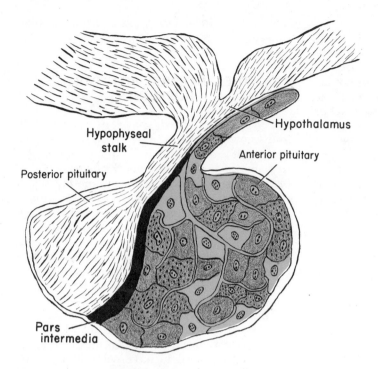

Figure 34–2.   The pituitary gland.

ens, and the soft tissues, including the heart, liver, tongue, and all the other internal organs, increase in size. In other words, growth hormone is exactly what its name implies; it causes the person to grow.

Function of Growth Hormone After Adolescence.   Growth hormone continues to function after adolescence as before, that is, to promote protein synthesis and formation of other cellular elements. However, by this age, most of the body's bones have grown as much as they can, so that continued growth of body stature ceases. A few bones, though, especially those of the lower jaw and nose, do continue to grow, increasing the prominence of these structures with age even in the normal person.

Basic Metabolic Effects of Growth Hormone.   Growth hormone is known to have the following basic effects on the metabolic processes of the body:

1. Increased rate of protein synthesis in all cells of the body
2. Decreased rate of carbohydrate utilization by all or most cells
3. Increased mobilization of fats and use of fats for energy

The most conspicuous influence of growth hormone is to promote growth, which occurs mainly because the hormone causes widespread protein synthesis. Yet, it also has effects on both carbohydrates and fat metabolism, conserving carbohydrates while at the same time using up the fat stores.

Stimulation of Growth of Cartilage and Bone—Role of "Somatomedin."   The growth of cartilage and bone is not a direct effect of growth hormone on these structures. Instead, growth hormone causes the liver,

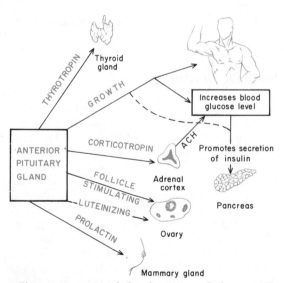

Figure 34–3.   Metabolic functions of the anterior pituitary hormones.

and perhaps the kidney to a lesser extent, to form a substance called somatomedin. This substance then acts directly on the cartilage and bone to make them grow. Somatomedin is necessary for deposition of collagen and ground substance, both of which are essential for growth of the cartilage and bone.

The reason that growth stops at adolescence is that the growth cartilages that separate the ends, or "epiphyses," of the long bones from the shafts of these bones exhaust their potential for growth at this age. Consequently, the epiphyses then unite directly with the shafts of the bones at the point where the growth cartilages had previously existed. Increased linear growth of these bones therefore cannot occur beyond this time.

Stimulation of Protein Deposition by Growth Hormone. Aside from promoting the growth of cartilage and bone, growth hormone also causes growth of most other tissues as well because of its effect of increasing the formation of protein. The precise mechanism or mechanisms by which it stimulates protein formation are not clear, but it is known that growth hormone has the following effects, all of which could be important in increasing the total body protein:

1. Increased transport of amino acids through cell membranes, thereby providing an appropriate intracellular supply of the building blocks from which proteins are formed.

2. Increased formation of RNA, which in turn causes greater formation of protein molecules.

3. Activation of the ribosomes to raise the rate at which they synthesize protein.

4. Decreased breakdown of proteins once they have been formed, thus allowing proteins to accumulate in greater quantities in the cells.

Effect of Growth Hormone on Fat and Carbohydrate Metabolism. Growth hormone causes release of fatty acids from adipose tissue and thus increases the overall utilization of fat by energy-consuming processes of the body. At the same time, it increases the utilization of carbohydrate in most cells, so that glycogen begins to accumulate in the cells and uptake of glucose from the blood decreases. Therefore, growth hormone enhances the circulating fats in the blood because of their increased release from the fat storage areas, and it also augments the circulating glucose in the blood because of diminished utilization of the glucose.

Unfortunately, the mechanisms by which growth hormone causes these effects on both fats and carbohydrates are still unknown. However, they are important to the overall metabolism of the body mainly because they tend to decrease the deposition of fat stores at the same time that they enhance the protein and carbohydrate stores.

Control of Growth Hormone Secretion. The secretion of growth hormone by the anterior pituitary gland is controlled by the hypothalamus, which secretes a *growth hormone–releasing factor* that then acts on the anterior pituitary gland to cause growth hormone secretion. This will be discussed more fully later in the chapter. However, various factors can affect the hypothalamus to cause increased or decreased secretion of growth hormone.

The rate of growth hormone secretion varies markedly from day to day depending on the metabolic needs of the body, which is quite contrary to the earlier belief that its secretion was constant in children, as a continuous stimulus to growth, but absent in adults. One situation that causes marked increase in growth hormone secretion is generalized nutritional deficiency. It is presumed that this in some way results from depressed formation of proteins, which in turn has an effect on the hypothalamus to trigger growth hormone secretion. The increased growth hormone then helps to promote deposition of new protein stores, thereby relieving cellular protein deficiency. In this way, the growth hormone mechanism could act as a controller of the protein stores of some or all the cells in the body.

Abnormalities of Growth Hormone Secretion. A decrease in growth hormone secretion causes *dwarfism,* while an increase causes overgrowth of the person, which can result in *giantism* or in the condition called *acromegaly.*

DWARFS. A person whose anterior pituitary gland fails to secrete growth hormone fails to grow. This causes the so-called "pituitary" type of dwarf (though most dwarfs have small stature because of their heredity rather than because of anterior pituitary failure). The pituitary type of dwarf remains childlike in all physical respects. His organs as well as bones fail to grow, and his degree of sexual development remains that of a young child even after he has reached the age of an adult. He may never grow to more than twice the height of a newborn baby.

GIANTS. Secretion of excess growth hormone, if it occurs before adolescence, can cause a giant. After adolescence the growing parts of most bones have "fused" and are no longer capable of growing longer regardless of the amount of growth hormone available. Giantism almost always results from a tumor of the acid-staining cells of the pituitary gland, this tumor secreting tremendous quantities of growth hormone.

ACROMEGALY. If a growth hormone–secreting tumor develops after adolescence, the excessive secretion of growth hormone cannot cause increased height; nevertheless, it can still cause enlargement of the soft tissues. It can also cause thickening of the bones, and a few of the bones, the so-called "membranous bones," can even continue to grow. As a result, disproportionate growth occurs in some parts of the body, resulting in the condition called *acromegaly*. In acromegaly the lower jaw, which is formed of membranous bone, grows excessively long, causing the chin sometimes to protrude a half-inch or more beyond the remainder of the face. The nose and lips also enlarge, and the internal organs such as the tongue, the liver, and the various glands enlarge. The bones of the feet and hands, unlike most bones of the body, continue to grow, so that the hands and feet become very large. In short, an acromegalic person is the same as a giant, except that failure of his already "fused" long bones to grow causes his height to be normal while other features of his body exhibit excessive, disproportionate growth.

THYROTROPIN

Thyrotropin, also known as *thyrotropic hormone,* is another one of the hormones secreted by the anterior pituitary gland. It controls the amount of hormone secreted by the thyroid gland by increasing the number of thyroid cells, the size of these cells, and also their rate of thyroxine production. When the anterior pituitary fails to secrete thyrotropin, the thyroid gland becomes so incapacitated that it secretes almost no measurable amount of hormone. In other words, the thyroid gland is almost completely controlled by thyrotropin. Further relationships of the thyroid gland to thyrotropin are discussed later in the chapter in connection with thyroxine.

CORTICOTROPIN

A third hormone secreted by the anterior pituitary, *corticotropin,* also known as *adrenocorticotropic hormone* or *ACTH,* controls secretion of adrenocortical hormones by the adrenal cortices in much the same manner as thyrotropin controls secretion by the thyroid gland. Corticotropin increases both the number of cells in the adrenal cortex and their degree of activity, resulting in increased output of adrenocortical hormones. The relationship of corticotropin to the different adrenocortical hormones is discussed in the following chapter.

PROLACTIN

Prolactin is a hormone secreted by the anterior pituitary gland during the last few days of pregnancy and during the entire period of milk production after birth of the baby. This hormone stimulates both breast growth and secretory functions of the breasts. These effects will be discussed in relation to reproduction in Chapter 38.

THE GONADOTROPIC HORMONES

The functions of the two gonadotropic hormones are summarized briefly in the following paragraphs; their relationship to sexual function will be presented in detail in Chapter 37.

Follicle-Stimulating Hormone. In the female, follicle-stimulating hormone initiates growth of *follicles* (fluid chambers) in the ovaries. In each of these a single ovum develops in preparation for fertilization. This hormone also helps to cause the ovaries to secrete *estrogens,* one of the female sex hormones. In the male, follicle-stimulating hormone stimulates growth of the germinal epithelium in the testes, thus promoting the development of sperm that can then fertilize the female ovum.

Luteinizing Hormone. In the female, luteinizing hormone joins with follicle-stimulating hormone to cause estrogen secretion. It also causes the follicle to rupture, allowing the ovum to pass into the abdominal cavity and then through a fallopian tube, within which time fertilization may take place. And it helps to cause the corpus luteum to secrete *progesterone*. In the male, luteinizing hormone causes the testes to secrete the male sex hormone *testosterone*.

REGULATION OF ANTERIOR PITUITARY SECRETION – THE HYPOTHALAMIC-HYPOPHYSEAL PORTAL SYSTEM

The anterior pituitary gland is a highly vascular organ which receives blood supply from two sources: (1) the usual arteriolar source and (2) the so-called *hypothalamic-hypophyseal portal system,* which is illustrated in Figure 34–4. The hypothalamus, a part of the brain lying immediately beneath the thalamus, controls many of the automatic functions of the body. After the blood passes through the capillaries of the hypothalamus, particularly through the lower part called the *median eminence,* it leaves by way of small *hypothalamic-hypophyseal portal veins* which course down the anterior surface of the pituitary stalk to the anterior pituitary gland. At this point, the veins dip into the gland where the blood flows through large numbers of *venous sinuses* that bathe the anterior pituitary cells.

The hypothalamus secretes a series of different neurosecretory substances called *hypothalamic releasing* and *inhibitory factors.* These factors are secreted into the blood of the hypothalamic-hypophyseal portal system and are then transported to venous sinuses in the anterior pituitary gland where they regulate secretion of the various anterior pituitary hormones. The six factors most important for control of anterior pituitary secretion are:

1. *Thyrotropin-releasing factor* (TRF), which causes release of thyrotropin, also called thyroid-stimulating hormone.

2. *Corticotropin-releasing factor* (CRF), which causes release of corticotropin, also known as ACTH or adrenocorticotropic hormone.

3. *Growth hormone–releasing factor* (GRF), which causes release of growth hormone.

4. *Luteinizing hormone–releasing factor* (LRF), which causes release of luteinizing hormone.

5. *Follicle-stimulating hormone–releasing factor* (FRF), whch causes release of follicle-stimulating hormone.

6. *Prolactin inhibitory factor* (PIF), which causes inhibition of prolactin secretion.

Note that five of these factors cause release

Figure showing labeled diagram with: Hypothalamus, Optic chiasma, Artery, Primary capillary plexus, Hypothalamic-hypophyseal portal vessels, Sinuses, Anterior pituitary gland, Vein, Posterior pituitary gland, Median eminence, Mammillary body.

Figure 34–4. The hypothalamic-hypophyseal portal system.

(secretion) of the respective anterior pituitary hormones. In the absence of these five releasing factors, the five hormones are secreted in only minute quantities by the anterior pituitary gland.

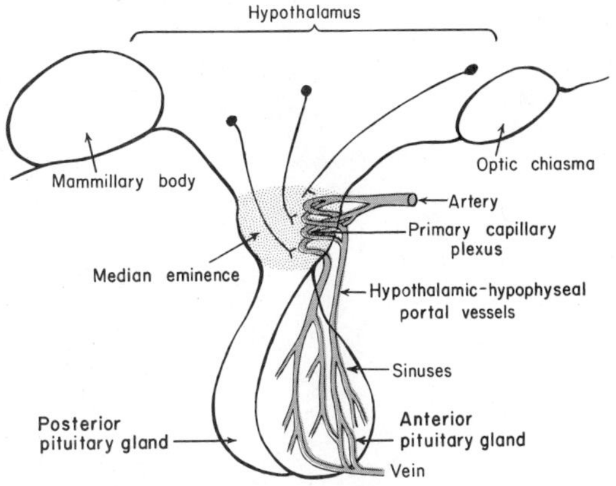

On the other hand, there is one important inhibitory factor secreted by the hypothalamus, prolactin inhibitory factor. In the absence of this factor, the anterior pituitary gland secretes an excess of prolactin.

Because of the powerful effects of these releasing and inhibitory factors on anterior pituitary secretion, most control of the anterior pituitary gland is carried out through stimulation or inhibition of the many neuronal control centers in the hypothalamus that in turn regulate the output of the hypothalamic releasing and inhibitory factors. For instance, it was pointed out earlier in this chapter that protein deficiency in some way excites the hypothalamus to induce increased secretion of growth hormone by the anterior pituitary gland. Later in this chapter we will see that thyroid hormone secretion also is controlled almost entirely by stimulation or inhibition of hypothalamic centers that in turn control thyrotropin secretion by the anterior pituitary gland. And in succeeding chapters we shall see that control of many of the other endocrine hormones is effected through a similar hypothalamic-pituitary system.

THE POSTERIOR PITUITARY HORMONES

Though the posterior pituitary gland is located immediately behind the anterior pituitary gland, the two glands do not have any

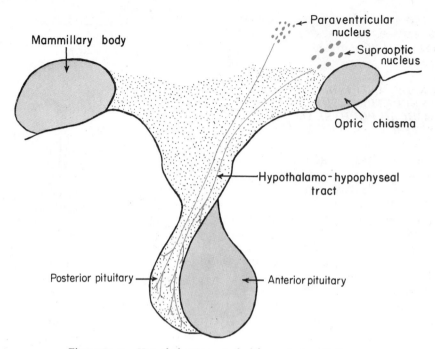

Figure 34–5.   Hypothalamic control of the posterior pituitary.

proved direct relationship. Also, in a sense, the posterior pituitary gland is not a gland at all, because it only stores hormones rather than secreting them. It stores the two hormones *antidiuretic hormone* (also called vasopressin) and *oxytocin*. Antidiuretic hormone has a very important function of helping to control water excretion by the kidney, which was discussed in detail in Chapter 18. Oxytocin stimulates contraction of muscle in the uterus and breasts. Therefore, it plays important roles in birth of the baby and in expelling milk from the mother's breast. These functions of oxytocin will be discussed in Chapter 38.

Both antidiuretic hormone and oxytocin are secreted by neuronal cells in the anterior hypothalamus and then are conducted through nerve axons to the posterior pituitary gland where both hormones are stored. Figure 34–5 illustrates the anatomy of this mechanism. Antidiuretic hormone is formed in the neurons of the *supraoptic nucleus*. Then, over a period of hours and days, this hormone flows slowly down the axoplasmic center of the nerve fibers by way of the hypothalamic-hypophyseal nerve fiber tract and is stored at the tips of these nerve fibers in the posterior pituitary gland. Oxytocin, on the other hand, is formed in the neurons of the *paraventricular*

nucleus and transported in a similar manner to the posterior pituitary gland.

Once these two hormones have reached the posterior pituitary gland, they remain there until nerve impulses are transmitted from either the supraoptic nucleus to cause release of antidiuretic hormone or from the paraventricular nucleus to cause release of oxytocin.

Though both of these hormones are discussed in greater detail elsewhere in this text, let us briefly review their functions here as well.

Antidiuretic Hormone and Control of Water Reabsorption in the Tubules of the Kidney. Antidiuretic hormone prevents the body fluids from becoming too concentrated in the following way: When the concentration of sodium and other osmotically active substances in the extracellular fluids rises, the increased osmotic pressure of these fluids causes special neurons in the supraoptic nucleus called *osmoreceptors,* to send impulses to the posterior pituitary gland. The impulses cause antidiuretic hormone to be released from the gland. This in turn passes by way of the blood to the collecting tubules of the kidney and causes increased quantities of water to be reabsorbed while allowing large quantities of sodium and other tubular solutes to pass on into the urine. The mechanism by which antidiuretic hormone increases the water reabsorption is to increase the pore size in the collecting ducts enough for water molecules to diffuse through, but not large enough for most other substances in the tubular fluid to pass through. Thus, the water of the body fluids is conserved, while the sodium and other solutes gradually decrease. In this way, overconcentration of the extracellular fluids is corrected.

PRESSOR FUNCTION OF ANTIDIURETIC HORMONE. Antidiuretic hormone is also frequently called *vasopressin* because injection of very large quantities of purified hormone causes the arterial pressure to rise. Also, even normal amounts of circulating antidiuretic hormone in the blood to a slight extent keep the arterial pressure from falling. Therefore, antidiuretic hormone plays at least some part in daily control of arterial pressure. And, when the circulatory system is subjected to very severe stress, such as during debilitating surgical operations, it is likely that antidiuretic hormone is then very important to help maintain the arterial pressure in a normal range.

Oxytocin. An "oxytocic" agent is a substance that causes the uterus to contract. This is one of the primary effects of *oxytocin*. This hormone is secreted in moderate quantities during the latter part of pregnancy and in especially large quantities at the time that the baby is born. It probably aids in the expulsion of the baby from the uterus. It is well known that a mother whose oxytocin-secreting mechanism has been destroyed has considerable difficulty in delivering her baby. This will be discussed further in relation to childbirth in Chapter 38.

EFFECT OF OXYTOCIN ON MILK EJECTION. Oxytocin also has an important function in helping the mother to provide milk for the newborn infant as follows: Milk formed by the *glandular cells* of the breasts is secreted into the *alveoli* of the breasts and remains there until the baby begins to nurse. For approximately the first minute after nursing begins the baby receives essentially no milk, but the suckling stimulus excites the mother's nipple and transmits nerve signals upward through the spinal cord and finally to the hypothalamus where they cause secretion of oxytocin. This hormone in turn flows by way of the blood to the breast where it causes many small *myoepithelial cells* surrounding the alveoli to contract, thereby squeezing the milk into the ducts so that the baby can removing it by suckling.

THYROXINE

Formation of Thyroxine by the Thyroid Gland. Figure 34–6 illustrates the basic structure of the thyroid gland, showing it to comprise large follicles. The cells lining the follicles secrete thyroxine to the interior of the follicles where it is temporarily stored.

The mechanism of thyroid secretion is illustrated in Figure 34–7, which shows that iodine is first absorbed into the follicular cell in the form of ionic iodine (I^-). The follicular cell converts the iodide ion to elemental iodine. At the same time, the cell also secretes a protein called *thyroglobulin* into the follicle. The elemental iodine reacts with the thyroglobulin both before being released into the follicle and after release to convert much of the amino acid tyrosine in the molecule of the thyroglobulin into *thyroxine*. The chemical reaction for this process is shown on the next page.

Another hormone very similar to thyroxine but with one less iodine in its structure, triiodothyronine, is formed in smaller amounts at the same time that thyroxine is formed. It has almost exactly the same effects in the body as thyroxine except that it acts several times as rapidly as thyroxine. Because the greater proportion of the thyroid hormone is thyroxine, we shall discuss its functions in the subsequent paragraphs, but it should be remembered that smaller amounts of triiodothyro-

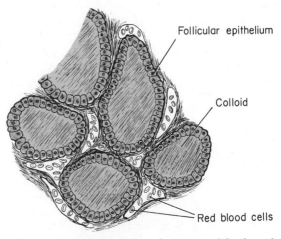

Figure 34–6. Cross-section of a portion of the thyroid gland, showing colloid material stored in the follicles.

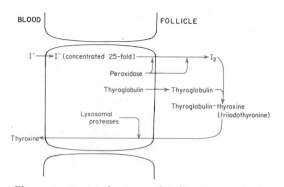

Figure 34–7. Mechanisms of iodine transport, thyroxine formation, and thyroxine release into the blood. (Triiodothyronine formation and release parallels that of thyroxine.)

nine function in an almost parallel manner to the thyroxine.

Storage and Release of Thyroxine. Thyroxine formed in the above manner usually remains stored in the follicle for several weeks as part of the thyroglobulin molecule before being released back through the follicular cells into the circulating blood. At the time of release, thyroxine is broken away from the thyroglobulin by the action of *proteases* released into the follicle by lysosomes in the thyroid cells. Then the thyroxine diffuses backward through the thyroid cells and enters the blood.

In the blood the thyroxine immediately combines with a plasma protein from which it is released over a period of several days to the tissue cells. Thus, this overall process assures a steady, slow flow of thyroxine into the tissues. Because of this slowness, some of the effects of thyroxine are still apparent as long as 6 to 8 weeks after it is first formed in the thyroid gland.

All the preceding reactions related to thyroxine formation are under the control of another hormone, *thyrotropin,* secreted by the anterior pituitary gland; this will be discussed in more detail in following sections.

MECHANISM OF ACTION OF THYROXINE

Effect of Thyroxine on Cellular Metabolism and on the Cellular Enzymes. Thyroxine steps up the rate of metabolism of all cells. Though the precise means by which this is accomplished are not known, studies have shown that tissues exposed to thyroxine develop greatly enhanced quantities of most of their enzymes. To date, at least 13 different cellular enzymes have been shown to be greatly increased under the influence of thyroxine, and because enzymes are the regulators of chemical reactions in the cells it is quite easy to understand how this could raise the metabolic rates of the cells. Because thyroxine affects all tissues of the body, it regulates the overall rate of activity of all functions of the body.

We still do not know the intracellular mechanism by which thyroxine augments the quantities of enzymes. However, since so many different enzymes are involved, and since the rate of enzyme formation is controlled primarily by genetic mechanisms of the cell, it is presumed that thyroxine activates specific genes to increase these enzymes. It also increases the number and sizes of the mitochondria, but whether this is a primary or a secondary effect is unknown.

Effects of Thyroxine on Specific Functions of the Body. EFFECT ON TOTAL BODY METABOLISM. Total lack of thyroxine production by the thyroid gland decreases the metabolic rate to about one-half normal. On the other hand, secretion of very large quantities of thyroxine can increase the rate of metabolism to as much as two times normal. Therefore overall, the thyroid gland can change the rate of metabolism as much as fourfold. Measurements of the basal metabolic rate can be used to estimate the degree of activity of the thyroid gland.

Thyroxine causes the body to burn its available carbohydrates very rapidly, and then to make additional deep inroads on the stores of fats. Therefore, a person with excess production of thyroxine usually loses weight, sometimes very rapidly. On the other hand, a person with less than normal production of thyroxine often develops extreme obesity.

EFFECT ON THE CARDIOVASCULAR SYSTEM. Thyroxine affects the cardiovascular system in two ways. First, because the metabolic rate rises, all the tissues of the body require increased quantities of nutrients. This leads to vasodilatation in all the tissues, and causes the heart to pump greater quantities of blood than usual. Second, thyroxine has a direct effect on the heart, increasing its rate of metabolism as well as its rate and forcefulness of contraction. These effects also help to boost the cardiac output.

EFFECT ON THE NERVOUS SYSTEM. Thyroxine greatly increases the activity of the nervous system. The reflexes become very excitable with excess thyroxine, but very sluggish with diminished thyroxine. Thyroxine increases a person's degree of wakefulness, while lack of thyroxine sometimes makes him sleep as much as 12 to 15 hours per day.

A special effect of thyroxine on the nervous system is to cause a continuous tremor of the muscles. The tremor is usually very fine but rapid, having a frequency of 10 to 20 times per second, which is considerably faster than the tremor caused by basal ganglia or cerebellar disease.

EFFECT ON THE GASTROINTESTINAL TRACT. Thyroxine increases the motility of the gastrointestinal tract and promotes copious flow of digestive juices. If these activities are enhanced sufficiently, diarrhea may develop. On the other hand, lack of thyroxine causes the opposite effects — sluggish motility and greatly diminished secretion — resulting in constipation. Excess production of thyroxine also causes a voracious appetite because of the rapid rate of metabolism. The person eats a tremendous amount of food, digests it rapidly, absorbs large quantities of nutrients, but metabolizes these as rapidly as they become available.

REGULATION OF THYROXINE PRODUCTION

Earlier in the chapter it was pointed out that thyroxine production is regulated almost entirely by *thyrotropin* from the anterior pituitary gland. In turn, as illustrated in Figure 34–8, the rate of secretion of thyrotropin is regulated by *thyrotropin-releasing factor* secreted by the hypothalamus. Therefore, to describe the regulation of thyroxine production, we need only to discuss the factors that regulate secretion of thyrotropin-releasing factor.

The primary stimulus controlling the rate of secretion of thyrotropin-releasing factor is the level of metabolism in the body. If the metabolism falls to a low value, the rate of secretion of thyrotropin-releasing factor increases automatically, in turn increasing the secretion of thyrotropin and consequently of thyroxine. The thyroxine then raises the metabolic rate of the body back toward normal.

On the other hand, if the body's rate of metabolism rises above normal, the hypothalamus decreases its secretion of thyrotropin-releasing factor, and an opposite sequence of events reduces the secretion of thyroxine, thereby reducing the metabolic rate back toward normal. Thus, as demonstrated in Fig-

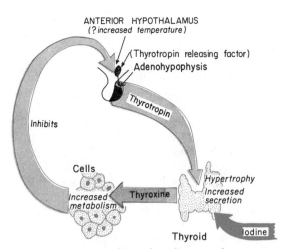

Figure 34–8. Interrelationships between the anterior pituitary gland and the thyroid gland for regulating the metabolic rate.

ure 34–8, the thyrotropin-thyroxine mechanism ordinarily acts as a feedback system to regulate body metabolism at the normal mean value.

Effect of Cold Weather on Thyroxine Production. When a person or an animal is exposed to very severe cold, the hypothalamus secretes greater quantities of thyrotropin-releasing factor. Over a period of 3 to 4 weeks, the thyroid gland gradually enlarges, and the rate of thyroxine secretion increases. The basal metabolic rate can be boosted by as much as 20 to 30 per cent by this mechanism, which helps to warm the body and partially compensates for the cooling effect of the weather.

ABNORMALITIES OF THYROID SECRETION

Hyperthyroidism. Failure of the anterior pituitary–thyroid regulatory system to function properly frequently leads to greatly increased production of thyroxine, sometimes to as much as 15 times normal. Most frequently the increased production is caused by a substance called *long-acting thyroid stimulator* (LATS) that is believed to be an antibody formed by the immune system and that has specific reactivity for the thyroid cells. The excess production of thyroxine resulting from this stimulation of the thyroid gland then causes the rate of thyrotropin secretion by the hypothalamus usually to be reduced rather than elevated.

In a few hyperthyroid persons, overproduction of thyroxine is caused by a small *thyroid adenoma,* a tumor of the thyroid gland that secretes thyroxine independently of control by the anterior pituitary gland or other stimuli. In either event the excess production of thyroxine causes *hyperthyroidism.*

In hyperthyroidism the basal metabolic rate rises very high — sometimes to as much as twice normal — the person loses weight, he develops diarrhea, he becomes highly nervous and tremulous, his heart rate is greatly increased, and the heart often beats so hard that he feels it palpitating in his chest. The state of hyperthyroidism is often so severe and so prolonged that it actually "burns out" the tissues, leading to degenerative processes in many parts of the body. One of the most common areas of degeneration is the muscle of the heart itself.

The usual methods for treating hyperthyroidism are (1) administration of a drug that suppresses thyroid function or destroys the thyroid gland or (2) surgical removal of a major portion of the gland. Administration of *propylthiouracil* blocks the chemical reaction of tyrosine and iodine to form thyroxine, and hyperthyroidism can often be controlled with this drug.

EXOPHTHALMOS (PROTRUSION OF THE EYEBALLS). Most patients with severe hyperthyroidism also develop protruding eyeballs, a condition called *exophthalmos* (Fig. 34–9). It is believed that thyroxine has little to do with the protrusion of the eyeballs but instead that this is caused by an abnormal hormone called *exophthalmos-producing substance* that is secreted in large quantities by the anterior pituitary gland in place of thyrotropin. The cause of this interplay between exophthalmos-producing substance and thyrotropin is not understood, though exophthalmos is indeed a very unfortunate and common side effect of hyperthyroidism. Exophthalmos-producing substance causes edema, excessive growth, and degeneration of the tissues behind the eyeballs. Because exophthalmos results at least partially from increased quantity of tissue behind the eyes, elimination of the hyperthyroidism will not eliminate all the exophthalmos, which remains throughout life thereafter.

Hypothyroidism. Diminished production of thyroxine is called *hypothyroidism.* A person can live for many years with complete lack of thyroxine production, but the rate of metabolism in all his tissues is decreased to about one-half normal. He is extremely lethargic, sleeping sometimes as much as 12 to 15 hours a day. He usually is constipated, his mental reactions are sluggish, and he often becomes fat. In addition to increased deposition of fat throughout his body, a gelatinous mixture of mucoprotein and extracellular fluid is deposited in the spaces between the cells, giving an edematous appearance. For this reason the condition is often called *myxedema.*

Goiter. In hyperthyroidism the overactive thyroid gland usually enlarges two- to threefold, and is then called a *goiter.* In hypothyroidism the gland frequently enlarges also, and here again the enlarged gland is called a goiter. Therefore, the term goiter is not synonymous with either hyper- or hypothyroidism but instead means simply enlargement of the thyroid gland.

Hypothyroidism is usually caused by some abnormality of the thyroid gland that makes it impossible for the gland, even when stimulated by thyrotropin, to secrete enough thyroxine. Yet the poorly secreting gland enlarges more and more in a futile attempt to produce adequate quantities of thyroxine, and large amounts of colloid substance containing almost no thyroxine are secreted into the follicles. For this reason this type of enlarged thyroid gland is called a *colloid goiter.*

Sometimes a colloid goiter becomes 15 times as large as the normal thyroid gland, weighing 500 or more grams and occupying a space in the neck or upper chest as large as one-half liter. Obviously, a gland this large can obstruct breathing and swallowing.

ENDEMIC GOITER. Persons residing in areas of the world where the food contains

Figure 34–9. A hyperthyroid person with exophthalmos.

very little iodine cannot produce an adequate quantity of thyroxine. As a result, their metabolic rates fall below normal, and this enhances the output of thyrotropin which in turn stimulates the thyroid gland in an attempt to produce increased quantities of thyroxine. Unfortunately, even this stimulus cannot enhance the output of thyroxine when iodine is lacking. But the anterior pituitary gland continues to produce large amounts of thyrotropin, so that the thyroid continues to enlarge, becoming progressively filled with colloid that contains almost no thyroxine. The enlarged gland is called an *endemic goiter* because everyone in the iodine-deficient geographic region develops an enlarged gland. Endemic goiter was formerly widely prevalent in those regions of the world, such as the Great Lakes region of the United States and the Swiss Alps, where iodine is not present in the soil. More recently, however, a small amount of iodine has been added to most commercial table salts, so that now an inadequate intake of iodine is very rare.

REFERENCES

Berson, S. A., and Yalow, R. S.: Peptide Hormones. New York, American Elsevier Publishing Company, 1973.

Catt, K. J., and Dufau, M. L.: Peptide hormone receptors. *Ann. Rev. Physiol.,* 39:529, 1977.

Daughaday, W. H., Herington, A. C., and Phillips, L. S.: The regulation of growth by endocrines. *Ann. Rev. Physiol.,* 37:211, 1975.

Edelman, I. S., and Ismail-Beigi. F.: Thyroid thermogenesis and active sodium transport. *Recent Prog. Horm. Res.,* 30:235, 1974.

Fain, J. N., and Butcher, F. R.: Cyclic nucleotides in mode of hormone action. *Intern. Rev. Physiol.,* 16:241, 1977.

Fisher, L. (ed.): Neuroendocrine Integration: Basic and Applied Aspects. New York, Raven Press, 1975.

Greer, M. A., and Haibach, H.: Thyroid secretion. *In* Greep, R. O., and Astwood, E. B. (eds.): Handbook of Physiology. Sec. 7, Vol. 3. Baltimore, The Williams & Wilkins Company, 1974, p. 135.

Jard, S., and Bockaert, J.: Stimulus-response coupling in neurophyophyseal peptide target cells. *Physiol. Rev.,* 55:489, 1975.

Kostyo, J. L., and Isaksson, O.: Growth hormone and the regulation of somatic growth. *Intern. Rev. Physiol.,* 13:255, 1977.

Krulich, L., and Fawcett, C. P.: The hypothalamic hypophysiotropic hormones. *Intern. Rev. Physiol.,* 16:35, 1977.

Martin, J. B.: Regulation of the pituitary-thyroid axis. *In* MTP International Review of Science: Physiology. Vol. 5, Baltimore, University Park Press, 1974, p. 67.

Seif, S. M., and Robinson, A. G.: Localization and release of neurophysins. *Ann. Rev. Physiol.,* 40:345, 1978.

Sterling, K., and Lazarus, J. H.: The thyroid and its control. *Ann. Rev. Physiol.,* 39:349, 1977.

Tixier-Vidal, A., and Farquhar, M. G.: The Anterior Pituitary. New York, Academic Press, Inc., 1975.

Vale, W., Rivier, C., and Brown, M.: Regulatory peptides of the hypothalamus. *Ann. Rev. Physiol.,* 39:473, 1977.

QUESTIONS

1. Explain the role of cyclic AMP as an intracellular hormone mediator.
2. Explain how hormones sometimes control cellular function by activating genes.
3. How does growth hormone stimulate growth of cartilage and bone?
4. How does growth hormone promote protein deposition?
5. What are the causes of dwarfs, giants, and acromegaly?
6. List the six most important anterior pituitary hormones.
7. What are the hypothalamic releasing and inhibitory factors?
8. What functions do antidiuretic hormone and oxytocin subserve?
9. Give the cellular and chemical mechanisms for the formation of thyroxine.
10. What are the effects of thyroxine on cellular metabolism, and what is the mechanism of these effects?
11. What are the causes and effects of hyperthyroidism and hypothyroidism?

35 ADRENOCORTICAL HORMONES, INSULIN, AND GLUCAGON

ADRENOCORTICAL HORMONES

The adrenal glands, one of which is shown in cross-section in Figure 35–1, are flattened structures located immediately above each of the two kidneys. These glands produce hormones of two entirely different types. The *medulla* of each gland is part of the sympathetic nervous system, and it secretes *epinephrine* and *norepinephrine* in response to sympathetic stimuli; this was described in Chapter 28. The outer portion of the adrenal gland is the *cortex*, which secretes *adrenocortical hormones*. These hormones are all chemically similar — they are all steroids — but they can be divided into three different categories on the basis of their functions: *mineralocorticoids, glucocorticoids,* and *androgens*. The mineralocorticoids control the kidney excretion of sodium and potassium; the glucocorticoids help to control the metabolism of protein, fat, and glucose; and the androgens cause masculinizing effects.

MINERALOCORTICOIDS — ALDOSTERONE

The function of mineralocorticoids is to regulate ions, particularly potassium, but also sodium, in the extracellular fluids.

The adrenal cortex secretes at least three different hormones that can be classified as mineralocorticoids: aldosterone, corticosterone, and minute quantities of deoxycorticosterone. However, aldosterone accounts for at least 95 per cent of the total mineralocorticoid activity.

The function of aldosterone in relation to electrolyte absorption from the kidney tubules has already been presented in Chapter 18; its overall function can be summarized briefly as follows:

Effect on Sodium. A direct effect of aldosterone is to increase the rate of sodium reabsorption by the renal tubular epithelium. When large quantities of aldosterone are se-

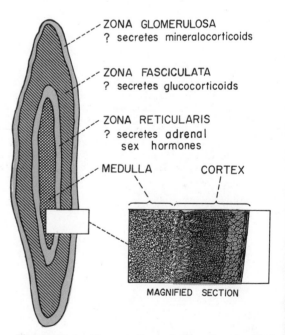

ZONA GLOMERULOSA
? secretes mineralocorticoids

ZONA FASCICULATA
? secretes glucocorticoids

ZONA RETICULARIS
? secretes adrenal
sex hormones

MEDULLA   CORTEX

MAGNIFIED SECTION

Figure 35–1. Cross-section of the adrenal gland, showing the three zones of the adrenal cortex that are believed to secrete the three types of adrenocortical hormones.

creted, essentially all of the sodium entering the glomerular filtrate is reabsorbed into the blood, and almost no sodium passes into the urine. On the other hand, when minute amounts of aldosterone are secreted, much less sodium is reabsorbed, and as much as 20 to 30 grams of sodium may appear in the urine each day. Therefore, it is evident that aldosterone is essential to prevent rapid depletion of sodium in the body. Thus, aldosterone secretion represents a method for controlling the total quantity of sodium in the extracellular fluids.

BASIC MECHANISM BY WHICH ALDOSTERONE INCREASES SODIUM REABSORPTION. Aldosterone does not cause an immediate increase in sodium reabsorption. Instead, 45 minutes to an hour after aldosterone has been secreted, the various enzymes required for sodium reabsorption begin to increase in the tubular epithelial cells, and these reach a maximum level within 3 or more hours. Therefore, it is believed that aldosterone activates the genetic mechanism of the cell to increase production of the specific enzymes required for sodium reabsorption. After disappearance of the aldosterone, these enzymes decrease back to a low level within another 2 to 4 hours.

Effect on Potassium. Aldosterone also has a direct effect on potassium transport through the tubular epithelium. However, instead of increasing the rate of potassium reabsorption it causes a marked increase in rate of secretion of potassium into the tubules and, therefore, a marked excretion of potassium into the urine.

This increase in potassium secretion is probably caused by two separate effects of the aldosterone. First, aldosterone directly stimulates the active transport process for potassium secretion, increasing this secretion at the same time that it enhances sodium reabsorption. Second, the marked absorption of positively charged sodium ions caused by aldosterone creates a state of electronegativity in the distal and collecting tubules that is strong enough actually to pull positive potassium ions from the extracellular fluids into the tubular fluids.

Thus, aldosterone has a very powerful effect of increasing excretion of potassium into the urine and consequently of decreasing the potassium concentration in the extracellular fluid.

Effect on Chloride Ion. The electronegativity that occurs in the renal tubules when sodium is absorbed also affects chloride ion transport through the tubular membrane. The chloride ions, being themselves electronegative, are repelled from the negative tubular fluids into the surrounding extracellular fluid. Thus, a secondary effect of mineralocorticoid action is to decrease the amount of chloride lost in the urine and to increase the quantity of chloride ions in the extracellular fluid. The net result is greater sodium chloride (salt) in the extracellular fluid and reduced potassium.

Effect on Water Reabsorption from the Tubules and on Extracellular Fluid Volume. The great increase in sodium and chloride reabsorption from the tubules results in an increase in the reabsorption of water as well. The reason for this is simply an osmotic effect that can be explained as follows: Transport of the sodium and chloride ions from the tubules into the extracellular fluid diminishes the concentrations of these substances in the tubular fluid and thus also reduces the tubular osmotic pressure. Therefore, an osmotic gradient develops across the tubular membrane to cause greatly increased absorption of water by osmosis into the peritubular fluids.

The increase in both sodium chloride and water in the extracellular fluids often raises the extracellular fluid volume as much as 10 to 20 per cent.

Effect on Blood Volume and on Circulatory Dynamics. The excess extracellular fluid enters both the interstitial spaces and the blood, increasing both the interstitial fluid volume and the blood volume. The greater blood volume forces a larger than normal quantity of blood toward the heart, which boosts the *cardiac output* a small amount; and this in turn often causes *elevated arterial pressure* as well. Therefore, excess aldosterone is sometimes a cause of high blood pressure.

REGULATION OF ALDOSTERONE SECRETION

The regulation of aldosterone secretion has already been discussed in Chapter 18 in connection with the effects of aldosterone on renal function. However, let us review this regulation briefly.

Effect of Potassium Concentration on Aldosterone Secretion. The most potent long-term stimulator of aldosterone secretion in experimental animals is a rise in potassium ion concentration in the plasma. The increased secretion of aldosterone in turn causes the kidneys to excrete large amounts of potassium from the body, thereby reducing the potassium concentration in the extracellular fluid back toward normal. Research studies have shown that this is a very powerful mechanism whereby the body regulates the long-term level of potassium concentration in the extracellular fluid.

Effect of Decreased Body Sodium and Decreased Extracellular Fluid Volume on Aldosterone Secretion. When an animal's sodium intake is greatly decreased, the rate of aldosterone secretion increases markedly during the next few days. On the other hand, an acute reduction in sodium concentration in the extracellular fluid does not have an immediate effect of increasing aldosterone secretion such as that caused by a rise in potassium ion concentration. Because there is no immediate effect of diminished sodium on aldosterone secretion, it is not certain why a long-term deficiency of sodium leads to greater aldosterone secretion. Some research workers have suggested that depressed sodium ion concentration has a delayed effect of (a) causing secretion of a pituitary hormone that controls aldosterone secretion, or (b) causing the kidney to activate the renin-angiotensin system which in turn directly influences the adrenal gland to increase aldosterone secretion.

However, it is also possible that diminished sodium increases aldosterone secretion indirectly rather than directly. That is, diminished sodium intake leads to greatly reduced extracellular fluid volume. It is known that other factors besides sodium deficiency that decrease the extracellular fluid volume also increase aldosterone secretion. Therefore, it may be that the adrenal gland in some unknown way boosts aldosterone secretion in response to diminished extracellular fluid volume. For instance, decreased cardiac output and reduced tissue blood flow may be the stimuli for this effect.

Effect of Angiotensin on Aldosterone Secretion. When blood flow through the kidneys decreases, the kidneys cause the formation of renin and angiotensin by mechanisms discussed in Chapter 14. One of the effects of angiotensin is to stimulate the adrenal gland to augment its secretion of aldosterone. However, this increase in aldosterone is very marked for only 6 to 8 hours, and thereafter is much less. Therefore, this mechanism seems to be especially important for causing transient increases in aldosterone secretion.

Effect of Corticotropin on Aldosterone Secretion. We shall note in the following paragraphs that the anterior pituitary hormone corticotropin (also called adrenocorticotropin or ACTH) has a very powerful effect of increasing the secretion of glucocorticoid hormones by the adrenal cortex. This hormone also causes a slight to moderate increase in the secretion of aldosterone. However, under most conditions, the rate of aldosterone secretion is controlled mainly by the potassium ion concentration in the circulating fluids and by the circulatory status of the person.

GLUCOCORTICOIDS — CORTISOL

The functions of the glucocorticoids are not nearly so well understood as those of the mineralocorticoids. However, many of the body's metabolic systems become greatly deranged when glucocorticoids are absent, and the person becomes unable to resist almost any trauma or disease condition that tends to destroy tissues. Therefore, the most important function of glucocorticoids is to enhance resistance to physical "stress," though the means by which this is effected are yet very vague.

Several different adrenocortical hormones exhibit glucocorticoid activity, but by far the most abundant one is *cortisol* (also known as *hydrocortisone*). Other much less important glucocorticoids are corticosterone and cortisone.

Effect of Cortisol on Glucose Metabolism. The earliest discovered effect of glucocorticoids is their ability to increase the concentration of glucose in the blood. This is caused by two different functions of cortisol. First, cortisol depresses utilization of glucose by the tissue cells. Consequently, the glucose that enters the extracellular fluid, instead of being utilized fully by the tissue cells, collects in the fluid.

Second, cortisol causes the liver cells to convert proteins and the glycerol portion of fats into glucose, a process called *gluconeogenesis*. It is believed that this results partly because cortisol increases the liver enzymes that promote gluconeogenesis. However, gluconeogenesis also results because cortisol has a special effect, one that will be discussed further in subsequent paragraphs, of causing amino acids to be mobilized from the protein stores of the body and fat to be mobilized from the fat stores. The excess supply in the blood of amino acids and glycerol (derived from the fat) provides material that the liver can convert into glucose.

Gluconeogenesis is very important during starvation because it supplies a continual source of glucose in the blood even when the person is not ingesting glucose. This glucose is necessary especially to provide nutrition for the neurons because they can use only glucose for energy.

Effect of Cortisol on Protein Metabolism. One of the most important effects of cortisol is that of decreasing the quantity of protein in most tissues of the body, with the exception of the liver. It does this both by suppressing the rate of protein formation in the nonliver cells and by causing breakdown of some of the proteins already present in the cells into amino acids and then release of these into the blood.

The blood concentration of amino acids increases considerably under the influence of cortisol, both because of the decreased use of amino acids by the cells to form proteins and because of release of amino acids from the cells into the blood. This is an important effect because it makes amino acids available for use wherever in the body they are needed.

And a final effect of cortisol on protein metabolism is to *increase* the rate of protein formation in the liver cells. Also, since it is the liver cells that produce most of the plasma proteins, cortisol increases the quantity of plasma proteins.

Effect of Cortisol on Fat Metabolism. Cortisol mobilizes fat from the fat depots in much the same manner that it mobilizes amino acids from the cells. The net result is a decrease in the amount of fat in the storage areas and increased use of fat for energy and other purposes. It is believed, for example, that the mobilization of fat from the storage areas during periods of starvation is caused mainly by increased production of cortisol.

During periods of rapid fat mobilization, the liver splits much of the fat into *keto acids*. If these are used immediately by the cells for energy, they cause no significant physiologic effect, but if they are not used immediately, their concentration in the extracellular fluid can become great enough to cause acidosis. This is occasionally one of the untoward effects of excessive cortisol secretion.

Effect of Cortisol on Lysosome Stabilization. When cells are damaged either because of trauma or because of disease, the lysosomes in the cytoplasm begin to break open, releasing their digestive enzymes into the interior of the cell. These enzymes in turn autolyze internal structures of the cells, which diminishes cellular function or at times actually kills the cells. Therefore, it is often important to prevent this process. Cortisol has a special ability to stabilize the membranes of lysosomes and thereby to prevent their dissolution. Fortunately, in disease states, large quantities of cortisol are almost invariably secreted into the blood. Therefore, it is believed that one of the most important functions of cortisol is to help a person resist the devastating effects of some diseases by preventing rupture of the lysosomes inside the cells.

Regulation of Cortisol Secretion. Figure 35–2 summarizes the regulation of adrenal secretion of cortisol and other glucocorticoids. The mechanism can be explained as follows: The primary stimulus that initiates glucocorticoid secretion is called "stress," which includes almost any type of damage to the body. For instance, a painful contusion of some part of the body, a broken bone, severe damage to large tissue areas by some disease condition, or any other destruction of parts of the body usually sets off a sequence of events that

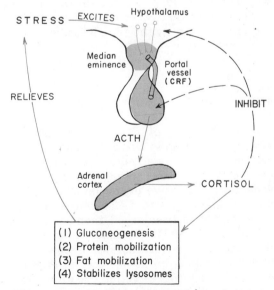

Figure 35–2. Mechanism for regulation of glucocorticoid secretion.

leads to cortisol secretion. The stress probably causes these reactions by initiating nerve impulses that are transmitted from the periphery into the hypothalamus. The hypothalamus then secretes the substance *corticotropin-releasing factor* which passes by way of the hypothalamic-hypophyseal portal system into the anterior pituitary gland. Here, this factor causes the cells of the gland to secrete *corticotropin* which flows in the blood to the adrenal cortex where it elicits *cortisol* secretion. The cortisol then mobilizes protein and fat from all over the body and also causes gluconeogenesis. The increased availability of amino acids, fats, and glucose in the blood helps in the repair of the damage, thus attenuating the initial stimulus that had set off the sequence of events leading to cortisol secretion. And, in addition, the cortisol prevents breakdown of the lysosomes, thus also preventing destruction of the tissues.

ANDROGENS

Androgens are hormones that cause the development of male sex characteristics, as we shall see in Chapter 37. Though the male testes are the primary source of these hormones, the adrenal cortex also secretes minute quantities, quantities so small that in the normal person they have no significant effect. However, an androgen-producing tumor of the adrenocortical cells develops occasionally which secretes very large quantities of male sex hormones that cause serious masculinizing effects even in the female body. This is explained below.

ABNORMALITIES OF ADRENOCORTICAL SECRETION

Hyposecretion of Adrenocortical Hormones — Addison's Disease. The adrenal cortices are occasionally destroyed by disease, or sometimes they simply atrophy. And, sometimes, excessive overstimulation of the adrenal gland by stress also causes it to become, first, greatly enlarged, then hemorrhagic, and finally replaced almost completely by fibrous tissue. In each instance hyposecretion of adrenocortical hormones occurs, or rarely complete lack of secretion, leading to a serious condition called *Addison's disease*.

Complete failure of the adrenal cortices usually causes death within 3 to 5 days unless the person is appropriately treated. This early death is caused by the lack of aldosterone, because the kidney's sodium reabsorptive and potassium secretory mechanisms are highly dependent on this hormone. Without adequate sodium reabsorption, the extracellular fluid volume decreases so greatly within only a few days that death ensues. This effect can be offset to some extent by having the afflicted person eat extra quantities of salt, and it can be overcome entirely by having him ingest extremely minute amounts of one of the mineralocorticoids, for instance, 0.1 mg. of fludrocortisone, a synthetic mineralocorticoid, by mouth each day.

Even if the life of the person with Addison's disease is saved by administering a mineralocorticoid, he still remains unable to resist stress, and even a slight respiratory infection may prove lethal. He also has little energy. Usually, therefore, to provide completely adequate function of the body, treatment with cortisol (about 30 mg. daily) or some other glucocorticoid is necessary in addition to the mineralocorticoid.

Hypersecretion of Adrenocortical Hormones. Hypersecretion by the adrenal cortex can result either directly from a tumor in one part of an adrenal gland, or indirectly from increased production of corticotropin by the anterior pituitary, this in turn stimulating the adrenal cortices.

The effects of hyperadrenalism depend upon which part of the adrenal gland is secreting the excessive quantities of hormone. If it is the inner zone of the adrenal cortex, the *zona reticularis,* then excessive quantities of androgens are secreted, and the person, even a child or a female, develops masculine characteristics such as growth of hair on the face, deepening of the voice, sometimes baldness, changes in portions of the female sexual organs to resemble the sexual organs of the male, atrophy of the female breasts, and considerably enhanced muscular development.

If the middle zone of the cortex, the *zona fasciculata,* secretes excess hormones, there are symptoms of excess cortisol secretion. These include excessive mobilization of proteins and fats from their storage areas. The mobilization of proteins causes weakness of the muscles and sometimes weakness of other protein structures such as the fibers that hold the tissues together beneath the skin. This allows laxity of the skin and predisposes to tears in the subcutaneous tissues that can be observed as long, linear, *purplish striae.* The excess mobilization of proteins and fats also increases gluconeogenesis, and raises the blood glucose concentration, sometimes enough to cause a very high blood sugar level, a condition known as *adrenal diabetes*.

If the increased secretion occurs in the outer layer of the adrenal cortex, the *zona glomerulosa,* the effects are those of excessive aldosterone secretion. The concentration of potassium decreases, the blood volume increases, cardiac output increases, and the arterial pressure rises moderately.

INSULIN

Secretion of Insulin by the Pancreas. The pancreas is a long gland that lies immediately beneath the stomach, a picture of which was shown in Figure 30–11 in Chapter 30. Also, a microscopic picture of a small section of the pancreas is shown in Figure 35–3. It is composed of two different types of tissue. One type is the *acini,* which secrete digestive juices into the intestines; this was discussed in Chapter 30. The other type is the *islets of Langerhans,* which secrete hormones directly into the blood; this secretion into the blood is an *endocrine* function of the pancreas, while the secretion of digestive juices is an *exocrine* function.

The islets of Langerhans are composed of two different types of cells, the *alpha* and *beta cells*. The beta cells secrete the hormone *insulin,* while the alpha cells secrete the hormone *glucagon,* which is discussed later in the chapter. Both of these hormones are small proteins.

FUNCTION OF INSULIN IN THE BODY

The basic function of insulin is to control glucose metabolism in the body. It does this by increasing the rate of glucose transport through the cellular membrane. The pores of the membrane are much too small for the glucose molecule to enter the cell by the process of simple diffusion. Instead, glucose must be carried through the membrane combined with a chemical carrier as shown in Figure 35–4. This is a process called *facilitated diffusion,* which was discussed in Chapter 8.

In the absence of insulin, only a small amount of glucose can combine with the carrier to be transported into the cells, but in the presence of normal amounts of insulin the

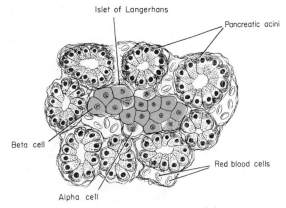

Figure 35–3. Microscopic structure of the pancreas.

transfer is accelerated as much as 3- to 5-fold, and in the presence of large amounts of insulin the rapidity of glucose transfer is increased as much as 15- to 20-fold. Therefore, insulin controls the rate of glucose metabolism in the body by controlling the entry of glucose into the cells.

Effect of Insulin on Blood Glucose. Because insulin accelerates the rate of glucose transfer from the extracellular fluids to the interior of the cells, the concentration of glucose in the blood and extracellular fluids becomes diminished. Conversely, lack of insulin secretion causes glucose to dam up in the blood instead of entering the cells. Complete lack of insulin usually produces a rise in blood glucose concentration from a normal value of 90 mg. per 100 ml. up to about 350 mg. per

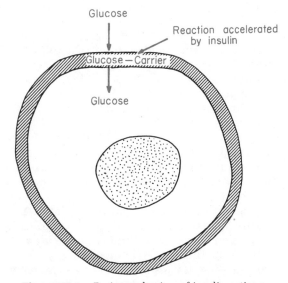

Figure 35–4. Basic mechanism of insulin action.

100 ml. On the other hand, a great excess of insulin can decrease the blood glucose to about 25 mg. per 100 ml., which is about one-fourth normal.

Effect of Insulin and Insulin Lack on Fat Metabolism. Insulin has almost as profound effects on fat metabolism as on glucose metabolism. However, these effects probably occur mainly secondarily to the carbohydrate effects in the following manner: Whenever large amounts of glucose are available, insulin causes some of this glucose to be transported into the fat cells. Products of glucose metabolism, especially *acetic acid* and *alpha-glycerophosphate,* then promote fat storage. The acetic acid is polymerized into fatty acid which reacts with the glycerophosphate to form neutral fat, thereby synthesizing fat. Conversely, in the absence of insulin, very little glucose enters the fat cells, which means that appropriate products are not available to cause fat storage. Instead, an exactly opposite effect occurs, namely, release of fatty acids into the blood.

Therefore, insulin has an effect on fat metabolism essentially opposite to its effect on carbohydrate metabolism. That is, in the presence of insulin carbohydrates are utilized preferentially and excess carbohydrate is stored as fat, whereas in the absence of insulin fatty acids are mobilized and utilized in place of carbohydrates.

Effect of Insulin on Protein Metabolism. Insulin is almost as potent as growth hormone in causing protein deposition in cells. It accomplishes this by both direct and indirect effects on protein metabolism.

The direct effects of insulin on protein metabolism are threefold:

1. Insulin increases the rate of transport of most of the amino acids through the cell membranes, thereby increasing the quantities of amino acids available in the cells for synthesizing proteins.

2. Insulin increases the formation of RNA in cells.

3. Insulin increases the formation of proteins by the ribosomes.

Thus, insulin has a strong direct effect of promoting protein formation in all or most cells of the body.

The indirect effect of insulin on protein metabolism results from its promotion of glucose utilization by the cells. When glucose is available to be utilized for energy, a *protein-sparing* effect is provided, because carbohydrates are then used in preference to proteins.

On the other hand, in the absence of insulin this protein-sparing effect is lost, so that large quantities of protein instead of carbohydrates are utilized along with fats for energy.

EFFECT OF INSULIN ON GROWTH. Because insulin promotes the formation of proteins, as well as makes available large amounts of energy from carbohydrates, this hormone has a powerful effect on growth. Indeed, its absence in an animal causes almost as much stunting of growth as does the absence of growth hormone from the anterior pituitary gland.

REGULATION OF INSULIN SECRETION

When the blood glucose level rises, the pancreas begins secreting insulin within a few minutes. This is caused by a direct effect of glucose on the beta cells in the pancreatic islets, causing them to produce greatly increased quantities of insulin. The insulin in turn causes the excess glucose to be transported into the cells where it can be used for energy, stored as glycogen, or converted into fat.

Thus, the insulin mechanism provides a feedback means for controlling the concentration of glucose in the blood and extracellular fluids. Too high a glucose level initiates insulin secretion, which then causes increased utilization of glucose and return of the blood glucose level back to or toward normal. Conversely, when the blood glucose level falls too low, the rate of insulin secretion decreases, and glucose is now conserved in the body fluids until its level returns to normal.

DIABETES

Diabetes is the disease that results from failure of the pancreas to secrete insulin. It is caused by degeneration of the beta cells of the islets of Langerhans, but the cause of degeneration is usually not known. In lower animals certain chemical compounds, notably *alloxan,* can cause degeneration of these cells, and excess dietary carbohydrates in some instances can so overwork the secretory function of the beta cells that they actually "burn out." Therefore, it is possible that some persons develop diabetes secondary to exposure to abnormal chemicals or secondary to eating excess quantities of carbohydrates. However, much diabetes is caused by inheritance from a parent or some preceding ancestor, that is, by hereditary degeneration of the beta cells. This often occurs very early in life and then lasts for the lifetime of the person. Finally, many middle-aged or old persons develop immunity to insulin. That is, the immune system produces antibodies that destroy the insulin before it can become effective.

Pathophysiologic Effects of Diabetes. The primary abnormality in diabetes is failure to utilize adequate quantities of glucose for energy. This causes the blood glucose level to rise, often to as high as three times normal and rarely as high as 10 times normal. Large quantities of glucose are lost into the urine because the kidney tubules cannot reabsorb all that enters them in the glomerular filtrate each minute. The excess tubular glucose also creates a tremendous amount of osmotic pressure in the tubules, and this diminishes the reabsorption of water. As a result, the diabetic person loses large quantities of water as well as glucose into the urine. In extreme cases the excess output of urine causes extracellular dehydration, which can in itself be very harmful.

Failure of the diabetic person to utilize glucose for energy deprives him of a major portion of the energy in his food. He loses weight and becomes weakened because of excess burning of his fat and protein stores. As a result of the nutrient deficiency in diabetes, the diabetic person usually becomes very hungry, so that he often eats voraciously even though the carbohydrate portion of his food contributes little to his nutrition.

KETOSIS AND DIABETIC COMA. The extremely rapid metabolism of fats in diabetes sometimes increases the quantity of keto acids in the extracellular fluids to as high as 20 to 30 milliequivalents per liter, which is 25 to 50 times normal. On occasion this becomes sufficient to make the pH of the body fluids fall from its normal value of 7.4 to as low as 7.0, or rarely to as low as 6.9. This degree of acidosis is incompatible with life for more than a few hours. The person breathes extremely rapidly and deeply to blow off carbon dioxide, which helps to offset the metabolic acidosis, but despite this the acidosis often becomes severe enough to cause coma. And unless the person is treated he usually dies in less than 24 hours. Treatment requires immediate administration of large quantities of insulin. Sometimes glucose is administered along with the insulin to help promote the shift from fat to glucose metabolism. Intravenous administration of alkaline solutions can also be of great benefit for two reasons:

First, alkaline solutions neutralize some of the acidosis, and, second, they correct the dehydration that often accompanies severe diabetic coma.

Treatment of Diabetes with Insulin. The diabetic person can usually be treated quite adequately by daily injections of insulin. Two principal types of insulin available for this purpose are *crystalline zinc insulin* and *protamine zinc insulin*. The duration of action of crystalline zinc insulin after injection is 4 to 6 hours, and that of protamine zinc insulin is 24 to 30 hours. Usually a person who has severe diabetes must take an injection of crystalline zinc insulin at mealtimes when his blood glucose concentration is likely to rise very high temporarily, and he must take an injection of protamine zinc insulin each morning to provide a steady rate of glucose inflow into his cells throughout the day.

Atherosclerosis in Diabetes. Prolonged diabetes usually leads to early development of atherosclerosis, and this subsequently causes heart disease, kidney damage, cerebral vascular accidents, or other circulatory disorders. The probable reason for the development of atherosclerosis is that, even with the best possible treatment of diabetes, glucose metabolism can never be maintained at a sufficiently high level to prevent some excess fat metabolism, and cholesterol deposition in the walls of the blood vessels is always an unfortunate accompaniment of rapid fat metabolism. Because of this, the person who develops diabetes in childhood usually has a shortened life, regardless of how well he is treated.

HYPERINSULINISM

Hyperinsulinism occasionally develops because of either overtreatment of a diabetic person with insulin or too much secretion of insulin by a pancreatic islet tumor. In either case low blood glucose concentration ensues. This in turn causes overexcitability of the brain at first and then coma. The neurons require a constant supply of glucose because they cannot utilize significant amounts of fats or proteins for energy. Furthermore, the rate of glucose uptake by the neurons, unlike that of other cells, is dependent mainly on the blood glucose concentration rather than on the amount of insulin available. Whenever excess insulin is available, the blood glucose becomes very low so that the neurons no longer receive the amounts of glucose needed to maintain their metabolism. This causes them first to become excessively excitable, and later depressed. In the excitement stage convulsions may occur, but in the depressed stage the person develops coma not unlike the coma that occurs in untreated diabetes. Indeed, it is sometimes a problem to diagnose the cause of coma in a diabetic. It may result from too little insulin secretion, in which case it is diabetic coma, or from too much treatment with insulin, in which case the abnormality is hyperinsulinism.

GLUCOSE TOLERANCE TEST

A method often used for testing the ability of the pancreas to secrete insulin is the so-called glucose tolerance test, illustrated in Figure 35–5. In the normal person the blood glucose concentration at the beginning of the test is less than 100 mg. per cent. At this time 50 grams of glucose is ingested, and during the ensuing half-hour or more the blood glucose rises to a value of 130 to 160 mg. per cent. In less than one hour, however, large quantities of insulin will have been produced by the pancreas, causing the blood glucose concentration to fall very rapidly, reaching normal once again about two and one-half hours after the ingestion of glucose. Then the concentration actually falls below normal, which is called the *hypoglycemic response,* because of continuing action of the insulin that has been secreted.

A diabetic person whose pancreas cannot secrete adequate quantities of insulin often exhibits a hyperglycemic response in which the blood glucose concentration not only begins at an excessively high level but also rises much more than in the normal person. And the blood glucose level may require as long as 5 or more hours to return to the original level. These effects are shown in the middle portion in Figure 35–5. Also, because no insulin has been secreted, the glucose level never shows the hypoglycemic response.

To the right in Figure 35–5, the effect of hyperinsulinism on the glucose tolerance curve is illustrated. Note that the normal fasting level of glucose is low, and even ingestion of a large excess of glucose still will not elevate the glucose concentration to a high level.

GLUCAGON

The alpha cells of the islets of Langerhans secrete a hormone called *glucagon.* Many of the functions of glucagon are opposite to those of insulin, while others complement the functions of insulin.

Glucagon raises the blood glucose level; insulin reduces it. On the other hand, both insulin and glucagon increase the availability of glucose to the cells for their utilization. Glucagon does this by mobilizing glucose from the liver, while insulin does so by in-

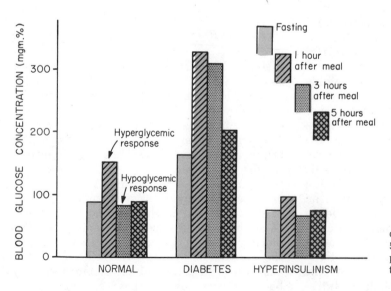

Figure 35–5. Effect on blood glucose concentration of administering 50 gm. of glucose to, first, a normal person, second, a diabetic person, and, third, a person with hyperinsulinism.

creasing the transport of glucose into the cells. For instance, during heavy exercise both of these hormones work together to promote utilization of glucose by the muscles.

Basic Mechanisms of Glucagon Function. Glucagon elevates the blood glucose concentration in two ways: First, it increases the breakdown of liver glycogen into glucose, making this available for transport into the blood. Glucagon achieves this effect by activating the enzyme *adenylcyclase,* which in turn increases the quantity of cyclic AMP in the liver cells. The cyclic AMP then causes *glycogenolysis* (breakdown of the glycogen to glucose).

Second, glucagon increases *gluconeogenesis* (conversion of proteins to glucose) by the liver. It does this mainly by mobilizing proteins from many tissues of the body and then promoting uptake of amino acids into the liver as well as conversion of the amino acids into glucose.

The blood glucose concentration can rise as much as 20 per cent within a few minutes after injection of glucagon.

Control of Glucagon Secretion, and Function of Glucagon in the Body. Glucagon secretion is controlled in almost exactly the opposite manner to the control of insulin. That is, when the blood glucose concentration falls below normal, the pancreas begins to secrete increased quantities of glucagon into the blood. Indeed, when the blood glucose concentration falls as low as 60 mg. per 100 ml. of blood (about 30 per cent below the normal resting level), the pancreas literally pours glucagon into the blood. This effect of low blood glucose concentration on glucagon secretion results from direct stimulation of the *alpha* cells in the islets of Langerhans. The glucagon in turn causes almost immediate release of glucose from the liver, thereby rapidly increasing blood glucose concentration back up toward the normal level of 90 to 100 mg. per 100 ml.

Thus, the glucagon mechanism, like the insulin mechanism, helps to regulate the blood glucose concentration, but with one difference: The glucagon mechanism acts to keep the blood glucose concentration from falling too low while insulin keeps it from rising too high. Thus, the glucagon mechanism is especially activated during severe exercise and during starvation, both of which tend to decrease blood glucose.

An especially important function of glucagon is to keep the glucose concentration high enough to prevent hypoglycemic convulsions or hypoglycemic coma, which were discussed above.

REFERENCES

Azarnoff, D. L.: Steroid Therapy. Philadelphia, W. B. Saunders Company, 1975.

Brodish, A., and Lymangrover, J. R.: The hypothalamic-pituitary adrenocortical system. *Intern. Rev. Physiol., 16*:93, 1977.

Dorfman, R. I.: Steroid Hormones. New York. American Elsevier Publishing Company, 1974.

Gerich, J. E., Charles, M. A., and Grodsky, G. M.: Regulation of pancreatic insulin and glucagon secretion. *Ann. Rev. Physiol., 38*:353, 1976.

Gorden, P., Gavin, J. R., 3rd., Kahn, C. R., Archer, J. A., Lesniak, M., Hendricks, C., Neville, D. M., Jr., and Roth, J.: Application of radioreceptor assay to circulating insulin, growth hormone, and to their tissue receptors in animals and man. *Pharmacol. Rev., 25*:179, 1973.

Gorski, J., and Gannon, F.: Current models of steroid hormone action: a critique. *Ann. Rev. Physiol., 38*:425, 1976.

Havel, R. J.: Lipid transport and the availability of insulin. *Horm. Metab. Res., Suppl. 4*:51, 1974.

Leung, K., and Munck, A.: Peripheral actions of glucocorticoids. *Ann. Rev. Physiol., 37*:245, 1975.

Metabolic effects of glucocorticoids in man. *Nutr. Rev., 32*:301, 1974.

O'Malley, B. W., and Schrader, W. T.: The receptors of steroid hormones. *Sci. Amer., 234(2)*:32, 1976.

Pilkis, S. J., and Park, C. R.: Mechanism of action of insulin. *Ann. Rev. Pharmacol., 14*:365, 1974.

Sherwin, R., and Felig, P.: Glucagon physiology in health and disease. *Intern. Rev. Physiol., 16*:151, 1977.

Unger, R. H., and Orci, L.: Physiology and pathophysiology of glucagon. *Physiol. Rev., 56*:788, 1976.

Unger, R. H., Dobbs, R. E., and Orci, L.: Insulin, glucagon, and somatostatin secretion in the regulation of metabolism. *Ann. Rev. Physiol., 40*:307, 1978.

QUESTIONS

1. Review the effects of aldosterone on electrolyte and water metabolism.
2. Discuss the important factors that regulate aldosterone secretion.
3. What are the metabolic effects of cortisol on carbohydrates, fats, and proteins?

4. What is the relationship of cortisol to function of the lysosomes?
5. What role does "stress" play in the regulation of cortisol secretion?
6. Give the effects of insulin on glucose metabolism.
7. Explain the effects of insulin on protein metabolism and on growth.
8. What role does insulin play in the control of blood glucose concentration?
9. What are the effects of the disease diabetes on cellular metabolism, and why do ketosis and diabetic coma frequently occur?
10. How do the functions of glucagon differ from those of insulin?

CALCIUM METABOLISM, BONE, PARATHYROID HORMONE, AND PHYSIOLOGY OF TEETH

CALCIUM METABOLISM

The adult human body contains about 1200 grams of calcium, at least 99 per cent of which is deposited in the bones, but a very small and extremely important portion of the calcium is dissolved in the blood plasma and interstitial fluid. Approximately one-half of that in the plasma is ionized, and the remaining half is bound with the plasma proteins. It is the ionized calcium that diffuses into the interstitial fluid and enters into chemical reactions.

Functions of Calcium Ions. EFFECT ON CELL MEMBRANES AND THE NERVOUS SYSTEM. One of the principal functions of calcium ions is their effect on the cell membrane. In unicellular animals, calcium decreases the permeability and increases the strength of the membrane, and without calcium the membrane becomes very friable (easily ruptured).

In the human being this basic effect of calcium on the membrane is not so obvious, but some of the secondary effects of its action on the cellular membrane are very important. For instance, a *decreaee* in calcium ion concentration to only one-half the normal level causes the membranes of nerve fibers to become very leaky to sodium ions and, therefore, to become partially depolarized and to transmit repetitive and uncontrolled impulses. These may occur so rapidly that they actually cause spasm of the skeletal muscles, a condition called *tetany*.

On the other hand, a great increase in the concentration of calcium ions depresses the neuronal activity, especially in the central nervous system. This presumably occurs because the membranes will not depolarize with normal ease.

EFFECT OF CALCIUM IONS ON THE HEART. A second effect of decreased calcium ion concentration is weakness of the heart muscle. Decreased calcium causes the duration of cardiac systole to decrease, and the heart dilates excessively during diastole. Usually, however, the person dies of tetany before his heart function is greatly impaired by calcium deficiency. An excess of calcium promotes overcontraction of the heart, causing the muscle to contract much too forcefully during systole and not to relax satisfactorily during diastole. Fortunately, this effect of excess calcium can be demonstrated only in experiments, for the calcium concentration never rises to the level required to cause this effect even in disease conditions.

These effects of high and low calcium concentrations on the heart can be explained by the basic mechanism of muscle contraction, which was discussed in Chapter 10. When the cardiac impulse passes over the cardiac muscle, a small amount of calcium ions is released into the muscle fibers, part of the ions coming from the longitudinal tubules and part through the cell membrane from the extracellular fluid. And it is these calcium ions that cause the contractile process. When small amounts of calcium are available, the intensity of contraction is reduced, whereas excess calcium in the extracellular fluids causes overcontraction of the heart.

EFFECT OF CALCIUM IONS ON BLOOD COAGULATION. Another important function of calcium ions is to promote blood coagulation. It will be recalled from the discussion of blood coagulation in Chapter 7 that calcium enters into most of the chemical reactions of the clotting process. Fortunately, though, the calcium ion concentration only rarely falls low enough or rises high enough to cause serious abnormalities of clotting.

REACTION OF CALCIUM IONS WITH PHOSPHATE IONS. Finally, an extremely important function of calcium ions is to react with phosphate ions to form bone salts, a function that is discussed in detail in subsequent parts of this chapter.

REACTION OF CALCIUM IONS WITH PHOSPHATE IONS

Calcium (Ca^{++}) and phosphate $HPO_4^{--})$ ions react together to form *calcium phosphate* $(CaHPO_4)$, a relatively insoluble compound. The mathematical product of the concentrations of calcium and phosphate ions in a solution can never be greater than a critical value called the *solubility product,* or they will precipitate in the form of calcium phosphate crystals. Therefore, the greater the concentration of calcium in the solution, the less can be the concentration of phosphate; or the greater the concentration of phosphate, the less can be the concentration of calcium.

Therefore, calcium and phosphate are inextricably related to each other in the deposition and reabsorption of bone, because the salts of bone are mainly calcium phosphate compounds. Every time calcium is deposited, phosphate is deposited also; and every time bone is reabsorbed, both calcium and phosphate are absorbed into the body fluids at the same time.

Figure 36–1 illustrates the absorption, utilization, and excretion of calcium and phosphate. Note that both calcium and phosphate are absorbed from the intestines and then are excreted mainly by way of the kidneys. Vitamin D plays an especially significant role in promoting calcium absorption from the intestines, and parathyroid hormone is important both in controlling the rate of calcium absorption from the intestines and in controlling the rate of calcium excretion by the kidneys. And another potent effect of parathyroid hormone is to regulate the absorption of calcium from bones; this determines the rate of turnover of calcium between the blood and the bones.

In general, phosphate absorption, excretion, and bone turnover follow essentially the same pathways as those for calcium. However, it is the calcium ion concentration that is controlled by vitamin D and parathyroid hormone, while the metabolism of phosphate is regulated mainly secondarily to what happens to the calcium ions, as we shall see in succeeding paragraphs.

FUNCTION OF VITAMIN D IN CALCIUM AND PHOSPHATE ABSORPTION

In the absence of vitamin D, almost no calcium is absorbed from the intestinal tract, but in the presence of this vitamin calcium

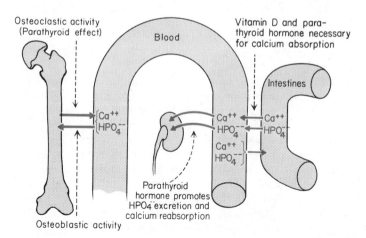

Figure 36–1. Absorption, utilization, and excretion of calcium and phosphate.

absorption can be very great. And, when the rate of calcium absorption is increased, so also is the rate of phosphate absorption, usually. The reason for this is that movement of calcium out of the intestines releases large amounts of phosphates from insoluble or un-ionized calcium phosphate compounds of the food. Since phosphate is highly absorbable by the intestinal mucosa, it simply follows along when the calcium is absorbed.

Source of Vitamin D.   There are several different vitamin D compounds, but the most important of these is the substance *cholecalciferol.* This is usually formed in the skin of animals by ultraviolet irradiation of 7-dehydrocholesterol, a fatty substance in the skin. Consequently, a person sufficiently exposed to sunlight needs no vitamin D in his diet. If he does not receive sufficient exposure to sunlight, then generally vitamin D must be provided in his food. Usually only foods of animal origin contain vitamin D. A cow exposed to sunlight forms vitamin D continually in its skin, and some of it is secreted into the milk. Also, most animals store vitamin D in great quantities in the liver; therefore, liver is usually an excellent source of this vitamin. An artificial source of vitamin D is irradiated milk, for milk contains steroids which can be converted into vitamin D by irradiation with ultraviolet light.

Basic Mechanism of Function of Vitamin D.   Vitamin D itself has almost no direct effect on increasing calcium absorption from the intestines. Instead, it must first be converted into an active product, *1,25-dihydroxycholecalciferol.* It is this compound that directly influences the intestinal epithelium to promote calcium absorption. The conversion of vitamin D to 1,25-dihydroxycholecalciferol occurs in the following steps, which are also illustrated in Figure 36–2.

1. The liver first converts the vitamin D *(cholecalciferol)* into *25-hydroxycholecalciferol.* As the quantity of this substance builds up in the liver, it inhibits further conversion of vitamin D, thus providing a feedback regulation of the amount of 25-hydroxycholecalciferol available at any one time.

2. The kidney converts the 25-hydroxycholecalciferol into 1,25-dihydroxycholecalciferol, which is the final active product. However, for this conversion to take place, *parathyroid hormone* is required. This is one major mechanism, perhaps the most important, by which parathyroid hormone exerts its effects on calcium metabolism, as we shall see in greater detail later in the chapter.

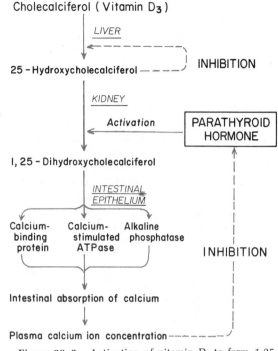

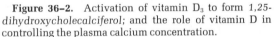

Figure 36–2.   Activation of vitamin D$_3$ to form *1,25-dihydroxycholecalciferol;* and the role of vitamin D in controlling the plasma calcium concentration.

MECHANISM BY WHICH 1,25-DIHYDROXY-CHOLECALCIFEROL INCREASES CALCIUM ABSORPTION.   Figure 36–2 illustrates three different mechanisms by which 1,25-dihydroxycholecalciferol enhances the absorption of calcium from the gut. Probably the most important of these is the effect of this "hormone" to increase the quantity of a special protein in the intestinal epithelium, called *calcium-binding protein,* that combines with calcium and causes it to be transported through the epithelium. Though the precise mechanism of calcium transport is yet unknown, it is abundantly clear that the role played by 1,25-dihydroxycholecalciferol is an essential part of the transport mechanism.

It should also be noted in Figure 36–2 that once the plasma calcium ion concentration becomes too high, the rate of formation of parathyroid hormone is inhibited, which then decreases the rate of activation of vitamin D, thereby shutting off the absorption of calcium from the intestines. This is one of the important mechanisms for control of the plasma calcium ion concentration, a subject that will be discussed in much greater detail after we have considered the relationship of bone to calcium metabolism.

BONE AND ITS FORMATION

Figure 36–3 illustrates the general structure of a long bone, showing an articular surface at each end where it is jointed with the other bones, and a hollow shaft that is excellent for resisting mechanical stresses. To the right in the figure is shown a greatly magnified microscopic cross-section of bone; the white portion of the figure represents deposits of bone salts while the black areas represent spaces that contain blood vessels and tissue fluids. Bone, like other tissue, is continually supplied with an adequate flow of nutrients in the blood.

Chemical Composition of Bone. Bone is composed of a strong protein *matrix* having a consistency almost like that of leather and of *salts* deposited in this matrix to make it hard and nonbendable. By far the major portion of the salts of bone have the following approximate chemical composition:

$$[Ca_3(PO_4)_2]_3 \cdot Ca(OH)_2$$

This is the chemical formula for *hydroxyapatite,* a hard marblelike compound. Small amounts of calcium carbonate ($CaCO_3$) are also present but probably are not of primary importance. The protein matrix prevents the bone from breaking when tension is applied, and the salts prevent the bone from crushing when pressure is applied. The matrix, therefore, is analogous to steel in reinforced concrete structures, and the salts are analogous to the concrete itself.

DEPOSITION OF BONE

Figure 36–4 illustrates both bone deposition and bone absorption. Note on the upper surface of the bone and also in some of the cavities a type of cell called *osteoblasts*. These are the cells that deposit bone. Bone deposition occurs in two stages as follows:

First, the osteoblasts secrete a protein substance that polymerizes to become very strong *collagen fibers*. These constitute by far the major portion of the matrix for the new bone.

Second, after the protein matrix has been deposited, calcium salts precipitate in the matrix, making it the hard structure that we know to be bone. Deposition of the salts requires (1) combination of calcium and phosphate to form calcium phosphate, $CaHPO_4$, and (2) slow conversion of this compound into hydroxyapatite over a period of several more weeks or months. However, the concentrations of calcium and phosphate in the extracellular fluids are normally not sufficient to cause automatic precipitation of calcium phosphate crystals. It is believed that the newly formed collagen fibers of the bone matrix have a special affinity for calcium phosphate, causing calcium phosphate crystals to deposit even though the mathematical product of calcium ion and phosphate ion concen-

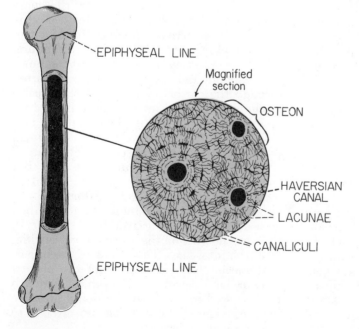

Figure 36–3. Structure of bone.

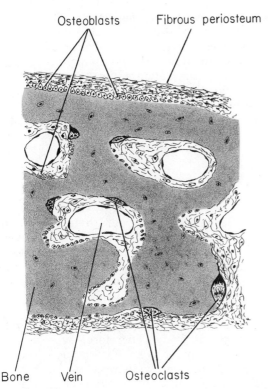

Figure 36-4. Osteoblastic deposition of bone, and osteoclastic reabsorption of bone.

trations in the surrounding fluids is not as great as the solubility product.

Regulation of Bone Deposition. Deposition of bone is regulated partially by the amount of strain applied to bone. That is, the greater the weight applied and the greater the bending of the bone the more active are the osteoblasts, the reason for which is yet unknown. Nevertheless, a bone subjected to continuous and excessive loads usually grows thick and strong, while bones not used at all, such as the bones of a leg in a plaster cast, waste away.

Another important cause of bone production is a break in a bone. Injured osteoblasts at the site of the break become extremely active and proliferate in all directions, secreting large quantities of protein matrix to cause deposition of new bone. As a result, the break is normally repaired within a few weeks.

REABSORPTION OF BONE

Figure 36-4 also shows, in addition to osteoblasts, several giant cells, called *osteoclasts*, each of which contains many nuclei. These cells are present in almost all the cavities of the bone, and have the ability to cause bone reabsorption. They do this probably by secreting enzymes or other substances, or both, that digest the protein matrix and that also help to dissolve the bone salts so that they will be absorbed into the surrounding fluid. Thus, as a result of osteoclastic activity, both calcium and phosphate are released into the extracellular fluid, while the bone is literally eaten away.

Balance Between Osteoclastic Reabsorption and Osteoblastic Deposition of Bone. Osteoclastic reabsorption of bone occurs all the time in multiple small cavities throughout the bone as shown in Figure 36-4, but this is offset by continued osteoblastic deposition of new bone. Indeed, new bone is usually being deposited on the opposite side of the same cavity. The strength of the bone depends on the relative rates of the two processes. If the rate of osteoblastic activity is greater than that of osteoclastic activity, the bone will be increasing in strength. This occurs in athletes and in others who subject their bones to excessive strain. On the other hand, osteoclastic activity is usually greater than osteoblastic activity when the bones are out of use, thereby causing the bones to become weakened.

One might wonder why bone is continually reabsorbed and new bone deposited in its place. However, in those persons in whom this does not occur, as is true of a few disease conditions, the bone becomes very brittle and then breaks easily. Therefore, it is believed that this continual turnover of bone keeps the tensile strength of the matrix strong. That is, new and strong collagen fibers take the place of old and weakened fibers. It is especially evident that the bones become more and more brittle in older people, in whom this process of bone reabsorption and redeposition occurs progressively more slowly with advancing age.

An interesting effect of continual absorption and redeposition of bone is the tendency for crooked bones to become straight over a period of years. Compression of the inner curvature of a bent bone seems to promote rapid deposition of new bone, while stretching of the outer curvature seems to promote reabsorption. As a result, the inside of the curvature becomes filled with new bone while the outside is absorbed. In a child a broken bone may initially heal with many degrees of angulation, and yet will become essentially straight within a few years.

PARATHYROID HORMONE AND ITS REGULATION OF CALCIUM METABOLISM

Secretion by the Parathyroid Glands. Parathyroid hormone, also called *parathormone,* is a small protein secreted by the parathyroid glands. It causes release of calcium salts from bones as well as increased calcium absorption from the intestine and kidney tubules. And its primary function is to regulate the concentration of ionic calcium in the extracellular fluids.

Normally there are four separate very minute parathyroid glands lying respectively behind the four poles of the thyroid gland. Each of these glands weighs only 0.02 gram. A histologic cross-section of a parathyroid gland is shown in Figure 36–5, illustrating two types of cells in the gland, *chief cells* and *oxyphil cells.* The parathyroid glands of some animals contain only chief cells, for which reason it is assumed that the chief cells are the ones that secrete parathyroid hormone. The oxyphil cells may be the same as the chief cells but in a different state of activity.

EFFECT OF PARATHYROID HORMONE ON INCREASING THE EXTRACELLULAR FLUID CONCENTRATION OF CALCIUM IONS

Parathyroid hormone increases the extracellular fluid calcium ion concentration by two different basic mechanisms, one of which is a very powerful and rapidly acting mechanism while the other is much more important for long-term regulation.

The short-term mechanism is based on the ability of parathyroid hormone to cause bone

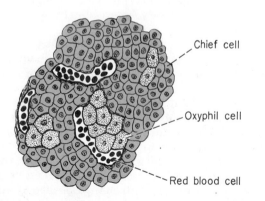

Figure 36–5. Histologic structure of a parathyroid gland.

Chief cell

Oxyphil cell

Red blood cell

absorption while the long-term mechanism is based on the ability of parathyroid hormone to increase absorption of calcium both from the intestine and from the kidney tubules.

Increase in Calcium Ion Concentration When Parathyroid Hormone Causes Bone Absorption. Parathyroid hormone has a very potent effect of activating the osteoclastic cells in bone. These cells, which literally eat their way into bone, become extremely active within minutes after the administration of parathyroid hormone. Both their number and size increase, and histologic study shows rapid invasion of the bone by the osteoclasts. When large amounts of parathyroid hormone are secreted in the body over a period of several months, the osteoclasts become so plentiful and large that they can actually eat entirely through a bone and cause it to fracture even with normal usage.

Bones contain about 800 times as much calcium as that found in all of the extracellular fluids of the body. Therefore, even a very slight amount of bone absorption causes a very large percentage increase in the calcium ion concentration of the extracellular fluid. For this reason, the calcium ion concentration of the blood can be greatly elevated within hours after injecting parathyroid hormone. Such a rise in calcium concentration is illustrated by the upper curve in Figure 36–6, showing a maximum increase in the calcium concentration 8 hours after parathyroid hormone injection.

Therefore, if ever the calcium ion concentration of the blood falls too low, greater secretion of parathyroid hormone can correct this within a few hours by increasing the rate of bone resorption.

Increase in Calcium Ion Concentration Caused by Parathyroid Hormone Stimulation of Intestinal and Kidney Tubular Absorption of Calcium Ions. Earlier in the chapter it was pointed out that parathyroid hormone is one of the factors that activates vitamin D. The active form of vitamin D then promotes rapid absorption of calcium from the intestinal tract. A similar effect is believed to occur in the kidney tubules as well. Therefore, increased quantities of parathyroid hormone enhance the input of calcium ions to the extracellular fluid from the intestines while at the same time decreasing excretion of calcium by the kidneys. However, the usual daily turnover of calcium by the intestines and the kidneys is only about 1 gram. Therefore, this is a very slow mechanism. Yet, it is the mech-

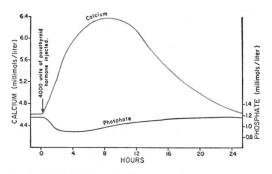

Figure 36–6.   Effect on plasma calcium and phosphate concentrations of injecting 4000 units of parathyroid hormone into a human being.

anism that determines the long-term fluctuating balance between intake and output of calcium. And, for this reason, it is also the mechanism that in the long run determines the total amount of calcium in the body, both in the bones and in the extracellular fluids.

Effect of Parathyroid Hormone on Extracellular Fluid Phosphate Concentration.   Parathyroid hormone has a special effect on phosphate metabolism that it does not have on calcium metabolism: It causes *increased* kidney excretion of phosphate ions. Therefore, at the same time that parathyroid hormone increases the concentration of calcium ions in the body fluids, it tends to *decrease* the phosphate concentration a small amount. This effect is illustrated in Figure 36–6, which shows that the phosphate ion concentration decreases by about 20 per cent in the extracellular fluids within 2 to 4 hours after injection of a large dose of parathyroid hormone.

This decrease in phosphate ion concentration reduces the likelihood that calcium will precipitate with phosphate at unwanted places in the body, an effect that does occur in the blood, muscle, tendons, and elsewhere when both the calcium and phosphate concentrations rise too high at the same time.

REGULATION OF PARATHYROID SECRETION, AND FEEDBACK CONTROL OF CALCIUM ION CONCENTRATION

Almost any factor that decreases the calcium ion concentration of the extracellular fluid will cause the parathyroid glands to secrete increased quantities of parathyroid hormone. And if the decrease in calcium ion concentration continues for a long period of time,

the parathyroid glands will often enlarge as much as tenfold, thereby multiplying the formation of parathyroid hormone many times.

Thus, the parathyroid hormone system plays an essential role in both hour-by-hour and year-by-year control of calcium ion concentration in the extracellular fluid. That is, when the ionic level of calcium falls too low, parathyroid hormone is secreted and calcium ions are absorbed from the bone, the intestines, and the kidneys to increase calcium ion concentration level in the extracellular fluid. In this way, a relatively constant concentration of calcium ions is maintained in the extracellular fluids at all times.

The Value of Parathyroid Hormone to the Body.   If the calcium ion concentration falls more than 50 per cent below normal, the person develops immediate tetany and dies because of spasm of his respiratory muscles. On the other hand, if the calcium concentration becomes too great, he is likely to have rather severe mental or cardiac disturbances. Therefore, it is necessary that the calcium ion concentration remain almost exactly constant all the time. Of course, the bones have almost a thousand times the calcium in all the extracellular fluids, so that the slight amount of calcium mobilized from the bones from day to day usually is not missed. The main value of parathyroid hormone, then, is to regulate calcium ion concentration in the extracellular fluid, so that the other functions of this ion besides that of bone deposition can continue normally all the time. The bone, on the other hand, can wait several months if necessary until adequate calcium is available to replenish its lost stores.

ABNORMALITIES OF PARATHYROID HORMONE SECRETION

Hypersecretion of Parathyroid Hormone.   Occasionally a parathyroid tumor develops in one of the glands, and extreme secretion of parathyroid hormone causes tremendous overgrowth of the osteoclasts. Sometimes these cells grow so large and so numerous that they cause large honeycomb cavities in the bone, and even combine to form large masses that resemble tumors. The result is often bones so weakened that even walking on a leg can cause it to break. Indeed, most hyperparathyroid persons first become aware of their disease through a broken bone.

Hypersecretion also increases the calcium

ion concentration in the body fluids, but if the hypersecretion is not too great, the level of calcium ion will not rise enough to cause untoward effects. Rarely, though, the level becomes great enough to cause precipitation of calcium phosphate in tissues other than the bone, such as the lungs, muscles, and heart, sometimes leading to death within a few days.

Hyposecretion of Parathyroid Hormone. The most common cause of deficient parathyroid secretion is surgical removal of all the parathyroid glands. This usually occurs inadvertently when a surgeon is removing the thyroid gland, because of the close proximity of the parathyroid glands to this other gland.

Loss of parathyroid secretion allows the calcium ion concentration to fall so low that tetany occurs within about 3 days, and unless the person is treated he dies almost immediately. However, sufficient calcium to restore normal function can be mobilized from the bone within a few hours by administration of parathyroid hormone or certain of the vitamin D compounds that, in extremely large doses, can cause calcium absorption from the bones.

A rare person has hereditary hyposecretion of parathyroid hormone. Usually the degree of hyposecretion is not sufficient to cause tetany, but it may be sufficient to cause chronically depressed osteoclastic activity in the bones. This results in brittle bones, probably because without osteoclastic activity the constant absorption and reformation of new bone ceases, and the protein matrix becomes old and brittle and is not replaced as often as needed to maintain adequate bone strength.

RICKETS

Rickets is a disease caused by prolonged calcium deficiency. Lack of dietary calcium for a short time will never cause rickets, for when the calcium ion concentration in the extracellular fluid falls below normal, large quantities of parathyroid hormone are secreted and calcium is automatically absorbed from the bones, thereby reestablishing an adequate calcium ion concentration. However, if insufficient calcium is absorbed from the gut for many months, then all or most of the calcium in the bones will have been absorbed, and little more is available. The calcium ion concentration of the extracellular fluids then

falls to very low values. At this point the person has fully developed rickets, the two major effects of which are (1) depletion of calcium salts from the bones and, therefore, severe weakening of the bones, often culminating in fractures or deformed bones, and (2) tetany caused by the diminished extracellular fluid calcium ion concentration.

The most frequent cause of rickets is a deficiency of vitamin D rather than lack of calcium in the diet. The usual child living in temperate climates receives far too little sunlight during the winter months to provide the amount of vitamin D needed for absorption of calcium from the gut. Fortunately, the stores of vitamin D in the child's liver are usually sufficient to provide adequate calcium absorption for the early months of winter. Therefore, rickets usually develops in the early spring, before the child has been exposed to the sun again.

CALCITONIN — A CALCIUM-DEPRESSING HORMONE

A recently discovered hormone called *calcitonin* causes exactly the opposite effect on blood calcium ion concentration to that caused by parathyroid hormone, that is, decrease in calcium ion concentration rather than increase. Calcitonin is secreted by the thyroid gland and therefore is sometimes called "thyrocalcitonin." Calcitonin decreases calcium ion concentration in three distinct ways:

1. Within minutes it greatly suppresses the activity of the osteoclasts and therefore reduces the rate of release of calcium from the bone into the blood.

2. It increases osteoblastic activity within about 1 hour, and this causes greater deposition of calcium in the bone and therefore removes calcium from the extracellular fluid.

3. It causes a prolonged reduction in the rate of formation of new osteoclasts.

The action of calcitonin is different from that of parathyroid hormone in a very important respect: It begins to act almost immediately, whereas the effect of parathyroid hormone is hardly observable for several hours after its injection.

Secretion of calcitonin is greatly enhanced when blood calcium concentration rises above normal. The calcitonin in turn causes some of the blood calcium to deposit in the bones, thereby returning the calcium ion concentration toward normal. Obviously, therefore, the calcitonin mechanism is a rapidly acting feedback mechanism, even more rapid than that of the parathyroid mechanism, that helps to stabilize calcium ion concentration in the extracellular fluids. However, it has far less effect quantitatively than does the parathyroid mechanism, and its long-term effect is even less significant, so that for practical purposes one can consider the long-term regulation of calcium ion concentration to be almost entirely a parathyroid function.

PHYSIOLOGY OF TEETH

The teeth cut, grind, and mix food. To perform these functions, the jaws have extremely powerful muscles capable of providing an occlusive force of as much as 50 to 100 pounds between the front teeth and as much as 150 to 200 pounds between the jaw teeth. Also, the upper and lower teeth are provided with projections and facets that interdigitate so that each set of teeth fits with the other. This fitting is called *occlusion,* and it allows even small particles of food to be caught and ground between the tooth surfaces.

FUNCTION OF THE DIFFERENT PARTS OF TEETH

Figure 36–7 illustrates a lengthwise section of a tooth, showing its major functional parts, *enamel, dentine, cementum,* and *pulp.* The tooth can also be divided into the *crown,* which is the portion that protrudes above the gum into the mouth, and the *root,* which is the portion in the bony socket of the jaw. The collar between the crown and the root, where the tooth is surrounded by the gum, is called the *neck.*

Dentine. The main body of the tooth is composed of dentine, which has a very strong bony structure. Dentine is made up principally of calcium salts of phosphate and carbonate (hydroxyapatite crystals) embedded in a strong meshwork of *collagen fibers.* In other words, the principal constituents of dentine are very much the same as those of bone. The major difference is its structure, for dentine does not contain any osteoblasts, osteoclasts, or spaces for blood vessels or nerves. Instead, it is deposited and nourished by a layer of cells called *odontoblasts,* which line the inner surface of the pulp cavity.

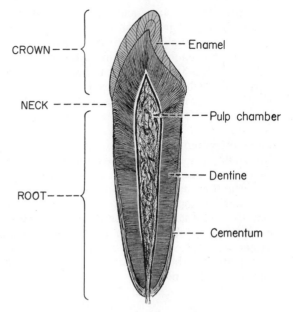

Figure 36–7. Functional parts of a tooth.

The calcium salts in dentine make it extremely resistant to compressional forces, while the collagen fibers make it tough and resistant to tensional forces that might result when the teeth are struck by solid objects.

Enamel. The outer surface of the tooth is covered by a layer of enamel that is formed prior to eruption of the tooth by special epithelial cells called *ameloblasts.* Once the tooth has erupted, no more enamel is formed. Enamel is composed of very small crystals of hydroxyapatite embedded in a fine meshwork of extremely strong fibers composed of a protein similar to the keratin of hair. The smallness of the crystalline structure of the calcium compounds makes the enamel extremely hard, much harder than dentine. Also, the protein meshwork makes enamel very resistant to acids, enzymes, and other corrosive agents, because this protein of enamel is one of the most insoluble and resistant proteins known.

Cementum. Cementum is a bony substance secreted by the *periodontal membrane* that lines the tooth socket. It forms a thin layer between the tooth and the inner surface of the socket. Many collagen fibers pass directly from the bone of the jaw, through the periodontal membrane, and into the cementum. It is these collagen fibers and the cementum that hold the tooth in place. When the teeth are exposed to excessive strain, the layer of cementum increases in thickness and strength. It also increases in thickness and strength with age, causing the teeth to become progressively more firmly seated in the jaws as one reaches adulthood and beyond.

Pulp. The inside of each tooth is filled with pulp, which is composed of connective tissue with an abundant supply of nerves, blood vessels, and lymphatics. The cells lining the surface of the pulp cavity are the odontoblasts, which, during the formative years of the tooth, lay down the dentine but at the same time encroach more and more on the pulp cavity, making it smaller. In later life the dentine stops growing and the pulp cavity remains essentially constant in size. However, the odontoblasts are still viable and send projections into small *dentinal tubules* that penetrate all the way through the dentine; these are of importance for providing nutrition.

DENTITION

Human beings and most other mammals develop two sets of teeth during a lifetime. The first teeth are called the *deciduous teeth* or *milk teeth,* and they number 20 in the human being. These erupt between the seventh month and second year of life and last until the sixth to the thirteenth year. After each deciduous tooth is lost, a permanent tooth replaces it, and an additional 8 to 12 molars appear posteriorly in the jaw, making the total number of permanent teeth 28 to 32, depending on whether the person finally grows his 4 *wisdom teeth,* which do not appear in everyone.

Formation of Teeth. Figure 36–8 shows the formation and eruption of teeth. Figure 36–8A shows protrusion of the oral epithelium into the *dental lamina,* this to be followed by the development of a tooth-producing organ. The upper epithelial cells form ameloblasts which secrete the enamel on the outside of the tooth. The lower epithelial cells grow upward to form a pulp cavity and also to form the odontoblasts that secrete dentine. Thus, enamel is formed on the outside of the tooth, and dentine is formed on the inside, developing an early tooth as illustrated in Figure 36–8B.

Eruption of Teeth. During early childhood, the teeth begin to protrude upward from the jawbone through the oral epithelium into the mouth, as in Figure 36–8C. The cause of eruption is unknown, though several theories have been offered. One of these assumes that an increase of the material inside the pulp cavity of the tooth causes much of it to be extruded downward through the root canal, and that this pushes the tooth upward. However, a more likely theory is that the bone underneath the tooth hypertrophies progressively, and in so doing shoves the tooth forward.

Development of Permanent Teeth. During embryonic life a tooth-forming organ also develops in the dental lamina for each permanent tooth that will be needed after the deciduous teeth are gone. These tooth-producing organs slowly form the permanent teeth throughout the first 6 to 20 years of life. When each permanent tooth becomes fully formed, it, like the deciduous tooth, pushes upward through the bone of the jaw. In so doing, it erodes the root of the deciduous tooth and eventually causes it to loosen and fall out.

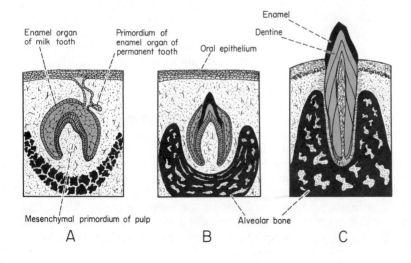

Figure 36–8. (A) Primordial tooth organ. (B) The developing tooth. (C) The erupting tooth. (Modified from Bloom and Fawcett: A Textbook of Histology, 8th ed. W. B. Saunders Co.)

Soon thereafter the permanent tooth erupts to take the place of the original one.

Effect of Metabolic Factors on Tooth Development. The rate of development and the speed of eruption of teeth can be accelerated by both thyroid and growth hormones. Also, the deposition of salts in the early forming teeth is affected considerably by various factors of metabolism, such as the availability of calcium and phosphate in the diet, the amount of vitamin D present, and the rate of parathyroid hormone secretion. When all these factors are normal, the dentine and enamel will be healthy. When these factors are deficient, the calcification of the teeth also may be defective, so that the teeth will be abnormal throughout life.

MINERAL EXCHANGE IN TEETH

The salts of teeth, like those of bone, are composed of calcium phosphate and calcium carbonate salts bound together in a single very hard crystalline substance. New calcium salts are constantly being deposited, while old salts are being reabsorbed from the teeth, as also occurs in bone. However, experiments indicate that this deposition and reabsorption occurs in the dentine and cementum, while very little occurs in the enamel.

The mechanism by which minerals are deposited in and reabsorbed from the dentine is probably the following: The small processes of the odontoblasts protruding into the tubules of the dentine are believed to be capable of absorbing salts and then providing new salts to take the place of the old.

DENTAL ABNORMALITIES

The two most common dental abnormalities are *malocclusion* and *caries*. Malocclusion means failure of the facets of the upper and lower teeth to interdigitate properly, while caries means erosions of the teeth.

Malocclusion. Malocclusion is usually the result of a hereditary abnormality that causes the teeth of one jaw to grow in an abnormal direction. When malocclusion occurs, the teeth cannot perform adequately their normal grinding or cutting action. Occasionally malocclusion causes abnormal displacement of the lower jaw in relation to the upper jaw, causing such undesirable effects as pain in the mandibular joint or deterioration of the teeth.

The orthodontist can often correct malocclusion by applying long-term gentle pressure against the teeth with appropriate braces. The gentle pressure causes absorption of alveolar jaw bone on the compressed side of the tooth, and deposition of new bone on the tensional side of the tooth. In this way the tooth gradually moves to a new position as directed by the applied pressure.

Caries. It is generally agreed by all research investigators of dental caries that caries result from the action of bacteria on the teeth. The first event in the development of caries is the deposit of *plaque,* a film of precipitated products of saliva and food, on the teeth. Large numbers of bacteria inhabit this plaque and are readily available to cause caries. However, these bacteria depend to a great extent on carbohydrates for their food. When carbohydrates are available, their metabolic systems are strongly activated and they also multiply. In addition, they form acids, particularly lactic acid, and proteolytic enzymes. The enzymes presumably digest the protein matrix of the enamel and dentine while the acid causes absorption of the calcium salts.

Because of the dependence of the caries bacteria on carbohydrates, it has frequently been taught that eating a diet high in carbohydrate content will lead to excessive development of caries. However, it is not the quantity of carbohydrate ingested but instead the frequency with which it is eaten that is important. If eaten in many small parcels throughout the day, such as in the form of candy, the bacteria are supplied with their preferential metabolic substrate for many hours of the day and the development of caries is extreme. If eaten in large amounts only at mealtimes, the extensiveness of the caries is greatly reduced.

Some teeth are more resistant to caries than others. Studies show that teeth formed in children who drink water containing small amounts of fluorine develop enamel that is more resistant to caries than the enamel in children who drink water not containing fluorine. Fluorine does not make the enamel harder than usual, but instead it is said to inactivate proteolytic enzymes before they digest the protein matrix of the enamel. Regardless of the precise means by which fluorine protects the teeth, it is known that small amounts of fluorine deposited in enamel make teeth about three times as resistant to caries as are teeth without fluorine.

REFERENCES

Borle, A. B.: Calcium and phosphate metabolism. *Ann. Rev. Physiol., 36*:361, 1974.

David, D. S.: Calcium metabolism in renal failure. *Am. J. Med., 58*:48, 1975.

DeLuca, H. F.: The kidney as an endocrine organ involved in the function of vitamin D. *Am. J. Med., 58*:39, 1975.

Gray, T. K., Cooper, C. W., and Munson, P. L.: Parathyroid hormone, thyrocalcitonin, and the control of mineral metabolism. *In* MTP International Review of Science: Physiology. Vol. 5. Baltimore, University Park Press, 1974, p. 239.

Hancox, N. M.: Biology of Bone. Cambridge. Eng., Cambridge University Press, 1972.

Harrison, H. E., and Harrison, H. C.: Calcium. *Biomembranes, 48*:793, 1974.

Kreitzman, S. N.: Enzymes and dietary factors in caries. *J. Dent. Res., 53*:218, 1974.

Norman, A. W.: 1.25-Dihydroxyvitamin D_3: a kidney-produced steroid hormone essential to calcium homeostasis. *Am. J. Med., 57*:21, 1974.

Raisz, L. G., Mundy, G. R., Dietrich, J. W., and Canalis, E. M.: Hormonal regulation of mineral metabolism. *Intern. Rev. Physiol., 16*:199, 1977.

Rasmussen, H., Bordier, P., Kurokawa, K., Nagata, N., and Ogata, F.: Hormonal control of skeletal and mineral homeostasis. *Am. J. Med., 56*:751, 1974.

Seltzer, S., and Bender, I. B.: The Dental Pulp. 2nd ed. Philadelphia, J. B. Lippincott Company, 1975.

Tada, M., Yamamoto, T., and Tonomura, Y.: Molecular mechanism of active calcium transport by sarcoplasmic reticulum. *Physiol. Rev., 58*:1, 1978.

Wheeler, R. C.: Dental Anatomy, Physiology and Occlusion. 5th ed. Philadelphia, W. B. Saunders Company, 1975.

QUESTIONS

1. What are the functions of calcium in cellular metabolism and in the formation of bone?
2. How do phosphate ions interact with calcium ions?
3. Explain the role of vitamin D in calcium absorption, and give the schema for its function.
4. Review the composition of bone and its deposition.
5. What is the difference between osteoclastic and osteoblastic functions of bone cells?
6. In what ways does parathyroid hormone increase blood calcium concentration?
7. Discuss the relative roles of parathyroid hormone and calcitonin in the control of blood calcium concentration.
8. Explain the effects of hyper- and hyposecretion of parathyroid hormone.
9. What are the parts of teeth, and how do they resemble bone?
10. How does mineral exchange occur in teeth.?

SEXUAL FUNCTIONS OF THE MALE AND FEMALE, AND THE SEX HORMONES

37

The male and female play equal parts in initiating reproduction and in determining the hereditary characteristics of the baby — the male provides the *sperm* and the female the *ovum*. Combination of a single sperm with a single ovum forms a *fertilized ovum* that can grow into an *embryo,* then into a *fetus,* and eventually into a newborn baby. The sexual and endocrine functions of both the male and the female for initiating the process of reproduction are discussed in the present chapter, and reproduction itself is considered in the following chapter.

MALE SEXUAL FUNCTIONS

Figure 37–1 illustrates the male sexual organs, the most important of which are the testes, the prostate, and the penis. The functions of these are described in subsequent sections of the chapter.

THE SPERM

Characteristics of Sperm. The *testes* of the male produce billions of *spermatozoa,* also called *sperm.* Figure 37–2 shows the structure of a sperm, which itself is a single cell constructed principally of a *head* and a *tail.* The head is composed mainly of the nucleus of the cell, with only a very thin cytoplasmic and membrane layer around its surface. Most of the cytoplasm of the sperm is aggregated into the tail, and this is covered by a long extension of the cellular membrane.

The nucleus in the head of the sperm carries the genes of the male, while the tail provides motility. The tail moves back and forth *(flagellar movement),* propelling the sperm forward. Normal sperm move in a straight line at a velocity of about 1 to 4 mm. per minute. This movement allows them to move up the female genital tract in quest of the ovum.

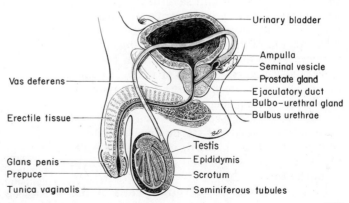

Figure 37–1. The male sexual organs. (Modified from Bloom and Fawcett: A Textbook of Histology, 10th ed. W. B. Saunders Co.)

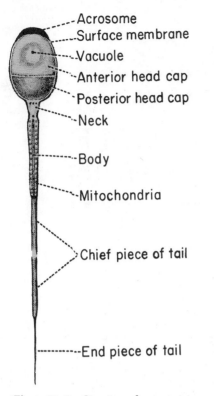

--- Acrosome
--- Surface membrane
--- Vacuole
--- Anterior head cap
--- Posterior head cap
--- Neck
--- Body
--- Mitochondria
--- Chief piece of tail
--- End piece of tail

Figure 37–2.   Structure of a sperm.

Formation   of   Sperm — Spermatogenesis.   The testes are composed of many thousands of small tubules, called *seminiferous tubules,* which form the sperm. A cross-section of a typical seminiferous tubule is illustrated in Figure 37–3A, and the formation of sperm in a small section of a tubule, which is called *spermatogenesis,* is shown in Figure 37–3B. The epithelioid cells lining the outer wall of the seminiferous tubules are called the *male germinal epithelium,* and it is from these that all sperm are formed.

The earliest form of cell in the development of spermatozoa is the *spermatogonium.* As this cell divides through several generations, the newly forming cells move toward the lumen of the tubule. The first stage in the development of the sperm is the *primary spermatocyte;* this divides again to form two *secondary spermatocytes,* each of which in turn divides to form two *spermatids.* The spermatid changes into a sperm by, first, losing some of its cytoplasm, second, reorganizing the chromatin material of its nucleus to form a compact head, and, third, collecting the remaining cytoplasm at one end of the cell to form the tail.

Located among the germinal cells of the tubule are many large cells called *sustentacular cells.* These lie adjacent to the developing spermatocytes, and the spermatocytes remain attached to them until the head and tail of the sperm are formed. It is believed that the sustentacular cells secrete substances that are needed by the developing sperm, but their exact function is not known.

DIVISION OF CHROMOSOMES DURING SPERM FORMATION.   When a primary spermatocyte divides to form two secondary spermatocytes, the cell division is not of the usual type, for reproduction of the chromosomes does not occur. Instead, each of the 23 pairs of chromosomes simply splits apart, allowing 23 unpaired chromosomes to pass into one of the two secondary spermatocytes, and the other 23 to pass into the other. In other words, only half of the genes enter each secondary spermatocyte, and, similarly, only half of the genes from the parent cell are in each respective sperm. A similar division of genes occurs in the ovum. Thus, each parent makes an equal contribution to hereditary characteristics of the child.

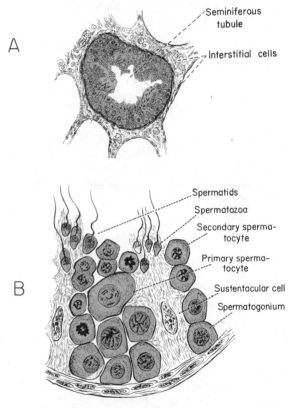

A

--- Seminiferous tubule
--- Interstitial cells

B

--- Spermatids
--- Spermatozoa
--- Secondary spermatocyte
--- Primary spermatocyte
--- Sustentacular cell
--- Spermatogonium

Figure   37–3.   (A) Cross-section of seminiferous tubule. (B) Development of spermatozoa by the germinal epithelium. (Modified from Arey: Developmental Anatomy, 7th ed. W. B. Saunders Co.)

SEX DETERMINATION BY THE SPERM. Whether a baby will be male or female is determined by the sperm, for the following reasons. One of the 23 pairs of chromosomes in all human cells is called the sex pair because these chromosomes determine whether the person will be male or female. One type of sex chromosome is known as a *Y chromosome* or a *male chromosome.* Another is known as an *X chromosome* or a *female chromosome.* A female has two X chromosomes. A male has one X chromosome and one Y chromosome. When the spermatocytes divide in the testis to form sperm, the X-Y pair of sex chromosomes splits; the Y chromosome passes into one sperm and the X chromosome passes into another. Since the unfertilized ovum contains only a single X chromosome, if a sperm containing an X chromosome combines with an ovum, the fertilized ovum then has a pair of X chromosomes. This combination leads to formation of a female. On the other hand, combination of a sperm containing a Y chromosome with an ovum creates an X-Y pattern in the fertilized ovum and causes the development of a male child.

Maturation and Storage of Sperm. Sperm are formed continually in the seminiferous tubules of the testes. After formation they pass first into the *epididymis* where they usually remain for several days. As long as the sperm are still in the seminiferous tubules they are unable to fertilize the ovum, but after remaining in the epididymis for as long as 18 hours they develop the power of motility and also become capable of fertilization, a process called *maturation.* Some of the sperm remain stored in the epididymis until the time of ejaculation, but others pass on into the vas deferens and even into the ampulla of the vas deferens where they are stored until ejaculated. In this storage state they can remain fertile for as long as 42 days.

THE MALE SEXUAL ACT

In addition to the testes, the male sexual organs include, as shown in Figure 37–1: the *vas deferens,* which conducts the sperm from the testis to the urethra; the *seminal vesicles,* which secrete a mucoid material into the upper end of the vas deferens; the *prostate gland,* which secretes a milky fluid also into the upper end of the vas deferens; the *urethra,* which conducts the sperm from the vas deferens to the exterior; and the *penis,* which pro-

vides the passageway into the female vagina. The penis in turn has two important parts for performance of the sexual act. These are the *glans,* which is the sensitive portion that elicits sexual excitement, and the *erectile tissue,* which surrounds the urethra and causes erection.

Erection and Lubrication. Stimulation of the glans of the penis by to-and-fro movement of the penis in the vagina causes sensory impulses to pass from the glans into the sacral portion of the spinal cord, and if the person is also in the appropriate psychic mood, reflex signals return through parasympathetic nerve fibers to the genital organs. These signals dilate the arteries that supply the erectile tissue of the penis and probably also constrict the veins. As a result, a tremendous amount of blood enters the erectile tissue under high pressure and blows it up like a balloon. This causes the penis to become greatly enlarged, become hard, and extend forward, which is the act of *erection.*

In addition to causing erection, the parasympathetic signals initiate mucus secretion by the *bulbourethral glands* located at the upper end of the urethra, and also by many small mucous glands along the course of the urethra. The mucus is expelled and helps to lubricate the to-and-fro movement of the penis in the vagina and, therefore allows the vagina to massage the penis during movement rather than to abrade it. The massaging effect creates the necessary sexual stimulus to cause ejaculation; an abrasive effect causes pain, which inhibits sexual desire and blocks completion of the sexual act.

Ejaculation. When the degree of sexual stimulation has reached a critical level, neuronal centers in the tip of the spinal cord send impulses through the sympathetic nerves to the male genital organs to initiate rhythmic peristalsis in the genital ducts. The peristalsis begins in the epididymis and then passes upward through the vas deferens, the seminal vesicles, the prostate gland, and the penis itself. This process, called *ejaculation,* moves sperm all the way from the epididymis out the tip of the penis into the vagina.

In addition to the sperm, mucus from the seminal vesicles and a milky serous fluid from the prostate are expelled during ejaculation. All these fluids including the sperm are called *semen,* and the total quantity at each ejaculation is usually about 3.5 ml., each milliliter containing about 120 million sperm, a total of about two-fifths billion sperm.

The milky fluid from the prostate, which is highly alkaline, neutralizes the acidic fluid from the testes, in this way immediately stimulating the sperm to action, for sperm are immobile in acidic media but very active in slightly alkaline media.

Male Sterility.   Approximately 1 male out of every 25 to 30 is sterile. The most frequent cause of this is previous infection in the male genital ducts, though occasionally the seminiferous tubules of the testes may have been partially or totally destroyed by mumps infection, typhus infection, x-ray irradiation, or nuclear radiation. Also, a few males have congenitally deficient testes that are incapable of producing normal sperm. The sperm may have two tails, two heads, or other less obvious abnormalities. These cannot fertilize an ovum, and when a large number of them occur in the ejaculate, it is usually an indication that all of the sperm are abnormal even though many appear to be completely normal.

Male sterility also can occur when the number of sperm in the ejaculate falls too low, even though all the sperm are normal. Usually a male is sterile when the number falls below about 75 million sperm in a single ejaculation. It is difficult to understand why sterility should exist in this instance, since only one sperm is required to fertilize the ovum. However, it is believed that the large number of sperm are necessary to provide enzymes or other substances that help the single fertilizing sperm to reach the ovum. The head of the sperm secretes *hyaluronidase* and several *proteinases* that probably help to disperse the granulosal cells that normally cover the surface of the ovum. This action then, theoretically, allows the sperm to attack the ovum.

HORMONAL REGULATION OF MALE SEXUAL FUNCTIONS

ROLE OF THE ANTERIOR PITUITARY GONADOTROPIC HORMONES

Puberty.   The testes of the child remain dormant until they are stimulated at the age of 10 to 14 by gonadotropic hormones from the pituitary gland. At that age, the hypothalamus begins to secrete gonadotropic hormone–releasing factors, which in turn cause the anterior pituitary gland to secrete both *follicle-stimulating hormone* and *luteinizing hormone.* These cause an upsurge of testicular

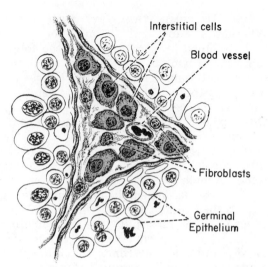

Figure 37–4.   The interstitial cells of the testis. (Modified from Bloom and Fawcell: A Textbook of Histology, 8th ed. W. B. Saunders Co.)

growth and function, causing the male sex life to begin. This stage of development is called *puberty.*

Follicle-Stimulating   Hormone.   Follicle-stimulating hormone causes proliferation of the cells in the germinal epithelium, promoting the formation of secondary spermatocytes from primary spermatocytes. Coincident with this proliferation of the germinal cells, small amounts of estrogens — female sex hormones — are produced locally in the seminiferous tubules. Therefore, it has been suggested that follicle-stimulating hormone might act primarily on the germinal epithelium to cause estrogen production, and that it may be the estrogens that make the germinal cells proliferate.

Luteinizing Hormone.   Luteinizing hormone causes the production of testosterone by the *interstitial cells,* located in the testes, as shown in Figure 37–4, in the interstices between the seminiferous tubules. Testosterone in turn is the major hormone responsible for development of male sexual characteristics.

FUNCTIONS OF TESTOSTERONE

Effect on Spermatogenesis.   Testosterone causes the testes to enlarge. Also, it must be present, along with follicle-stimulating hormone, before spermatogenesis will go to completion. Unfortunately, though, the precise action of testosterone in spermatogenesis is unknown.

Effect on Male Sex Characteristics. After a fetus begins developing inside its mother's uterus, its testes begin to secrete testosterone when it is only a few weeks old. This testosterone then helps the fetus to develop male sexual organs and male secondary characteristics. That is, it accelerates the formation of a penis, a scrotum, a prostate, the seminal vesicles, the vas deferens, and other male sexual organs. In addition, testosterone causes the testes to descend from the abdominal cavity into the scrotum; if the production of testosterone by the fetus is insufficient, the testes fail to descend, but instead remain in the abdominal cavity in the same manner that the ovaries remain in the abdominal cavity in the female.

Testosterone secretion by the fetal testes is caused by a hormone, called *chorionic gonadotropin,* that is formed in the placenta during pregnancy, as will be discussed in more detail in the following chapter. Immediately after birth of the child, loss of connection with the placenta removes this stimulatory effect, so that the testes then stop secreting testosterone. Consequently, the sexual characteristics cease developing from birth until puberty. At puberty the reinstitution of testosterone secretion causes the male sex organs to begin growing again. The testes, scrotum, and penis then enlarge about tenfold.

Effect on Secondary Sex Characteristics. In addition to the effects on the genital organs, testosterone exerts other general effects throughout the body to give the adult male his distinctive characteristics. It promotes *growth of hair* on his face, along the midline of his abdomen, on his pubis, and on his chest. On the other hand, it causes *baldness* in those male individuals who also have a hereditary predisposition to baldness. It increases the *growth of the larynx* so that the male, after puberty, develops a deeper pitch to his voice. It causes an *increase in the deposition of protein* in his muscles, bones, skin, and other parts of his body, so that the male adolescent becomes generally larger and more muscular than the female. Also, testosterone sometimes promotes abnormal secretion by the sebaceous glands of the skin, this leading to acne on the face of the postpuberal male.

FEMALE SEXUAL FUNCTIONS

Figure 37–5 shows the sexual organs of the female. The general function of these is the following: An *ovary* forms an *ovum,* which is

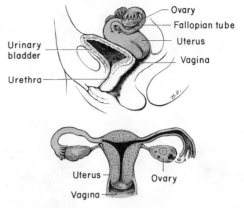

Figure 37–5. The female sexual organs.

then transported through a *fallopian tube* into the *uterus*. If a sperm fertilizes the ovum, the fertilized ovum develops in the uterus into a fetus and then into a child. The purpose of the present section is to explain the female sexual and endocrine functions that are necessary to initiate the process of reproduction.

THE OVUM

The ovaries, like the testes, have a germinal epithelium. However, this is not arranged in tubules as in the testes, but instead lies on the surface of each ovary. In the early stages of ovarian development, some of the cells from this germinal epithelium migrate into the mass of the ovary and after several divisions form large cells which become the *ova*. Approximately 400,000 ova in all are formed in the two ovaries, most of which are already present in the substance of the ovaries at birth and remain there in a dormant state until puberty. At this time, the gonadotropic hormones from the anterior pituitary gland begin promoting the release of ova one at a time, as explained below.

Each ovum in the ovary is surrounded by a layer of epithelioid cells called *granulosal cells,* and these along with the enclosed ovum are called a *primary follicle,* one of which is illustrated in the upper left-hand corner of Figure 37–6.

Growth of the Follicle and Ovulation. At puberty, when the ovaries begin to be stimulated by gonadotropic hormones from the adenohypophysis, a few of the primary follicles begin to enlarge. The granulosal cells proliferate, and adjacent stromal cells from the substance of the ovary also take on epithe-

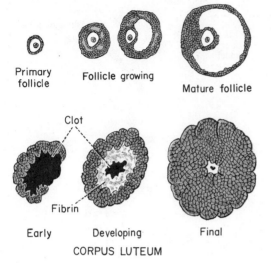

CORPUS LUTEUM

Figure 37-6. Growth of the follicle, and formation of the corpus luteum. (Modified from Arey: Developmental Anatomy, 7th ed. W. B. Saunders Co.)

lioid characteristics and join the growing follicle. The follicular cells then secrete fluid within the mass of follicular cells, creating a cavity called an *antrum* in the follicle, as also shown in Figure 37-6. The antrum grows until the follicle actually balloons outward on the surface of the ovary; several of these follicles are shown on the surface of the ovary to the right in Figure 37-5.

One of the follicles soon balloons outward more than the others and ruptures, and an ovum covered by a mass of granulosal cells is expelled into the abdominal cavity, which is the process called *ovulation.* As soon as this follicle ruptures, the remainder of the growing follicles suddenly begin to disappear without ever rupturing. Presumably the release of the follicular fluid from the ruptured follicle into the abdominal cavity and absorption of its hormones by the peritoneum in some way cause the involution of the other follicles. Regardless of the exact cause, only one ovum is usually expelled into the abdominal cavity each month, though rarely a second or even more ova are expelled before the remaining follicles start to involute. This is the major cause of multiple births.

DIVISION OF CHROMOSOMES IN THE OVUM. Figure 37-7 shows the development of the mature ovum from the germinal epithelium. After several divisions a germinal epithelial cell becomes a *primary oocyte,* which is the ovum of a primary follicle. Toward the end of growth of the follicle the nucleus of the primary oocyte divides, but without reproduc-

ing the chromosomes. Instead, each pair of chromosomes splits apart and 23 unpaired chromosomes remain in the ovum while the other 23 chromosomes are expelled in the so-called *first polar body.* The ovum now is called a *secondary oocyte,* and it has only half the genes of the mother. Then, approximately when the follicle ruptures, the secondary oocyte divides again; this time each of the unpaired chromosomes is reproduced, and half of them pass into the *second polar body* while half remain in the ovum. The ovum at this point is said to be mature, and it is now ready to be fertilized by a sperm.

Transport of the Ovum in the Fallopian Tube. Referring once again to Figure 37-5, it can be seen that the ends of the fallopian tubes lie in close proximity to the ovaries. Cilia on the epithelial surfaces of the fallopian tubes are always beating toward the uterus, causing fluids from the region of the ovary to move constantly out of the abdominal cavity into the tubes. The ovum, on being expelled into the abdominal cavity from the rupturing follicle, is thus transported into a fallopian tube along with this current of fluid flow. The lining of the fallopian tube, however, is pocked along its entire extent with large cavities that obstruct the movement of the ovum, so that usually 3 to 4 days are required for its passage to the uterus.

The ovum must be fertilized within 8 to 24 hours after it is released from the ovary or it

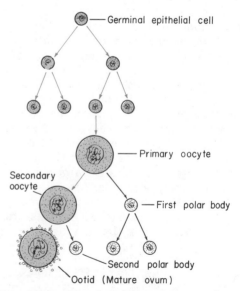

Figure 37-7. Formation of the mature ovum from the germinal epithelium. (Modified from Arey: Developmental Anatomy, 7th ed. W. B. Saunders Co.)

dies. Therefore, fertilization usually takes place in the upper portion of the fallopian tube, and the fertilized ovum begins to divide and to form the early stages of the embryo while it is still in the tube. The fallopian tube epithelium secretes substances that provide nutrition for the developing mass of cells. However, much if not most of the nutrition is supplied by the large amount of cytoplasm already in the ovum itself.

Female Sterility.   About 1 female out of 15 is sterile. As is true in the male, many of these instances of sterility are caused by previous infection. The infection sometimes blocks the fallopian tubes, or at other times involves the ovaries in a mass of scar tissue. The most common infection causing female sterility, as is true also in the male, is gonorrhea. Sterility is often caused also by an inborn failure of the ovaries to develop and expel ova into the abdominal cavity. Most frequently this results from insufficient stimulation by the anterior pituitary gonadotropic hormones. At other times it is caused by an excessively thick capsule on the surface of the ovary, or occasionally by congenitally abnormal ovaries.

THE FEMALE SEXUAL ACT

Part of the female's role in the sexual act is to help the male reach a sufficient degree of sexual stimulation for ejaculation to occur. She does this by providing appropriate conditions for maximum masculine stimulation, including especially female erection and the secretion of adequate lubricant.

Female Erection and Sexual Lubrication.   Located around the opening of the vagina are masses of erectile tissue exactly like that in the male penis. Appropriate psychic excitation in the female, plus sexual stimulation of the female genital organs, causes parasympathetic impulses to pass from the caudal spinal cord to this erectile tissue, making it swell and provide a tight though distensible opening into the vaginal canal.

In addition to erection the parasympathetic impulses cause two small glands called *Bartholin's glands,* located immediately inside the vagina on either side, to secrete large quantities of mucus. It is this mucus from the female that is mainly responsible for the lubrication that facilitates movement of the penis in the vagina during intercourse, though mucus from the male helps to a lesser extent.

The Female Climax.   Most of the sexual stimulation of the female is provided by the massaging action of the penis against the *female clitoris*. This structure is the homologue in the female of the masculine penis, and it is located at the anterior margin of the vulva immediately outside the vagina. The female, like the male, develops intense sexual stimulation, and when the degree of stimulation reaches sufficient intensity, the uterus and fallopian tubes begin rhythmic, upward-directed peristaltic contractions. This stage is called the *female climax*. The peristaltic contractions are believed to help propel sperm from the vagina into the fallopian tubes.

HORMONAL REGULATION OF FEMALE SEXUAL FUNCTIONS

RELATION OF THE ANTERIOR PITUITARY GONADOTROPIC HORMONES TO THE MONTHLY FEMALE CYCLE

The anterior pituitary gland of the female child, like that of the male child, secretes essentially no gonadotropic hormones until the age of 10 to 14 years. However, at that time the anterior pituitary gland begins to secrete two gonadotropic hormones. At first it secretes mainly *follicle-stimulating hormone,* which initiates the beginning of sexual life in the growing female child, but later it secretes *luteinizing hormone* as well, which helps to control the monthly female cycle. The rates of

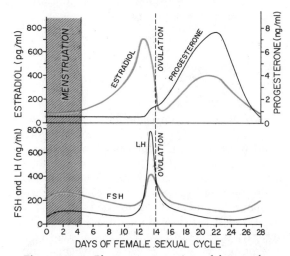

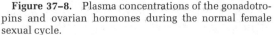

Figure 37–8.   Plasma concentrations of the gonadotropins and ovarian hormones during the normal female sexual cycle.

secretion of these hormones as well as of estrogen and progesterone during the different phases of the monthly female cycle are shown in Figure 37–8.

Follicle-Stimulating Hormone. It is follicle-stimulating hormone that causes a few of the primary follicles of the ovary to begin growing each month, promoting very rapid proliferation of the epithelioid cells surrounding the ovum. These cells then begin to secrete *estrogens,* one of the two major female sex hormones. Thus, the two functions of follicle-stimulating hormone are to cause proliferation of the ovarian follicular cells and to cause secretory activity by these cells. Consequently, the follicular cavities develop and the follicles enlarge greatly. As soon as the follicles grow to about one-half their maximum size, the anterior pituitary gland begins to secrete greatly increased quantities of *luteinizing hormone* along with the follicle-stimulating hormone.

Luteinizing Hormone. This hormone increases still more the rate of secretion by the follicular cells, and soon makes one follicle grow so large that it ruptures, expelling its ovum into the abdominal cavity. Simultaneously, the luteinizing hormone also causes the individual follicular cells to grow in size and to develop a fatty, yellow appearance. These cells are then known as *lutein cells,* and the entire mass of lutein cells develops within 1 to 2 days into an enlarged growth on the ovary called a *corpus luteum,* which means a "yellow body." This change of follicular cells into lutein cells is illustrated in the lower part of Figure 37–6. The corpus luteum then continues to secrete estrogens but also secretes large quantities of progesterone as well.

THE OVARIAN HORMONES – ESTROGEN AND PROGESTERONE

Two ovarian hormones, *estrogen* and *progesterone,* are responsible for sexual development of the female and also for the female monthly sexual changes. These hormones, like the adrenocortical hormones and the male hormone testosterone, are both steroid compounds, and they are formed mainly from the fatty substance cholesterol. Estrogen is actually several different hormones called *estradiol, estriol,* and *estrone,* but they have identical functions and almost but not exactly identical chemical structures. For this reason

they are considered together as if they were a single hormone.

Functions of Estrogen. Estrogen causes the cells in several parts of the body to *proliferate* — that is, to increase in number. For example, it causes the smooth muscle cells of the uterus to proliferate, making the female uterus, after puberty, become about two to three times larger than that of a child. Also, estrogen causes enlargement of the vagina, development of the *labia* surrounding the vagina, growth of hair on the pubis, broadening of the hips, conversion of the pelvic outlet into an ovoid shape rather than the funnel shape of the male, growth of the breasts, proliferation of the glandular elements of the breasts, and finally deposition of fatty tissues in characteristic female areas such as on the thighs and hips. In summary, essentially all of the characteristics that distinguish the female from the male are caused by estrogen, and the basic reason for the development of these characteristics is the ability of estrogen to promote proliferation of the respective cellular elements in certain regions of the body.

Estrogen also increases the growth rate of all bones immediately after puberty, but they cause the growing portions of the bone to "burn out" within a few years, so that growth then stops. As a result, the female grows very rapidly for the first few years after puberty and then ceases growing entirely. On the other hand, the male child continues to grow even beyond this time and grows to a taller height than the female —not because of more rapid growth but because of more prolonged growth.

Estrogen also has very important effects on the internal lining of the uterus, the *endometrium,* which is discussed below in relation to the menstrual cycle.

Functions of Progesterone. Progesterone has little to do with the development of the female sexual characteristics; instead, it is concerned principally with preparing the uterus for acceptance of a fertilized ovum and preparing the breasts for secretion of milk. In general, progesterone increases the degree of secretory activity of the glands of the breasts and also of the cells lining the uterine wall. This latter effect is also discussed below in relation to the menstrual cycle. Finally, progesterone inhibits the contractions of the uterus, and prevents the uterus from expelling a fertilized ovum that is trying to implant or a fetus that is developing.

REGULATION OF THE FEMALE SEXUAL CYCLE

Oscillation Between the Anterior Pituitary and Ovarian Hormones. The monthly female sexual cycle is caused by alternating secretion of follicle-stimulating and luteinizing hormones by the anterior pituitary gland and estrogens and progesterone by the ovaries, as illustrated in Figure 37–8. The cycle of events that causes this alternation is the following:

1. At the beginning of the monthly cycle, that is, when menstruation begins, the anterior pituitary gland begins to secrete increasing quantities of *follicle-stimulating hormone* along with smaller quantities of *luteinizing hormone*. These two together cause several follicles to grow in the ovaries as well as cause considerable secretion of *estrogen*.

2. The estrogen then is believed to have two sequential effects on the anterior pituitary gland secretion. First, it inhibits secretion of both follicle-stimulating hormone and luteinizing hormone, their secretory rates falling to a low ebb at about the tenth day of the cycle. Then, suddenly, the anterior pituitary gland begins to secrete very large quantities of both follicle-stimulating hormone and luteinizing hormone, but especially of luteinizing hormone. This phase of secretion is called the *luteinizing hormone surge*. It is this surge that causes very rapid final development of one of the ovarian follicles and causes it to rupture within about 2 days.

3. The process of ovulation, occurring at approximately the fourteenth day of the normal 28 day cycle, then leads to the development of the corpus luteum as described earlier in the chapter. The corpus luteum secretes large quantities of *progesterone* and considerable amounts of *estrogen*.

4. The estrogen and progesterone secreted by the corpus luteum inhibit the anterior pituitary gland once again and diminish greatly their rate of secretion of both follicle-stimulating hormone and luteinizing hormone. Without these hormones to stimulate it, the corpus luteum involutes so that the secretion of both progesterone and estrogen now falls to a very low ebb. It is at this time that menstruation begins, brought on by this sudden demise of estrogen and progesterone secretion.

5. At this time, the anterior pituitary gland, being no longer inhibited by estrogen and progesterone, begins to secrete large quantities of follicle-stimulating hormone once again, thus initiating the cycle of the next month. This process continues on throughout the entire reproductive life of the female.

The Reproductive Life of the Female, and the Menopause. Before puberty the anterior pituitary gland is unable to secrete any gonadotropic hormones. Consequently, until that time the alternation between anterior pituitary and ovarian hormones cannot take place. As soon as a sufficient quantity of follicle-stimulating hormone begins to be secreted, however, the reproductive life of the female begins.

After about 30 years of reproductive life, the ovaries "burn out" because all the primary follicles finally have grown into mature follicles and ruptured or involuted. Thereafter, the monthly cycles also cease because the ovaries no longer have any follicular cells to secrete estrogen and progesterone, even though the anterior pituitary gland continues to secrete large quantities of follicle-stimulating hormone for the remainder of the woman's life. This cessation of the monthly cycles occurs in the average woman at an age of about 45, and this stage of her life is called the *menopause*.

The Endometrial Cycle. It is evident from Figure 37–8 that during the first half of the monthly cycle the only *ovarian* hormone secreted in large quantity is estrogen. Then during the latter half both estrogen and progesterone are secreted.

Estrogen causes the endometrium — that is, the lining of the inside of the uterus — to grow in thickness. Both the epithelial cells on the surface and the deeper cells of the endometrium proliferate approximately threefold. Also, the glands of the endometrium increase greatly in depth and in tortuosity. These changes constitute the *proliferative phase* of endometrial development and are illustrated in the first portion of Figure 37–9. This phase occurs during the first 10 to 12 days of the monthly cycle after menstruation is over.

At approximately the fourteenth day of the cycle the corpus luteum begins secreting progesterone, which then causes further thickening of the endometrium while also initiating several special effects as follows: (1) The endometrial glands begin secreting a nutrient fluid that can be used by a fertilized ovum before it implants; (2) large quantities of

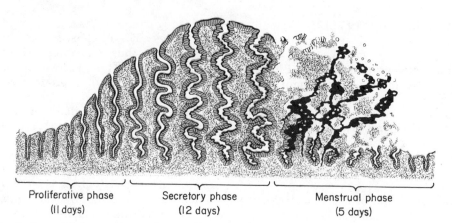

Proliferative phase   Secretory phase   Menstrual phase
(11 days)   (12 days)   (5 days)

Figure 37–9. Endometrial changes during the monthly sexual cycle, showing especially the process of menstruation.

fatty substances and glycogen deposit in the deeper cells of the endometrium; and (3) blood flow to the endometrium increases. In other words, the major function of progesterone is to make available an adequate supply of nutrients, already stored in the endometrium, for an embryo, should it implant in the endometrium and begin to develop.

Menstruation. If an ovum becomes fertilized and implanted in the uterus, a special hormone called *chorionic gonadotropin* (to be discussed more fully in the following chapter) is then released by the developing ovum itself to stimulate the corpus luteum. This causes it to continue to produce estrogen and progesterone, and pregnancy begins. However, if toward the end of the monthly cycle implantation of an ovum has not occurred, no such hormone appears, the corpus luteum dies, and the production of both estrogen and progesterone almost ceases. The sudden lack of these two hormones causes the blood vessels of the endometrium to become spastic so that blood flow to the surface layers of the endometrium ceases. As a result, much of the endometrial tissue dies and sloughs into the uterine cavity, a process called *menstruation*. Blood oozes from the denuded endometrial wall, causing a blood loss of about 50 ml. during the several days of menstruation. The necrotic endometrial tissue, plus the blood and much serous exudate from the denuded uterine surface, is gradually expelled for about 3 to 5 days.

The deepest pits of the endometrial glands remain intact during menstruation, despite the sloughing of all the surface layers of the endometrium. During the first few days after menstruation, new epithelium grows outward from the pits of the glands to cover the inner surface of the uterus. Then, under the influence of renewed estrogen production by the ovaries, the endometrial cycle begins again. These sequential changes during the entire cycle are shown in Figure 37–9.

The Period of Fertility During the Sexual Cycle. An ovum can be fertilized by a sperm for a period of 8 to 24 hours after ovulation. Also, sperm can live in the female genital tract usually for 24 to 48 hours. Therefore, for successful fertilization, sexual exposure must occur either shortly before ovulation so that sperm are already available when ovulation occurs, or within a few hours after ovulation.

Ovulation occurs almost exactly 14 days *before* menstruation begins. Therefore, in any woman who has a normal 28 day sexual cycle, ovulation will occur on the fourteenth day *after* the beginning of menstruation. However, many women, instead of having normal 28 day cycles, have cycles as short as 21 days or as long as 40 days. If these women are regular, they can still calculate that ovulation will occur 14 days *before* menstruation begins. However, in those women who are irregular, it becomes impossible to predict the time of ovulation.

To be on the safe side, one usually allows 4 to 5 days on either side of the calculated day of ovulation when using the rhythm method of contraception.

HORMONAL SUPPRESSION OF FERTILITY — THE LUTEINIZING HORMONE SURGE AND "THE PILL"

It has long been known that either estrogen or progesterone given in sufficient quantity during the early part of the monthly female cycle can prevent ovulation. It will be recalled from earlier discussions in this chapter that ovulation occurs approximately one and a half days after a massive surge of luteinizing hormone secretion by the anterior pituitary gland on the twelfth to thirteenth day of the normal cycle. It is believed that estrogen and progesterone suppress ovulation by inhibiting the anterior pituitary secretion of luteinizing hormone and thereby preventing the luteinizing hormone surge. This obviously prevents conception as well.

The problem in devising methods for hormonal suppression of ovulation has been to develop appropriate combinations of estrogen and progesterone that will suppress ovulation but that will not cause unwanted effects of these two hormones. For instance, too much of either of the hormones can cause abnormal menstrual bleeding patterns. Fortunately, several synthetic estrogens and progestins have been developed which have maximum capability to suppress the luteinizing hormone surge but at the same time will not affect the menstrual cycle severely. Appropriate combinations of these synthetic hormones are marketed as "the pill."

Yet, most hormone regimens for contraception still have undesirable side effects in the body, particularly salt retention, which can, over a period of years, lead to high blood pressure in many women. Therefore, successful control of conception in ways other than by use of the pill is much to be desired.

A recent advance that might eliminate many of the hazards of the pill is a small intrauterine device (IUD) that is made of Silastic and contains within a central cavity a highly concentrated quantity of progestin. When this device is inserted into the uterus, minute amounts of the progestin filter through the Silastic into the uterine cavity for a period of a year or more. This directly affects the endometrium to make it impossible for the ovary to implant, and yet there are no side effects elsewhere in the body. The device needs replacement only once a year.

REFERENCES

Arimura, A.: Hypothalamic gonadotropin-releasing hormone and reproduction. *Intern. Rev. Physiol.,* 13:1, 1977.

Austin, C. R.: Recent progress in the study of eggs and spermatozoa: insemination and ovulation to implantation. *In* MTP International Review of Science: Physiology. Vol. 8. Baltimore. University Park Press, 1974, p. 95.

Bedford, J. M.: Maturation, transport, and fate of spermatozoa in the epididymis. *In* Greep, R. O., and Astwood, E. B. (eds.): Handbook of Physiology. Sec. 7, Vol. 5. Baltimore, The Williams & Wilkins Company, 1975, p. 303.

Brenner, R. M., and West, N. B.: Hormonal regulation of the reproductive tract in female mammals. *Ann. Rev. Physiol.* 37:273, 1975.

Briggs, M. H., and Briggs, M.: Biochemical Contraception. New York. Academic Press, Inc., 1975.

Brown-Grant, K.: Physiological aspects of the steroid hormone–gonadotropin interrelationship. *Intern. Rev. Physiol.,* 13:57, 1977.

Davidson, J. M.: Neurohormonal basis of male sexual behavior. *Intern. Rev. Physiol.,* 13:225, 1977.

James, V. H. T., and Martini, L. (eds.): The Endocrine Function of the Human Testis. New York, Academic Press, Inc., 1974.

Owen, J. A., Jr.: Physiology of the menstrual cycle. *Am. J. Clin. Nutr.,* 28:333, 1975.

Phillips, D. M.: Spermiogenesis. New York, Academic Press, Inc., 1974.

Prasad, M. R. N., and Rajalakshmi, M.: Recent advances in the control of male reproductive functions. *Intern. Rev. Physiol.,* 13:153, 1977.

Segal, S. J.: The physiology of human reproduction. *Sci. Amer., 231(3)*:52, 1974.

Wilson, J. D.: Sexual differentiation. *Ann. Rev. Physiol., 40*:279, 1978.

QUESTIONS

1. Describe the formation of sperm, their maturation, and their functional characteristics.
2. Give the mechanisms of the male sexual act.
3. What initiates puberty in both the male and the female?

4. What are the roles of follicle-stimulating hormone and luteinizing hormone in the male?
5. In what ways does testosterone affect bodily function?
6. Trace the development of the ovum, its maturation, and the process of ovulation.
7. How is the ovum transported to the uterus?
8. In what ways does the female sexual act resemble the male sexual act, and in what ways is it dissimilar?
9. What are the effects of estrogen on the body?
10. What are the effects of progesterone on the body?
11. How do the gonadotropic hormones from the anterior pituitary gland interact with the ovarian hormones to cause the rhythmic monthly sexual cycle?
12. Describe the endometrial cycle and menstruation.
13. Explain the hormonal mechanisms for suppression of fertility.

REPRODUCTION AND FETAL PHYSIOLOGY

38

FERTILIZATION OF THE OVUM AND EARLY GROWTH

Entry of the Sperm into the Ovum. After the ovum is expelled from the ovary, it remains viable for 8 to 24 hours. During this time it usually moves approximately one-quarter of the distance down the fallopian tube toward the uterus. Obviously, then, fertilization must occur either in the abdominal cavity before the ovum enters the tube, or in the upper portion of one of the tubes.

Sperm travel at a velocity of only 1 to 4 mm. per minute, and the total length of a fallopian tube is approximately 15 cm. Therefore, the minimum time usually required for a sperm to reach the ovum is approximately 40 minutes.

The ovum, when released from the ovary, carries on its surface a large number of granulosal cells which collectively are called the *corona radiata,* shown in Figure 38–1. It is claimed, though not proved, that the corona radiata is dispersed by enzymes secreted by the sperm, especially *hyaluronidase* and several *proteinases.* These enzymes supposedly digest the protein linkages that hold the cells together, and allow the cells to break away from the ovum. In other words, the sperm supposedly removes this barrier so that it can attack and enter the ovum. However, the alkaline secretions of the fallopian tube also are believed to help remove the corona radiata.

After a single sperm has entered the ovum, the membrane of the ovum suddenly becomes impermeable to entry of additional sperm. And sperm that do reach the ovum are inac-tivated as they attempt to enter. This effect prevents more than one set of male chromosomes from combining with the chromosomes of the ovum.

Shortly after a sperm has entered the ovum, the head of the sperm begins to swell, forming a *male pronucleus,* as shown in Figure 38–1D; the original nucleus of the ovum is also still present and is called the *female pronucleus.* It will be recalled from the previous chapter that each of these pronuclei has only 23 *unpaired* chromosomes. However, the two pronuclei soon combine so that the fertilized ovum now has 46 *paired* chromosomes in a single nucleus, as do all other cells of the human body. A few hours later the chromosomes reproduce and split apart, initiating division of the ovum into two daughter cells.

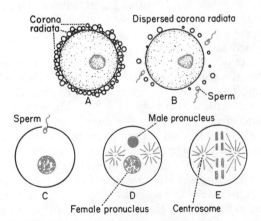

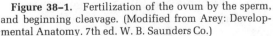

Figure 38–1. Fertilization of the ovum by the sperm, and beginning cleavage. (Modified from Arey: Developmental Anatomy. 7th ed. W. B. Saunders Co.)

CELL MULTIPLICATION AND DIVISION

Early Division of the Ovum. The first cleavage of the fertilized cell occurs approximately 30 hours after the sperm has entered the ovum, and the next few generations occur every 10 to 15 hours. By the time the ovum reaches the uterus, the total number of cells in the cellular mass is usually about 16 to 32. By this time the cells have actually begun to *differentiate,* which means that some of them have developed different characteristics from the others.

Differentiation. Figure 38–2 illustrates various stages of cleavage of the fertilized mammalian ovum, the darkened cells representing those that will eventually form the embryo and the lighter cells those that will eventually form the fetal membranes. Note the differences in size and other characteristics of the cells as cell division proceeds during the first few days after fertilization. This is the process of cell differentiation. Obviously, still many more changes in characteristics of the cells will take place before the final postnatal human being has been formed. Unfortunately, we do not know all the causes of differentiation, but let us discuss the theories of how the cells change their characteristics to form all the different organs and tissues of the body.

The earliest and simplest theory for explaining differentiation was that the genetic composition of the nucleus undergoes changes during successive generations of cells in such a way that one daughter cell develops one set of characteristics while another daughter cell develops entirely different characteristics. This theory probably explains some stages of differentiation, because highly differentiated cells, when grown in tissue culture, cannot be reverted all the way back to the original primordial state. Yet most cells can be reverted at least two or three steps backward in differentiation, which indicates that other factors besides simple changes in genetic potency probably also play roles in differentiation and might even play the dominant role.

The cytoplasm, in addition to the nucleus, is known also to play a role in differentiation, because cells without nuclei can divide and even differentiate to some extent for a few stages before the cells die. However, the long-term function of the cytoplasm in differentiation is probably controlled by the nucleus.

Embryologic experiments show that certain

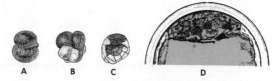

Figure 38–2. Differentiation of cells in the early embryo during gestation: (A) two blastomeres, (B) six blastomeres, (C) a hemisected early blastocyst, (D) a later blastocyst stage showing the early fetus at the top. (Modified from Arey: Developmental Anatomy, 7th ed. W. B. Saunders.)

cells in an embryo control the differentiation of adjacent cells. For instance, a central axis develops in the embryo, and it in turn, as a result of complex *inductions* in the surrounding tissues, causes formation of essentially all the organs of the body.

Another instance of induction occurs when the developing eye vesicles deep in the head come in contact with the outer layer of the head, the ectoderm, and cause it to thicken into a lens plate which folds inward to form the lens of the eye. It is possible that the entire embryo develops as a result of complex inductions, one part of the body affecting another part, and this part affecting still other parts.

Thus our understanding of cell differentiation is still hazy. We know many different control mechanisms by which differentiation *could* occur. Yet the overall basic controlling factors in cell differentiation are yet to be discovered; when learned, they will make a tremendous difference in our understanding of bodily development.

IMPLANTATION OF THE OVUM

Figure 38–3 shows the *blastocyst* stage of a developing human ovum one and one-half weeks after fertilization. It illustrates the method by which the mass of cells attaches itself to the inner wall of the uterus. At this stage the cells on the outside of the blastocyst, called *trophoblasts,* secrete large quantities of proteolytic enzymes that digest the endometrium, and then the trophoblasts phagocytize the digested products. Thus, they literally eat their way into the wall of the endometrium. The trophoblasts grow and divide very rapidly at the same time, and soon they and adjacent cells begin forming the placenta and

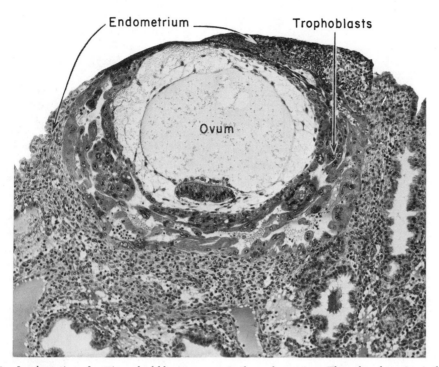

Figure 38-3.   Implantation of a 1½ week old human ovum in the endometrium. The colored portion is the developing fetus. (Courtesy of Dr. Arthur Hertig.)

fetal membranes while the embryo develops on the inside of the blastocyst.

NUTRITION OF THE FETUS IN THE UTERUS

THE TROPHOBLASTIC PHASE OF NUTRITION

During the first few weeks after implantation of the ovum, the placenta and its blood supply are not sufficiently formed to supply the fetus with nutrients. During this phase, nutrition is provided only by trophoblastic digestion and phagocytosis of the endometrium. It will be recalled that prior to implantation of the ovum, the endometrium stores large quantities of proteins, lipid materials, and glycogen. Also, small quantities of iron and vitamins are stored awaiting phagocytosis by the developing ovum. Even so, the embryo can obtain nutrition in this manner for only the first few weeks of its development. By that time the placenta will have developed to a stage such that it can begin supplying nutrition. Figure 38-4 shows that the trophoblastic period of phagocytic nutrition lasts until approximately the twelfth week of pregnancy.

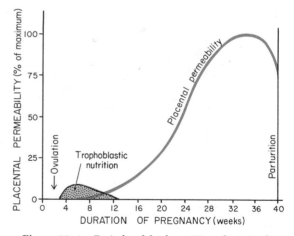

Figure 38-4.   Periods of fetal nutrition: first, trophoblastic phagocytosis of the endometrium lasting for the first few weeks of pregnancy and, second, diffusion through the placenta during the remainder of pregnancy.

PLACENTAL NUTRITION OF THE FETUS

Figure 38-5 shows growth of the fetus in the uterus and also shows the *placenta* and the *fetal membranes.* The fetal membranes attach to the entire inner surface of the uter-

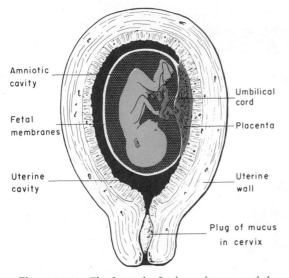

Figure 38-5.   The fetus, the fetal membranes, and the placenta in the pregnant uterus.

us, forming a cavity called the *amniotic cavity*. The fetus floats freely in *amniotic fluid* that fills the amniotic cavity.

The placenta covers about one-sixth the surface of the uterus, and the fetus is connected with it by means of the *umbilical cord*. The function of the umbilical cord is to transport blood from the baby to the placenta and then back again to the baby. In this way the baby's blood picks up nutrition from the placenta and transports it into the developing child.

Figure 38-6 illustrates the gross organization of the placenta and, at the bottom, a cross-section of a *placental villus*. The mother's arterial blood flows into large placental chambers called *placental sinuses*. The fetal portion of the placenta is composed of many small cauliflowerlike growths that project into the placental sinuses. Each of these in turn is covered with tremendous numbers of small villi containing blood capillaries. The baby's blood flows into the capillaries of the villi where it receives nutrients from the mother's blood, and then it passes back through the umbilical veins to the fetus.

The cross-section of a villus in the lower part of Figure 38-6 shows the close proximity of the mother's blood in the placental sinuses to the baby's blood in the fetal capillaries. The mother's blood flows so rapidly through the sinuses that the sinus blood maintains very high concentrations of the nutrients needed by the fetus.

Diffusion of Nutrients Through the Villi.   Nutrients pass through the placental villi into the fetal blood almost entirely by the process of diffusion. The gaseous pressure of oxygen in the blood of the maternal sinuses is normally 50 to 60 mm. Hg, whereas the gaseous pressure of oxygen in the fetal capillaries is only 30 mm. Hg. Because of this pressure difference, oxygen simply diffuses through the tissues of the villus from the mother's blood to the baby's blood. Likewise, glucose, amino acids, fats, many of the vitamins, and most of the minerals are present in greater concentration in the mother's blood than in the baby's blood because the baby utilizes these substances almost as rapidly as they enter his blood. As a consequence, all of them diffuse into the baby's blood and thereby supply the fetus with nutrition.

Figure 38-4 shows the progressive increase in placental permeability as pregnancy proceeds. The total amount of nutrients that can be transported through the placenta reaches a maximum 32 to 36 weeks after pregnancy begins, about 6 weeks before the baby is born.

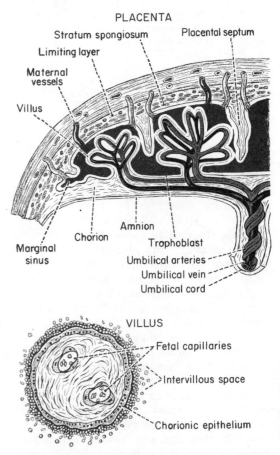

Figure 38-6.   Gross anatomic structure of the placenta and microscopic cross-section of a placental villus. (Modified from Goss: Gray's Anatomy of the Human Body. Lea & Febiger; and from Arey: Developmental Anatomy, 7th ed. W. B. Saunders Co.)

At this point the placental tissues begin to grow old and degenerate. As a result, even though the fetus still requires progressively increasing quantities of nutrients, the ability of these nutrients to enter the baby becomes lessened. Fortunately, birth of the baby occurs soon, and the baby assumes independent existence.

EXCRETION THROUGH THE PLACENTA BY DIFFUSION. In addition to the diffusion of nutrients into the baby, excretory products diffuse through the placenta from the fetal blood to the maternal blood. For instance, the baby's metabolism continually forms large quantities of carbon dioxide, urea, uric acid, creatinine, phosphates, sulfates, and other normal excretory products. The concentration of each of these substances rises in the baby's blood until it is greater than the concentration in the mother's blood. Then, because of this reverse concentration gradient, these substances diffuse backward through the villi, and the mother excretes them through her kidneys, lungs, and gastrointestinal tract.

Active Transport Through the Villi. The chorionic epithelium that covers the surface of the placental villi is formed from the trophoblastic cells of the early fetus. These cells continue their phagocytic activity throughout the life of the placenta, so that, in addition to the large amount of nutrients that pass by diffusion through the placental membrane, small amounts of nutrients are actively transported through the villi all through the period of gestation. In the first few weeks of placental development this aids greatly in increasing the amount of amino acids, fatty substances, and some minerals that can be supplied to the baby, but after 12 to 20 weeks the amount of nutrients supplied in this manner becomes insignificant compared with that supplied by the simple process of diffusion.

THE HORMONES OF PREGNANCY

In the previous chapter, the vital roles played by the different sex hormones in the initiation of reproduction became clear. However, hormones play an equally important, if not more important, role in helping the processes of pregnancy be completed. Most of these hormones are secreted by the placenta itself. Two of these are *estrogen* and *progesterone,* the same female sex hormones secreted by the ovaries during the normal female monthly cycle. However, two others that are also important and even necessary for pregnancy to proceed are *chorionic gonadotropin* and *human placental lactogen*. These hormones affect both the mother and the fetus. In the mother, they help to control changes in the uterus and breasts that are necessary to carry the fetus to birth and also to provide milk production. And they help to regulate the development of the fetus itself, especially the sexual organs.

SECRETION OF ESTROGEN DURING PREGNANCY

During the normal menstrual cycle moderate quantities of estrogen are secreted first by the cells lining the developing ovarian follicles and then by the corpus luteum that forms from the ruptured follicle. If the ovum becomes fertilized and implants in the uterus, the corpus luteum, instead of degenerating as it normally does at the end of the normal monthly cycle, grows even larger and produces increasing amounts of estrogens for an additional 4 to 5 months. By this time, the placental tissues have begun to form extreme quantities of estrogen, the amount progressively increasing to a maximum immediately before birth, as shown in Figure 38–7. The peak production reaches several times as much as the production during a normal monthly sexual cycle.

Functions of Estrogen During Pregnancy. In the mother, estrogen causes (1) rapid proliferation of the uterine musculature; (2) greatly increased growth of the vascular system supplying the uterus; (3) enlargement of

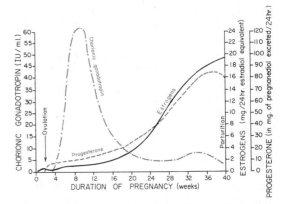

Figure 38–7. Rates of secretion of estrogens, progesterone, and chorionic gonadotropin at different stages of pregnancy.

the external sex organs and of the vaginal opening, providing an appropriately enlarged passageway for birth of the baby; and (4) probably also some relaxation of the pelvic ligaments that allows the pelvic opening to stretch as the baby is born.

In addition to the effects on the reproductive organs, estrogen also causes the breasts to grow rapidly. The ducts, especially, enlarge, and the glandular cells proliferate. Finally, estrogen causes a pound or more of extra fat to deposit in the breasts.

Another effect of estrogen not yet completely understood is the effect on the fetus itself. It is believed that estrogen causes much of the rapid proliferation of fetal cells and also aids in the differentiation of some of these cells into special organs. In particular estrogen is believed to control development of some of the female sex characteristics.

SECRETION OF PROGESTERONE DURING PREGNANCY

Figure 38–7 also shows the rates of progesterone secretion during pregnancy. Progesterone, like estrogen, is secreted in moderate quantities by the corpus luteum during the early part of pregnancy. At approximately the sixteenth week of gestation the placenta begins producing very large quantities of progesterone as well as estrogen, reaching a maximum production toward the end of pregnancy of as much as ten times that produced by the corpus luteum.

Functions of Progesterone During Pregnancy. The first function of progesterone during pregnancy is to make increased quantities of nutrients available to the early endometrium for use by the developing ovum. It does this by causing the endometrial cells to store glycogen, fat, and amino acids. In addition, progesterone has a strong inhibitory effect on the uterine musculature, causing it to remain relaxed throughout pregnancy. It is believed that this effect allows pregnancy to continue until the fetus is large enough to be born and live an independent existence.

Progesterone complements the effects of estrogens on the breasts. It causes the glandular elements to enlarge further and to develop a secretory epithelium, and it promotes deposition of nutrients in the glandular cells, so that when milk production is required the appropriate materials will be available.

SECRETION AND FUNCTIONS OF CHORIONIC GONADOTROPIN DURING PREGNANCY

If the corpus luteum degenerates or is removed from the ovary at any time during the first 3 to 4 months of pregnancy, the loss of estrogen and progesterone secretion by this corpus luteum causes the fetus to stop developing and to be expelled within a few days. For this reason it is important that the corpus luteum remain active at least during the first third and preferably during the first half of pregnancy. Beyond that time removal of the corpus luteum does not affect pregnancy because the placenta by then is secreting many times as much estrogens and progesterone as the corpus luteum.

It will be recalled from the previous chapter that the corpus luteum normally degenerates and is absorbed at the end of each female monthly cycle. To keep the corpus luteum intact when the ovum implants, a special hormone, a small glycoprotein called *chorionic gonadotropin,* is secreted by the developing fetal tissues — by the trophoblasts. This hormone has almost exactly the same properties as luteinizing hormone. It not only keeps the corpus luteum from involuting, but also actually stimulates it so greatly that it enlarges severalfold during the first 2 to 4 months of pregnancy.

Chorionic gonadotropin begins to be formed from the day that the trophoblasts implant in the uterine endometrium. Its concentration is highest approximately at the tenth week of pregnancy, as shown in Figure 38–7. Thus, the concentration is greatest at the time of pregnancy when it is essential to prevent involution of the corpus luteum. In the middle and latter parts of pregnancy the secretion of chorionic gonadotropin falls to very low values. Its only known function at this time of pregnancy is to stimulate production of testosterone by the testes of the male fetus, which was discussed in the previous chapter and which plays an important role in the development of the male fetus.

SECRETION AND FUNCTIONS OF HUMAN PLACENTAL LACTOGEN

Recently, a hormone called *human placental lactogen* has been discovered. This is a small protein that begins to be secreted about the fifth week of pregnancy, with progressively increasing secretion throughout the remainder of pregnancy.

Human placental lactogen has two types of effects: First, it has actions very similar to those of growth hormone. It causes deposition of protein throughout the body in the same way as growth hormone. It also has the same effects as growth hormone on glucose metabolism and other metabolic functions of the cells. This growth-hormonelike function of human placental lactogen probably plays an important role in promoting growth of the fetus as well as in development of many of its organs.

Second, human placental lactogen specifically promotes growth of the breasts and actually causes milk production after the breasts have been appropriately prepared by estrogen and progesterone. These are very much the same effects that the hormone prolactin, secreted by the anterior pituitary gland, has on the breasts after birth of the baby, as we shall discuss later in the chapter.

FETAL PHYSIOLOGY

In general, the physiology of the growing fetus during the last 3 months of pregnancy is not far different from that of the normal child. All the organs of the body assume their final anatomic form, with minor exceptions, by 4 to 5 months after the beginning of pregnancy, and most of them can function almost normally by 6 months after the beginning of pregnancy. For example, long before birth the kidneys of the fetus become at least partially active, the gastrointestinal tract imbibes and absorbs fluid from the amniotic cavity, breathing is attempted even though it cannot be accomplished because the fetus is immersed in amniotic fluid, the heart pumps blood quite normally from approximately the third month on, and the metabolic systems operate very much the same as they do after birth.

GROWTH OF THE FETUS

Figure 38–8 depicts the increase in length and weight of the fetus during the 40 weeks of pregnancy. The length of the fetus increases almost directly in proportion to its age, while the weight increases in proportion to the third power of the age. The weight is almost infini-

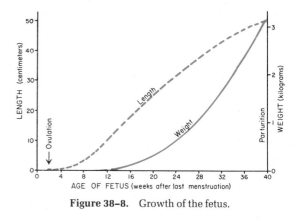

Figure 38–8.   Growth of the fetus.

tesimal until the twelfth to sixteenth week, but at that time it begins to increase extremely rapidly. Two months before birth the fetus usually weighs about one-half its birth weight, and one month prior to birth it weighs approximately three-fourths its birth weight. In other words, the major gain in weight occurs during the last 2 to 3 months. This is an especially important factor when one is considering the appropriate nutrition for the mother during pregnancy, because essentially all of the nutrients required by the baby are needed during the last 3 months. However, for several months prior to that time, growth of the uterus, placenta, and fetal membranes requires additional nutrients.

Special Nutrients Required by the Fetus.   The fetus requires especially large quantities of iron, calcium, phosphorus, amino acids, and vitamins. Iron begins to be used to form red blood cells within the first weeks of development of the fetus. Much of this early iron enters the fetus by active absorption from the endometrium by the trophoblasts. Throughout the remainder of pregnancy, still larger quantities of iron diffuse through the placental membrane to be used by the liver, spleen, and bone marrow for production of the fetus' blood.

Calcium is needed to ossify the bones. During the first two-thirds of pregnancy the fetal bones contain mainly organic matrix and almost no calcium salts. During the last 3 months of fetal development, ossification occurs very rapidly, approximately one-half occurring in the final month. It is at this time, therefore, that the mother needs an especially abundant amount of milk or other foods containing calcium and phosphate.

The large quantities of amino acids and vitamins required by the fetus provide the necessary building materials for growth of the fetal tissues. During the last 3 months of pregnancy, a mother can become depleted of protein and vitamins if these are not in the diet in adequate amounts. Also, brain development in the fetus is highly dependent on these nutrients.

THE FETAL NERVOUS SYSTEM

The major anatomic parts of the nervous system are formed during the first few months of fetal growth, but complete function of this system is not reached even by the time the child is born. A premature baby always exhibits signs of poor nervous system function. As an example, one of the most difficult problems in treating a premature baby is to keep his body temperature normal, because the hypothalamic temperature control centers of a 6- or 7-month fetus usually have not developed sufficiently to regulate body temperature. The child must be placed in an incubator for several weeks or sometimes a month or more until his nervous system has developed to a greater degree.

By the time of birth most, if not all, of the peripheral nerve fibers are completely developed, but in the central nervous system the deposition of myelin around many of the large nerve fibers is far from complete. Though myelin is not always necessary for nerve fibers to function, it has been inferred from this lack of complete myelination that certain portions of the central nervous system probably are far from optimum function at the time of birth. On the other hand, all the neurons that will ever be formed by the child are believed to be present by birth. This means that every time a neuron is destroyed thereafter, no new neuron will take its place.

CHANGES IN THE CIRCULATION AT BIRTH

Figure 38–9 illustrates the circulatory system in the fetus, showing very important special vessels that disappear after birth. Special *umbilical vessels* allow blood to flow into and out of the placenta. The blood returning from the placenta through the umbilical vein enters the fetal circulatory system through a special vessel called the *ductus venosus,* which passes directly through the liver. Also, blood from the portal veins of the fetus flows

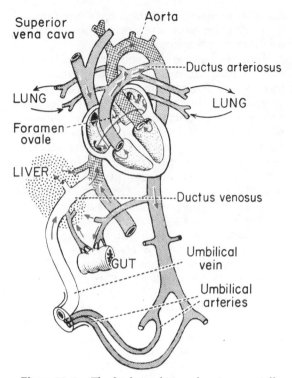

Figure 38–9. The fetal circulation, showing especially the ductus arteriosus, ductus venosus, and foramen ovale. (Modified from Arey: Developmental Anatomy, 7th ed. W. B. Saunders.)

through the ductus venosus, by-passing the liver entirely and emptying directly into the general venous system.

On reaching the heart, fetal blood by-passes the nonaerated lungs by two routes. First, it can flow from the right atrium through an opening called the *foramen ovale* directly into the left atrium, thus by-passing both the right ventricle and the lungs. Second, the remainder of the blood enters the right ventricle and is pumped into the pulmonary artery. However, most of this blood, instead of going through the lungs and left heart, passes through a vessel called the *ductus arteriosus* directly into the aorta.

Thus, by utilizing the ductus venosus, the foramen ovale, and the ductus arteriosus, blood flow through the liver and lungs occurs only to a minor extent in the fetus. This is a means for conserving the energy of the fetal heart, because the functions normally required of the fetal liver are performed instead by the mother's liver, and the fetal lungs cannot aerate the blood until birth.

After birth all three of these openings soon close. The ductus venosus and the ductus arteriosus become gradually occluded during the first few days or weeks of life. Presumably the ductus venosus becomes occluded because it no longer carries the tremendous blood flow from the umbilical vein, and the lesser blood flow from the portal system is not sufficient to keep it open. The ductus arterosus probably closes because the functioning lungs increase the oxygen in the blood which in turn has a direct effect on the wall of the ductus to cause increased contraction of the ductus muscle.

The foramen ovale is covered by a thin valve on the left atrial side. As long as the pressure in the right atrium is greater than that in the left atrium, which is the case before birth of the fetus, blood flows from right to left, by-passing the right ventricle and lungs. However, when the lungs expand, the right heart can then pump blood so easily through the lungs that the right atrial pressure falls to about 2 mm. Hg less than the left atrial pressure. This backward pressure differential thereafter keeps the valve of the foramen ovale closed. In about two-thirds of all persons the valve gradually adheres to the opening so that it becomes permanently closed; in about one-third it does not, but, regardless of which is true, blood ceases to flow through it after birth.

Thus, because of a series of changes in the dynamics of the circulation, blood begins to flow through both the liver and lungs immediately after birth, even though in the fetus the heart had conserved its energy by by-passing these two organs.

PHYSIOLOGY OF THE MOTHER DURING PREGNANCY

The mother's physiology is changed during pregnancy in several ways. First, accessory changes occur in her reproductive organs and breasts to provide for development of the fetus and to provide nutrition for the newborn child. Second, all her metabolic functions are increased to supply sufficient nutrition to the growing fetus. Third, tremendous production of certain hormones by the placenta during pregnancy causes many side effects not directly associated with reproduction.

Changes in Weight. The pregnant mother gains an average of about 22 pounds (approximately 10 kg.) during pregnancy. In general, this gain is accounted for in the following manner: fetus, 7 pounds; uterus, 2 pounds;

placenta and membranes, 2 pounds; breasts, 1.5 pounds; and the remainder, about 9.5 pounds, fat and increased quantities of extracellular fluid and blood. The amount of increase in fat and body fluids varies tremendously from one mother to another, depending upon her eating habits, especially salt and fat intake, and upon the amounts of hormones secreted during pregnancy.

Changes in Metabolism. The mother's metabolic rate in general rises approximately in proportion to her increase in weight, plus perhaps an additional 5 to 10 per cent. Much of this increase is occasioned simply by the greater amount of energy required for the mother to carry the growing load. However, rapid growth of the fetus also demands heightened activity of most of the mother's functions, such as rapid intermediary metabolism in her liver, rapid pumping of blood by her heart, increased respiration, and increased digestion and assimilation of food.

Changes in the Body Fluids and Circulation. The female sex hormones and extra adrenocortical hormones produced during pregnancy cause the mother usually to gain about 5 to 7 pounds of fluid, or, in other words, about 3 liters. About 0.5 liter of this is in the plasma, and another 0.5 liter is red blood cells, making a total gain in blood volume of about 1 liter. About one-third of the extra blood is needed to fill the sinuses of the placenta, but the other two-thirds of a liter collects in the circulation, causing blood to flow toward the heart with far greater ease than usual. As a result, the cardiac output becomes roughly 30 per cent more than normal.

During birth of the baby, the mother loses an average of 200 to 300 ml. of blood as the placenta separates from the uterus. This ordinarily causes no physiologic inconvenience because of the extra blood that had been stored during pregnancy. After birth, loss of the estrogen and other steroid hormones produced by the placenta causes the kidneys to excrete most of the remainder of the excess fluid and salt in the next few days.

BIRTH OF THE BABY (PARTURITION)

Duration of Pregnancy and Onset of Parturition. The duration of pregnancy, between the time of the last menstrual period until birth of the baby, is normally 40 weeks, though occasionally surviving babies are born as early as 28 weeks or as late as 46 weeks. Approximately 90 per cent of all babies are born within 10 days before or after the 40-week interval.

The reason for the relatively constant duration of pregnancy has never been completely understood. Presumably, growth of the baby and of the placenta to a certain size and state of maturity initiates birth. The probable factors that cause the onset of parturition are the following:

When the baby becomes large, pressure of its body against the uterus stretches the uterine musculature, which in turn initiates uterine contractions. In addition to this, movements of the baby in the uterus — such as the feet and hands striking the uterine wall — also initiate contractions. Obviously, the larger the baby becomes, the more likely are these initiated contractions to become strong enough to cause birth.

Perhaps equally as important as the mechanical factors, however, are hormonal factors. The concentration of *progesterone* secreted by the placenta begins to decrease a few weeks prior to birth, and, since progesterone normally inhibits the uterus, this change allows an increase in uterine contractions. On the other hand, the concentration of *estrogen* increases up until birth, and this increases the activity of the uterus, in contrast to the inhibition caused by progesterone. These two hormonal factors probably account for much of the progressively increasing contractility of the uterus shortly before birth.

A third factor possibly helping to initiate parturition is an increase in the secretion of *oxytocin* by the hypothalamic–posterior pituitary system shortly before the end of pregnancy. This hormone causes extreme contractility of the uterine musculature, and absence of its secretion in animals usually makes parturition difficult.

Mechanism of Parturition. Uterine musculature, like almost all smooth muscle, undergoes rhythmic contractions all the time. However, because of the influence of progesterone, these contractions are so weak during the early months of pregnancy that they can hardly be noted. During the last 3 months of pregnancy, they increase steadily and reach extreme intensity a few hours before birth. These very strong contractions wedge the head against the cervix, then slowly over a period of hours stretch the cervical ring and vaginal canal, and expel the baby. This period, from onset of the strong contractions

until the baby is born, is called the period of *labor,* and the actual expulsion of the child is called *parturition*.

Approximately 19 times out of 20 the portion of the baby that pushes against the cervix is the head, and this acts as a wedge to open the cervical and vaginal canals. In most of the remaining cases the buttocks are the presenting portion of the baby, though occasionally a leg, a shoulder, or even the side of the baby may be against the cervical canal. In all of these instances the cervix cannot be wedged open nearly as effectively as by the head, and birth of the baby is considerably more difficult. When the head comes first, the remainder of the body slips through the vaginal canal within a few seconds after the head is born; when the head is not the presenting part, the body portion of the baby may be born relatively easily, but the head, the largest part of the child, then has difficulty in passing through the canal. One of the problems of bottom-first birth is often a period of interrupted umbilical blood supply to the baby because the cord can become compressed against the wall of the delivery canal for a minute or more by the large head.

Theoretical Cause of the Sudden Onset of Labor. The reason the rhythmic uterine contractions suddenly become intense enough to cause labor is yet undetermined, but the following explanation has been offered as a possibility.

Irritation of the cervix is known to cause a muscular reaction over the entire uterus, making it become very excitable and causing its contractions to increase. It is presumed that when the natural contractions of the uterus reach a certain level of intensity, which occurs at the end of approximately 9 months of pregnancy, they push the baby's head against the cervix, which stretches it and thereby makes the entire uterus contract more forcefully. This contraction in turn pushes the baby's head into the cervix still more and initiates still further intensity of uterine contraction. Thus, a vicious cycle develops, as illustrated in Figure 38–10, and the uterine contractions become stronger and stronger until finally the baby is expelled.

The increased uterine excitability caused by stretching or irritating the cervix is believed to result from two mechanisms. First, cervical stimulation is believed to transmit impulses upward through the uterine musculature itself, thus resulting in uterine contraction. Second, in lower animals, and there-

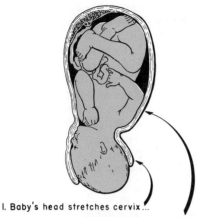

1. Baby's head stretches cervix...
2. Cervical stretch excites fundic contraction...
3. Fundic contraction pushes baby down and stretches cervix some more...
4. Cycle repeats over and over again...

Figure 38-10. A postulated mechanism for initiation of labor.

fore presumably in man, cervical irritation has been shown to transmit nerve impulses to the hypothalamus, causing greater secretion of *oxytocin,* which in turn increases the contractility of the uterus, as was explained in Chapter 34.

PRODUCTION OF MILK (LACTATION)

HORMONAL CONTROL OF LACTATION – PROLACTIN

The secretion of the *estrogen* and *progesterone* after puberty in a girl begins to prepare the breasts for lactation. The breasts enlarge and the glandular elements begin to develop. However, the degree of glandular development is relatively slight compared with that achieved during pregnancy. The tremendous quantities of estrogen and *human placental lactogen* secreted by the placenta during pregnancy cause rapid development of the glands in the breasts, and the large quantities of progesterone as well as the human placental lactogen change the glandular cells into actual secreting cells. By the time the baby is born, the breasts will have reached a degree of development capable of producing milk. Yet the estrogen and progesterone, despite their developmental effects on the breasts, inhibit the actual formation of milk until after the baby is born.

Figure 38-11 shows the changes in *estrogen, progesterone,* and *prolactin* production at the time of parturition. Sudden expulsion of the placenta from the uterus removes the source of estrogen and progesterone (as well as that of human placental lactogen) so that the availability of these hormones falls immediately almost to zero. Throughout pregnancy, the production of prolactin by the anterior pituitary gland has been inhibited by the placental estrogen and progesterone, but now that these placental hormones are gone, the anterior pituitary gland begins to secrete large quantities of prolactin.

Prolactin has a direct stimulatory effect on the glands of the breast, causing them to enlarge, to develop additional glandular elements, and to produce milk. In other words, it takes over the function of the human placental lactogen that had been produced by the placenta. In addition, it directly stimulates all the secretory cells of the breast to cause copious production of milk.

Control of Prolactin Secretion by the Hypothalamus. It will be recalled from Chapter 34 that the hypothalamus controls the secretion of essentially all the anterior pituitary hormones. Normally, it controls the secretion of prolactin by secreting an inhibitory substance, *prolactin inhibitory factor*. Except for the period following birth of a baby, this inhibitory factor keeps the anterior pituitary secretion of prolactin mainly inhibited. In the absence of this inhibitory factor, the anterior pituitary gland has a natural tendency to secrete large quantities of prolactin.

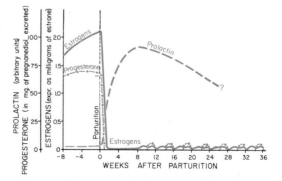

Figure 38-11. Changes in rates of secretion of estrogens, progesterone, and prolactin at parturition and during the succeeding weeks after parturition, showing especially the rapid increase in prolactin secretion immediately after parturition.

When the breasts are filled with milk, nerve signals from the swollen breasts to the hypothalamus cause large quantities of prolactin inhibitory factor to be formed, and this inhibits the secretion of prolactin, thereby also inhibiting further production of milk. However, if the breasts are prevented from overfilling with milk — that is, if the baby continues to remove the milk — the hypothalamus will not form prolactin inhibitory factor, and therefore the anterior pituitary gland will continue to produce prolactin. Thus, this feedback mechanism automatically controls the amount of milk formed by the breasts in accord with the food demands of the baby. Then, when the baby ceases to remove milk from the mother's breast, large quantities of prolactin inhibitory factor begin to appear in the hypothalamus, and in a few days to a few weeks the breasts cease lactating entirely.

As long as the mother continues to lactate and the anterior pituitary gland continues to produce prolactin, something about the prolactin production depresses the output of large quantities of follicle-stimulating and luteinizing hormones by the anterior pituitary gland. Therefore, during the first few months after parturition, the ovaries of a lactating mother frequently will not ovulate, and the rate of formation of estrogen and progesterone remains reduced. However, after several months, the normal cycle reappears even though lactation continues. If the mother does not lactate after birth, her next ovulation appears in about 4 weeks, and her next menstrual cycle in about 6 weeks.

Milk Ejection — Role of Oxytocin.   When a baby sucks on the nipples, it usually obtains no milk from the breasts for approximately the first 45 seconds to a minute. Then suddenly milk appears in the ducts of both breasts even though suckling occurs on only one breast, indicating that some general phenomenon has occurred to cause milk to flow toward the nipples.

Experiments have shown that suckling causes sensory nerve signals to pass first into the spinal cord, then upward through the brain stem, and finally into the hypothalamus to stimulate the production of oxytocin. This then circulates through the blood to the breasts, where it causes *myoepithelial cells* surrounding the alveoli to contract and expel the milk collected in the alveoli into the ducts leading to the nipples. This process is called *milk ejection*.

This milk ejection mechanism can be adversely affected tremendously by psychic factors. For example, a mother's fear that she might not be able to nurse her baby can actually keep her from doing so. Also, disturbances caused by other children in the family or by overly concerned relatives may lead to difficulty in milk ejection and cause the breasts to fail to empty. Failure to empty in turn allows the anterior pituitary gland to cease its production of prolactin, and the breasts stop secreting milk.

COMPOSITION OF MILK

Milk contains the usual substances needed for energy and growth by the baby. These include two types of protein, *casein* and *lactalbumin*; an easily digested sugar, *lactose*, composed of one molecule of glucose and one of galactose; and large quantities of fatty substances such as *butter fats, cholesterol,* and *phospholipids*.

In addition, milk contains small quantities of vitamins and large quantities of calcium phosphate. Yet, it has a conspicuous lack of iron. This lack usually is not detrimental to the baby in early life, because a sufficient quantity of this mineral is stored in the fetal liver prior to birth to continue the formation of hemoglobin for about 2 months. Beyond this time, however, iron must be in the baby's diet, or he will develop progressive anemia.

Table 38–1 shows the relative compositions of human and cow's milk. When cow's milk is substituted for human milk, the baby's metabolic systems must adjust to a large increase in ions — to the phosphates in particular —

Table 38–1.   **Composition of Milk in Per Cent**

	HUMAN	Cow
Water	88.5	87.0
Fat	3.3	3.5
Sugar	6.8	4.8
Casein	0.9	2.7
Lactalbumin and other protein	0.4	0.7
Electrolytes	0.2	0.7

which can occasionally lead to temporarily serious ion disturbances in the baby. Also, it is desirable to fortify cow's milk with extra quantities of easily digested sugar such as pure glucose (dextrose).

EFFECT OF LACTATION ON THE MOTHER

Production of milk by the mother in some ways is as great a drain on her metabolic systems as pregnancy itself. Particularly does she lose tremendous amounts of stored proteins and fats during lactation. Also, if large quantities of calcium phosphate are not in her diet, her parathyroid glands become greatly overactive. This causes reabsorption of her bones, releasing the necessary calcium and phosphate for milk formation. As a result, her bones are likely to become weakened. Fortunately, though, the amount of calcium phosphate stored in the bones of the average mother is tremendous in comparison with the amount that the baby will need during the first few months of life. Therefore, only when the mother already has some degree of decalcification will this loss cause her any difficulty.

Frequently mothers develop extensive dental caries during pregnancy, which has often been ascribed to the loss of calcium phosphate from the teeth. Experiments, however, have shown that no significant amount of calcium phosphate leaves the teeth, but that the caries are probably caused by enhanced bacterial growth in the mouth during pregnancy.

REFERENCES

Battaglia, F. C., and Meschia, G.: Principal substrates of fetal metabolism. *Physiol. Rev., 58*:499, 1978.

Biggers, J. D., and Borland, R. M.: Physiological aspects of growth and development of the preimplantation mammalian embryo. *Ann. Rev. Physiol., 38*:95, 1976.

Bradley, R. M., and Mistretta, C. M.: Fetal sensory receptors. *Physiol. Rev., 55*:352, 1975.

Brenner, R. M., and West, N. B.: Hormonal regulation of the reproductive tract in female mammals. *Ann Rev. Physiol. 37*:273, 1975.

Cockburn, F., and Drillien, C. M.: Neonatal Medicine. Philadelphia, J. B. Lippincott Company, 1975.

Daughaday, W. H. Herington, A. C., and Phillips, L. S.: The regulation of growth by endocrines. *Ann. Rev. Physiol., 37*:211, 1975.

Epel, D.: The program of fertilization. *Sci. Amer., 237(5)*:128, 1977.

Goldberg, V. J., and Ramwell, P. W.: Role of prostaglandins in reproduction. *Physiol. Rev., 55*:325, 1975.

Hytten, F. E., and Leitch, I.: The Physiology of Human Pregnancy. Philadelphia, J. B. Lippincott Company, 1974.

Miller, R. K., and Berndt, W. O.: Mechanisms of transport across the placenta: an in vitro approach. *Life Sci., 16*:7, 1975.

Nathanielsz, P. W.: Endocrine mechanisms of parturition. *Ann. Rev. Physiol., 40*:411, 1978.

Smith, C. A., and Nelson, N. M.: The Physiology of the Newborn Infant. Springfield, Ill., Charles C Thomas, Publisher, 1974.

Villee, D. B.: Human Endocrinology: A Developmental Approach. Philadelphia, W. B. Saunders Company, 1975.

Yoshinaga, K.: Hormonal interplay in the establishment of pregnancy. *Intern. Rev. Physiol., 13*:201, 1977.

QUESTIONS

1. Describe the fertilization of the ovum and early growth of the fetus.
2. Explain the process of implantation and early nutrition of the fetus.
3. Describe the function of the placenta for nutrition of the fetus.
4. What are the functions of estrogen during pregnancy?
5. What are the functions of progesterone during pregnancy?
6. What are the functions of chorionic gonadotropin during pregnancy?
7. What are the functions of human placental lactogen during pregnancy?
8. What are some of the important nutritional needs of the growing fetus?
9. Describe the changes in the circulation of the fetus at birth.
10. Explain the following changes in the mother during pregnancy: changes in weight, changes in metabolism, and changes in body fluids and circulation.
11. What is the mechanism of parturition, and what factors cause it to begin?
12. Explain the functions of estrogen, progesterone, human placental lactogen, prolactin, and oxytocin in the process of milk production.

INDEX